P9-CPV-838

Study Guide

GENERAL CHEMISTRY

EBBING

FIFTH • EDITION

LARRY K. KRANNICH
UNIVERSITY OF ALABAMA AT BIRMINGHAM

JOAN I. SENYK

HOUGHTON MIFFLIN COMPANY • BOSTON • TORONTO
Geneva, Illinois • Palo Alto • Princeton, New Jersey

ISBN: 0-395-75928-5

23456789–CS–99 98 97 96

CONTENTS

PREFACE

This study guide has been written to complement the fifth edition of *General Chemistry* by Darrell Ebbing. It gives the student both information about studying and help in mastering general chemistry. We present a complete and graduated approach to problem solving, with an emphasis on self-testing. The study guide is closely tied to Ebbing's *General Chemistry* through extensive cross-referencing. A student who misses an answer on any of the diagnostic tests, post-tests, or unit exams in this study guide is told exactly what to go back and study in the text. By using vivid descriptive language and examples relevant to the student experience, we have made every effort to engage the student and to share with him or her our own deep involvement with the study of chemistry.

The introduction "To the Student" in the study guide describes how to approach chemistry: what to do before, during, and after lectures; how to use the textbook and study guide; how to master concepts; and how to solve problems in specific steps.

Each study guide chapter includes a list of terms and definitions; a diagnostic test with answers; a summary of chapter topics including operational skills for each subject area, worked-out solutions to exercises in the text, hints for mastering specific chapter material, and additional explanations complementing the text presentation; study questions related to in-depth essays on particular chemicals with answers; a set of additional problems with answers; and a post-test with answers. The solutions to exercises are worked in thinking format — using the headings *Wanted*, *Given*, *Known*, and *Solution* — to help students learn the mental steps in problem solving. Eight unit exams with answers are provided, each covering material presented in the previous two to four text chapters. Test and unit exam answers in this study guide are keyed to text chapter sections and operational skills.

In all solutions, one extra digit has been retained in intermediate calculations and answers are reported to the correct number of significant digits and, in most cases, in scientific notation. The dimensional-analysis method is used whenever applicable in problem solving. We have found that because students succeed in problem solving with this method, its use is a major factor in student motivation and involvement in learning.

Larry K. Krannich wishes to thank his wife, Beverley, for her encouragement and support. Joan I. Senyk would like to thank James D. Crum of California State University at San Bernardino for his superior teaching and encouragement to pursue the field of chemistry.

L. K. K.
J. I. S.

TO THE STUDENT

You've signed up to take general chemistry! You've probably heard a lot about it: "It's really tough!" "It's a flunk-out course." "It's dull and boring." But maybe you were fortunate enough to have been excited by chemistry in high school, or even in elementary school. If so, you know that you are involved with chemistry every minute of your life. Your body is the most marvelous chemical factory in the universe, and everything you come in contact with — the food you eat, the book you are reading, the movie you saw Saturday night — relates to chemistry in some way. For chemistry is the study of the material world — what it's made of and how its individual pieces relate to each other.

The problem with chemistry, as with anything worth studying, is that you can't learn it by osmosis — by sitting back and letting it ooze in. What secrets our universe will reveal are available only to those who are prepared to discover them. And you can be prepared, if you want to be.

Good study habits are the key to success in learning chemistry. You must allot sufficient time on a regular basis to study. But how you spend this time is of great importance. Merely reading is a very poor way to study. Watching while others show you how to solve problems is a pure waste of your time. You will learn only by digging in — by memorizing, by working problems, and by continually testing yourself to find what you still haven't mastered. Seek help only when you are stuck. You will find that learning chemistry gets easier as you get into the swing of studying and as you become more familiar with the vocabulary. Success breeds success, and enjoyment.

Your textbook and this study guide have both been written to help you take an active role in mastering chemistry. The formats of the textbook and the study guide are designed to encourage good study habits. The following paragraphs describe how the textbook and study guide are set up to help you and how you can best use them.

USING YOUR TEXTBOOK

The author of your textbook has done many things to help you learn chemistry. Each chapter presents terms and concepts in the best way for your understanding and retention of the topics discussed. The list of contents at the beginning of each chapter gives you a general idea of what the chapter covers. The most important terms and their definitions appear in boldface type in the text. Italics emphasize other important terms and statements. Key concepts highlighted in color are important in problem solving. Notes in the margin present additional information that is helpful or interesting.

Within each chapter are worked-out Examples. You will also find Exercises for you to try as you go through the chapter. Doing these exercises will help you grasp a concept as soon as you meet it. At the end of each exercise is a reference to end-of-chapter problems on the same topic. Answers to the exercises and to odd-numbered problems at the end of each chapter are in the appendix of the text so you can find out right away if you have mastered the concept being tested. Worked-out solutions to the exercises appear in the study guide, in case you miss an answer.

At the end of each chapter is a list of Important Terms and, for most chapters, a list of Key Equations. Next to each important term is the textbook section number in which the term is explained. You should memorize these terms and definitions, and the sooner the better. An important step in learning chemistry is to learn the vocabulary. You cannot speak a new language if you do not know the meaning of the words. The definitions are listed in this study guide to help you, and they also appear in the glossary at the end of the textbook.

Next you will find a Chapter Summary. This is a condensed presentation of facts and concepts developed in the chapter. You should read the summary even before reading the chapter to establish in your mind some of the goals of the chapter material. Read it again when you finish the text material to help refresh your mind and as a quick review of these important topics.

Following the summary is a list of Operational Skills. This list summarizes the problem-solving skills that are essential for you to master. Use it as a checklist to see if you are ready to be tested on the material in the chapter. If you have difficulty with any of these skills, you may need to do more work on the topic or to get help from your instructor. Note that each operational skill is keyed to a specific example within the chapter, so you can go back and study a worked-out example if any of the operational skills gives you trouble.

At the end of each chapter are Review Questions and many Problems. There are two of each kind of problem, to give you lots of practice in each skill you need to master. The answer to the first problem in each pair is given at the back of the text. Most of the problems are grouped under topic headings. A set of Unclassified Problems gives problems without topic headings, just as they would appear on a test or quiz. A final set of Cumulative-Skills Problems integrates skills from earlier chapters with the material just covered. These problems present particularly interesting and challenging opportunities to sharpen your problem-solving skills.

At the end of the book are a number of appendices containing valuable information, including answers to the in-chapter exercises and to the odd-numbered end-of-chapter problems. There is also a glossary of all the Important Terms in the text. Familiarize yourself with what is there.

USING THIS STUDY GUIDE

This study guide supplements your textbook. Each chapter corresponds to a textbook chapter. After every two to four chapters, a Unit Exam (with answers and references to text sections and operational skills) tests material covered in the preceding chapters. The study guide should be used after you have read your textbook chapter and heard the lecture on the material in it.

In each study guide chapter, you will first find Chapter Terms and Definitions. This section gives definitions for the Important Terms listed at the end of the textbook chapter and the italicized terms in the chapter. (Italicized terms are starred in the study guide.) Terms are listed in the order they appear in the chapter. Memorize the definitions of at least the Important Terms as quickly as you can. You will find knowing them a great asset to your study.

There is a Diagnostic Test and a Post-Test in each study guide chapter. These tests are designed to help you find out what topics you need to study. After you take the Chapter Diagnostic Test, turn to the answer section immediately following it. You will see the correct answers, along with a number giving the text section where the topic is covered and, where appropriate, another number giving the operational skill on which the test question is based. If you got a wrong answer, you know where to go back and study.

After the Diagnostic Test is a section-by-section Summary of Chapter Topics. In each section, you will find the operational skill(s) covered in that section, many helpful comments and hints, and the worked-out solutions to the exercises in that section of the text. Many solutions are worked out in steps to help you learn to analyze problems. (The four steps used in problem solving are explained below under "Two Hints for Success.") Also included are Questions for Study, with answers, related to the essays on specific chemicals found in selected chapters in the text.

A set of Additional Problems, with answers, is provided in each chapter for drill and practice of basic skills. Solutions are provided through Chapter 10 and for the more difficult problems thereafter. Do these problems after you have worked your way through the chapter if you feel you need additional practice. Or, you might wish to work them in preparation for an exam.

The Chapter Post-Test is at the end of the study guide chapter. Take it when you think you are finished studying. The answers follow, along with the numbers of the text sections where the topics are covered and the numbers of the operational skills used. If you missed any answers, go back and study the appropriate sections of the text and study guide and practice the relevant operational skills.

Each Unit Exam covers the set of two to four chapters preceding it. Check your answers; then go back and review the text sections and operational skills for any questions you missed.

STEPS IN STUDYING CHEMISTRY

Now that you know how your text and study guide work together to help you study, here is a way of studying for your chemistry course. This method requires regular effort but is much more effective than trying to learn everything right before an exam.

Before the Lecture

Before you go to a lecture, read over the text material to be covered. Do not expect to understand the material completely from this one reading. Your textbook is well written, but scientific material is much more difficult to read than material from other subject areas. One reason for this is that you are being introduced to many new terms. This initial reading will prepare your mind to get far more out of the lecture.

During the Lecture

Take the best set of notes that you can. Some lecturers write down the important things and it's easy to take notes. Others don't, and you then must write down key words and fill in the gaps later.

After the Lecture

As soon after the lecture as possible, go over your notes and fill in words and ideas that are still fresh in your mind, so that when you review your notes in the future they are understandable. Check with your textbook or your lecture professor on anything that isn't clear. Most professors cover the more important and more difficult material in their lectures, so a good set of notes is very important.

Do the Examples, Exercises, and Problems

Now you are ready to go back to the textbook. Read through a section. Study the example. Then work the accompanying exercise and check the answer in the text. Now try the problems referenced after the exercise and look up the answers.

Take the Diagnostic Test

When you have worked through the exercises and problems in your textbook, turn to this study guide and take the Chapter Diagnostic Test. Allow yourself no more than one hour and do not use any notes. Work out all problems in detail, as this helps you solve the problems correctly and is essential to your retention of the material. After you have completed the Diagnostic Test, turn to the answers to find out how you did.

Analyze the Diagnostic Test Results

If your answers to all questions were correct, you are very familiar with the concepts presented in this chapter. You can now (a) read through the chapter or solve the additional problems in this study guide to reinforce your knowledge, (b) take the Post-Test, or (c) proceed to the next chapter.

If you missed any Diagnostic Test questions, study the text and study guide sections that cover the topics of these questions. Be sure that you have memorized both the definitions of the important terms and other facts that provide the link between what you want for an answer and what you are given to work with in a problem. Also refer to the operational skills for these sections. You may need to spend more time mastering the necessary skills to work the problems.

If a problem stumps you, review your definitions. Also refer to the information on problem solving that follows here. If you still cannot work the problem, give up on it for now. You are probably missing an important concept that your lecturer or teaching assistant can usually point out to you in less than five minutes. If you can't solve a problem within 20 minutes, you waste valuable time in struggle. Go on to another section or, if the missing concept is crucial, get help or put chemistry away until you can get help.

Don't hesitate to see your lecture professor and/or teaching assistant for help. In addition, find out if any audio-visual self-study materials or programmed instructional units are available for your use. If so, use the units that cover the topics you have the most trouble with. Sometimes reading the discussion of a particular subject in a different textbook helps to make that subject a little clearer. Some students hire a tutor. Check with the chemistry department office for a list of tutors.

Complete the Additional Problems and Take the Post-Test

When you feel you have mastered the material, review the questions you missed. Work the additional problems in the study guide and analyze your results. If you missed any questions, review the appropriate text and study guide sections. Once again review the important terms and operational skills. Then take the Chapter Post-Test and analyze your results. It is this kind of "follow-through effort" that will bring you your desired result — learning chemistry.

TWO HINTS FOR SUCCESS

Memorizing

Many students balk at this. They say that memorizing is not learning, that thinking and understanding are what learning is all about. Of course, this notion is not entirely wrong, but the truth is that without necessary definitions and data in your memory, you have nothing to think with. Constant use does eventually commit facts to memory, but this takes time and much repetition. In many cases it is better to work at memorizing. Each study guide chapter will help you decide what to memorize.

The best way to memorize is to test yourself continually. Put items to be memorized on a set of 3" × 5" index cards. For instance, in memorizing a list of definitions, put the word on one side and the definition on the other. Then flip through the cards, putting the cards with words you know in one pile and those with words you don't know in the other pile. Work on the ones you don't know. You may prefer to list terms and definitions to learn on the left- and right-hand sides of a sheet of paper, respectively. Cover one side with a blank sheet of paper and go down the list, moving the blank sheet to reveal the correct answer once you have guessed.

Acronyms (words in which each letter stands for a word) are useful tools to help you remember things. For example, in studying electrochemistry, you will learn that in voltaic cells the anode is negative. A useful acronym is VAN, the first letter of voltaic, anode, and negative.

Problem Solving

There is nothing magical in being able to come up with a solution to a problem. It is a skill to be mastered. There are four basic steps in problem solving.

The headings used in the study guide for these four steps are *Wanted*, *Given*, *Known*, and *Solution*. Many of the exercises worked in the study guide follow these steps to help you master the skill of problem solving.

Step 1: *Wanted*   First write down what you want for an answer. To do this, look for a key word in the problem statement. Write the word and write the units of the quantity, if specific units are indicated. For example, if a problem asks you to solve for the mass in grams of a given element in a sample of a compound, write

Wanted: mass in grams.

Step 2: *Given*   Next write *Given* and list all of the information given in the problem for you to work with.

Step 3: *Known*   The third step is the key: brainstorm. Write down everything you can think of that is related to what is given and to what is wanted. Write down the definitions of any given terms, mathematical expressions, and relationships that come to mind. *What you must come up with is a way to relate what is given to what is wanted.* If you don't see a solution after doing all this, you are probably missing a key relationship. Review the related text section. If you can't find the missing concept, get help.

Step 4: *Solution*   Finally, go through the calculations to solve the problem. If you are using a formula, first solve it algebraically before you plug in numerical values. As a rule, report answers in scientific notation. (Exceptions to this rule are noted in Chapter 1 of the text.)

The best of luck to you in your study!

CHAPTER 1   CHEMISTRY AND MEASUREMENT

CHAPTER TERMS AND DEFINITIONS

Numbers in parentheses after definitions give the text sections in which the terms are explained. Starred terms are italicized in the text. Where a term does not fall directly under a text section heading, additional information is given for you to locate it.

experiment   observation of natural phenomena carried out in a controlled manner so that the results can be duplicated and rational conclusions obtained   (1.2)

law   concise statement or mathematical equation about a fundamental relationship or regularity of nature   (1.2)

hypothesis   tentative explanation of some regularity of nature   (1.2)

theory   tested explanation of basic natural phenomena   (1.2)

scientific method*   the creative process of understanding the physical world that involves hypothesis formation, experimentation, and modification of theory followed by more experimentation   (1.2)

mass   quantity of matter in a material   (1.3)

matter   whatever occupies space and can be perceived by our senses   (1.3)

law of conservation of mass   total mass remains constant during a chemical change (chemical reaction)   (1.3)

weight*   force of gravity exerted on an object   (1.3)

fluids*   liquids and gases: kinds of matter that flow easily and change their shapes in response to slight outside forces   (1.4)

compressibility*   property of a gas enabling it to be pressed into a smaller space   (1.4)

expansibility*   property of a gas enabling it to spread out to fill a given space   (1.4)

solid   form of matter characterized by rigidity; has fixed volume and shape   (1.4)

liquid   form of matter that is a relatively incompressible fluid; has fixed volume but no fixed shape   (1.4)

gas   form of matter that is an easily compressible fluid; has neither fixed volume nor fixed shape   (1.4)

vapor*   the gaseous state of any matter that normally exists as a liquid or solid   (1.4)

states of matter   the three forms in which matter exists: solid, liquid, and gas   (1.4)

physical change   change in the form of matter but not in its chemical identity   (1.4)

distillation*   process in which a liquid is vaporized and then condensed; used to separate substances that differ in volatility   (1.4)

condenser*   cooled tube in which vapor changes back into liquid   (1.4)

receiver*   container in which distilled material is collected   (1.4)

chemical change (chemical reaction)   change in which one or more kinds of matter are transformed into a new kind of matter or several new kinds of matter   (1.4)

physical property   characteristic of a material that can be observed without changing its chemical identity   (1.4)

chemical property   characteristic of a material involving its ability to undergo chemical change   (1.4)

substance   kind of matter that cannot be separated into other kinds of matter by any physical process   (1.4)

element   kind of matter composed of only one chemically distinct type of atom; substance that cannot be chemically decomposed into simpler substances   (1.4)

compound   kind of matter composed of atoms of two or more elements chemically combined in fixed proportions; substance that can be chemically decomposed into two or more elements   (1.4)

law of definite proportions (law of constant composition)   a pure compound, whatever its source, always contains definite or constant proportions of the elements by mass   (1.4)

mixture   material that can be separated by physical means into two or more substances   (1.4)

heterogeneous mixture   mixture that consists of physically distinct parts, each with different properties   (1.4)

homogeneous mixture (solution)   mixture that is uniform in its properties throughout given samples   (1.4)

phase   one of several different homogeneous materials present in the portion of matter under study   (1.4)

chromatography*   technique for separating a mixture that is based on the partitioning of the components of the mixture between stationary and mobile phases   (Instrumental Methods: Separation of Mixtures by Chromatography)

paper chromatography*   chromatographic procedure in which components of a mixture are dissolved in a solution and carried up a paper strip at different rates   (Instrumental Methods: Separation of Mixtures by Chromatography)

column chromatography*   chromatographic procedure in which a mixture is placed on a stationary column and washed down with solvent   (Instrumental Methods: Separation of Mixtures by Chromatography)

chromato-*   color   (Instrumental Methods: Separation of Mixtures by Chromatography)

gas chromatography (GC)*   chromatographic procedure in which a mixture of gases or highly volatile liquids is separated using a carrier gas passing through a solid, or viscous liquid on a solid support, packed in a column   (Instrumental Methods: Separation of Mixtures by Chromatography)

carrier* unreactive gas that flows through the solid column in a gas chromatograph (Instrumental Methods: Separation of Mixtures by Chromatography)

retention time* time it takes for a substance to travel through the column to the detector in a gas chromatograph   (Instrumental Methods: Separation of Mixtures by Chromatography)

chromatogram* chart recording of peaks corresponding to the passage of different substances by the detector in a gas chromatograph   (Instrumental Methods: Separation of Mixtures by Chromatography)

unit fixed standard of measurement   (1.5)

precision agreement among measured values of a quantity   (1.5)

accuracy agreement of a measured value with the true value   (1.5)

significant figures those digits in a measured number (or result of a calculation with measured numbers) that include all certain digits plus a final one having some uncertainty   (1.5)

number of significant figures number of digits reported for the value of a measured or calculated quantity, indicating the precision of the value   (1.5)

scientific notation method of writing numbers in the form $A \times 10^{n}$, where A is a number with a single nonzero digit to the left of the decimal point, and n is an integer, or whole number   (1.5)

exact number number that arises when we count items or sometimes when we define a unit   (1.5)

rounding dropping nonsignificant digits in a calculation result and adjusting the last digit reported   (1.5)

livre* eighteenth-century unit of measurement for the pound, equivalent to 9216 grains (1.6, margin note)

metric system* decimal set of weights and measures   (1.6)

International System (SI) group of metric units internationally accepted in 1960 as the standard units of scientific measurement   (1.6)

SI base units SI units from which all others can be derived   (1.6)

SI prefix prefix used in the International System to indicate a power of 10   (1.6)

meter (m) SI base unit of length (about 39 inches)   (1.6)

angstrom (Å) traditional non-SI unit of length (1 Å = 10^{-10} m)   (1.6)

kilogram (kg) SI base unit of mass (about 2.2 lb)   (1.6)

second (s) SI base unit of time; unit on absolute temperature scale   (1.6)

Celsius scale temperature scale in general scientific use, in which there are exactly 100 units between the freezing and boiling points of water   (1.6)

kelvin (K) SI base unit of temperature   (1.6)

absolute temperature* temperature reported using a scale on which the lowest point that can be attained theoretically is zero   (1.6)

area* length times width (1.7)

speed* rate of change of distance with time (1.7)

SI derived unit unit obtained by combining SI base units (1.7)

volume* length times width times height (1.7)

liter (L) traditional unit of volume (1 L = 1 dm^3) (1.7)

density mass per unit volume (1.7)

dimensional analysis (factor-label method) method of calculation that includes the units for quantities and treats them as numbers (1.8)

conversion factor ratio equivalent to 1 that converts a quantity expressed in one unit to a quantity expressed in another unit (1.8)

CHAPTER DIAGNOSTIC TEST

For questions 1–3, write the letter of the one best answer.

1.   Chemistry is classified as a scientific endeavor, since it
     (a)   investigates physical changes in matter.
     (b)   is based on a close relationship between theory and experiment.
     (c)   is founded on a molecular classification of matter.
     (d)   depends on a quantitative relationship between atoms and molecules.
     (e)   is continually expanding into new areas in an attempt to explain natural phenomena.

2.   The derived unit of pressure in SI units would be

     (a) $\dfrac{g}{cm\ s^2}$ (c) $\dfrac{g}{m\ s^2}$ (e)   either (a) or (d)

     (b) $\dfrac{kg}{cm\ s^2}$ (d) $\dfrac{kg}{m\ s^2}$

3.   An irregularly shaped object has a mass of 2.3×10^{-2} g and a density of 3.4×10^3 kg/m^3. The volume of this object would be calculated to be:

     (a) $\dfrac{m^3}{3.4 \times 10^3\ kg} \times \dfrac{1\ kg}{10^3\ g} \times 2.3 \times 10^{-2}\ g$

     (b) $\dfrac{3.4 \times 10^3\ kg}{m^3} \times \dfrac{10^3\ g}{1\ kg} \times \dfrac{1}{2.3 \times 10^{-2}\ g}$

(c) $\dfrac{m^3}{3.4 \times 10^3 \text{ kg}} \times \dfrac{1 \text{ kg}}{10^{-3} \text{ g}} \times 2.3 \times 10^{-2} \text{ g}$

(d) $\dfrac{3.4 \times 10^3 \text{ kg}}{m^3} \times \dfrac{10^3 \text{ g}}{1 \text{ kg}} \times 2.3 \times 10^{-2} \text{ g}$

(e) none of the above.

4. The number of significant figures in a numerical quantity tells us the _precision_ of the measurement.

5. Absolute temperature is measured in units called _Kelvins_ .

6. Match each value in the left-hand column with its expression in scientific notation in the right-hand column.

b	(1) 201		(a)	1.4916×10^4
c	(2) 0.0000750		(b)	2.01×10^{-2}
e	(3) 14.916×10^{-5}		(c)	7.50×10^{-5}
f	(4) 0.750×10^4		(d)	2.01×10^2
d	(5) 0.0201		(e)	1.4916×10^{-4}
a	(6) 14,916		(f)	7.50×10^3

7. Give the number of significant figures in each of the following numbers:

(a) 1745.0 – 5   (c) 2.89×10^4 – 3   (e) 6200 – 4 ↑ uncertain.

(b) 0.0156 – 3   (d) 1.000×10^{-3} – 4

8. Which of the following is (are) correctly rounded off?

(a) 74.63 to two significant figures equals 75.

(b) 2.4501 to two significant figures equals 2.4.

(c) 0.093374 to four significant figures equals 0.09337.

9. Write the correct values on the lines.

(a) 1 pg = 10^{-12} g   (b) 1 g = 10^9 ng   (c) 1 kg = 10^3 g

10. Perform the following arithmetic operations and report the answers in scientific notation with the correct number of significant figures:

(a) $\dfrac{0.00215}{(74.31)(1.434 \times 10^5)}$ (b) $0.0076 + 24.6(0.000064)$ 1.6×10^{-3}

$= 2.102 \times 10^{-10}$ $= 9.2 \times 10^{-3}$

11.   Make the following conversions. Observe significant figures.

   (a) 6.45×10^4 g to pounds   (d)   91.2°F to °C

   (b) 3.2×10^2 m to inches   (e)   14.7 lb/in^2 to g/m^2

   (c)   4.5 gal to L

12.   In an electrolysis experiment, 0.0004 cm of copper was plated onto an electrode.
      What was the thickness of the copper in nm? $0.0004 cm \left(\frac{1m}{100cm}\right)\left(\frac{10^9 nm}{1m}\right) = 4\times10^{-15}$

      $\boxed{4\times10^3 nm}$

13.   The mass of the moon is 7.345×10^{22} kg. Its volume is 2.1991×10^{25} cm^3. What
      is the density of the moon in g/cm^3? $3.340 g/cm^3$

14.   There is growing concern today over depletion of the ozone layer in our atmosphere.
      The mass of this protective layer has been estimated at 2×10^9 metric tons. If the
      density of ozone is 2.144×10^{-3} g/cm^3, what is the volume corresponding to the
      estimated mass?   (1 metric ton = 1000 kg) 2×10^9 metric tons $\times \left(\frac{1000 kg}{1 metric ton}\right)\left(\frac{1000 g}{1 kg}\right)\left(\frac{cm^3}{2.144\times10^{-3}g}\right)$

      $= 9.328 \times 10^{17} cm^3$

ANSWERS TO CHAPTER DIAGNOSTIC TEST

If you missed an answer, study the text section and operational skill given in parentheses
after the answer.

1.   b   (1.2)   2.   d   (1.6, 1.7)   3.   a   (1.7, 1.8, Op. Sk. 5)

4.   precision   (1.5)   5.   kelvins   (1.6)

6.   (1) d,   (2) c,   (3) e,   (4) f,   (5) b,   (6) a   (Appendix A)

7.   (a) 5,   (b) 3,   (c) 3,   (d) 4,   (e)   uncertain   (1.5)

8.   a, c   (1.5, Op. Sk. 2)   9.   (a) 10^{-12},   (b) 10^9,   (c) 10^3 (1.6)

10.   (a) 2.02×10^{-10},   (b) 9.2×10^{-3} (1.5, Op. Sk. 2)

11.   (a)   142 lb   (1.8, Op. Sk. 2, 6)   (d)   32.9°C   (1.6, Op. Sk. 2, 3)

      (b) 1.3×10^4 in   (1.8, Op. Sk. 2, 6)   (e) 1.03×10^7 g/m^2 (1.8, Op.
                                                               Sk. 2, 6)

      (c)   17 L   (1.8, Op. Sk. 2, 6)

12. 4×10^3 nm   (1.6, Op. Sk. 2, 6)

13. 3.340 g/cm^3 (1.7, 1.8, Op. Sk. 4, 6)

14. 9×10^{17} cm^3 (1.7, 1.8, Op. Sk. 5, 6)

SUMMARY OF CHAPTER TOPICS

1.1   MODERN CHEMISTRY: A BRIEF GLIMPSE

1.2   EXPERIMENT AND EXPLANATION

In Sections 1.1 and 1.2, you are introduced to modern chemistry — a quantitative science where the final test of truth is repeatable experimental observation. It is important to understand the difference between laws made by governments, which any person may choose to break, and observed natural laws. You cannot choose to break the law of gravity! Can you think of any theories? Examples are the theory of evolution, from biology, and Einstein's theory of relativity, from physics. You will learn about chemistry's atomic theory in Chapter 2.

1.3   LAW OF CONSERVATION OF MASS

Operational Skill

1.   Using the law of conservation of mass. Given the masses of all substances in a chemical reaction except one, calculate the mass of this one substance   (Example 1.1).

> **Exercise 1.1**   You place 1.85 grams of wood in a vessel with 9.45 grams of air and seal it. Then you heat the vessel strongly so that the wood burns. After the experiment, you weigh the ash that remains after the wood burns and find that its mass is 0.28 grams. What is the mass of wood converted to gas by the end of the experiment?
>
> *Wanted:*   mass of gases (in grams) that were formed when the wood burns.
>
> *Given:*   1.85 grams of wood, 0.28 grams of ash. The gases are lost when the wood burns to give the ash.

Solution: 1.85 grams wood
              – 0.28 grams ash
              ‾‾‾‾‾‾‾‾‾‾‾‾‾‾
              1.57 grams gases

1.4   MATTER: PHYSICAL STATE AND CHEMICAL CONSTITUTION

Be sure that you understand the difference between a material and a substance. A material is any sample of matter that has mass and occupies space. A substance is a type of material that is unique. It has its own chemical formula and composition and its own specific properties. There are two types of substances: elements and compounds.

Exercise 1.2   Potassium is a soft, silvery-colored metal melting at 64°C. It reacts vigorously with water, with oxygen, and with chlorine. Identify all of the physical properties given in this description. Identify all of the chemical properties given.

Known: definitions of physical properties and chemical properties.

Solution: three physical properties — soft, silvery-colored, melting point 64°C; four chemical properties — it is a metal (its chemical identity), it reacts vigorously with water, it reacts vigorously with oxygen, and it reacts vigorously with chlorine.

1.5   MEASUREMENT AND SIGNIFICANT FIGURES

Operational Skill

2.   Using significant figures in calculations.   Given an arithmetic setup, report the answer to the correct number of significant figures and round it properly (Example 1.2).

Only since the wide use of calculators has the subject of significant figures become important. Previously, when a slide rule was used for calculations, there was little chance to report too many figures. Remembering the rules for reporting significant figures is helpful, but the best way to learn the process is to practice it as much as possible. Note that exact numbers have no effect on the number of significant figures you should report in the answer to a calculation. The number of significant figures in a calculation result will depend only on the numbers of significant figures in quantities having uncertainties.

Scientific notation greatly simplifies chemical calculation. Remember that in converting a number to scientific notation we are not changing its value. We are merely changing the expression of its value.

   Students often have trouble determining the sign of the power of 10 when writing numbers in scientific notation. When we move the decimal point to the left or right, we are in effect dividing or multiplying the original number by 10 for each place we move it. We must then indicate the opposite operation with the power of 10. For example, 49012 equals 4.9012×10^4. We moved the decimal four places to the left, indicating division by 10^4. Therefore, we must multiply the smaller number by 10^4. As another example, putting 0.0036024 in scientific notation we get 3.6024×10^{-3}. This time we moved the decimal to the right, indicating multiplication by 10^3. Therefore, we must indicate division by 10^3. We do this by multiplying by 10^{-3}.

   Get used to writing any numbers less than 0.01 and greater than 999 in scientific notation. Remember that any number written in scientific notation shows the precision in measurement of the number (significant figures). Only the digits in the number, not the exponent, are significant.

   As an exercise, write the following numbers in scientific notation, referring to Appendix A for help:

   (a)   6093   (b)   4218   (c)   0.00413   (d)   0.00006987

The answers are   (a) 6.093×10^3,   (b) 4.218×10^3,   (c) 4.13×10^{-3},
(d) 6.987×10^{-5}.

A point worth remembering is that any number to the zero power is 1, so $10^0 = 1$. You will see this again in later chapters.

   Exercise 1.3   Give answers to the following arithmetic setups. Round to the correct number of significant figures.

   (a) $\dfrac{5.61 \times 7.891}{9.1}$ (b) $8.91 - 6.435$

   (c) $6.81 - 6.730$ (d) $38.91 \times (6.81 - 6.730)$

Solution: (a)   4.9. The calculator result is 4.864671429, but we must report only two digits, as 9.1 has but two. Following rule 1 of the rounding procedure, we increase 8 to 9 because the leftmost digit to be dropped is 6.   (b)   2.48. The calculator result is 2.475, but we report only two decimal places, as 8.91 has but two. Following rule 1 of the rounding procedure, we increase 7 to 8 and drop the 5.   (c)   0.08. The calculator value is correct, since we must report only two decimal places, as 6.81 has but two.   (d)   3. We do the computation in parentheses first:

$$38.91 \times (0.0\underline{8}) = \underline{3}.1128$$

Since 0.08 has but one significant figure, we round off according to rule 2 of the rounding procedure. Thus, our answer has one digit.

1.6   SI UNITS

Operational Skill

3.   Converting from one temperature scale to another.   Given a temperature reading on one scale, convert it to another scale — Celsius, Kelvin, or Fahrenheit (Example 1.3).

Can you imagine what it would be like to go to a party where everybody spoke to you in a different language and you had to have an interpreter to understand what was said? This is what it was like in the scientific world before the use of SI units. It is crucial that you memorize the metric prefixes *mega-, kilo-, deci-, centi-, milli-, micro-, nano-,* and *pico-* and their symbols. Also memorize the SI base units for the base quantities mass, length, time, and temperature. 10^{10} 10^3 10 10^{-2} 10^{-3} 10^{-6} 10^{-9}

10^{-12}

Exercise 1.4   Express the following quantities using an SI prefix and a base unit. For instance, 1.6×10^{-6} m $= 1.6\ \mu$m. A quantity such as 0.000168 g could be written 0.168 mg or 168 µg.

(a) 1.84×10^{-9} m   (c) 7.85×10^{-3} g   (e)   0.000732 s

(b) 5.67×10^{-12} s   (d) 9.7×10^{3} m   (f)   0.000000000154 m

Solution: (a)   1.84 nm (nanometer)
              (b)   5.67 ps (picosecond)
              (c)   7.85 mg (milligram)
              (d)   9.7 km (kilometer)
              (e)   0.732 ms (millisecond) or 732 µs (microsecond)
              (f)   0.154 nm (nanometer) or 154 pm (picometer)

Exercise 1.5   (a)   A person with a fever has a temperature of 102.5°F. What is this temperature in degrees Celsius?   (b)   A cooling mixture of dry ice and isopropyl alcohol has a temperature of -78°C. What is this temperature in kelvins?

Solution: (a) $°C = \dfrac{°F - 32}{1.8} = \dfrac{102.5 - 32}{1.8} = \dfrac{70.5}{1.8} = 39.2°C$

              (b) $K = °C + 273.15 = -78 + 273.15 = 195\ K$

1.7   DERIVED UNITS

Operational Skills

4.   Calculating the density of a substance.   Given the mass and volume of a substance, calculate the density   (Example 1.4).

5.   Using the density to relate mass and volume.   Given the mass and density of a substance, calculate the volume; or given the volume and density, calculate the mass (Example 1.5).

>   **Exercise 1.6**   A piece of metal wire has a volume of 20.2 cm^3 and a mass of 159 g. What is the density of the metal? The metal is either manganese, iron, or nickel, which have densities of 7.21 g/cm^3, 7.87 g/cm^3, and 8.90 g/cm^3, respectively. From which metal is the wire made?
>
>   *Solution:* $d = \dfrac{m}{V} = \dfrac{159 \text{ g}}{20.2 \text{ cm}^3} = 7.87$ g/cm^3
>
>   The object is made of iron.

>   **Exercise 1.7**   Ethanol (grain alcohol) has a density of 0.789 g/cm^3. What volume of ethanol must be poured into a graduated cylinder to equal 30.3 g?
>
>   *Solution:*   Since $d = \dfrac{m}{V}$,
>
> $$V = \frac{m}{d} = 30.3 \; \cancel{g} \times \frac{\text{cm}^3}{0.789 \; \cancel{g}} = 38.4 \text{ cm}^3$$

1.8   UNITS AND DIMENSIONAL ANALYSIS (FACTOR-LABEL METHOD)

Operational Skill

6.   Converting units.   Given an equation relating one unit to another (or a series of such equations), convert a measurement expressed in one unit to a new unit   (Examples 1.6, 1.7, and 1.8).

One of the most useful methods for solving problems in all fields of science is dimensional analysis. This problem-solving technique cannot be overstressed. If you are not totally familiar with this technique, review Section 1.8 of your textbook. The time you now spend learning and mastering this method will reward you tenfold during your science education.

As you review this section in your textbook, pay particular attention to the fact that in every problem and/or example the unit is always written with the number. These units are used to "set up" the problem correctly and to assure you of getting the correct answer. Get into the habit now of using units as part of the problem as well as part of the answer.

It is a good idea to memorize the relationships in text Table 1.5. You will find this won't take much effort if you give yourself lots of practice working the conversion problems. You will eventually remember the factors just from using them.

Exercise 1.8   The oxygen molecule (the smallest particle of oxygen gas) consists of two oxygen atoms a distance of 121 pm apart. How many millimeters is this distance?

Solution: We have been given the distance in picometers and must convert to millimeters:

$$121 \text{ pm} \times \frac{10^{-12} \text{ m}}{1 \text{ pm}} \times \frac{1 \text{ mm}}{10^{-3} \text{ m}} \quad = \quad 121 \text{ pm} \times \frac{10^{-12} \text{ m}}{1 \text{ pm}} \times \frac{1 \text{ mm}}{10^{-3} \text{ m}}$$

$$\underbrace{\qquad}_{\substack{\text{converts} \\ \text{pm to m}}} \underbrace{\qquad}_{\substack{\text{converts} \\ \text{m to mm}}}$$

$$= 1.21 \times 10^{-7} \text{ mm}$$

Note that we report the answer in scientific notation.

Exercise 1.9   A crystal is constructed by stacking small, identical pieces of crystal, much as you construct a brick wall by stacking bricks. A unit cell is the smallest such piece from which a crystal can be made. The unit cell of a crystal of gold metal has a volume of 67.6 $\mathring{A}^3$. What is this volume in cubic decimeters?

Solution: We have been given the volume in $\mathring{A}^3$ and must convert to dm^3:

$$67.6 \, \mathring{A}^3 \times \left(\frac{10^{-10} \text{ m}}{1 \, \mathring{A}} \right)^3 \times \left(\frac{10 \text{ dm}}{1 \text{ m}} \right)^3 \quad = \quad 67.6 \, \mathring{A}^3 \times \frac{10^{-30} \text{ m}^3}{\mathring{A}^3} \times \frac{10^3 \text{ dm}^3}{\text{m}^3}$$

$$\underbrace{\qquad}_{\substack{\text{converts} \\ \mathring{A}^3 \text{ to m}^3}} \underbrace{\qquad}_{\substack{\text{converts} \\ \text{m}^3 \text{ to dm}^3}}$$

$$= 6.76 \times 10^{-26} \text{ dm}^3$$

Exercise 1.10 Using the definitions 1 in = 2.54 cm and 1 yd = 36 in (both exact), obtain the conversion factor for yards to meters. How many meters are there in 3.54 yd?

Solution: $\dfrac{yd}{m} = \dfrac{1 \ yd}{36 \ \cancel{in}} \times \dfrac{1 \ \cancel{in}}{2.54 \ \cancel{cm}} \times \dfrac{1 \ \cancel{cm}}{10^{-2} \ m} = 1.093613298$ yd/m

$$\underbrace{\qquad\qquad}_{\substack{\text{converts} \\ \text{to yd/cm}}} \quad \underbrace{\qquad\qquad}_{\substack{\text{converts} \\ \text{to yd/m}}}$$

Because all conversion factors are exact, this number is exact to an infinite number of figures. We will use a conversion factor of 1.094 in the next calculation:

$$3.54 \ yd = 3.54 \ \cancel{yd} \times \dfrac{1 \ m}{1.094 \ \cancel{yd}} = 3.24 \ m$$

$$\underbrace{\qquad\qquad}_{\substack{\text{converts} \\ \text{yd to m}}}$$

ADDITIONAL PROBLEMS

1. A 5.00-g sample of magnesium ribbon burns in oxygen. When the bright blue-white flame subsides, a white ash is left. If the ash weighs 8.29 g, what mass of oxygen combined with the magnesium during the reaction? $5g \ Mg^{+O} \rightarrow 8.29g$
$$O = 8.29 - 5 = 3.29$$

2. Perform the following calculations and report the answers to the correct number of significant figures:

   (a) $\dfrac{3.9 \times 4.87}{2.412} = 7.9$

   (b) $8.941 + 2.11 = 11.05$

   (c) $4.785 - 2.1003 = 2.685$

   (d) $10.56 - 17.8 \times 0.04 = 9.89$
   $\underset{0.712}{\qquad}$

3. Round the following numbers to the designated number of significant figures, and express the answer in scientific notation:

   (a) 0.004977 to three significant figures — 4.98×10^{-3}
   (b) 13.955 to four significant figures — $1.396 \times 10^{+1}$
   (c) 200,143,000 to three significant figures — 2.00×10^8
   (d) -0.19223×10^6 to one significant figure — -2×10^5
   (e) 0.00000088852932956 to eight significant figures — 8.8852933×10^{-7}

4. Express each of the following quantities using the most appropriate SI prefix and base unit. Give the abbreviation, such as dm for decimeter, for each.

    (a) 4.3×10^{-6} g $= 4.3 \, \mu g$ (c) 8.7×10^{-9} s $- 8.7 \, ns$

    (b) 6.8×10^{-3} L $= 6.8 \, mL$ (d) 3.82×10^{3} m $- 3.82 \, km$

5. Express each of the following as an SI base unit. (Use scientific notation.)

    (a)   2.5 g   (b)   9.8 ps   (c)   4.7 Mg   (d)   2.54 cm

    $= 2.5 \times 10^{-3} kg$ $= 9.8 \times 10^{-12} s$ $4.7 \times 10^{3} kg$ $2.54 \times 10^{-2} m$

6. Convert:

    (a)   5.89 kg to mg   (c)   6.28 Å to cm   (e)   98.6°F to kelvins

    (b)   4.01 µL to mL   (d)   −40°C (exact) to °F

7. How many milliliters of soft drink are in a 12-oz can of Classic Coke? (32 fluid oz = 1 qt)

8. The density of iron is 7.86 g/cm^3. Express this in units of oz/in^3.

9. Suppose your car averages 45 miles per gallon of gasoline when traveling on the interstate highway. If your tank contains 12 gallons of gasoline, can you make the trip from Atlanta, Georgia, to New Orleans, Louisiana, a distance of 772 km, without stopping for gas?

10. A thermos bottle contains liquid nitrogen, which boils at − 323°F. Is the bottle cold enough to contain liquid oxygen, solid oxygen, or both? The melting point of oxygen is − 219°C; the boiling point of oxygen is − 183°C.

11. The Hope diamond, probably the most famous diamond in the world, weighs 45.52 carats. Its density is 3.51 g/cm^3. What is the volume of this blue gem? (1 carat (c) = 0.200 g)

12. A soft plastic material weighing 1.0 lb is shaped into a cylinder 13.5 cm in height and 6.8 cm in diameter. Will this cylinder sink or float when dropped into water? Assume that the density of water is 1.0 g/mL.

13. The densest known form of matter is the metal osmium, with a density of 22.48 g/cm^3. A grapefruit with a diameter of 10.50 cm weighs about 3/4 lb. How many pounds does a sphere of osmium of this size weigh?

ANSWERS TO ADDITIONAL PROBLEMS

If you missed an answer, study the text section and operational skill given in parentheses after the answer.

1. 8.29 g residue − 5.00 g magnesium ribbon = 3.29 g oxygen   (1.3, Op. Sk. 1)

2. (a) 7.9   (b) 11.05   (c) 2.685   (d) $10.56 - (.7) = 9.9$
    (1.5, Op. Sk. 2)

3. (a) 4.98×10^{-3} (c) 2.00×10^{8} (e) 8.8852933×10^{-7} (1.5)
    (b) 1.396×10^{1} (d) -2×10^{5}

4. (a) 4.3 micrograms (μg)   (c) 8.7 nanoseconds (ns)
    (b) 6.8 milliliters (mL)   (d) 3.82 kilometers (km)   (1.6)

5. (a) $2.5 \ \cancel{g} \times \dfrac{10^{-3} \ \text{kg}}{1 \ \cancel{g}} = 2.5 \times 10^{-3} \ \text{kg}$

    (b) $9.8 \ \cancel{ps} \times \dfrac{10^{-12} \ \text{s}}{1 \ \cancel{ps}} = 9.8 \times 10^{-12} \ \text{s}$

    (c) $4.7 \ \cancel{Mg} \times \dfrac{10^{3} \ \text{kg}}{1 \ \cancel{Mg}} = 4.7 \times 10^{3} \ \text{kg}$

    (d) $2.54 \ \cancel{cm} \times \dfrac{10^{-2} \ \text{m}}{1 \ \cancel{cm}} = 2.54 \times 10^{-2} \ \text{m}$ (1.6)

6. (a) $5.89 \ \cancel{kg} \times \dfrac{10^{3} \ \cancel{g}}{1 \ \cancel{kg}} \times \dfrac{1 \ \text{mg}}{10^{-3} \ \cancel{g}} = 5.89 \times 10^{6} \ \text{mg}$

    (b) $4.01 \ \cancel{\mu L} \times \dfrac{10^{-6} \ \cancel{L}}{1 \ \cancel{\mu L}} \times \dfrac{1 \ \text{mL}}{10^{-3} \ \cancel{L}} = 4.01 \times 10^{-3} \ \text{mL}$

    (c) $6.28 \ \cancel{\overset{\circ}{A}} \times \dfrac{10^{-10} \ \cancel{m}}{1 \ \cancel{\overset{\circ}{A}}} \times \dfrac{1 \ \text{cm}}{10^{-2} \ \cancel{m}} = 6.28 \times 10^{-8} \ \text{cm}$ (1.6, 1.8, Op. Sk. 6)

    (d) $\left(-40°\cancel{C} \times \dfrac{9°\text{F}}{5°\cancel{C}} \right) + 32°\text{F} = -40°\text{F}$ (exact)

(e) $(98.6°\cancel{F} - 32°\cancel{F}) \times \dfrac{5°C}{9°\cancel{F}}$ $= 37.0°C$

          $37.0°C + 273.15$ $= 310.2\ K$ (1.6, Op. Sk. 3)

7.   Calculate the number of milliliters (mL) represented by 12 fluid oz.

$$12\ \cancel{oz} \times \frac{1\ \cancel{qt}}{32\ \cancel{oz}} \times \frac{0.9464\ \cancel{L}}{1\ \cancel{qt}} \times \frac{1\ mL}{10^{-3}\ \cancel{L}} = 3.5 \times 10^2\ mL$$

    (1.8, Op. Sk. 6)

8. $\dfrac{7.86\ \cancel{g}}{cm^3} \times \dfrac{1\ oz}{28.35\ \cancel{g}} \times \dfrac{(2.54\ cm)^3}{1\ in^3} = 4.54\ oz/in^3$ (1.8, Op. Sk. 6)

9.   Yes. Calculate the number of kilometers the car will travel on 12 gallons of gasoline.

$$12\ \cancel{gal} \times \frac{45\ \cancel{mile}}{\cancel{gal}} \times \frac{1.609\ km}{1\ \cancel{mile}} = 8.7 \times 10^2\ km$$

The distance from Atlanta to New Orleans is 772 km; therefore, you can make the trip without stopping for gas.   (1.8, Op. Sk. 6)

10.   The bottle is cold enough to contain liquid oxygen but not solid oxygen.

$$°C = (-323°\cancel{F} - 32°\cancel{F}) \times \frac{5°C}{9°\cancel{F}} = -197°C \quad (1.6,\ Op.\ Sk.\ 3)$$

11.   Since $d = \dfrac{m}{V}$, $V = \dfrac{m}{d}$:

$$V = 45.52\ \cancel{c} \times \frac{0.200\ \cancel{g}}{1\ \cancel{c}} \div \frac{3.51\ \cancel{g}}{cm^3}$$

$$V = 2.59\ cm^3 \quad (1.7,\ 1.8,\ Op.\ Sk.\ 5,\ 6)$$

12. The cylinder will float, because its density is less than that of water;
$V = h\pi r^{2-}, r = \frac{d}{2}$

$$\text{Density} = \frac{\text{mass}}{\text{volume}} = \frac{1.0 \text{ lb}}{13.5 \text{ cm} \times \pi \times \left(\frac{6.8 \text{ cm}}{2}\right)^2} \times \frac{453.6 \text{ g}}{\text{lb}}$$

$$= 0.93 \text{ g/cm}^3 \quad (1.7, \text{ Op. Sk. 4})$$

13. Since $d = \frac{m}{V}$, $m = dV$; $V = \frac{4}{3}\pi r^3$, $r = \frac{d}{2}$

$$m = \frac{22.48 \text{ g}}{\text{cm}^3} \times \frac{4}{3} \times \pi \times \left(\frac{10.50}{2}\right)^3 \text{ cm}^3 \times \frac{1 \text{ lb}}{453.6 \text{ g}} = 30.04 \text{ lb}$$

(1.7, 1.8, Op. Sk. 5, 6)

CHAPTER POST-TEST

1. In the scientific method, one experiment can disprove a theory if the experiment is repeatable. True/False

2. The law of conservation of mass states that ___ mass reactants = mass products ___.
    _____.

3. Which of the following could be classified as a possible theory? (There may be more than one answer.)
    (a) Animals have a special sense to warn them of upcoming natural disasters such as earthquakes.
    (b) Water boils at 100°C at 1 atm pressure.
    (c) Sodium reacts with chlorine to form sodium chloride.
    (d) One inch equals 2.54 centimeters.
    (e) Molecules are composed of atoms bound together.

4. The derived unit of volume in SI units is

    (a) cm^3 (c) L   (e) none of the above.

    (b) mL   (d) m^3

5. Which of the following equals the volume of 469 ft^3 of gas in L?

(a) $469 \text{ ft}^3 \times \dfrac{12 \text{ in}^3}{1 \text{ ft}^3} \times \dfrac{2.54 \text{ cm}^3}{1 \text{ in}^3} \times \dfrac{1 \text{ mL}}{1 \text{ cm}^3} \times \dfrac{1 \text{ L}}{10^3 \text{ mL}}$

(b) $469 \text{ ft}^3 \times \dfrac{12 \text{ in}^3}{1 \text{ ft}^3} \times \dfrac{2.54 \text{ cm}^3}{1 \text{ in}^3} \times \dfrac{1 \text{ mL}^3}{1 \text{ cm}^3} \times \dfrac{10^{-3} \text{ L}}{1 \text{ mL}^3}$

(c) $469 \text{ ft}^3 \times \left(\dfrac{12 \text{ in}}{1 \text{ ft}}\right)^3 \times \left(\dfrac{2.54 \text{ cm}}{1 \text{ in}}\right)^3 \times \dfrac{1 \text{ mL}}{1 \text{ cm}^3} \times \dfrac{10^{-3} \text{ L}}{1 \text{ mL}}$

(d) $469 \text{ ft}^3 \times \left(\dfrac{12 \text{ in}}{1 \text{ ft}}\right)^3 \times \left(\dfrac{2.54 \text{ cm}}{1 \text{ in}}\right)^3 \times \dfrac{1 \text{ mL}}{1 \text{ cm}^3} \times \dfrac{1 \text{ L}}{10^{-3} \text{ mL}}$

6. Express the following numbers in scientific notation:
(a) -0.00042 (c) 19,758   (e) 0.1603×10^{10}
(b) 37.4×10^{-6} (d) 250,000,000

7. Which of the following give(s) the correct number of significant figures for the respective number?
(a) 104.710 has six significant figures.
(b) 0.000305 has six significant figures.
(c) -91.20×10^{-4} has four significant figures.
(d) 5.62 has three significant figures.
(e) 0.00000001 has eight significant figures.

8. Round off the following numbers to the designated number of significant figures:
(a) 46,574 to three significant figures
(b) 6.5535 to four significant figures
(c) 0.009374 to three significant figures

9. Which of the following answers give(s) the correct number of significant figures for the indicated arithmetic calculation?

(a) $\dfrac{146 \,(0.015)}{746 + 0.094} = 2.935 \times 10^{-3}$

(b) $(64.4)^{-1} (0.00342) + 1.43 \times 10^{-5} = 6.74 \times 10^{-5}$

10. Complete the following conversions. Observe significant figures.

    (a) 17.3 in = _____ cm   (d) 145°C = _____ °F
    (b) 185 lb = _____ kg   (e) 55 mi/h = _____ m/s
    (c) 100$\underline{0}$ mL = _____ gal

11. The freezing point of ethyl alcohol is -117.6°C. Using this information, indicate whether each of the following statements is true or false:

    (a) This is colder than -179.7°F.   (b) This is colder than 175 K.

12. Light travels through space at the speed of 1.86×10^5 mi/s. Using this information, answer the following questions:
    (a) What is the speed of light in cm/s?
    (b) A light-year is the distance light travels in one year. How many miles are there in one light-year?

13. A jeweler was asked to determine the identity of the metal in a gold-colored lapel pin. The volume of the pin was determined to be 0.8721 cm^3, and its mass was determined to be 14.790 g. Calculate the density of the metal and determine if the pin was made from brass (density 8.5 g/cm^3), 14-karat gold (density 15 g/cm^3), or 18-karat gold (density 17 g/cm^3).

14. The planet Jupiter is often referred to as a failed star because its density, 1.33 g/cm^3, is about that of our sun. If the volume of Jupiter is 1.44×10^{30} cm^3, by what factor is its mass less than the mass of the sun (approximately 2×10^{30} kg)?

ANSWERS TO CHAPTER POST-TEST

If you missed an answer, study the text section and operational skill given in parentheses after the answer.

1. True. (1.2)   2. Mass remains constant during a chemical change. (1.3)

3. a, e (1.2)   4. d (1.7)   5. c (1.7, 1.8, Op. Sk. 5, 6)

6. (a) -4.2×10^{-4},   (b) 3.74×10^{-5},   (c) 1.9758×10^4,

   (d) 2.5×10^8,   (e) 1.603×10^9 (Appendix A)

7.   a, c, d   (1.5)

8.   (a) 4.66×10^{4},   (b)   6.554,   (c)   0.00937   (1.5)

9.   b   (1.5, Op. Sk. 2)

10. (a)   43.9 cm,   (b)   83.9 kg,   (c)   0.2642 gal,

     (d)   293°F,   (e)   25 m/s   (1.6, 1.8, Op. Sk. 3, 6)

11. (a)   False,   (b)   True   (1.6, Op. Sk. 3)

12. (a) 2.99×10^{10} cm/s,   (b) 5.87×10^{12} mi   (1.7, 1.8, Op. Sk. 6)

13. $d = 16.96$ g/cm^{3}; the metal is 18-karat gold   (1.7, Op. Sk. 4)

14. $m = 1.92 \times 10^{27}$ kg; 1000 times less massive than our sun
     (1.7, 1.8, Op. Sk. 5, 6)

CHAPTER 2   ATOMS, MOLECULES, AND IONS

CHAPTER TERMS AND DEFINITIONS

Numbers in parentheses after definitions give the text sections in which the terms are explained. Starred terms are italicized in the text. Where a term does not fall directly under a text section heading, additional information is given for you to locate it.

atomic theory   explanation of the structure of matter in terms of different combinations of very small particles   (2.1)

atom   minute particle of which matter is composed; the smallest part of an element that can enter into chemical reaction   (2.1)

compound   type of matter composed of atoms of two or more elements chemically combined in fixed proportions   (2.1)

chemical reaction   rearrangement of atoms present in reacting substances to give new chemical combinations present in the substances formed   (2.1)

law of multiple proportions   when two elements form more than one compound, the masses of one element in these compounds for a fixed mass of the other element are in ratios of small whole numbers   (2.1)

atomic symbol   one- or two-letter notation used to represent an atom corresponding to a particular element   (2.1)

nucleus   positively charged central core of an atom; contains most of the atom's mass   (2.2)

electron   very light, negatively charged particle that exists in the region around the positively charged nucleus   (2.2)

cathode*   negative electrode   (2.2)

anode*   positive electrode   (2.2)

cathode rays*   rays that originate from the cathode, or negative electrode, in a gas-discharge tube   (2.2)

coulomb (C)*   unit of electric charge   (2.2)

nuclear model*   most of the mass of an atom is concentrated in a positively charged center, called the nucleus, around which negatively charged electrons move   (2.2)

proton   nuclear particle having a positive charge equal to $+e$ (e being the charge on an electron) and a mass more than 1800 times that of an electron   (2.3)

atomic number (Z)   number of protons in an atomic nucleus; identifies the element   (2.3)

element   substance whose atoms all have the same atomic number   (2.3)

neutron   neutral particle of mass almost identical to that of a proton, but without electric charge   (2.3)

mass number (A)   total number of protons and neutrons in a nucleus   (2.3)

nuclide   particular nucleus characterized by a definite atomic number and mass number   (2.3)

nuclide symbol*   symbol for a nuclide in which the mass number is written as a superscript and the atomic number as a subscript on the left of the symbol for the element   (2.3)

isotopes   atoms whose nuclei have the same atomic number but different mass numbers   (2.3)

mass spectrometer*   instrument used to determine atomic mass   (2.4)

atomic mass unit (amu)   mass unit equal to exactly one-twelfth the mass of a carbon-12 atom   (2.4)

atomic weight   average atomic mass for the naturally occurring element, expressed in atomic mass units   (2.4)

mass spectrum*   chart recording from the mass spectrometer that shows the relative numbers of atoms for various masses   (2.4)

fractional abundance   fraction of the total number of atoms that is composed of a particular isotope   (2.4)

periodic table   tabular arrangement of elements in rows and columns, highlighting the regular repetition of properties of the elements   (2.5)

period (of periodic table)   elements in any one horizontal row of the periodic table   (2.5)

group (of periodic table)   elements in any one column of the periodic table   (2.5)

main-group (representative) elements*   elements in the A groups of the periodic table   (2.5)

transition elements*   elements in the B groups of the periodic table   (2.5)

inner-transition elements*   two rows of elements at the bottom of the periodic table   (2.5)

lanthanides*   first of the two rows of inner-transition elements   (2.5)

actinides*   second of the two rows of inner-transition elements   (2.5)

alkali metals*   elements in Group IA of the periodic table   (2.5)

halogens*   elements in Group VIIA of the periodic table   (2.5)

metal   substance or mixture that has a characteristic luster or shine, and is generally a good conductor of heat and electricity; elemental metals are to the left of the staircase line on the periodic table   (2.5)

malleable*   able to be hammered into sheets   (2.5)

ductile*   able to be drawn into wire   (2.5)

nonmetal   element to the right of the staircase line on the periodic table; exhibits characteristics different from those of metals   (2.5)

metalloid (semimetal)   element bordering the staircase line on the periodic table; exhibits both metallic and nonmetallic properties   (2.5)

semiconductors*   elements that, when pure, are poor conductors of electricity at room temperature but become good conductors at higher temperatures   (2.5)

doping*   adding small amounts of other elements to pure semiconductor elements to make them very good electrical conductors   (2.5, margin note)

chemical formula   notation that uses atomic symbols with numerical subscripts to convey the relative proportions of atoms of the different elements in the substance   (2.6)

molecule   definite group of atoms that are chemically bonded together and, as a group, electrically neutral   (2.6)

molecular substance*   substance composed of molecules all of which are alike   (2.6)

molecular formula   gives the exact number of different atoms of an element in a molecule   (2.6)

structural formula*   chemical formula that shows which atoms are bonded to one another in a molecule   (2.6)

ion   electrically charged particle obtained from an atom or chemically bonded group of atoms by addition or removal of one or more electrons   (2.6)

anion   negatively charged ion   (2.6)

cation   positively charged ion   (2.6)

ionic compound   compound composed of cations and anions   (2.6)

crystal*   solid having a regular three-dimensional arrangement of either ions, atoms, or molecules   (2.6)

formula unit   group of atoms or ions explicitly symbolized in the chemical formula   (2.6)

chemical nomenclature   systematic naming of chemical compounds   (2.7)

organic compounds   compounds that contain carbon combined with other elements, such as hydrogen, oxygen, and nitrogen   (2.7)

inorganic compounds   compounds composed of elements other than carbon   (2.7)

monatomic ion   ion formed from a single atom   (2.7)

Stock system*   system for naming compounds in which a Roman numeral within parentheses follows the first-named element to indicate its charge or oxidation number   (2.7)

oxidation state* or **oxidation number***   hypothetical charge assigned in accordance with certain rules; denoted with a Roman numeral following the name of the metal atom   (2.7 margin note)

-ous*   in an older system of nomenclature, a suffix added to the stem name of an element to indicate the cation of lower charge; also indicates the oxoacid with fewer oxygen atoms   (2.7)

-ic*   in an older system of nomenclature, a suffix added to the stem name of an element to indicate the cation of higher charge; indicates the oxoacid with more oxygen atoms; also indicates an acid solution obtained from binary compounds of hydrogen and nonmetals   (2.7)

-ide*   suffix added to the stem name of the element to name monatomic anions or the more electronegative element in binary compounds   (2.7)

polyatomic ion   ion consisting of two or more atoms chemically bonded together and carrying a net electric charge   (2.7)

oxoanion (oxyanion)*   anion composed of oxygen with another element, which is the central element   (2.7)

acid*   molecular compound that can yield one or more hydrogen ions, H^+, and an anion for each acid molecule when the acid dissolves in water   (2.7)

oxoacid   acid containing hydrogen, oxygen, and another element   (2.7)

-ate*   suffix denoting the oxoanion with the greater number of oxygen atoms   (2.7)

-ite*   suffix denoting the oxoanion with the lesser number of oxygen atoms   (2.7)

hypo-*   prefix denoting the oxoacid or oxoanion with the least number of oxygen atoms in the series   (2.7)

per-*   prefix denoting the oxoacid or oxoanion with the greatest number of oxygen atoms in the series   (2.7)

acid anions*   anions that have hydrogen atoms they can lose as hydrogen ions, H^+ (2.7)

di-*   Greek prefix meaning two   (2.7)

thio-*   prefix meaning an oxygen in the root ion name has been replaced by a sulfur atom (2.7)

binary compound   compound composed of only two elements   (2.7)

hydro-*   prefix added to the stem name of the nonmetal to name the acid solution obtained from binary compounds of hydrogen and nonmetals   (2.7)

hydrate   compound that contains water molecules weakly bound in its crystals   (2.7)

chemical equation   symbolic representation of a chemical reaction in terms of chemical formulas   (2.8)

reactant   starting substance in a chemical reaction; appears to the left of the arrow in a chemical equation   (2.8)

product   substance that results from a chemical reaction; appears to the right of the arrow in a chemical equation   (2.8)

coefficient*   number that appears in front of a formula in a chemical equation and gives the relative number of molecules or formula units of a substance involved in the reaction   (2.8)

(*g*)*   phase label placed after a formula in a chemical equation to indicate that the substance is a gas   (2.8)

(*l*)*   phase label placed after a formula in a chemical equation to indicate that the substance is a liquid   (2.8)

(*s*)*   phase label placed after a formula in a chemical equation to indicate that the substance is a solid   (2.8) (*aq*)*   phase label placed after a formula in a chemical equation to indicate that the substance is in aqueous (water) solution   (2.8)

catalyst*   substance that speeds up a reaction without undergoing any net change itself (2.8)

balanced*   describes a chemical equation having correct coefficients   (2.9)

balancing by inspection*   trial-and-error method of balancing a chemical equation by writing appropriate coefficients until there is the same number of any one elemental atom on each side of the arrow   (2.9)

electrolysis*   process of using electricity to provide energy for decomposing a stable substance   (A Substance That Matters: Sodium [a Reactive Metal])

*chloros**   Greek word meaning "greenish-yellow"   (A Substance That Matters: Chlorine   [a Reactive Nonmetal])

CHAPTER DIAGNOSTIC TEST

1. Dalton's atomic theory postulated that matter
    (a) is in continuous motion.
    (b) is continuous (infinitely divisible) in nature.
    (c) changes in mass when heated to combustion.
    (d) can exist in three states — gas, liquid, and solid.
    (e) is composed of small particles called atoms.

2. Robert Millikan's ___oil-drop___ experiment, in conjunction with Thomson's value for *m/e* of the electron, allowed an accurate calculation of the mass of the electron.

3. In a mass spectrograph, the natural isotopes of iron were observed to have the following atomic masses (and percentage abundances):   53.94 (5.84%), 55.94 (91.68%), 57.94 (2.17%), and 57.93 (0.310%). From these data, give the average atomic weight of iron.   At wt = 5587.2759
    = 55.9 amu

4. Explain the difference in meaning between the symbols H, H_2, and H_2O.

5. Complete the following statements about aluminum nitrate, $Al(NO_3)_3$.

    (a) One formula unit of aluminum nitrate contains ___1___ Al atom(s).
    (b) One formula unit of aluminum nitrate contains ___3___ N atom(s).
    (c) One formula unit of aluminum nitrate contains ___9___ O atom(s).

6. A nucleus consists of 32 protons and 41 neutrons. What is the nuclide symbol for this nucleus?
    $^{73}_{32}Ge$

7. In a sample of hydrogen gas there are 4.92×10^{18} hydrogen atoms. How many methane molecules (formula CH_4) could be formed? 1.23×10^{18} molecules $\left(\dfrac{4.92 \times 10^{18} \text{ H atoms}}{4} \right)$

8. When the ions Na^+ and CO_3^{2-} chemically combine, the compound sodium carbonate (called soda ash) is formed. Keeping in mind that chemical formulas are written as electrically neutral species, write the correct formula for sodium carbonate. Na_2CO_3

9. Write the molecular formulas for the molecules having the following structural formulas. C_3H_6 CNH_4Cl N_2F_4

(a)
$$\begin{array}{c} H \\ | \\ H{-}C{-}C{=}C \end{array} \begin{array}{c} H \\ \\ H \end{array}$$
with H on the left carbon top and bottom

(b)
$$H{-}C{-}N \begin{array}{c} H \\ Cl \end{array}$$
with H on left carbon

(c)
$$\begin{array}{c} F \\ F \end{array} N{-}N \begin{array}{c} F \\ F \end{array}$$

(d) $F{-}O{-}O{-}F$ O_2F_2

10. Match each term in the left-hand column with a descriptive example in the right-hand column.

  f   (1) compound   (a) Sn
  d   (2) Rutherford   (b) characterized by Z and A
  a   (3) atomic symbol   (c) 1/12 the mass of a C-12 atom
  e   (4) J. J. Thomson   (d) gold-foil experiment
  b   (5) nuclide   (e) cathode-ray-tube experiments
  c   (6) amu   (f) methane (CH_4)

11. On the following diagram of the periodic table, indicate the regions where non-metals, metals, and metalloids are to be found.

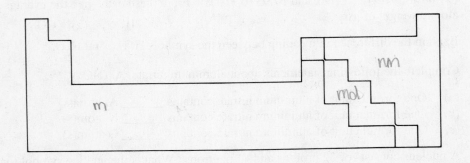

12. Fluorine, chlorine, and bromine are ___nonmetals___ (metals/metalloids/non-metals) that belong to the Group VIIA family, commonly referred to as the ___halogens___.

13. Carbonic acid, H_2CO_3, exists only in aqueous solution and is formed when CO_2 dissolves in water. This is what gives carbonated drinks their "sparkling" taste. Give the formula and name of the oxoanion of carbonic acid.

$$CO_3^{2-}$$

14. Tell whether you would expect the following compounds to be ionic or molecular in nature: (a) XeF_4, (b) CS_2, (c) NaI.

ionic molecular   molecular   ionic

15. Complete the following chart with the appropriate numbers or symbols.

symbol	C	Al^{3+}	H ?	S^{2-} ?	Ag^{1+} ?
atomic number	? 6	13 ?	1	16 ?	47
protons	? 6	13 ?	1 ?	16	47 ?
neutrons	? 8	14	0 ?	16 ?	61
mass number	A = 14	27 ?	1	32	108 ?
electrons	? 6	10 ?	1	18	46

16. Write the formula for each of the following substances:

   (a) zinc hydrogen carbonate
   (b) copper(II) dichromate dihydrate
   (c) manganous hydroxide
   (d) bismuth nitride
   (e) plumbous iodide
   (f) periodic acid

17. Write the name of each of the following substances. Where appropriate, give the Stock system name and the common name.

   (a) $Cr(NO_3)_3 \cdot 9H_2O$ (c) Fe_2S_3 (e) $Co(CN)_2$ (g) CO

   (b) $AlPO_4$ (d) $PbSO_3$ (f) P_2O_5

18. Fill in the blanks, referring to the following:

$$Fe_3O_4 + 4H_2 \longrightarrow 3Fe + 4H_2O$$

The above expression is called a chemical __(a)__ . It describes a chemical __(b)__ in which a chemical __(c)__ occurs in the identity of the reacting molecules. Write a description of the information it gives you: __(d)__ .

ANSWERS TO CHAPTER DIAGNOSTIC TEST

If you missed an answer, study the text section and operational skill given in parentheses after the answer.

1.   e   (2.1)   2.   oil-drop   (2.2)   3.   55.9 amu   (2.4, Op. Sk. 2)

4.   H is the symbol for the element hydrogen. It represents one atom of hydrogen. H_2 represents a molecule of the element hydrogen, which is made up of two atoms of hydrogen. H_2O represents a molecule of a compound (two or more *different* elements combined together) that contains two atoms of hydrogen and one atom of oxygen.   (2.6)

5.   (a)   1,   (b)   3,   (c)   9   (2.6)

6. $^{73}_{32}$Ge   (2.3, Op. Sk. 1)

7. 1.23×10^{18} molecules   (2.6)   8. Na_2CO_3 (2.6, Op. Sk. 3)

9.   (a) C_3H_6,   (b) CH_4NCl,

      (c) N_2F_4,   (d) O_2F_2 (2.6)

10.   (1) f (2.1), (2) d (2.2), (3) a (2.1), (4) e (2.2), (5) b (2.3), (6) c (2.4)

11.

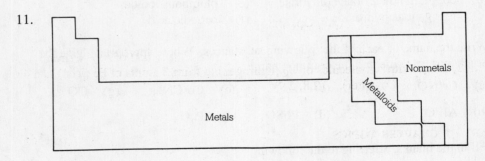

Metals

Metalloids

Nonmetals

(2.5)

12.   nonmetals, halogens   (2.5)

13. $CO_3{}^{2-}$ is the carbonate ion.   (2.7, Op. Sk. 6)

14. (a) molecular   (b) molecular   (c) ionic   (2.7, Op. Sk. 5)

15.

symbol			H	S^{2-}	Ag^+
atomic number	6	13		16	
protons	6	13	1		47
neutrons	8		0	16	
mass number		27			108
electrons	6	10			

(2.3, Op. Sk. 1)

16. (a) $Zn(HCO_3)_2$ (c) $Mn(OH)_2$ (e) PbI_2

(b) $CuCr_2O_7 \cdot 2H_2O$ (d) BiN   (f) HIO_4 (2.7, Op. Sk. 4)

17. (a) chromium(III) nitrate nonahydrate
(b) aluminum phosphate
(c) iron(III) sulfide or ferric sulfide
(d) lead(II) sulfite or plumbous sulfite
(e) cobalt(II) cyanide or cobaltous cyanide
(f) diphosphorus pentoxide
(g) carbon monoxide   (2.7, Op. Sk. 4)

18. (a) equation, (b) reaction, (c) change, (d) 1 formula unit of Fe_3O_4 (iron oxide) reacts with 4 molecules of H_2 (hydrogen) to form 3 atoms of Fe (iron) and 4 molecules of H_2O (water).   (2.8, 2.9)

SUMMARY OF CHAPTER TOPICS

The definitions presented in this chapter are central to the language of chemistry. Work at mastering them as soon as possible. The descriptions of the terms *atom, element,* and *compound* are theoretical explanations that form the basis of our understanding of chemistry. Besides knowing their definitions, it is equally important that you know how atoms, elements, and compounds behave in laboratory work. Both theoretical and practical descriptions are given in the list of chapter terms and definitions.

To keep ion names and charges on the tip of your tongue, use text Tables 2.4 and 2.5 and study guide Table 2.1 to make flip cards on 3" x 5" index cards. Write the name of the ion or element on one side and the symbol (with charge, if an ion) on the other side. Flip through the cards, putting the ones you don't know in a separate pile. Work on the ones you don't know in your spare time.

2.1   ATOMIC THEORY OF MATTER

2.2   THE STRUCTURE OF THE ATOM

In the cathode ray tube, the electrode through which the electrons enter the tube, the cathode, is designated the negative electrode. The electrode at which the electrons leave the tube, the anode, is electrically positive with respect to the cathode. In Chapter 19 you will learn more about electricity as it relates to chemistry.

J. J. Thomson established the mass-to-charge ratio of the electron and Robert Millikan determined the electric charge on an electron. Shortly thereafter, Ernest Rutherford, on the basis of his gold-foil experiment, postulated a nuclear model of the atom.

2.3   NUCLEAR STRUCTURE; ISOTOPES

Operational Skill

1.   Writing nuclide symbols.   Given the number of protons and neutrons in a nucleus, write its nuclide symbol   (Example 2.1).

The atomic number (the number of protons in the nucleus) tells what element we are dealing with. Every calcium atom has 20 protons in its nucleus. The calcium atom loses two electrons to become the Ca^{2+} ion. This charged atom still has 20 protons in the nucleus. The bromine atom gains one electron to become the bromide ion, Br^{-}, with 36 electrons. This charged atom still has 35 protons in its nucleus.

Exercise 2.1   A nucleus consists of 17 protons and 18 neutrons. What is its nuclide symbol?

Known: The complete symbol includes the symbol for the element, the mass number, and the atomic number. Atomic number = number of protons; mass number = number of protons + number of neutrons. (See table on inside back cover of the text.)

Solution: Atomic number = 17; mass number = 17 + 18 = 35. The symbol is $^{35}_{17}Cl$.

The atomic numbers and average atomic weights of the elements are given in the periodic table on the inside front cover of the text. Each element is represented by a square with the atomic symbol in it. The number above the atomic symbol is the atomic number.

2.4 ATOMIC WEIGHTS

Operational Skill

2. Determining atomic weight from isotopic masses and fractional abundances. Given the isotopic masses (in atomic mass units) and fractional isotopic abundances for a naturally occurring element, calculate its atomic weight (Example 2.2).

If you look at text Table 2.2, you will see that the relative masses of protons and neutrons are not exactly one amu. Thus, the mass number of an isotope (the sum of the number of protons and neutrons) will always be a whole number, but the mass of the isotope (in amu) will not be. You will notice from Example 2.2 that we can get the mass number by rounding off the isotopic mass to a whole number.

To take an average, we usually add the given values and divide by the number of them. For instance, the average of 5, 7, and 9 is 5 + 7 + 9 = 21/3 = 7. This method gives us the correct answer because each value has equal weight or representation. We could have gotten the same average, 7, by taking 1/3 of each value, and then adding them:

$$\frac{1}{3}(5) + \frac{1}{3}(7) + \frac{1}{3}(9) = \frac{5}{3} + \frac{7}{3} + \frac{9}{3} = \frac{21}{3} = 7$$

This second method exemplifies a weighted average. We must use this method for determining average relative atomic weights because there is never an equal number of atoms of each isotope in any naturally occurring sample of an element. We could give the amount of each isotope as a percent of the total, but it is more useful to give the value in decimal form, which we call the fractional abundance. For example, in Example 2.2 the fractional abundance of Cr-50 is 0.0435. This means that 4.35% of the atoms in any naturally occurring sample of Cr are the Cr-50 isotope. You should memorize this method of finding atomic weights.

Exercise 2.2 Chlorine consists of the following isotopes:

Isotope	Mass (amu)	Fractional Abundance
Chlorine-35	34.96885	0.75771
Chlorine-37	36.96590	0.24229

What is the atomic weight of chlorine?

Known: The atomic weight is the weighted average of isotopic masses.

Solution: 34.96885 amu × 0.75771   = 26.49̲62   amu
                   36.96590 amu × 0.24229   =   8.9564̲7 amu

Atomic weight of chlorine = 35.453   amu

Although the actual masses of atoms are known, the relative atomic masses (called atomic weights) are much easier to use in calculations. For example, the actual average atomic mass of a calcium atom is 6.656×10^{-23} g, whereas its relative average atomic mass (atomic weight) is 40.08 amu. Be sure you understand the difference between the two and the relationship between them.

2.5   PERIODIC TABLE OF THE ELEMENTS

There are many periodic phenomena in our experience. The phases of the moon are periodic; every 28 days we see a full moon. The seasons of the year are periodic; there is a regular repetition of winter, spring, summer, and fall.

Exercise 2.3   By referring to the periodic table (Figure 2.14 or inside front cover of the text), identify the group and period to which each of the following elements belongs. Then decide whether the element is a metal, nonmetal, or metalloid.
(a) Se   (b) Cs   (c) Fe   (d) Cu   (e) Br

Known: Elements to the left of the periodic table staircase line with characteristic properties are metals; elements to the right of the line with characteristic properties are nonmetals; elements bordering the line with properties of metals and nonmetals are metalloids.

Solution: (a)   Selenium is a nonmetal. It is to the right of the staircase line, in Period 4 and Group VIA. (b)   Cesium is a metal. It is to the left of the line, in Period 6 and Group IA, the alkali metals. (c)   Iron is a metal. It is in Period 4 and Group VIIIB. (d)   Copper is a metal. It is in Period 4 and Group IB. (e)   Bromine is a nonmetal. It is in Period 4 and Group VIIA, the halogens.

2.6   CHEMICAL FORMULAS; MOLECULAR AND IONIC SUBSTANCES

Operational Skill

3.   Writing an ionic formula, given the ions.   Given the formulas of a cation and an anion, write the formula of the ionic compound of these ions   (Example 2.3).

> **Exercise 2.4**   Potassium chromate is an important compound of chromium (see Figure 2.19). It is composed of K^+ and CrO_4^{2-} ions. Write the formula of the compound.
>
> *Known:*   Compounds must be electrically neutral.
>
> *Solution:*   There must be two K^+ ions to provide two positive charges to neutralize the 2- charge on the chromate ion. The formula is K_2CrO_4.

2.7   NAMING SIMPLE COMPOUNDS

Operational Skills

4.   Writing the name of a compound from its formula, or vice versa.   Given the formula of a simple compound (ionic, binary molecular, acid, or hydrate) write the name (Examples 2.4, 2.6, and 2.9), or vice versa   (Examples 2.5, 2.7, and 2.10).

5.   Writing the name and formula of an anion from the acid.   Given the name and formula of an oxoacid, write the name and formula of the oxoanion; or from the name and formula of the oxoanion, write the formula and name of the oxoacid   (Example 2.8).

You must learn the element names and symbols and the ion names, formulas, and charges to master naming and writing the formulas of compounds. Refer to text Tables 2.4 and 2.5 or to study guide Table 2.1. Make flip cards to help you.

It will also be useful for you to know the formulas and names of common acids. They are given in Table 2.7 in the text and Table 2.2 in this study guide.

> **Exercise 2.5**   Write the names of the following compounds:
> (a)   CaO   (b) $PbCrO_4$
>
> (a)   *Known:*   CaO is a binary compound, will end in *-ide*. Ca (Group IIA) forms the 2+ ion; thus, the oxygen ion is O^{2-}.
>
> *Solution:*   calcium oxide.

(b)   *Known:*   The chromate ion has a charge of 2-. The lead ion is thus Pb^{2+}.

     Solution: lead(II) chromate, or plumbous chromate.

Exercise 2.6   A compound has the name thallium(III) nitrate. What is the formula? (The symbol of thallium is Tl.)

Known: Tl (Group IIIA) has a charge of 3+; nitrate is NO_3^-; the compound must be neutral.

Solution: $Tl(NO_3)_3$.

Table 2.1 Charges, Formulas, and Names of Some Common Ions

Cations (+ charge)			Anions (− charge)		
Charge	Formula	Name	Charge	Formula	Name
1+	NH_4^+	ammonium	1−	$C_2H_3O_2^-$	acetate
	Cs^+	cesium		Br^-	bromide
	Cu^+	copper(I); cuprous		Cl^-	chloride
	H^+	hydrogen		ClO^-	hypochlorite
	Li^+	lithium		ClO_2^-	chlorite
	K^+	potassium		ClO_3^-	chlorate
	Rb^+	rubidium		ClO_4^-	perchlorate
	Ag^+	silver		CN^-	cyanide
	Na^+	sodium		$H_2PO_4^-$	dihydrogen phosphate
				F^-	fluoride
2+	Ba^{2+}	barium		H^-	hydride
	Be^{2+}	beryllium		HCO_3^-	hydrogen carbonate; bicarbonate
	Cd^{2+}	cadmium			
	Ca^{2+}	calcium		HSO_4^-	hydrogen sulfate; bisulfate
	Co^{2+}	cobalt(II); cobaltous		HSO_3^-	hydrogen sulfite; bisulfite
	Cu^{2+}	copper(II); cupric		OH^-	hydroxide
	Fe^{2+}	iron(II); ferrous		I^-	iodide
	Pb^{2+}	lead(II); plumbous		NO_3^-	nitrate
	Mg^{2+}	magnesium		NO_2^-	nitrite
	Mn^{2+}	manganese(II); manganous		MnO_4^-	permanganate
	Hg_2^{2+}	mercury(I); mercurous	2−	CO_3^{2-}	carbonate
	Hg^{2+}	mercury(II); mercuric		$C_2O_4^{2-}$	oxalate
	Ni^{2+}	nickel(II); nickelous		CrO_4^{2-}	chromate
	Sr^{2+}	strontium		$Cr_2O_7^{2-}$	dichromate
	Sn^{2+}	tin(II); stannous		HPO_4^{2-}	monohydrogen phosphate
	Zn^{2+}	zinc		O^{2-}	oxide
				O_2^{2-}	peroxide
3+	Al^{3+}	aluminum		SO_4^{2-}	sulfate
	Bi^{3+}	bismuth		S^{2-}	sulfide
	Cr^{3+}	chromium(III); chromic		SO_3^{2-}	sulfite
	Fe^{3+}	iron(III); ferric		$S_2O_3^{2-}$	thiosulfate
4+	Sn^{4+}	tin(IV); stannic	3−	AsO_4^{3-}	arsenate
				N^{3-}	nitride
				PO_4^{3-}	phosphate

Table 2.2   Formulas and Names of Some Common Acids

Formula	Name	Formula	Name
$HF(aq)$	hydrofluoric acid	$HClO_2$	chlorous acid
$HCl(aq)$	hydrochloric acid	$HClO_3$	chloric acid
$HBr(aq)$	hydrobromic acid	$HClO_4$	perchloric acid
$HI(aq)$	hydroiodic acid	HNO_3	nitric acid
$H_2S(aq)$	hydrosulfuric acid	HNO_2	nitrous acid
$HCN(aq)$	hydrocyanic acid	H_2SO_4	sulfuric acid
$HC_2H_3O_2$	acetic acid	H_2SO_3	sulfurous acid
H_2CO_3	carbonic acid	H_3AsO_4	arsenic acid
$HClO$	hypochlorous acid	H_3PO_4	phosphoric acid

On the following two pages are practice grids for writing chemical formulas and names for ionic and covalent compounds. The answers follow on the next two pages. (This is a naming exercise; some of the named compounds do not exist.)

Formula-Nomenclature Practice Grid for Ionic Compounds

Cation \ Anion	Fluoride (-1)	Chloride (-1)	Oxide (-2)	Sulfide (-2)	Sulfate SO_4^{-2}	Sulfite SO_3^{-2}	Carbonate CO_3^{-2}	Chlorate ClO_3^-
Sodium (+1)	NaF	$NaCl$	Na_2O	Na_2S	Na_2SO_4	Na_2SO_3	Na_2CO_3	$Na(ClO_3)$
Potassium (+1)	KF	KCl	K_2O	K_2S	K_2SO_4	K_2SO_3	K_2CO_3	$KClO_3$
Magnesium (+2)	MgF_2	$MgCl_2$	MgO	MgS	$MgSO_4$	$MgSO_3$	$MgCO_3$	$Mg(ClO_3)_2$
Calcium (+2)	CaF_2	$CaCl_2$	CaO	CaS	$CaSO_4$	$CaSO_3$	$CaCO_3$	$Ca(ClO_3)_2$
Aluminum (+3)	AlF_3	$AlCl_3$	Al_2O_3	Al_2S_3	$Al_2(SO_4)_3$	$Al_2(SO_3)_3$	$Al_2(CO_3)_3$	$Al(ClO_3)_3$
Chromium(II) (+2)	CrF_2	$CrCl_2$	CrO	CrS	$CrSO_4$	$CrSO_3$	$CrCO_3$	$Cr(ClO_3)_2$
Iron(II) (+2)	FeF_2	$FeCl_2$	FeO	FeS	$FeSO_4$	$FeSO_3$	$FeCO_3$	$Fe(ClO_3)_2$
Iron(III) (+3)	FeF_3	$FeCl_3$	Fe_2O_3	Fe_2S_3	$Fe_2(SO_4)_3$	$Fe_2(SO_3)_3$	$Fe_2(CO_3)_3$	$Fe(ClO_3)_3$
Copper(I) (+1)	CuF	$CuCl$	Cu_2O	Cu_2S	Cu_2SO_4	Cu_2SO_3	Cu_2CO_3	$Cu(ClO_3)$
Copper(II) (+2)	CuF_2	$CuCl_2$	CuO	CuS	$CuSO_4$	$CuSO_3$	$CuCO_3$	$Cu(ClO_3)_2$
Silver (+1)	AgF	$AgCl$	Ag_2O	Ag_2S	Ag_2SO_4	Ag_2SO_3	Ag_2CO_3	$AgClO_3$
Ammonium (+1)	NH_4F	NH_4Cl	$(NH_4)_2O$	$(NH_4)_2S$	$(NH_4)_2SO_4$	$(NH_4)_2SO_3$	$(NH_4)_2CO_3$	$(NH_4)ClO_3$

This grid is designed to help you learn to name a compound when the formula is given, and to write the formula when the name of the compound is given. To use this grid effectively:

(1) read aloud the name of the compound as you write the formula, and

(2) once all formulas are written, cover up the names of the cations and anions and recite the names of the compounds from the formulas.

Note: This is a naming exercise; some of the named compounds do not exist.

Formula-Nomenclature Practice Grid for Molecular Compounds

	Fluorine		Chlorine		Oxygen		Sulfur	
	Formula	*Name*	*Formula*	*Name*	*Formula*	*Name*	*Formula*	*Name*
Boron(III)	BF_3	Boron trifluoride						
Silicon(IV)	SiF_4	Silicon tetrafluoride						
Phosphorus(V)	PF_5	phosphorus pentafluoride						
Tungsten(VI)	WF_6	tungsten hexafluoride						

This grid is designed to help you learn to name molecular compounds. To use this grid effectively:

(1)   write the formula for the compound formed between the element in the far left column (which is written as if it were an ion) and the nonmetal element listed across the top of the grid, and

(2)   write the name using Greek number prefixes.

Answers

Formula-Nomenclature Practice Grid for Ionic Compounds

Anion / Cation	Fluoride	Chloride	Oxide	Sulfide	Sulfate	Sulfite	Carbonate	Chlorate
Sodium	NaF	$NaCl$	Na_2O	Na_2S	Na_2SO_4	Na_2SO_3	Na_2CO_3	$NaClO_3$
Potassium	KF	KCl	K_2O	K_2S	K_2SO_4	K_2SO_3	K_2CO_3	$KClO_3$
Magnesium	MgF_2	$MgCl_2$	MgO	MgS	$MgSO_4$	$MgSO_3$	$MgCO_3$	$Mg(ClO_3)_2$
Calcium	CaF_2	$CaCl_2$	CaO	CaS	$CaSO_4$	$CaSO_3$	$CaCO_3$	$Ca(ClO_3)_2$
Aluminum	AlF_3	$AlCl_3$	Al_2O_3	Al_2S_3	$Al_2(SO_4)_3$	$Al_2(SO_3)_3$	$Al_2(CO_3)_3$	$Al(ClO_3)_3$
Chromium(II)	CrF_2	$CrCl_2$	CrO	CrS	$CrSO_4$	$CrSO_3$	$CrCO_3$	$Cr(ClO_3)_2$
Iron(II)	FeF_2	$FeCl_2$	FeO	FeS	$FeSO_4$	$FeSO_3$	$FeCO_3$	$Fe(ClO_3)_2$
Iron(III)	FeF_3	$FeCl_3$	Fe_2O_3	Fe_2S_3	$Fe_2(SO_4)_3$	$Fe_2(SO_3)_3$	$Fe_2(CO_3)_3$	$Fe(ClO_3)_3$
Copper(I)	CuF	$CuCl$	Cu_2O	Cu_2S	Cu_2SO_4	Cu_2SO_3	Cu_2CO_3	$CuClO_3$
Copper(II)	CuF_2	$CuCl_2$	CuO	CuS	$CuSO_4$	$CuSO_3$	$CuCO_3$	$Cu(ClO_3)_2$
Silver	AgF	$AgCl$	Ag_2O	Ag_2S	Ag_2SO_4	Ag_2SO_3	Ag_2CO_3	$AgClO_3$
Ammonium	NH_4F	NH_4Cl	$(NH_4)_2O$	$(NH_4)_2S$	$(NH_4)_2SO_4$	$(NH_4)_2SO_3$	$(NH_4)_2CO_3$	NH_4ClO_3

Answers

Formula-Nomenclature Practice Grid for Molecular Compounds

	Fluorine		Chlorine		Oxygen		Sulfur	
	Formula	*Name*	*Formula*	*Name*	*Formula*	*Name*	*Formula*	*Name*
Boron(III)	BF_3	boron trifluoride	BCl_3	boron trichloride	B_2O_3	diboron trioxide	B_2S_3	diboron trisulfide
Silicon(IV)	SiF_4	silicon tetrafluoride	$SiCl_4$	silicon tetrachloride	SiO_2	silicon dioxide	SiS_2	silicon disulfide
Phosphorus(V)	PF_5	phosphorus pentafluoride	PCl_5	phosphorus pentachloride	P_2O_5	diphosphorus pentoxide	P_2S_5	diphosphorus pentasulfide
Tungsten(VI)	WF_6	tungsten hexafluoride	WCl_6	tungsten hexachloride	WO_3	tungsten trioxide	WS_3	tungsten trisulfide

Exercise 2.7 Name the following compounds: (a) Cl_2O_6, (b) PCl_3, (c) PCl_5.

Known: For binary compounds we name the first element and add a suffix to the stem name of the second element. For elements forming more than one compound, we use the Greek prefixes listed in Table 2.6 in the text, omitting *mono-* for a first-named element.

Solution: (a) The text list of compounds in which the Greek prefixes are used shows that chlorine and oxygen form more than one compound. The name would be dichlorine hexoxide. (b) phosphorus trichloride. (c) phosphorus pentachloride.

Exercise 2.8 Give formulas for the following compounds: (a) carbon disulfide, (b) sulfur trioxide.

Known: We change the names of the elements to symbols and translate the prefixes to subscripts.

Solution: (a) CS_2 (b) SO_3

Exercise 2.9 What is the name and formula of the anion corresponding to perbromic acid, $HBrO_4$?

Known: We remove H^+ from the acid to get the ion. We name the ion by replacing *-ic* in the acid name with *-ate*.

Solution: The ion is BrO_4^-, and it is the perbromate ion.

Exercise 2.10 Washing soda has the formula $Na_2CO_3 \cdot 10H_2O$. What is the chemical name of this substance?

Known: The formula indicates the compound is a hydrate. We name hydrates from the anhydrous compound, followed by the word *hydrate* with a Greek prefix to indicate the associated number of water molecules.

Solution: The name is sodium carbonate decahydrate.

Exercise 2.11 Photographers' hypo, used to fix negatives during the development process, is sodium thiosulfate pentahydrate. What is the chemical formula of this compound?

Known: The anhydrous compound is composed of sodium ions (Na^+) and thio-sulfate ions $(S_2O_3^{2-})$; pentahydrate means there are five associated molecules of water.

Solution: The formula is $Na_2S_2O_3 \cdot 5H_2O$.

2.8   WRITING CHEMICAL EQUATIONS

2.9   BALANCING CHEMICAL EQUATIONS

Operational Skill

6.   Balancing simple equations.   Given the formulas of the reactants and products in a chemical reaction, obtain the coefficients of the balanced equation   (Example 2.11).

Balancing equations does not have to be a difficult process. The principle behind it is the law of conservation of matter. Usually, balancing equations is a trial-and-error process. Always check your final answer to be sure that you have the same number of atoms of each element on each side of the arrow.

Exercise 2.12   Find the coefficients that balance the following equations.

(a) $O_2 + PCl_3 \longrightarrow POCl_3$

(b) $P_4 + N_2O \longrightarrow P_4O_6 + N_2$

(c) $As_2S_3 + O_2 \longrightarrow As_2O_3 + SO_2$

(d) $Ca_3(PO_4)_2 + H_3PO_4 \longrightarrow Ca(H_2PO_4)_2$

Solution: (a) $O_2 + 2PCl_3 \longrightarrow 2POCl_3$

(b) $P_4 + 6N_2O \longrightarrow P_4O_6 + 6N_2$

(c) $2As_2S_3 + 9O_2 \longrightarrow 2As_2O_3 + 6SO_2$

(d) $Ca_3(PO_4)_2 + 4H_3PO_4 \longrightarrow 3Ca(H_2PO_4)_2$

A Substance That Matters: SODIUM (a Reactive Metal)

Questions for Study

1.   How is sodium metal stored in the laboratory? Explain.

2.   List some physical and chemical properties of sodium metal.

3.   When was the discovery of sodium announced? By whom? How did he prepare the element?

4.   What is sodium used for?

Answers to Questions for Study

1.   Sodium metal is stored covered with a liquid such as kerosene to keep the sodium metal from reacting with the oxygen and water vapor in the air.

2.   Physical properties of sodium include that it is soft enough to be cut with a very dull knife; it has a bright, silvery luster; the density is 0.968 g/cm^3, which means that it will float on water; and it melts at 98°C, below the boiling point of water. Chemical properties include that it is a metal, although it does not occur as the metal in nature because it is highly reactive with oxygen and violently reactive with water.

3.   The announcement of the discovery of sodium was made on November 19, 1807, by British chemist Humphry Davy. He prepared the element by electrolysis of sodium hydroxide.

4.   Sodium is used in the manufacture of many chemicals, including pharmaceuticals and dyes.

A Substance That Matters: CHLORINE (a Reactive Nonmetal)

Questions for Study

1.   Chlorine gas was first discovered by the Swedish chemist Karl Wilhelm Scheele in 1774. However, its present name was given to the substance by Humphry Davy. Why did he rename the substance? What is the basis for the name?

2.   How was the liquefaction of chlorine gas discovered? How are chlorine and many other gases liquefied?

3.   List some physical and chemical properties of chlorine.

4.   What are some commercial uses of chlorine?

Answers to Questions for Study

1. Davy renamed the substance because its original name, oxymuriatic acid, had been given on the basis of the incorrect assumption that it was a compound of oxygen. Its present name, chlorine, from the Greek word *chloros*, was given for its greenish-yellow color.

2. The liquefaction of chlorine gas was discovered by observation of drops of yellow oil inside a closed vessel in which a chlorine-releasing substance had been heated, greatly increasing the pressure. When the vessel was broken open for examination, the liquid vaporized explosively. Chlorine and many other gases are liquefied by increasing the pressure on them.

3. Some physical properties of chlorine include that it is a yellowish-green gas with a suffocating odor, and it is much more dense than air. Some chemical properties include that it is a nonmetal, a biological disinfectant at low concentrations, a biological poison at higher concentrations, and it reacts with many other substances.

4. Some commercial uses include as a water disinfectant, a bleach, and in the preparation of chlorinated organic compounds, including plastics for bottles and packaging film.

ADDITIONAL PROBLEMS

1. Supply the missing information in the following table.

symbol	C	$?S^{2-}$	Al^{3+}
atomic number	? 6	? 16	? 13
number of protons	? 6	16	? 13
number of neutrons	8	16	? 14
number of electrons	? 6	18	? 10
mass number	? 14	? 32	27

2. Explain the nature of isotopes of elements in terms of the subatomic particles of matter.

3. Write the symbol for:
    (a) a nucleus containing 8 protons and 8 neutrons $^{16}_{8}O$
    (b) the carbon-14 nucleus $^{14}_{6}C$

4. A sample of neon always contains the three isotopes of neon: Ne-20, Ne-21, and Ne-22. The natural abundances of these isotopes are 90.92%, 0.257%, and 8.82%,

respectively. Their isotopic masses are 19.99244 amu, 20.99395 amu, and 21.99138 amu, respectively. Calculate the atomic weight of neon.

5. Write the formula for the compound of each of the following ion pairs:

   (a) Ca^{2+} and H^-

   (b) Al^{3+} and OH^-

   (c) Mg^{2+} and PO_4^{3-}

   (d) NH_4^+ and SO_4^{2-}

6. Classify each of the following elements as a metal, a metalloid, or a nonmetal: S, Sr, Co, N, Ga, I, Ar, Si, Li, V.

7. Complete the following chart with the appropriate numbers or symbols.

symbol	P	?	H	Mg^{2+}	?
atomic number	?	11	?	?	29
protons	?	?	?	?	?
neutrons	?	12	2	?	?
mass number	31	?	?	24	64
electrons	?	10	?	?	29

8. Write the formula for each of the following substances:

   (a) sodium sulfite heptahydrate
   (b) iron(III) chloride
   (c) calcium fluoride
   (d) sulfuric acid

   (e) silicon tetrabromide
   (f) barium hydrogen sulfite
   (g) perchloric acid
   (h) dinitrogen tetroxide

   — with #'s in ( ).

9. Name each of the following compounds. Where appropriate, give both the Stock system name and the common name.

   (a) $SrBr_2 \cdot 6H_2O$

   (b) $MnCl_2$

   \ with ous & ic
   (c) $HBrO_3$

   (d) NO_2

   (e) LiH

   (f) BCl_3

10. For each of the following, write the symbol and name for the corresponding oxoacid or oxoanion. *\ anion consisting of O & another element*

    (a) HClO, hypochlorous acid

    (b) PO_4^{3-}, phosphate ion

    (c) HNO_3, nitric acid

    (d) SO_3^{2-}, sulfite ion

    'acid consisting of H, O & another element

    oxoacid.

    oxoanion

11.   Determine which of the following statement(s) is (are) incorrect with respect to the following chemical equation:

$$H_2 + Cl_2 \longrightarrow 2HCl$$

(a)   Two atoms of hydrogen occur on both sides of the equation.
(b)   The chemical reaction described by this equation consists of a rearrangement of the atoms of hydrogen and chlorine to give hydrogen chloride (HCl).
(c)   Since one molecule of H_2 and one molecule of Cl_2 react, two molecules of product must form.
(d)   This equation describes a chemical change.

12.   Balance each of the following equations:

$$\_\ Na + \_\ O_2 \longrightarrow \_\ Na_2O$$

$$\_\ As + \_\ H_2 \longrightarrow \_\ AsH_3$$

$$\_\ Ba(OH)_2 + \_\ HNO_3 \longrightarrow \_\ Ba(NO_3)_2 + \_\ H_2O$$

$$\_\ Al + \_\ H_2SO_4 \longrightarrow \_\ Al_2(SO_4)_3 + \_\ H_2$$

$$\_\ C_3H_8 + \_\ O_2 \longrightarrow \_\ CO_2 + \_\ H_2O$$

13.   A solution of lead sulfate reacts with a solution of sodium chloride to form solid lead chloride and aqueous sodium sulfate. Write the balanced equation for this reaction, including state (phase) labels.

ANSWERS TO ADDITIONAL PROBLEMS

If you missed an answer, study the text section and operational skill given in parentheses after the answer.

1.

symbol	C	S^{2-}	Al^{3+}
atomic number	6	16	13
number of protons	6	16	13
number of neutrons	8	16	14
number of electrons	6	18	10
mass number	14	32	27

(2.3)

2. Isotopes are atoms of the same element that have the same number of protons and same number of electrons but different numbers of neutrons. Therefore, isotopes have different masses. (2.3)

3. (a) $^{16}_{8}O$ (b) $^{14}_{6}C$ (2.3, Op. Sk. 1)

4. The atomic weight is the weighted average of isotopic masses.

$$19.99244 \text{ amu} \times 0.9092 \ = 18.1\underline{7}71 \text{ amu}$$
$$20.99395 \text{ amu} \times 0.00257 = 0.054\underline{0} \text{ amu}$$
$$21.99138 \text{ amu} \times 0.0882 \ = \underline{1.9\underline{3}96 \text{ amu}}$$

Atomic weight of neon $= 20.1707 = 20.17$ amu   (2.4, Op. Sk. 2)

5. (a) CaH_2 (c) $Mg_3(PO_4)_2$

    (b) $Al(OH)_3$ (d) $(NH_4)_2SO_4$ (2.6, Op. Sk. 3)

6. The metals are Sr, Co, Ga, Li, and V. Si is a metalloid. The nonmetals are S, N, I, and Ar.   (2.5)

7.

symbol		Na$^+$			Cu
atomic number	15		1	12	
protons	15	11	1	12	29
neutrons	16			12	35
mass number		23	3		
electrons	15		1	10	

(2.3)

8. (a) $Na_2SO_3 \cdot 7H_2O$ (e) $SiBr_4$

   (b) $FeCl_3$ (f) $Ba(HSO_3)_2$

   (c) CaF_2 (g) $HClO_4$

   (d) H_2SO_4 (h) N_2O_4 (2.7, Op. Sk. 4)

9. (a) strontium bromide hexahydrate
   (b) manganese(II) chloride, manganous chloride
   (c) bromic acid
   (d) nitrogen dioxide
   (e) lithium hydride
   (f) boron trichloride   (2.7, Op. Sk. 4)

10. (a) ClO^-, hypochlorite ion   (c) NO_3^-, nitrate ion

    (b) H_3PO_4, phosphoric acid   (d) H_2SO_3, sulfurous acid   (2.7, Op. Sk. 6)

11. c   (2.9)

12. $4Na + O_2 \longrightarrow 2Na_2O$

    $2As + 3H_2 \longrightarrow 2AsH_3$

    $Ba(OH)_2 + 2HNO_3 \longrightarrow Ba(NO_3)_2 + 2H_2O$

    $2Al + 3H_2SO_4 \longrightarrow Al_2(SO_4)_3 + 3H_2$

    $C_3H_8 + 5O_2 \longrightarrow 3CO_2 + 4H_2O$ (2.9, Op. Sk. 7)

13. $PbSO_4(aq) + 2NaCl(aq) \longrightarrow PbCl_2(s) + Na_2SO_4(aq)$

(2.6, 2.7, 2.8, 2.9, Op. Sk. 6, 7)

CHAPTER POST-TEST

1.   Indicate whether each of the following statements is true or false. If a statement is false, change it so it is true.
   (a)   According to Dalton's atomic theory, compounds are kinds of matter composed of atoms of two or more elements. True/False: _____

   _____ .

   (b)   A chemical symbol is used to designate elements, and formulas are used to designate formula units. True/False: _____

   _____ .

   (c)   A molecular formula does not indicate the arrangement of atoms in a molecule. True/False:

   _____ __ .

   (d) $KClO_3$, Na_2S, and BF_3 are ionic compounds. True/False: _____

   _____ .

2.   If you have one dozen formula units of $Mg(OH)_2$, how many hydrogen atoms do you have?

3.   The structure of sulfuric acid, the most widely used chemical, is

$$H-O-\overset{\displaystyle O}{\underset{\displaystyle O}{\overset{|}{\underset{|}{S}}}}-O-H$$

Write the molecular formula for sulfuric acid. (*Hint:*   The order of atoms is H, S, then O.)

4.   Which one of the following statements correctly describes the difference between a structural and a molecular formula?
   (a)   The structural formula indicates the composition of the molecule and the spatial arrangement of the atoms, and the molecular formula shows how the different atoms bond with each other.
   (b)   The structural formula represents the simplest composition of a molecule, and the molecular formula represents the actual composition.

(c)   The structural formula shows how the atoms are bonded together in a mole-
      cule, and the molecular formula shows the atomic composition of the
      molecule.
(d)   None of the above are correct statements.

5.   Write the symbol and name for the corresponding oxoacid or oxoanion.

(a) BrO_2^-, bromite ion   (c) IO_3^-, iodate ion

(b) HNO_2, nitrous acid   (d) $C_2H_3O_2^-$, acetate ion

6.   Complete the following chart with the appropriate numbers or symbols.

symbol	Cr	?	Br^-	?	?
atomic number	?	26	?	?	24
protons	?	?	?	92	?
neutrons	28	?	?	143	29
mass number	?	56	80	?	?
electrons	?	24	?	92	24

7.   Classify each of the following elements as a metal, a nonmetal, or a metalloid: Mg,
     Cu, Pb, As, Cl.

8.   Which of the following is an incorrect statement? Rewrite that statement so it is
     correct.
     (a)   Na and Cs are in the same group in the periodic table.
     (b)   The nonmetallic elements are on the far right side of the periodic table.
     (c)   Most of the known chemical elements are classified as metals.
     (d)   Elements in the same period of the periodic table have similar properties.
     (e)   Fluorine is classified as a nonmetallic element.

9.   Write the formula for each of the following:
     (a)   barium peroxide   (d)   ammonium iodate
     (b)   silver chlorite   (e)   stannous nitrite
     (c)   strontium oxalate monohydrate   (f)   phosphoric acid

10. Write the name of each of the following. Where appropriate, give the Stock system name and the common name.

    (a) $KClO_4$

    (b) Hg_2Br_2

    (c) $Mg(C_2H_3O_2)_2 \cdot 4H_2O$

    (d) CaH_2

    (e) $NaMnO_4$

    (f) HNO_3

11. Fill in the blanks, referring to the following:

    $$2ZnS(s) + 3O_2(g) \xrightarrow{\Delta} 2SO_2(g) + 2ZnO(s)$$

    The above equation is a symbolic representation of the ___(a)___ between two ___(b)___ of ___(c)___ in the ___(d)___ state and three ___(e)___ of oxygen ___(f)___ to produce two ___(g)___ of gaseous ___(h)___ and two formula units of solid ___(i)___. The triangle over the arrow indicates that the reactants are ___(j)___ to produce the products.

12. Balance the following equations:

    (a) $\_ P + \_ O_2 \longrightarrow \_ P_4O_{10}$

    (b) $\_ Ag + \_ NiSO_4 \longrightarrow \_ Ag_2SO_4 + \_ Ni$

    (c) $\_ C_6H_{14} + \_ O_2 \longrightarrow \_ CO_2 + \_ H_2O$

    (d) $\_ HF + \_ SiO_2 \longrightarrow \_ SiF_4 + \_ H_2O$

    (e) $\_ NaHCO_3 \longrightarrow \_ Na_2CO_3 + \_ H_2O + \_ CO_2$

ANSWERS TO CHAPTER POST-TEST

If you missed an answer, study the text section and operational skill given in parentheses after the answer.

1. (a) True. (2.1)   (b) True. (2.1, 2.6)   (c) True. (2.6)

    (d) False. $KClO_3$ and Na_2S are ionic compounds; BF_3 is a molecular compound. (2.7)

2. two dozen (2.1)   3. H_2SO_4 (2.6)   4. c (2.6)

5.   (a) $HBrO_2$, bromous acid   (c) HIO_3, iodic acid

     (b) NO_2^-, nitrite ion   (d) $HC_2H_3O_2$, acetic acid   (2.7, Op. Sk. 5)

6.

symbol		Fe^{2+}		U	Cr
atomic number	24		35	92	
protons	24	26	35		24
neutrons		30	45		
mass number	52			235	53
electrons	24		36		

                                                      (2.3)

7.   Mg, Cu, and Pb are metals, Cl is a nonmetal, and As is a metalloid.   (2.5)

8.   d. Elements in the same *group* of the periodic table have similar properties. (2.5)

9.   (a) BaO_2 (d) NH_4IO_3

     (b) $AgClO_2$ (e) $Sn(NO_2)_2$

     (c) $SrC_2O_4 \cdot H_2O$ (f) H_3PO_4 (2.7, Op. Sk. 4)

10.   (a) potassium perchlorate
      (b) mercury(I) bromide or mercurous bromide
      (c) magnesium acetate tetrahydrate
      (d) calcium hydride
      (e) sodium permanganate
      (f) nitric acid   (2.7, Op. Sk. 4)

11.   (a) chemical reaction   (b) formula units   (c) zinc sulfide   (d) solid
      (e) molecules   (f) gas   (g) molecules
      (h) sulfur dioxide   (i) zinc oxide   (j) heated   (2.8, 2.9)

12.   (a) $4P + 5O_2 \longrightarrow P_4O_{10}$

     (b) $2Ag + NiSO_4 \longrightarrow Ag_2SO_4 + Ni$

(c) $2C_6H_{14} + 19O_2 \longrightarrow 12CO_2 + 14H_2O$

(d) $4HF + SiO_2 \longrightarrow SiF_4 + 2H_2O$

(e) $2NaHCO_3 \longrightarrow Na_2CO_3 + H_2O + CO_2$

(2.9, Op. Sk. 6)

CHAPTER 3   CHEMICAL REACTIONS: AN INTRODUCTION

CHAPTER TERMS AND DEFINITIONS

Numbers in parentheses after definitions give the text sections in which the terms are explained. Starred terms are italicized in the text. Where a term does not fall directly under a text section heading, additional information is given for you to locate it.

solution*   homogeneous mixture of dissolved substances in a solvent such as water (3.1, introductory section)

ionic theory of solutions*   proposed by Svante Arrhenius; certain substances produce freely moving ions when they dissolve in water, and these ions conduct an electric current in an aqueous solution   (3.1)

electrolyte   substance that dissolves in water to give an electrically conducting solution; ionic solid that dissolves in water   (3.1)

nonelectrolyte   substance that dissolves in water to give a nonconducting or very poorly conducting solution   (3.1)

strong electrolyte   electrolyte that exists in solution almost entirely as ions   (3.1)

weak electrolyte   electrolyte that dissolves in water to give a relatively small percentage of ions   (3.1)

molecular equation   chemical equation in which the reactants and products are written as if they were molecular substances, even though they may actually exist in solution as ions   (3.2)

complete ionic equation   chemical equation in which you represent strong electrolytes (such as soluble ionic compounds) as separate ions in the solution   (3.2)

spectator ion   ion in an ionic equation that does not take part in the reaction   (3.2)

net ionic equation   ionic equation from which spectator ions have been canceled   (3.2)

solubility*   ability of a substance to dissolve in a solvent   (3.3)

precipitate   insoluble solid compound formed during a chemical reaction in solution (3.3)

exchange (or metathesis) reaction   reaction between compounds that, when written as a molecular equation, appears to involve the exchange of parts between the two reactants   (3.3)

acid–base indicator   dye used to distinguish between acidic and basic solutions by means of the color changes it undergoes in these solutions   (3.4)

acid (Arrhenius)   substance that produces hydrogen ions, H^+, when it dissolves in water   (3.4)

base (Arrhenius)   substance that produces hydroxide ions, OH^-, when it dissolves in water   (3.4)

proton-transfer reactions*   alternative view of acid–base reactions in which a proton, H^+, is transferred from an acid to a base in an acid–base reaction   (3.4)

acid (Brønsted–Lowry)   species (molecule or ion) that donates a proton to another species in a proton-transfer reaction   (3.4)

base (Brønsted–Lowry)   species (molecule or ion) that accepts a proton in a proton-transfer reaction   (3.4)

aq*   means that water molecules are associated with the species   (3.4)

hydronium ion*   represented by the formula $H_3O^+(aq)$ (3.4)

strong acid   acid that ionizes completely in water, forming a strong electrolyte   (3.4)

weak acid   acid that only partly ionizes in water, forming a weak electrolyte   (3.4)

strong base   base that is present in aqueous solution entirely as ions, one of which is OH^- (it is a strong electrolyte)   (3.4)

weak base   base that is only partly ionized in water, forming a weak electrolyte   (3.4)

neutralization reaction   reaction of an acid and a base that results in an ionic compound and possibly water   (3.4)

salt   ionic compound that is a product of a neutralization reaction   (3.4)

monoprotic acids*   acids that have only one acidic hydrogen atom per acid molecule   (3.4)

polyprotic acid   acid that yields two or more acidic hydrogens per molecule   (3.4)

acid salts*   salts that have acidic hydrogen atoms and can undergo neutralization with bases   (3.4)

oxidation number (oxidation state)   actual charge of the atom, if it exists as a monoatomic ion, or else a hypothetical charge assigned to the atom in the substance by simple rules   (3.5)

oxidation–reduction reaction (redox reaction)   reaction in which electrons are transferred between species or in which atoms change oxidation number   (3.5)

oxidized*   algebraic gain in oxidation number   (3.5)

reduced   algebraic decrease in oxidation number   (3.5)

half-reaction   one of two parts of an oxidation–reduction reaction, one part of which involves a loss of electrons (or increase of oxidation number) and the other a gain of electrons (or decrease of oxidation number)   (3.5)

oxidation   half-reaction in which there is a loss of electrons by a species (or an increase of oxidation number of an atom)   (3.5)

reduction   half-reaction in which there is a gain of electrons by a species (or a decrease in the oxidation number of an atom)   (3.5)

oxidizing agent   species that oxidizes another species; it is itself reduced   (3.5)

reducing agent   species that reduces another species; it is itself oxidized   (3.5)

combination reaction   reaction in which two substances combine to form a third
         substance   (3.5)

decomposition reaction   reaction in which a single compound reacts to give two or more
         substances   (3.5)

displacement (single-replacement) reaction   reaction in which an element reacts with a
         compound, displacing an element from it   (3.5)

combustion reaction   reaction in which a substance reacts with oxygen, usually with the
         rapid release of heat to produce a flame   (3.5)

skeleton equation*   equation that gives the essential information about a reaction   (3.6)

muriatic acid*   one name for hydrochloric acid (A Material That Matters: Hydrochloric
         Acid [a Strong Acid])

muria*   Latin for "seawater"; hydrochloric acid was originaly produced from sea salt (A
         Material That Matters: Hydrochloric Acid [a Strong Acid])

CHAPTER DIAGNOSTIC TEST

1.   What is the Arrhenius concept of an acid?

2.   Classify each of the following equations as either precipitation, acid–base, or
     oxidation–reduction by type of reaction:

     (a) $4Na + O_2 \longrightarrow 2Na_2O$

     (b) $2As + 3H_2 \longrightarrow 2AsH_3$

     (c) $Ba(OH)_2 + 2HNO_3 \longrightarrow Ba(NO_3)_2 + 2H_2O$

     (d) $2Al + 3H_2SO_4 \longrightarrow Al_2(SO_4)_3 + 3H_2$

     (e) $CdSO_4 + (NH_4)_2S \longrightarrow CdS + (NH_4)_2SO_4$

3.   Which of the following are oxidation–reduction reactions?

     (a) $C_3H_8 + 5O_2 \longrightarrow 3CO_2 + 4H_2O$

     (b) $I_2 + 2KCl \longrightarrow Cl_2 + 2KI$

(c) $K_2SO_4 + Ba(NO_3)_2 \longrightarrow BaSO_4 + 2KNO_3$

(d) $2CaSO_4 + C \longrightarrow 2CaO + 2SO_2 + CO_2$

4.   Classify each of the following as a strong acid, weak acid, strong base, or weak base:

(a) $H_3C_6H_5O_7$ (c) $Mg(OH)_2$ (e) HNO_2

(b) H_2SO_4 (d) $Cu(OH)_2$

5.   Write the net ionic equations for the following molecular equations:

(a) $LiOH(aq) + HCl(aq) \longrightarrow LiCl(aq) + H_2O(l)$

(b) $Pb(NO_3)_2(aq) + 2NaI(aq) \longrightarrow PbI_2(s) + 2NaNO_3(aq)$

6.   Determine the oxidation number of the underlined atom in each of the following compounds:

(a) $\underline{Cl}F_3$ (c) $Ca\underline{H}_2$ (e) $HC\underline{l}O_3$

(b) $Na_2\underline{S}_4O_6$ (d) $\underline{N}H_4Cl$

7.   Write the skeleton equation for the oxidation–reduction reaction below. Label the oxidizing and reducing agents, the oxidation and reduction parts of the equation, and the oxidized and reduced species.

$$Br_2(g) + SO_2(g) + 2H_2O(g) \longrightarrow 2HBr(g) + H_2SO_4(g)$$

8.   Balance the following equation using the half-reaction method:

$$NO + PbO_2 + H^+ \longrightarrow NO_3^- + Pb^{2+}$$

9.   Write the molecular equation and the net ionic equation for the neutralization of sodium hydroxide with sulfuric acid.

10.   Decide whether the two soluble compounds in each case will react to form a precipitate. If so, write the net ionic equation; if not, write NR.

(a) $(NH_4)_2CO_3 + MgCl_2$ (c) $KOH + Cu(NO_3)_2$

(b) $Na_2SO_4 + NiBr_2$ (d) $Ca(ClO_3)_2 + Li_3PO_4$

11.   Write a molecular equation for the preparation of lead(II) chloride from lead(II) sulfide.

ANSWERS TO CHAPTER DIAGNOSTIC TEST

If you missed an answer, study the text section and operational skill given in parentheses after the answer.

1. According to the Arrhenius concept, an acid is any hydrogen-containing substance that yields a hydrogen ion in a water solution.   (3.4)

2. (a) $4Na + O_2 \longrightarrow 2Na_2O$ Oxidation–reduction

   (b) $2As + 3H_2 \longrightarrow 2AsH_3$ Oxidation–reduction

   (c) $Ba(OH)_2 + 2HNO_3 \longrightarrow Ba(NO_3)_2 + 2H_2O$ Acid–base reaction

   (d) $2Al + 3H_2SO_4 \longrightarrow Al_2(SO_4)_3 + 3H_2$ Oxidation–reduction reaction

   (e) $CdSO_4 + (NH_4)_2S \longrightarrow CdS + (NH_4)_2SO_4$ Precipitation reaction

    (3.3, 3.4, 3.5, Op. Sk. 2)

3. (a), (b), (d)   (3.5, Op. Sk. 1)

4. (a) weak acid   (c) strong base   (e) weak acid   (3.4, Op. Sk. 3)
   (b) strong acid   (d) weak base

5. (a) $OH^-(aq) + H^+(aq) \longrightarrow H_2O(l)$

   (b) $Pb^{2+}(aq) + 2I^-(aq) \longrightarrow PbI_2(s)$ (3.3, 3.4, Op. Sk. 1, 2, 4)

6. (a) +3   (c) +2   (e) +5   (3.5, Op. Sk. 6)
   (b) +2.5   (d) −3

7.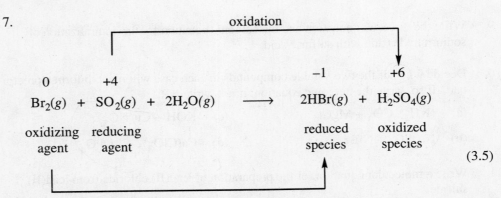

8.　$4H^+ + 2NO + 3PbO_2 \longrightarrow 2NO_3^- + 3Pb^{2+} + 2H_2O$　(3.5, Op. Sk. 6, 7)

9.　Molecular equation: $2NaOH(aq) + H_2SO_4(aq) \longrightarrow Na_2SO_4(aq) + 2H_2O(l)$

　　Net ionic equation:　$2OH^-(aq) + 2H^+(aq) \longrightarrow 2H_2O(l)$

$$\text{(3.4, Op. Sk. 1, 4)}$$

10.　(a)　$Mg^{2+}(aq) + CO_3^{2-}(aq) \longrightarrow MgCO_3(s)$　　　　(b)　NR

　　(c)　$2OH^-(aq) + Cu^{2+}(aq) \longrightarrow Cu(OH)_2(s)$

　　(d)　$3Ca^{2+}(aq) + 2PO_4^{3-}(aq) \longrightarrow Ca_3(PO_4)_2(s)$　(3.3, Op. Sk. 2)

11.　$PbS(s) + 2HCl(aq) \longrightarrow PbCl_2(s) + H_2S(g)$　(3.3, Op. Sk. 2, 5)

SUMMARY OF CHAPTER TOPICS

3.1　IONIC THEORY OF SOLUTIONS

3.2　MOLECULAR AND IONIC EQUATIONS

Operational Skill

1.　Writing net ionic equations.　Given a molecular equation, write the corresponding net ionic equation　(Example 3.1).

> **Exercise 3.1**　Write ionic and net ionic equations for each of the following molecular equations.
>
> (a)　$2HNO_3(aq) + Mg(OH)_2(s) \longrightarrow 2H_2O(l) + Mg(NO_3)_2(aq)$
>
> (b)　$Pb(NO_3)_2(aq) + Na_2SO_4(aq) \longrightarrow PbSO_4(s) + 2NaNO_3(aq)$
>
> *Known:*
>
> (a)　HNO_3 is a strong electrolyte and $Mg(NO_3)_2$ is a soluble ionic compound, which makes it a strong electrolyte. Because $Mg(OH)_2$ is an insoluble ionic compound, it is written as $Mg(OH)_2(s)$. Water, H_2O, is a nonelectrolyte, so

its molecular formula is used. The complete ionic equation contains all the ions and formulas. The net ionic equation results from canceling the spectator ions from the complete ionic equation.

(b) $Pb(NO_3)_2$, Na_2SO_4, and $NaNO_3$ are soluble ionic compounds, so are strong electrolytes and are written in ionic form. $PbSO_4$ is an insoluble ionic compound, so it is written as $PbSO_4(s)$. The complete ionic equation contains all the ions and formulas. The net ionic equation results from canceling the spectator ions from the complete ionic equation.

Solution:

(a)   Complete ionic equation:

$$2H^+(aq) + 2NO_3^-(aq) + Mg(OH)_2 \longrightarrow$$
$$2H_2O(l) + Mg^{2+}(aq) + 2NO_3^-(aq)$$

Net ionic equation:

$$2H^+(aq) + Mg(OH)_2 \longrightarrow 2H_2O(l) + Mg^{2+}(aq)$$

(b)   Complete ionic equation:

$$Pb^{2-}(aq) + 2NO_3^-(aq) + 2Na^+(aq) + SO_4^{2-} \longrightarrow$$
$$PbSO_4(s) + 2Na^+(aq) + 2NO_3^-(aq)$$

Net ionic equation:

$$Pb^{2-}(aq) + SO_4^{2-}(aq) \longrightarrow PbSO_4(s)$$

3.3   PRECIPITATION REACTIONS

Operational Skill

2.   Deciding whether precipitation occurs.   Using solubility rules, decide whether two soluble ionic compounds will react to form a precipitate. If they will, write the net ionic equation   (Example 3.4).

It is a good idea to memorize the groups of soluble compounds from Table 3.1 in the text. As a rule of thumb, Group IA and all ammonium (NH_4^+) compounds are soluble; nitrates are soluble; and hydroxides of Group IA, ammonium, and barium are soluble.

Exercise 3.2   You mix aqueous solutions of sodium iodide and lead acetate. If a reaction occurs, write the molecular equation and the net ionic equation. If no reaction occurs, write *NR* after the arrow.

Wanted: molecular equation, net ionic equation.

Given: written reactants.

Known: formulas NaI and Pb $(C_2H_3O_2)_2$; rules for writing formulas (Section 2.7); definitions of molecular and ionic equations; reaction occurs if a precipitate forms; solubility rules   (Table 3.1).

Solution: The reaction would be

$$NaI + Pb(C_2H_3O_2)_2 \longrightarrow NaC_2H_3O_2 + PbI_2 \quad \text{(unbalanced)}$$

Table 3.1 shows that PbI_2 is insoluble, so a reaction does occur. The molecular equation is

$$2NaI(aq) + Pb(C_2H_3O_2)_2(aq) \longrightarrow 2NaC_2H_3O_2(aq) + PbI_2(s)$$

The net ionic equation is

$$Pb^{2+}(aq) + 2I^-(aq) \longrightarrow PbI_2(s)$$

3.4   ACID–BASE REACTIONS

Operational Skill

3.   Classifying acids and bases as strong or weak.   Given the formula of an acid or base, classify it as strong or weak   (Example 3.3).

Exercise 3.3   Label each of the following as a strong or weak acid or base:
(a) H_3PO_4 (b)   HClO   (c) $HClO_4$ (d) $Sr(OH)_2$

Known: (a) H_3PO_4 is not listed in Table 3.3 in the text as a strong acid.

(b)   HClO is not listed in Table 3.3 as a strong acid.

(c) $HClO_4$ is listed among the strong acids in Table 3.3.

(d) $Sr(OH)_2$ is a Group IIA hydroxide.

Solution: (a)   weak acid   (b)   weak acid   (c)   strong acid   (d)   strong base

Operational Skill

4.   Writing an equation for a neutralization.   Given an acid and a base, write the molecular equation and then the net ionic equation for the neutralization reaction (Example 3.4).

> **Exercise 3.4**   Write the molecular equation and the net ionic equation for the neutralization of hydrocyanic acid, HCN, by lithium hydroxide, LiOH, both in aqueous solution.
>
> *Known:*   Formula of lithium hydroxide is LiOH; acid–base reactions produce water plus a salt; HCN is a weak acid, LiOH a strong base   (Table 3.3).
>
> *Solution:*   The molecular equation is
>
> $$HCN(aq) + LiOH(aq) \longrightarrow H_2O(l) + LiCN(aq)$$
>
> The net ionic equation is
>
> $$HCN(aq) + OH^-(aq) \longrightarrow H_2O(l) + CN^-(aq)$$

> **Exercise 3.5**   Write molecular and net ionic equations for the successive neutralizations of the acidic hydrogens of sulfuric acid with potassium hydroxide. (That is, write equations for the reaction of sulfuric acid with KOH to give the acid salt and for the reaction of the acid salt with more KOH to give potassium sulfate.)
>
> *Solution:*   The molecular equation for the neutralization of the first acidic hydrogen is
>
> $$H_2SO_4(aq) + KOH(aq) \longrightarrow H_2O(l) + KHSO_4(aq)$$
>
> The net ionic equation is
>
> $$H_2SO_4(aq) + OH^-(aq) \longrightarrow H_2O(l) + HSO_4^-(aq)$$
>
> The molecular equation for the second reaction is
>
> $$KHSO_4(aq) + KOH(aq) \longrightarrow H_2O(l) + K_2SO_4(aq)$$
>
> The net ionic equation is
>
> $$HSO_4^-(aq) + OH^-(aq) \longrightarrow H_2O(l) + SO_4^{2-}(aq)$$

Operational Skill

5. Writing an equation for a reaction with gas formation.   Given the reaction between a carbonate, sulfite, or sulfide and an acid, write the molecular and the net ionic equations   (Example 3.5).

> **Exercise 3.6**   Write the molecular equation and the net ionic equation for the reaction of calcium carbonate with nitric acid.
>
> *Known:*   definitions of molecular and net ionic equations; carbonates react with acids to produce unstable weak acids that decompose to give gaseous products; solubility rules from Table 3.1.
>
> *Solution:*   The exchange reaction is
>
> $$CaCO_3(s) + HNO_3(aq) \longrightarrow Ca(NO_3)_2(aq) + H_2CO_3(aq) \quad \text{(unbalanced)}$$
>
> The product carbonic acid, H_2CO_3, decomposes to form water and carbon dioxide gas. The molecular equation for the overall reaction is
>
> $$CaCO_3(s) + 2HNO_3(aq) \longrightarrow Ca(NO_3)_2(aq) + H_2O(l) + CO_2(g)$$
>
> The net ionic equation is
>
> $$CaCO_3(s) + 2H^+(aq) \longrightarrow Ca^{2+}(aq) + H_2O(l) + CO_2(g)$$

3.5 OXIDATION–REDUCTION REACTIONS

Operational Skill

6. Assigning oxidation numbers.   Given the formula of a simple compound or ion, obtain the oxidation numbers of the atoms, using the rules for assigning oxidation numbers (Table 3.5)   (Example 3.6).

Writing oxidation numbers of atoms in compounds is initially difficult for beginning students because of the negative charge on the electron. It makes sense to think a number decreases when items are lost. But since electrons are negatively charged, when they are lost, there are more protons on the species than electrons, and the oxidation number thus increases. When they are gained, there are more electrons to cancel the positive charge, and the oxidation number decreases. An acronym to help you remember these processes is "LEO goes GER": **L**oss of **E**lectrons is **O**xidation, **G**ain of **E**lectrons is **R**eduction.

Note that when we write oxidation numbers, we place the sign before the number, whereas when we write the charge on an ion, the sign follows the number.

Do not take the time to memorize the rules for determining oxidation numbers. Use them to work the examples, exercises, and problems, and you will learn them.

Exercise 3.7   Obtain the oxidation numbers of the atoms in (a) potassium dichromate, $K_2Cr_2O_7$, and (b) permanganate ion, MnO_4^-.

(a) $K_2Cr_2O_7$:

Known: $K = +1$, $O = -2$ (rules 2 and 3); sum of all oxidation numbers $= 0$ (rule 6).

Solution: K_2 Cr_2 O_7
            $+1$ $?$ -2

$$2x_K + 2x_{Cr} + 7x_O = 0$$

$$2(+1) + 2x_{Cr} + 7(-2) = 0$$

$$x_{Cr} = +6$$

(x = oxidation number of the subscript element.)

(b) MnO_4^-:

Known: $O = -2$ (rule 3); sum of all oxidation numbers $= -1$ (rule 6).

Solution: Mn O_4^-
            $?$ -2

$$x_{Mn} + 4x_O = -1$$

$$x_{Mn} + 4(-2) = -1$$

$$x_{Mn} = +7$$

An oxidizing agent, or a reducing agent, is a reactant that plays a role in a particular reaction. Zinc atom, Zn, can be only a reducing agent, as it can only lose electrons and be itself oxidized. Zn^{2+} ion can be only an oxidizing agent, as it can only take electrons and be itself reduced. Fe^{2+}, however, can be either. It can lose another electron to become Fe^{3+} and thus be a reducing agent, or it can gain two electrons to become the iron atom,

Fe, and thus be an oxidizing agent. What Fe^{2+} does depends on the other reactant as well as on the reaction conditions.

3.6 BALANCING OXIDATION–REDUCTION EQUATIONS

Operational Skill

7. Balancing equations by the half-reaction method. Given the skeleton equation for an oxidation–reduction equation, complete and balance it (Examples 3.7 and 3.8).

A simple way to balance the oxygens and hydrogens in a reaction in basic solution is to add an even number of hydroxide ions (2, 4, etc.) to the side needing oxygen and then to add half that number of water molecules to the other side. This is a trial-by-error method; that is, you can try adding two OH^- ions first, then if that doesn't work, add four, etc. It is a waste of time to try to memorize the rules for balancing oxidation–reduction equations. The rules are a great help in learning and reviewing how to balance them, but the way to become proficient at it is to balance enough reactions so the rules become second nature.

Exercise 3.8 Iodic acid, HIO_3, can be prepared by reacting iodine, I_2, with concentrated nitric acid. The skeleton equation is

$$I_2(s) + NO_3^-(aq) \longrightarrow IO_3^-(aq) + NO_2(g)$$

Balance this equation.

Given: the unbalanced ionic equation in acidic solution.

Known: the steps for balancing the equation (given in text).

Solution: 1. The oxidation numbers of all atoms are

$$\overset{0}{I_2} + \overset{+5}{N}\overset{-2}{O_3^-} \longrightarrow \overset{+5}{I}\overset{-2}{O_3^-} + \overset{+4}{N}\overset{-2}{O_2}$$

2. The two half-reactions are

Reduction: $NO_3^- \longrightarrow NO_2$

Oxidation: $I_2 \longrightarrow IO_3^-$

3.   To balance the reduction half-reaction:

(a)   N is balanced.
(b)   Balance O by adding H_2O to the right.
(c)   Balance H by adding $2H^+$ to the left.
(d)   Balance charge by adding e^- to the left. The balanced reduction half-reaction is

$$2H^+ + NO_3^- + e^- \longrightarrow NO_2 + H_2O$$

4.   To balance the oxidation half-reaction:

(a)   Balance I by adding a coefficient of 2 for IO_3^-.
(b)   Balance O by adding $6H_2O$ to the left.
(c)   Balance H by adding $12H^+$ to the right.
(d)   Balance charge by adding $10e^-$ to the right. The balanced oxidation half-reaction is

$$6H_2O + I_2 \longrightarrow 2IO_3^- + 12H^+ + 10e^-$$

5.   Combining the half-reactions, we multiple the reduction half-reaction by 10 to get an equal number of electrons in both half-reactions. Then we add to get the balanced reaction:

$$20H^+ + 10NO_3^- + \cancel{10e^-} \longrightarrow 10NO_2 + 10H_2O$$
$$6H_2O + I_2 \longrightarrow 2IO_3^- + 12H^+ + \cancel{10e^-}$$
$$\overline{20H^+ + 10NO_3^- + 6H_2O + I_2 \longrightarrow 10NO_2 + 10H_2O +}$$
$$2IO_3^- + 12H^+$$

or

$$8H^+ + 10NO_3^- + I_2 \longrightarrow 10NO_2 + 4H_2O + 2IO_3^-$$

Exercise 3.9   Balance the following equation using the half-reaction method.

$$H_2O_2 + ClO_2 \longrightarrow ClO_2^- + O_2 \quad \text{(basic solution)}$$

Given: the unbalanced reaction, in basic solution.

Known: the steps for balancing the equation (given in text).

Solution: 1.   The oxidation numbers of the atoms in the skeleton equation are

$$\overset{+1}{H_2}\ \overset{-1}{O_2}\ +\ \overset{+4}{Cl}\ \overset{-2}{O_2}\ \longrightarrow\ \overset{+3}{Cl}\ \overset{-2}{O_2^-}\ +\ \overset{0}{O_2}$$

2.   The two half-reactions are

Reduction: $ClO_2 \longrightarrow ClO_2^-$
Oxidation: $H_2O_2 \longrightarrow O_2$

3.   To balance the reduction half-reaction, only the charge needs to be balanced. We add $1e^-$ to the left, giving:

$$1e^- + ClO_2 \longrightarrow ClO_2^-$$

4.   To balance the oxidation half-reaction:

(a)   Balance O and H by adding $2H_2O$ to the right and $2OH^-$ to the left.

(b)   Balance charge by adding $2e^-$ to the right. The balanced oxidation half-reaction is:

$$2OH^- + H_2O_2 \longrightarrow O_2 + 2H_2O + 2e^-$$

5.   Combining the half-reactions, we multiply the reduction half-reaction by 2 to get an equal number of electrons in both half-reactions. Then we add to get the balanced reaction:

$$\cancel{2e^-} + 2ClO_2 \longrightarrow 2ClO_2^-$$
$$\underline{2OH^- + H_2O_2 \longrightarrow O_2 + 2H_2O + \cancel{2e^-}}$$
$$2ClO_2 + 2OH^- + H_2O_2 \longrightarrow 2ClO_2^- + O_2 + 2H_2O$$

Note that in the half-reaction method, when we write the half-reactions, we write the *species* containing the atoms that undergo reaction. It is very tempting to write only the atom with its oxidation state. Don't do this, because writing the species enables you to get the proper number of water molecules and hydroxide ions or H^+ ions, as well as to balance the reacting atoms.

You have now been introduced to both methods of balancing oxidation–reduction reactions. It is a good idea to be able to use both methods. However, if your instructor does not tell you otherwise, we suggest that you concentrate on learning the half-reaction method. It appears to take longer but in practice is less confusing. Later, the oxidation-number method will seem easier.

A Material That Matters: HYDROCHLORIC ACID (a Strong Acid)

Questions for Study

1.   What is hydrochloric acid? List several properties of hydrochloric acid.

2.   What is a common but older name for hydrochloric acid? What does this name derive from?

3.   Give chemical equations for the preparation of hydrogen chloride from sodium chloride. Classify each by type of chemical reaction.

4.   What is the major source of hydrogen chloride today?

5.   Write an equation for the reaction of iron with hydrochloric acid. What type of reaction is this?

6.   List several uses of hydrochloric acid.

Answers to Questions for Study

1.   Hydrochloric acid is an aqueous solution of hydrogen chloride, HCl, a colorless, corrosive gas. Properties include that HCl is a strong acid that is colorless when pure, and a concentrated solution of it fumes when exposed to air. It reacts with rust (Fe_2O_3) more readily than with metallic iron.

2.   A common, older name is *muriatic acid*. It derives from the Latin word *muria* (for "seawater"), because the original commercial preparation frequently used sea salt.

3. $NaCl(s) + H_2SO_4(aq) \xrightarrow{\Delta} NaHSO_4(s) + HCl(g)$ Exchange reaction

       $NaCl(s) + NaHSO_4(s) \xrightarrow{\Delta} Na_2SO_4(s) + HCl(g)$ Exchange reaction

4.   The major source of hydrogen chloride today is as a by-product of the preparation of chlorinated hydrocarbons, including certain plastics, insecticides, and refrigerants.

5. $Fe(s) + 2HCl(aq) \longrightarrow FeCl_2(aq) + H_2(g)$ Displacement reaction or oxidation–reduction

6. Uses of hydrochloric acid include the "pickling" of steel (removing surface rust before processing into finished goods), the production of corn syrup from corn starch, and the production of gelatin from bones.

A Material That Matters: HYDROGEN PEROXIDE (an Oxidizing and Reducing Agent)

Questions for Study

1. List some properties of hydrogen peroxide that are not possessed by water.

2. When blood is added to a solution of hydrogen peroxide, the solution bubbles furiously. What is the overall reaction that occurs?

3. Write the balanced equation for the oxidation of PbS to $PbSO_4$ by H_2O_2 in acidic solution.

4. Tetrahydroxochromate(III) ion, $Cr(OH)_4^-$, is obtained when Cr^{3+} ion is treated with a base. Write the balanced equation for the oxidation of $Cr(OH)_4^-$ ion to CrO_4^{2-} ion by H_2O_2 in basic solution.

5. Give an example of a reaction in which hydrogen peroxide functions as a reducing agent. Write the half-reaction for the oxidation of H_2O_2.

Answers to Questions for Study

1. Properties of hydrogen peroxide not possessed by water include a syrupy consistency and a pale blue color. Hydrogen peroxide may decompose explosively in the presence of small quantities of impurities, and its solutions decompose in the presence of various catalysts.

2. The overall reaction is: $2H_2O_2(l) \longrightarrow 2H_2O(l) + O_2(g)$

3. The equation is

$$PbS\,(s) + 4H_2O_2(aq) \longrightarrow PbSO_4(s) + 4H_2O(l)$$

4. The equation is

$$2OH^-(aq) + 2Cr(OH)_4^-(aq) + 3H_2O_2(aq) \longrightarrow 2CrO_4^{2-}(aq) + 8H_2O(l)$$

5. A reaction in which hydrogen peroxide functions as a reducing agent is

$$5H_2O_2(aq) + 2MnO_4^-(aq) + 6H^+(aq) \longrightarrow 5O_2(g) + 8H_2O(l) +$$
$$2Mn^{2+}(aq)$$

The half-reaction for the oxidation of H_2O_2 in basic solution is

$$2OH^- + H_2O_2 \longrightarrow O_2 + 2H_2O + 2e^-$$

ADDITIONAL PROBLEMS

1. Balance the following equations and classify each as either precipitation, acid–base, or oxidation–reduction.

    (a) __ $NaNO_3$ $\longrightarrow$ __ $NaNO_2$ + __ O_2

    (b) __ N_2 + __ H_2 $\longrightarrow$ __ NH_3

    (c) __ $BaCl_2$ + __ $AgNO_3$ $\longrightarrow$ __ $AgCl$ + __ $Ba(NO_3)_2$

    (d) __ K_2SO_4 + __ $CaCl_2$ $\longrightarrow$ __ $CaSO_4$ + __ KCl

    (e) __ KOH + __ HCN $\longrightarrow$ __ H_2O + __ KCN

2. Write the net ionic equation for each of the following:

    (a) $2(NH_4)_3PO_4(aq) + 3MgCl_2(aq) \longrightarrow Mg_3(PO_4)_2(s) + 6NH_4Cl(aq)$

    (b) $Na_2CO_3(aq) + CaS(aq) \longrightarrow CaCO_3(s) + Na_2S(aq)$

3. Classify each of the following as a strong acid, weak acid, strong base, or weak base:

    (a) $HClO_4$ (b) NH_3 (c) HCN (d) $Ba(OH)_2$

4. Write net ionic equations for the reactions that occur between solutions of the following compounds. Write *NR* for those cases in which there is no reaction.
    (a) potassium sulfide and barium nitrate
    (b) ammonium sulfate and sodium phosphate

(c)   magnesium chloride and mercury(II) nitrate

(d)   lead(II) acetate and sodium carbonate

5.   Write molecular and net ionic equations for reactions in aqueous solution of each of the following. Explain why these reactions go to completion.

(a)   sodium sulfite and nitric acid

(b)   ammonium sulfate and sodium hydroxide

(c)   barium carbonate and hydrogen chloride

(d)   sodium sulfide and sulfuric acid

6.   Write molecular and net ionic equations for each of the following reactions. Name the type of reaction, and explain why they go to completion.

(a)   solid calcium hydroxide with a solution of sulfuric acid

(b)   a solution of acetic acid with a solution of lithium hydroxide

7.   Write the oxidation number for the underlined atom in each of the following species:

(a) $\underline{Mn}O_4^-$ (b) $\underline{Cr}PO_4$ (c) $Na\underline{Cl}O_3$ (d) $Ba\underline{Cr}O_4$ (e) $HO\underline{Br}$

8.   Balance the following oxidation–reduction reactions:

(a) $AuCl_3(aq) + KI(aq) \longrightarrow AuCl(aq) + KCl(aq) + I_2(s)$

(b) $HNO_3(aq) + HI(aq) \longrightarrow NO(g) + I_2(s) + H_2O(l)$

(c) $NCl_3(l) + H_2O(l) \longrightarrow NH_3(aq) + HOCl(aq)$

(d) $SO_3^{2-}(aq) + H_2O(l) + CrO_4^{2-}(aq) \longrightarrow$

$$SO_4^{2-}(aq) + CrO_2^-(aq) + OH^-(aq)$$

ANSWERS TO ADDITIONAL PROBLEMS

If you missed an answer, study the text section and operational skill given in parentheses after the answer.

1.   (a) $2NaNO_3 \longrightarrow 2NaNO_2 + O_2$ Oxidation–reduction reaction

(b) $N_2 + 3H_2 \longrightarrow 2NH_3$ Oxidation–reduction reaction

(c) $BaCl_2 + 2AgNO_3 \longrightarrow 2AgCl + Ba(NO_3)_2$ Precipitation reaction

(d) $K_2SO_4 + CaCl_2 \longrightarrow CaSO_4 + 2KCl$ Precipitation reaction

(e) $2KOH + 2HCN \longrightarrow 2H_2O + 2KCN$ Acid–base reaction

(2.9, Op. Sk. 6; 3.3, 3.4, 3.5, Op. Sk. 2)

2.   (a) $2PO_4{}^{3-}(aq) + 3Mg^{2+}(aq) \longrightarrow Mg_3(PO_4)_2(s)$

(b) $CO_3{}^{2-}(aq) + Ca^{2+}(aq) \longrightarrow CaCO_3(s)$ (3.2, Op. Sk. 4)

3.   (a)   strong acid   (c)   weak acid
(b)   weak base   (d)   strong base   (3.4, Op. Sk. 3)

4.   (a) $S^{2-}(aq) + Ba^{2+}(aq) \longrightarrow BaS(s)$ (b)   *NR*   (c)   *NR*

(d) $Pb^{2+}(aq) + CO_3{}^{2-}(aq) \longrightarrow PbCO_3(s)$ (3.2, 3.3, Op. Sk. 1, 2)

5.   (a)   Molecular equation:

$Na_2SO_3(aq) + 2HNO_3(aq) \longrightarrow 2NaNO_3(aq) + H_2O(l) + SO_2(g)$

Net ionic equation:

$SO_3{}^{2-}(aq) + 2H^+(aq) \longrightarrow H_2O(l) + SO_2(g)$

(b)   Molecular equation:

$(NH_4)_2SO_4(aq) + 2NaOH(aq) \longrightarrow 2NH_3(g) + 2H_2O(l) + Na_2SO_4(aq)$

Net ionic equation:

$NH_4{}^+(aq) + OH^-(aq) \longrightarrow NH_3(g) + H_2O(l)$

(c)   Molecular equation:

$BaCO_3(aq) + 2HCl(aq) \longrightarrow BaCl_2(aq) + H_2O(l) + CO_2(g)$

Net ionic equation:

$CO_3{}^{2-}(aq) + 2H^+(aq) \longrightarrow H_2O(l) + CO_2(g)$

(d)   Molecular equation:

$$Na_2S(aq) + H_2SO_4(aq) \longrightarrow Na_2SO_4(aq) + H_2S(g)$$

Net ionic equation:

$$S^{2-}(aq) + 2H^+(aq) \longrightarrow H_2S(g)$$

These reactions go to completion because of the formation of gases that escape from the reaction mixtures.   (3.2, 3.4, Op. Sk. 1, 5)

6.   (a)   Molecular equation:

$$Ca(OH)_2(s) + H_2SO_4(aq) \longrightarrow CaSO_4(aq) + 2H_2O(l)$$

Net ionic equation:

$$Ca(OH)_2(s) + 2H^+(aq) \longrightarrow Ca^{2+}(aq) + 2H_2O(l)$$

(b)   Molecular equation:

$$HC_2H_3O_2(aq) + LiOH(aq) \longrightarrow H_2O(l) + LiC_2H_3O_2(aq)$$

Net ionic equation:

$$HC_2H_3O_2(aq) + OH^-(aq) \longrightarrow C_2H_3O_2^-(aq) + H_2O(l)$$

These reactions are neutralizations; they go to completion because a very weak electrolyte (in this case, water) is formed.   (3.2, 3.4, Op. Sk. 1, 4)

7.   (a) $+7$ (c) $+5$ (e) $+1$ (3.5, Op. Sk. 6)
      (b) $+3$ (d) $+6$

8.   (a) $AuCl_3(aq) + 2KI(aq) \longrightarrow AuCl(aq) + 2KCl(aq) + I_2(s)$

(b) $2HNO_3(aq) + 6HI(aq) \longrightarrow 2NO(g) + 3I_2(s) + 4H_2O(l)$

(c) $NCl_3(l) + 3H_2O(l) \longrightarrow NH_3(aq) + 3HOCl(aq)$

(d) $3SO_3^{2-}(aq) + H_2O(l) + 2CrO_4^{2-}(aq) \longrightarrow$

$$3SO_4^{2-}(aq) + 2CrO_2^-(aq) + 2OH^-(aq)$$

(3.5, Op. Sk. 6, 7)

CHAPTER POST-TEST

1. Classify each of the following reactions as precipitation, acid–base, or oxidtion–reduction:

    (a) $2SO_2 + 2NO_2 \longrightarrow 2SO_3 + 2NO$

    (b) $Al_2(SO_4)_3 + 3BaCl_2 \longrightarrow 3BaSO_4 + 2AlCl_3$

    (c) $Cd + CuSO_4 \longrightarrow CdSO_4 + Cu$

    (d) $CaCl_2 + (NH_4)_2CO_3 \longrightarrow CaCO_3 + 2NH_4Cl$

    (e) $H_3PO_4 + 2NaOH \longrightarrow Na_2HPO_4 + 2H_2O$

2. Classify each of the following as a strong acid, weak acid, strong base, or weak base: (a)   HI   (b) $Sr(OH)_2$ (c) $HC_9H_7O_4$ (d)   KOH

3. Write the net ionic equations for the following molecular equations:

    (a) $2Na_3PO_4(aq) + Fe_2(SO_4)_3(aq) \longrightarrow 2FePO_4(s) + 3Na_2SO_4(aq)$

    (b) $KCl(aq) + AgNO_3(aq) \longrightarrow KNO_3(aq) + AgCl(s)$

    (c) $CaCO_3(s) + 2HCl(aq) \longrightarrow H_2O(l) + CO_2(g) + CaCl_2(aq)$

    (d) $H_2SO_4(aq) + 2KOH(aq) \longrightarrow K_2SO_4(aq) + 2H_2O(l)$

4. Write the molecular and net ionic equations for the successive reactions of each acidic hydrogen of phosphoric acid with sodium hydroxide.

5. Aqueous solutions of potassium chloride and silver nitrate are mixed. If a reaction occurs, write the molecular and net ionic equations. If no reaction occurs, write *NR* after the arrow.

6. Write the molecular and net ionic equations for the reaction of ammonium chloride with sodium hydroxide.

7. Determine the oxidation number for the underlined atom in each of the following species:

    (a) $K_2\underline{Cr}O_4$ (b) $Na_2\underline{S}O_3$ (c) $Ba\underline{H}_2$ (d) $\underline{N}O_2^-$ (e) $\underline{Fe}(OH)_3$

8. Write the skeleton equation for the oxidation–reduction reaction below. Label the oxidizing and reducing agents and the oxidized and reduced species, and write the oxidation and reduction half-reactions.

$$2AuCl_4^-(aq) + 3Cu(s) \longrightarrow 2Au(s) + 8Cl^-(aq) + 3Cu^{2+}(aq)$$

9. Balance the following equation using the half-reaction method:

$$CN^- + MnO_4^- \longrightarrow CNO^- + MnO_2 \quad \text{(basic)}$$

10. Balance the following equation using the half-reaction method:

$$H_2SO_4 + HI \longrightarrow I_2 + SO_2 + H_2O$$

ANSWERS TO CHAPTER POST-TEST

If you missed an answer, study the text section and operational skill given in parentheses after the answer.

1. (a) $2SO_2 + 2NO_2 \longrightarrow 2SO_3 + 2NO$ Oxidation–reduction reaction

    (b) $Al_2(SO_4)_3 + 3BaCl_2 \longrightarrow 3BaSO_4 + 2AlCl_3$ Precipitation reaction

    (c) $Cd + CuSO_4 \longrightarrow CdSO_4 + Cu$ Oxidation–reduction reaction

    (d) $CaCl_2 + (NH_4)_2CO_3 \longrightarrow CaCO_3 + 2NH_4Cl$ Precipitation reaction

    (e) $H_3PO_4 + 2NaOH \longrightarrow Na_2HPO_4 + 2H_2O$
                 Acid–base reaction   (3.3, 3.4, 3.5, Op. Sk. 2, 6)

2. (a)   strong acid   (c)   weak acid
    (b)   strong base   (d)   strong base   (3.4, Op. Sk. 3)

3.   (a) $PO_4^{3-}(aq) + Fe^{3+}(aq) \longrightarrow FePO_4(s)$

     (b) $Ag^+(aq) + Cl^-(aq) \longrightarrow AgCl(s)$

     (c) $CaCO_3(s) + 2H^+(aq) \longrightarrow H_2O(l) + CO_2(g) + Ca^{2+}(aq)$

     (d) $H^+(aq) + OH^-(aq) \longrightarrow H_2O(l)$ (3.3, 3.4, Op. Sk. 2)

4.   (a)   Molecular equation:

$$NaOH(aq) + H_3PO_4(aq) \longrightarrow NaH_2PO_4(aq) + H_2O(l)$$

           Net ionic equation:

$$OH^-(aq) + H_3PO_4(aq) \longrightarrow H_2O(l) + H_2PO_4^-(aq)$$

     (b)   Molecular equation:

$$NaOH(aq) + NaH_2PO_4(aq) \longrightarrow Na_2HPO_4(aq) + H_2O(l)$$

           Net ionic equation:

$$OH^-(aq) + H_2PO_4^-(aq) \longrightarrow H_2O(l) + HPO_4^{2-}(aq)$$

     (c)   Molecular equation:

$$NaOH(aq) + Na_2HPO_4(aq) \longrightarrow Na_3PO_4(aq) + H_2O(l)$$

           Net ionic equation: $OH^-(aq) + HPO_4^{2-}(aq) \longrightarrow$
$$H_2O(l) + PO_4^{3-}(aq) \quad (3.3, 3.4, 3.5, \text{Op. Sk. 2, 4})$$

5.   Molecular equation: $KCl(aq) + AgNO_3(aq) \longrightarrow KNO_3(aq) + AgCl(s)$

     Net ionic equation: $Ag^+(aq) + Cl^-(aq) \longrightarrow AgCl(s)$
                                              (3.3, Op. Sk. 1, 2)

6.   Molecular equation:

$$NH_4Cl(aq) + NaOH(aq) \longrightarrow NH_3(g) + H_2O(l) + NaCl(aq)$$

     Net ionic equation:

$$NH_4^+(aq) + OH^-(aq) \longrightarrow NH_3(g) + H_2O(l)$$

                                              (3.4, Op. Sk. 5)

7. (a) $+6$ (c) -1 (e) $+3$ (3.4, Op. Sk. 6)
    (b) $+4$ (d) $+3$

8.

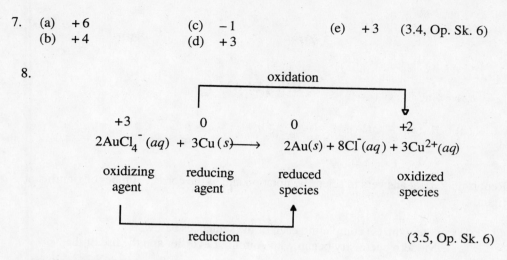

Reduction half-reaction:

$$3e^- + AuCl_4^-(aq) \longrightarrow Au(s) + 4Cl^-(aq)$$

Oxidation half-reaction:

$$Cu(s) \longrightarrow Cu^{2+}(aq) + 2e^-$$

9. $H_2O + 3CN^- + 2MnO_4^- \longrightarrow 3CNO^- + 2MnO_2 + 2OH^-$ (3.5, Op. Sk. 6)

10. $H_2SO_4 + 2HI \longrightarrow I_2 + SO_2 + 2H_2O$ (3.5, Op. Sk. 6)

UNIT EXAM 1

Report numerical answers in scientific notation and to the correct number of significant figures.

1. Write the word (s) completing each of the sentences below.
   (a) Modern chemistry began with Antoine Lavoisier and the use of the
   _____ .

   (b) A generalization about the behavior of nature that can be stated briefly is a
   _____ .

   (c) A new and untested explanation is a _____ .
   (d) The separation technique one could use to separate a solution of leaf pigments
   is _____ .

2. Perform the following arithmetic operations:

   (a) $31.149 + 123.2116$ (c) $(4.63 \times 10^8) \div (5.1694 \times 10^{-10})$

   (b) $(3.731 \times 10^{-4})^2$

3. Make the following conversions:
   (a) 4.87 pg to Mg   (c) 4.14 K to °F

   (b) 2.39×10^{-8} kL to dL   (d) 4.6×10^3 m to yd

4. What is the volume of 184 g of mercury?   (Density = 13.6 g/cm^3)

5. Classify each of the following as element (E), compound (C), homogeneous mixture (H), or heterogeneous mixture (M):

   _____ (a) ice cubes in water   _____ (d) sulfur

   _____ (b) water, H_2O _____ (e) rum

   _____ (c) an orange   _____ (f) iodine crystals

6. Match each term in the left-hand column with an appropriate expression in the right-hand column.

___ (1) metals
___ (2) ions
___ (3) molecular formula
___ (4) salts
___ (5) metalloids
___ (6) molecules
___ (7) strong acids
___ (8) strong bases

(a) $C_{16}H_{10}N_2$
(b) HCl, HNO_3, H_2SO_4
(c) As, B, Si, Ge
(d) 6.02×10^{23}
(e) KOH, $NaOH$, $Ba(OH)_2$
(f) $HC_2H_3O_2$, KNO_2, H_2CO_3
(g) $C_{18}H_{35}NO$, SF_6, Cl_2
(h) Al, K, Na, Cu, Zn
(i) KCl, Na_2SO_4, $Ni(NO_3)_2$
(j) NO_3^-, K^+, $C_2H_3O_2^-$

7. Write the formula for each of the following:

(a) magnesium nitride
(b) zinc phosphate
(c) manganese dioxide
(d) mercurous sulfate
(e) sodium oxalate
(f) cadmium acetate trihydrate

8. Write an acceptable name for each of the following:

(a) $(NH_4)_2S_2O_3$
(b) $K_2Cr_2O_7$
(c) PbO
(d) $CaCO_3 \cdot 6H_2O$
(e) $Ba(ClO_4)_2$
(f) SnF_2

9. Indicate whether each of the following is ionic or molecular in nature:

(a) OF_2
(b) $CaCl_2$
(c) KNO_3
(d) $SbBr_3$

10. Write the appropriate name and formula of an oxoanion or oxoacid for each of the following:

(a) nitric acid, HNO_3

(b) chlorate ion, ClO_3^-

(c) carbonic acid, H_2CO_3

(d) sulfite ion, SO_3^{2-}

11. Give some reasons for considering the elements Li, Na, K, Rb, and Cs as members of one group in the periodic table.

12.   Balance the following equations and classify each reaction as either precipitation, acid–base, or oxidation–reduction:

(a) $\_ MgCO_3 \xrightarrow{\Delta} \_ MgO + \_ CO_2$

(b) $\_ NaOH + \_ H_3PO_4 \longrightarrow \_ Na_3PO_4 + \_H_2O$

(c) $\_ AgNO_3 + \_ CuCl_2 \longrightarrow \_ AgCl + \_ Cu(NO_3)_2$

(d) $\_ Al + \_ Br_2 \longrightarrow \_ AlBr_3$

(e) $\_ Ba(OH)_2 + \_ HC_2H_3O_2 \longrightarrow \_ Ba(C_2H_3O_2)_2 + \_ H_2O$

13.   Write the net ionic equation for each reaction that will occur between the following soluble compounds to form a precipitate. If no reaction takes place, write *NR*.

(a) $NaClO_4 + Sr(C_2H_3O_2)_2$ (c) $KI + Ba(OH)_2$

(b) $(NH_4)_2SO_4 + SrCl_2$ (d) $Li_3PO_4 + Mg(NO_2)_2$

14.   Write the molecular and net ionic equations for the reaction in aqueous solution of sulfurous acid with excess lithium hydroxide.

15.   Write the oxidation number for each element in each of the following:

(a) KNO_3 (b) $Na_2Cr_2O_7$ (c) MgH_2

16.   Write the skeleton equation for the oxidation–reduction reaction below. Label the oxidizing and reducing agents and the oxidized and reduced species, and note the oxidation and reduction parts of the half-reactions.

$$2AuCl_4^-(aq) + 3Cu(s) \longrightarrow 2Au(s) + 8Cl^-(aq) + 3Cu^{2+}(aq)$$

17.   Balance the following equation using the half-reaction method:

$$ClO_3^- + S^{2-} \longrightarrow Cl^- + S \quad \text{(basic)}$$

ANSWERS TO UNIT EXAM 1

If you missed an answer, study the text section and operational skill given in parentheses after the answer.

1. (a) balance (1.3)
    (b) law (principle) (1.2)
    (c) hypothesis (1.2)
    (d) column chromatography (1.4, Instrumental Methods: Separation of Mixtures by Chromatography)

2. (a) 154.361 (b) 1.392×10^{-7} (c) 8.96×10^{17} (1.5, Op. Sk. 2)

3. (a) 4.87×10^{-18} Mg (c) $-452.22°F$

    (b) 2.39×10^{-4} dL (d) 5.0×10^{3} yd (1.6, 1.8, Op. Sk. 2, 3, 6)

4. 13.5 cm^3 (1.7, 1.8, Op. Sk. 2, 5)

5. (a) M (b) C (c) M (d) E (e) H (f) E (1.4)

6. (1) h (2.5) (2) j (2.6) (3) a (2.6) (4) i (2.6)

    (5) c (2.5) (6) g (2.6) (7) b (3.4) (8) e (3.4)

7. (a) Mg_3N_2 (d) Hg_2SO_4

    (b) $Zn_3(PO_4)_2$ (e) $Na_2C_2O_4$

    (c) MnO_2 (f) $Cd(C_2H_3O_2)_2 \cdot 3H_2O$ (2.7, Op. Sk. 4)

8. (a) ammonium thiosulfate
    (b) potassium dichromate
    (c) lead(II) oxide or plumbous oxide
    (d) calcium carbonate hexahydrate
    (e) barium perchlorate
    (f) tin(II) fluoride or stannous fluoride (2.7, Op. Sk. 4)

9. (a) molecular (b) ionic (c) ionic (d) molecular (2.7)

10. (a) nitrate ion, NO_3^- (c) carbonate ion, CO_3^{2-}

    (b) chloric acid, $HClO_3$ (d) sulfurous acid, H_2SO_3 (2.7, Op. Sk. 5)

11. The Group IA metals are all shiny, soft metals with low melting points. They all react vigorously with water to form hydroxides of the general formula MOH, where M corresponds to the Group IA metal. They also react vigorously with the halogens to form halides of the general formula MX, where M corresponds to the Group IA metal, and X to the Group VIIA nonmetal.   (2.5)

12. (a) balanced;   Oxidation–reduction reaction

    (b) $3NaOH + H_3PO_4 \longrightarrow Na_3PO_4 + 3H_2O$ Acid–base reaction

    (c) $2AgNO_3 + CuCl_2 \longrightarrow 2AgCl + Cu(NO_3)_2$ Precipitation reaction

    (d) $2Al + 3Br_2 \longrightarrow 2AlBr_3$ Combination reaction or oxidation–reduction

    (e) $Ba(OH)_2 + 2HC_2H_3O_2 \longrightarrow Ba(C_2H_3O_2)_2 + 2H_2O$ Acid–base reaction
    (2.9, Op. Sk. 6; 3.3, 3.4, 3.5, Op. Sk. 2)

13. (a) *NR*   (b) $SO_4^{2-}(aq) + Sr^{2+}(aq) \longrightarrow SrSO_4(s)$ (c) *NR*

    (d) $2PO_4^{3-}(aq) + 3Mg^{2+}(aq) \longrightarrow Mg_3(PO_4)_2(s)$ (3.2, 3.3, Op. Sk. 1, 2)

14. Molecular equation:

$$H_2SO_3(aq) + 2LiOH(aq) \longrightarrow 2H_2O(l) + Li_2SO_3(aq)$$

Net ionic equation:

$$H^+(aq) + OH^-(aq) \longrightarrow H_2O(l)$$ (3.4, Op. Sk. 4)

15. (a) $K = +1, N = +5, O = -2$
    (b) $Na = +1, Cr = +6, O = -2$
    (c) $Mg = +2, H = -1$ (3.4, Op. Sk. 6)

16.

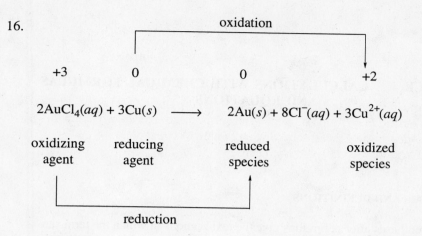

$$2AuCl_4(aq) + 3Cu(s) \longrightarrow 2Au(s) + 8Cl^-(aq) + 3Cu^{2+}(aq)$$

Reduction half-reaction:

$$3e^- + AuCl_4^-(aq) \longrightarrow Au(s) + 4Cl^-(aq)$$

Oxidation half-reaction:

$$Cu(s) \longrightarrow Cu^{2+}(aq) + 2e^-$$

17. $3H_2O + ClO_3^- + 3S^{2-} \longrightarrow Cl^- + 6OH^- + 3S$ (3.5, Op. Sk. 6)

CHAPTER 4   CALCULATIONS WITH CHEMICAL FORMULAS AND EQUATIONS

CHAPTER TERMS AND DEFINITIONS

Numbers in parentheses after definitions give the text sections in which the terms are explained. Starred terms are italicized in the text. Where a term does not fall directly under a text section heading, additional information is given for you to locate it.

vinaigre*   French for "sour wine," from which the name vinegar is derived   (4.1, introductory section)

molecular weight (MW)   sum of the atomic weights of all the atoms in a molecule   (4.1)

formula weight (FW)   sum of the atomic weights of all the atoms in a formula unit of a compound   (4.1)

mole concept*   idea of working with enormous quantities of tiny particles in groups called moles   (4.2)

mole (mol)   amount of substance that contains as many molecules or formula units as the number of atoms in exactly 12 g of carbon-12   (4.2)

Avogadro's number (N_A)   number of atoms in exactly 12 g of carbon-12; to three significant figures: 6.02×10^{23} (4.2)

molar mass (g/mol)   mass of one mole of a substance; atomic weight, or formula or molecular weight, expressed in grams   (4.2)

percentage composition   mass percentages of each element in a compound   (4.3, introductory section)

mass percentage   parts per hundred in terms of mass:

$$\text{Mass \% of A} = \frac{\text{mass of A in the whole}}{\text{mass of the whole}} \times 100\% \qquad (4.3)$$

empirical (simplest) formula   chemical formula of a substance with the smallest integer (whole number) subscripts   (4.5)

stoichiometry   calculation of the quantities of reactants and products involved in a chemical reaction   (4.6, introductory section)

limiting reactant (limiting reagent)   reactant that is entirely consumed when a reaction goes to completion   (4.8)

excess reactant*   reactant that is not completely consumed when a reaction goes to completion   (4.8)

theoretical yield   maximum amount of product that can be obtained in a reaction; calculated based on the limiting reagent   (4.8)

actual yield*   amount of product obtained from an experimental determination   (4.8)

percentage yield   actual yield (experimentally determined) of product from a reaction expressed as a percentage of the theoretical yield (calculated) of product:

$$\text{Percentage yield} = \frac{\text{actual yield}}{\text{theoretical yield}} \times 100\% \qquad (4.8)$$

solute*   substance dissolved, usually in a liquid, to form a solution   (4.9)

solvent*   component of a solution that is present in excess   (4.9)

concentration*   quantity of solute in a standard quantity of solution   (4.9)

dilute*   describing solution containing a low concentration of solute   (4.9)

concentrated*   describing solution containing a high concentration of solute; maximum concentration sold in a commercially available solution   (4.9)

molar concentration (molarity) (M)   number of moles of solute dissolved in one liter (cubic decimeter) of solution   (4.9)

initial molarity (M_i)*   molarity of a solution before dilution   (4.10)

final molarity (M_f)*   molarity of a solution after dilution   (4.10)

qualitative analysis*   identification of substances or species present in a material   (4.11, introductory section)

quantitative analysis   determination of the amount of a substance or species present in a material   (4.11, introductory section)

gravimetric analysis   type of quantitative analysis in which the amount of a species in a material is determined by converting the species to a product that can be isolated completely and weighed   (4.11)

titration   procedure for determining the amount of substance A by adding a carefully measured volume of a solution of known concentration of B until the reaction of A and B is just complete   (4.12)

volumetric analysis   method of analysis based on titration   (4.12)

buret*   glass tube graduated to measure the volume of liquid delivered from the stopcock   (4.12)

glacial*   icelike   (An Acid That Matters: Acetic Acid [a Weak Acid])

CHAPTER DIAGNOSTIC TEST

1. Determine whether each of the following statements is true or false. If a statement is false, change it so it is true.

    (a)   The balanced equation $CaCO_3 + H_2SO_4 \longrightarrow CaSO_4 + CO_2 + H_2O$ indicates that 1 mole of H_2SO_4 will produce 44 g CO_2. True/False:

    _____ .

    (b)   The limiting reactant in a chemical reaction determines the maximum quantity of any product obtainable from the reaction. True/False: _____

    _____ .

    (c)   The percentage yield of product indicates the minimum amount of product available from the chemical reaction. True/False: _____

    _____ .

2. Calculate:

    (a)   the mass in grams of a molecule of blood sugar, glucose, $C_6H_{12}O_6$;
    (b)   the number of molecules in 12.5 g of glucose;
    (c)   the mass percentage of the elements in glucose.

3. A 4.64-g sample of a compound containing only carbon, hydrogen, and oxygen was analyzed by combustion analysis. In all, 11.7 g CO_2 were collected. What was the percentage of carbon in the compound?

4. Determine the empirical formula of a compound that contains 89.7% bismuth and 10.3% oxygen.

5. Write the molecular formula for a compound with empirical formula C_2H_4O and a molecular weight of 88.0 amu.

6. Refer to the following reaction to answer the questions below:

    $$KClO_3 + 6KBr + 3H_2SO_4 \longrightarrow KCl + 3Br_2 + 3H_2O + 3K_2SO_4$$

    (a)   How many moles of H_2SO_4 are needed to react with 3.70 g $KClO_3$?

    (b)   If 305 g $KClO_3$ and 415 g KBr are mixed with excess H_2SO_4, what is the theoretical yield of Br_2?

    (c)   If 234 g Br_2 are recovered, what is the percentage yield in this reaction?

7. A stock solution of dilute nitric acid is 32 mass percentage HNO_3. Its density is 1.19 g/mL. What is the molarity of this solution?

8. Your lab experiment calls for 325 mL 0.250 M HCl. What volume of 6.00 M HCl must be used to make the desired solution?

9. How many milliliters of 0.100 M $Ca(OH)_2$ must be added to completely react with 45.0 mL of 0.150 M H_3PO_4? The reaction products are $Ca_3(PO_4)_2$ and water.

ANSWERS TO CHAPTER DIAGNOSTIC TEST

If you missed an answer, study the text section and operational skill given in parentheses after the answer.

1. (a) True (4.6), (b) True (4.8), (c) False. The percentage yield of product indicates the actual amount of product obtained from the chemical reaction. (4.8)

2. (a) 2.99×10^{-22} g (4.1, 4.2, Op. Sk. 1, 2)

   (b) 4.18×10^{22} molecules (4.2, Op. Sk. 4)

   (c) 40.0% C, 6.72% H, 53.3% O (4.4, Op. Sk. 5)

3. 68.8% (4.4, Op. Sk. 6)   4. Bi_2O_3 (4.5, Op. Sk. 8)

5. $C_4H_8O_2$ (4.5, Op. Sk. 9)   6. (a) 0.0905 mol H_2SO_4 (4.7, Op. Sk. 10)

                                        (b) 279 g Br_2 (4.8, Op. Sk. 1)

                                        (c) 83.9% (4.8)

7. 6.0 M HNO_3 (4.10, Op. Sk. 12, 13)   8. 13.5 mL (4.10, Op. Sk. 14)

9. 101 mL (4.12, Op. Sk. 17)

SUMMARY OF CHAPTER TOPICS

4.1   MOLECULAR WEIGHT AND FORMULA WEIGHT

Operational Skill

1.   Calculating the formula weight from a formula.   Given the formula of a
compound and a table of atomic weights, calculate the formula weight   (Example 4.1).

> **Exercise 4.1**   Calculate the formula weights of the following compounds, using
> a table of atomic weights. Give the answers to three significant figures:
>
> (a)   nitrogen dioxide, NO_2 (c)   sodium hydroxide, NaOH
>
> (b)   glucose, $C_6H_{12}O_6$ (d)   magnesium hydroxide, $Mg(OH)_2$

Solution: (a) $1 \times$ AW of N $=$ 14.0 amu
 $2 \times$ AW of O $= 2 \times 16.0 = $ <u>32.0 amu</u>
 FW of NO_2 $=$ 46.0 amu

 (b) $6 \times$ AW of C $= 6 \times 12.0 = $ 72.0 amu
 $12 \times$ AW of H $= 12 \times 1.0 = $ 12.0 amu
 $6 \times$ AW of O $= 6 \times 16.0 = $ <u>96.0 amu</u>
 FW of $C_6H_{12}O_6$ $=$ 180.0 amu
 $= 1.80 \times 10^{-2}$ amu

 (c) $1 \times$ AW of Na $=$ 23.0 amu
 $1 \times$ AW of O $=$ 16.0 amu
 $1 \times$ AW of H $=$ <u>1.0 amu</u>
 FW of NaOH $=$ 40.0 amu

 (d) $1 \times$ AW of Mg $=$ 24.3 amu
 $2 \times$ AW of O $= 2 \times 16.0 = $ 32.0 amu
 $2 \times$ AW of H $= 2 \times 1.0 = $ <u>2.0 amu</u>
 FW of $Mg(OH)_2$ $=$ 58.3 amu

4.2   THE MOLE CONCEPT

Operational Skills

2.   Calculating the mass of an atom or molecule.   Using the molar mass and
Avogadro's number, calculate the mass of an atom or molecule in grams   (Example 4.2).

3.   Converting moles of substance to grams, and vice versa.   Given the moles of a compound with a known formula, calculate the mass   (Example 4.3). Or, given the mass of a compound with a known formula, calculate the moles   (Example 4.4).

4.   Calculating the number of molecules in a given mass.   Given the mass of a sample of a molecular substance and its formula, calculate the number of molecules in the sample   (Example 4.5).

The mole is one of the most useful concepts in chemistry. It is important that you practice using it, and related terms, to be able to quickly say the number of particles that are in a mole of a substance and how much this number of particles weighs. For example, a mole of sodium consists of Avogadro's number (6.02×10^{23}) of sodium atoms and weighs 23.0 g (the atomic weight of sodium in amu's expressed in grams). Likewise, a mole of water consists of 6.02×10^{23} water molecules and weighs 18.0 g, and a mole of sodium chloride consists of 6.02×10^{23} formula units of NaCl and weighs 58.5 g.

If we know the mass of a mole of particles and the number of particles in a mole, we can calculate the mass of an individual particle, such as an atom. Example 4.2 in the text and Exercise 4.2, which follows, involve this concept.

Exercise 4.2   (a)   What is the mass in grams of a calcium atom, Ca?
(b)   What is the mass in grams of an ethanol molecule, C_2H_5OH?

(a)   *Wanted:*   mass of a Ca atom (g/atom).

Known: The atomic weight of Ca is 40.08 amu, so the molar mass is 40.08 g/mol; 1 mol Ca = 6.02×10^{23} atoms.

Solution: $\dfrac{40.08 \text{ g}}{\text{mol}} \times \dfrac{\text{mol}}{6.02 \times 10^{23} \text{ atoms}} = 6.66 \times 10^{-23}$ g/atom

(b)   *Wanted:*   mass of an ethanol molecule (g/molecule).

Given: Formula is C_2H_5OH.

Known: We can calculate the molar mass by adding the atomic weights; 1 mol = 6.02×10^{23} molecules.

Solution: First calculate the molar mass.

$$2 \text{ AW of C} = 2 \times 12.0 = 24.0 \text{ amu}$$
$$6 \text{ AW of H} = 6 \times 1.0 = 6.0 \text{ amu}$$
$$1 \text{ AW of O} = 1 \times 16.0 = \underline{16.0 \text{ amu}}$$
$$46.0 \text{ amu}$$

$$\text{molar mass} = 46.0 \text{ g/mol}$$

Then calculate the grams per molecule.

$$\frac{46.0 \text{ g}}{\cancel{\text{mol}}} \times \frac{\cancel{\text{mol}}}{6.02 \times 10^{23} \text{ molecules}} = 7.64 \times 10^{-23} \text{ g/molecule}$$

Exercise 4.3   Hydrogen peroxide, H_2O_2, is a colorless liquid. A concentrated solution of it is used as a source of oxygen for rocket propellant fuels. Dilute aqueous solutions are used as a bleach. Analysis of a solution shows that it contains 0.909 mol H_2O_2 in 1.00 L of solution. What is the mass of hydrogen peroxide in this volume of solution?

Wanted: mass of H_2O_2 (assume g).

Given: 0.909 mol H_2O_2/L solution.

Known: We can get the molar mass from the formula weight.

Solution: FW of H_2O_2 = $(2 \times 1.01) + (2 \times 16.0) = 34.\underline{0}2$ amu, keeping one extra digit.

$$\text{g } H_2O_2 = 0.909 \; \cancel{\text{mol}} \; H_2O_2 \times \frac{34.\underline{0}2 \text{ g}}{\cancel{\text{mol}}} = 30.9 \text{ g}$$

Exercise 4.4   Nitric acid, HNO_3, is a colorless, corrosive liquid used in the manufacture of nitrogen fertilizers and explosives. In an experiment to develop new explosives for mining operations, a 28.5-g sample of nitric acid was poured into a beaker. How many moles of HNO_3 are there in this sample of nitric acid?

Known: We need the molar mass.

Solution: FW of HNO_3 = $(1 \times 1.01) + (1 \times 14.0) + (3 \times 16.0) = 63.\underline{0}1$ amu, keeping one extra digit.

$$\text{mol } HNO_3 = 28.5 \; \cancel{\text{g}} \times \frac{1 \text{ mol}}{63.\underline{0}1 \; \cancel{\text{g}}} = 0.452 \text{ mol}$$

Exercise 4.5   Hydrogen cyanide, HCN, is a volatile, colorless liquid with the odor of certain fruit pits (such as peach and cherry pits). The compound is highly poisonous. How many molecules are there in 56 mg HCN, the average toxic dose?

Known: We can use the mole concept, molar mass, and N_A.

Solution: FW of HCN = 1.01 + 12.0 + 14.0 = 27.01 amu.

To find molecules of HCN:

$$56 \text{ mg HCN} \times \frac{\text{g HCN}}{10^3 \text{ mg HCN}} \times \frac{\text{mol HCN}}{27.01 \text{ g HCN}} \times \frac{6.02 \times 10^{23} \text{ molecules HCN}}{\text{mol HCN}}$$

$$= 1.2 \times 10^{21} \text{ molecules HCN}$$

4.3   MASS PERCENTAGES FROM THE FORMULA

Operational Skill

5.   Calculating the percentage composition from the formula.   Given the formula of a compound, calculate the mass percentages of the elements in it   (Example 4.6).

6.   Calculating the mass of an element in a given mass of compound.   Given the mass percentages of elements in a given mass of a compound, calculate the mass of any element   (Example 4.7).

Percentage means per 100. Thus, we can make a conversion factor from a given percentage. If a sample is 84% NaCl, the conversion factor would be

$$\frac{84 \text{ g NaCl}}{100 \text{ g sample}}$$

We specify "mass" percentage because we can also calculate volume percentage, which is percentage in terms of volume rather than mass. Mass percentage is used when working with solids, and volume percentage is used with liquids, which are more easily measured by volume.

Exercise 4.6   Ammonium nitrate, NH_4NO_3, which is prepared from nitric acid, is used as a nitrogen fertilizer. Calculate the mass percentages of the elements in ammonium nitrate (to three significant figures).

Wanted: mass % of elements in ammonium nitrate to three significant figures.

Given: Formula is NH_4NO_3.

Known: Mass % = $\dfrac{\text{mass of element A}}{\text{mass of whole}} \times 100\%.$

The "whole" we use is one mole of ammonium nitrate.

Solution: FW of NH_4NO_3 = $(2 \times 14.0) + (4 \times 1.01) + (3 \times 16.0)$

= 80.04 amu

Mass % N = $\dfrac{2 \times 14.0 \text{ g}}{80.04 \text{ g}} \times 100\%$ = 35.0%

Mass % H = $\dfrac{4 \times 1.01 \text{ g}}{80.04 \text{ g}} \times 100\%$ = 5.05%

Mass % O = $\dfrac{3 \times 16.0 \text{ g}}{80.04 \text{ g}} \times 100\%$ = 60.0%

The mass % of O can also be obtained by subtracting the sum of the mass percentages of N and H from 100%.

Exercise 4.7   How many grams of nitrogen, N, are there in a fertilizer containing 48.5 g of ammonium nitrate and no other nitrogen-containing compound? See Exercise 4.6 for the percentage composition of NH_4NO_3.

Wanted: grams of nitrogen.

Given: 48.5 g NH_4NO_3.

Known: Mass % N in NH_4NO_3 = 35.0%, from Exercise 4.6.

Solution: 48.5 g ammonium nitrate $\times \dfrac{35.0 \text{ g N}}{100 \text{ g ammonium nitrate}}$ = 17.0 g N

4.4   ELEMENTAL ANALYSIS: PERCENTAGES OF CARBON, HYDROGEN, AND OXYGEN

Operational Skill

7.   Calculating the percentages of C and H by combustion.   Given the masses of CO_2 and H_2O obtained from the combustion of a known mass of a compound of C, H, and O, compute the mass percentage of each element   (Example 4.8).

In this section you will learn to calculate mass percentages of elements in compounds just as analytical chemists do.

Exercise 4.8   A 3.87-mg sample of ascorbic acid (vitamin C) gives 5.80 mg CO_2 and 1.58 mg H_2O on combustion. What is the percentage composition of this compound (the mass percentage of each element)? Ascorbic acid contains only C, H, and O.

Wanted: percentage composition of vitamin C.

Given: Combustion of 3.87-mg sample gives 5.80 mg CO_2, 1.58 mg H_2O; ascorbic acid contains C, H, and O.

Known: definition of percentage composition; $1 \text{ g} = 10^3$ mg. All of the C in CO_2 and H in H_2O came from the sample; we first must find the mg of C and of H from the percentage compositions of CO_2 and H_2O (using dimensional analysis). To do this, we need the molecular weights of CO_2 and H_2O. Then we find the mg of O by subtracting mg (C + H) from mg of sample. We can then calculate the mass percentages.

Solution: Find molecular weights:

$$CO_2 = 12.0 + (2 \times 16.0) = 44.0 \text{ amu}$$

$$H_2O = (2 \times 1.01) + 16.0 = 18.0 \text{ amu}$$

Find mg C:

$$5.80 \times 10^{-3} \text{ g } CO_2 \times \frac{1 \text{ mol } CO_2}{44.0 \text{ g } CO_2} \times \frac{12.0 \text{ g C}}{1 \text{ mol } CO_2} \times \frac{10^3 \text{ mg C}}{1 \text{ g C}} = 1.5\underline{8}2 \text{ mg}$$

Find mg H:

$$1.58 \times 10^{-3} \text{ g } H_2O \times \frac{1 \text{ mol } H_2O}{18.0 \text{ g } H_2O} \times \frac{2.02 \text{ g H}}{1 \text{ mol } H_2O} \times \frac{10^3 \text{ mg H}}{1 \text{ g H}} = 0.17\underline{7}3 \text{ mg}$$

Then:

$$\text{Mass \% C} = \frac{1.5\underline{8}2 \text{ mg C}}{3.87 \text{ mg sample}} \times 100\% \quad = \quad 40.9\%$$

$$\text{Mass \% H} = \frac{0.17\underline{7}3 \text{ mg H}}{3.87 \text{ mg sample}} \times 100\% \quad = \quad 4.58\%$$

$$\text{Mass \% O} = 100\% - (40.9\% + 4.58\%) \quad = \quad 54.5\%$$

4.5   DETERMINING FORMULAS

Operational Skills

8.   Determining the empirical formula from percentage composition.   Given the masses of elements in a known mass of compound, or given its percentage composition, obtain the empirical formula   (Examples 4.9 and 4.10).

9.   Determining the molecular formula from percentage composition and molecular weight.   Given the empirical formula and molecular weight of a substance, obtain its molecular formula   (Example 4.11).

In this section, you see how to use percentage composition to obtain the simplest ratio of atoms in a compound (the empirical formula). Then, if the substance is molecular, you find how to use molecular-weight data from another experiment to determine the molecular formula.

Exercise 4.9   A sample of compound weighing 83.5 g contains 33.4 g of sulfur. The rest is oxygen. What is the empirical formula?

Wanted: empirical formula of the compound.

Given: 83.5 g of compound; 33.4 g sulfur; other element is oxygen.

Known: definition of empirical formula; get mass of oxygen by subtracting the two given masses; atomic weights of elements.

Solution: Calculate mass of oxygen:

83.5 g compound – 33.4 g sulfur = 50.1 g oxygen

Find moles of S:

$$33.4 \ \cancel{g \ S} \times \frac{1 \ \text{mol S}}{32.1 \ \cancel{g \ S}} = 1.040 \ \text{mol S}$$

Find moles of O:

$$50.1 \ \cancel{g \ O} \times \frac{1 \ \text{mol O}}{16.0 \ \cancel{g \ O}} = 3.131 \ \text{mol O}$$

Change to integers by dividing by 1.040, the smaller.

For S: $\dfrac{1.040}{1.040} = 1.000$ For O: $\dfrac{3.131}{1.040} = 3.011$

To two significant digits, the numbers are 1.0 and 3.0; thus, the empirical formula is SO_3.

Exercise 4.10   Benzoic acid is a white, crystalline powder used as a food preservative. The compound contains 68.8% C, 5.0% H, and 26.2% O, by mass. What is its empirical formula?

Wanted: empirical formula of benzoic acid.

Given: Compound is 68.8% C, 5.0% H, 26.2% O.

Known: Percent means per 100. In 100 g benzoic acid there are 68.8 g C, 5.0 g H, and 26.2 g O. Convert these to moles, divide by the smallest number, then, if necessary, adjust by multiplication to get whole numbers (integers).

Solution: Find moles C:

$$68.8 \text{ g C} \times \frac{1 \text{ mol C}}{12.0 \text{ g C}} = 5.7\underline{3}3 \text{ mol C}$$

Find moles H:

$$5.0 \text{ g H} \times \frac{1 \text{ mol H}}{1.01 \text{ g H}} = 4.\underline{9}5 \text{ mol H}$$

Find moles O:

$$26.2 \text{ g O} \times \frac{1 \text{ mol O}}{16.0 \text{ g O}} = 1.6\underline{3}8 \text{ mol O}$$

Change to integers by dividing by 1.6$\underline{3}$8 mol (the smallest).

For C: $\dfrac{5.7\underline{3}3 \text{ mol}}{1.6\underline{3}8 \text{ mol}} = 3.50$ For H: $\dfrac{4.\underline{9}5 \text{ mol}}{1.6\underline{3}8 \text{ mol}} = 3.0\underline{2}2$

For O: $\dfrac{1.6\underline{3}8 \text{ mol}}{1.6\underline{3}8 \text{ mol}} = 1.00$

To two significant digits these values are 3.5, 3.0, and 1.0, so we must multiply each by 2 to get 7.0, 6.0, and 2.0; the empirical formula is $C_7H_6O_2$.

Exercise 4.11   The percentage composition of acetaldehyde is 54.5% C, 9.2% H, and 36.3% O, and its molecular weight is 44 amu. Obtain the molecular formula of acetaldehyde.

Wanted: molecular formula of acetaldehyde.

Given: Compound is 54.5% C, 9.2% H, 36.3% O; MW = 44 amu.

Known: The molecular formula is some multiple of the empirical formula. In a 100-g sample there are 54.5 g C, 9.2 g H, and 36.3 g O. Solve for the empirical formula, then determine its formula weight and divide it into the molecular weight to get *n*. Multiply all subscripts in the empirical formula by *n*.

Solution: Solve for the empirical formula:

$$\text{mol C} = 54.5 \, \cancel{g \, C} \times \frac{1 \text{ mol C}}{12.0 \, \cancel{g \, C}} = 4.5\underline{4}2$$

$$\text{mol H} = 9.2 \, \cancel{g \, H} \times \frac{1 \text{ mol H}}{1.01 \, \cancel{g \, H}} = 9.\underline{1}1$$

$$\text{mol O} = 36.3 \, \cancel{g \, O} \times \frac{1 \text{ mol O}}{16.0 \, \cancel{g \, O}} = 2.2\underline{6}9$$

Divide by smallest value:

$$C = \frac{4.5\underline{4}2}{2.2\underline{6}9} = 2.0\underline{0}2 \qquad H = \frac{9.\underline{1}1}{2.2\underline{6}9} = 4.\underline{0}1 \qquad O = \frac{2.2\underline{6}9}{2.2\underline{6}9} = 1.00$$

Empirical formula = C_2H_4O

Empirical formula weight = $(2 \times 12.0) + (4 \times 1.01) + 16.0 = 44.\underline{0}1$ amu

$$n = \frac{44}{44.\underline{0}1} = 1.\underline{0}0$$

The molecular formula is C_2H_4O.

4.6   MOLAR INTERPRETATION OF A CHEMICAL EQUATION

Exercise 4.12   In an industrial process, hydrogen chloride, HCl, is prepared by burning hydrogen gas, H_2, in an atmosphere of chlorine, Cl_2. Write the chemical equation for the reaction. Below the equation, give the molecular, molar, and mass interpretations.

Solution:

$$H_2 \; + \; Cl_2 \; \longrightarrow \; 2HCl$$

$$1 \; H_2 \text{ molecule} \; + \; 1 \; Cl_2 \text{ molecule} \; \longrightarrow \; 2 \text{ HCl molecules}$$

$$1 \text{ mole of } H_2 \; + \; 1 \text{ mole of } Cl_2 \; \longrightarrow \; 2 \text{ moles of HCl}$$

$$2.02 \text{ g } H_2 \; + \; 71.0 \text{ g } Cl_2 \; \longrightarrow \; 2 \times 36.5 \text{ g HCl}$$

4.7   AMOUNTS OF SUBSTANCES IN A CHEMICAL REACTION

Operational Skill

10.   Relating quantities in a chemical equation.   Given a chemical equation and the amount of one substance, calculate the amount of another substance involved in the reaction   (Examples 4.12 and 4.13).

There are two things to remember in solving these types of problems.   (1) In order to do any stoichiometry problem, you must have the balanced equation. If the equation, or information to write it, is not given, you must look back through the chapter to find it. Without it you do not have the necessary relationships to solve the problem.   (2) You must use moles in your solution. Remember that the equation relates the reaction materials by moles. If the given information is not in moles, you must convert it to moles. Then you can use the "mole ratio" factor from the equation coefficients to get moles of the desired substance. This value can then be converted to the unit required for the answer.

> **Exercise 4.13**   Sodium is a soft, reactive metal that instantly reacts with water to give hydrogen gas and a solution of sodium hydroxide, NaOH. How many grams of sodium metal are needed to give 7.81 g of hydrogen by this reaction? (Remember to write the balanced equation first.)

Wanted: g Na metal.

Given: the word equation; the formula for NaOH; 7.81 g hydrogen produced.

Known: We can write the balanced equation using symbols and the given formula. Also, 1 mol Na = 23.0 g Na (molar mass of Na); 1 mol H_2 = 2.02 g H (molar mass of H_2). The reaction gives the conversion factor for moles H_2 to moles Na.

Solution: The equation is $2Na + 2H_2O \longrightarrow 2NaOH + H_2$. Thus, 2 moles of sodium atoms react to produce 1 mole of hydrogen molecules. Thus, the conversion factor is

$$2 \text{ mol Na (atoms)} \rightleftharpoons 1 \text{ mol } H_2 \text{ (molecules)}$$

In the calculation we must determine the moles of hydrogen produced to find the moles, then grams, of sodium that reacted. The calculation is

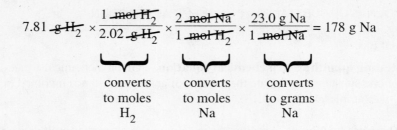

$$7.81 \; \cancel{\text{g H}_2} \times \frac{1 \; \cancel{\text{mol H}_2}}{2.02 \; \cancel{\text{g H}_2}} \times \frac{2 \; \cancel{\text{mol Na}}}{1 \; \cancel{\text{mol H}_2}} \times \frac{23.0 \; \text{g Na}}{1 \; \cancel{\text{mol Na}}} = 178 \; \text{g Na}$$

converts to moles H_2 converts to moles Na   converts to grams Na

Exercise 4.14   Sphalerite is a zinc sulfide (ZnS) mineral and an important commercial source of zinc metal. The first step in the processing of the ore consists of heating the sulfide with oxygen to give zinc oxide, ZnO, and sulfur dioxide, SO_2.

How many kilograms of oxygen gas combine with 5.00×10^3 g of zinc sulfide in this reaction? (You must first write the balanced chemical equation.)

Solution: The equation is

$$2ZnS + 3O_2 \longrightarrow 2ZnO + 2SO_2$$

The calculation is

$$5.00 \times 10^3 \; \cancel{\text{g ZnS}} \times \frac{1 \; \cancel{\text{mol ZnS}}}{97.5 \; \cancel{\text{g ZnS}}} \times \frac{3 \; \cancel{\text{mol O}_2}}{2 \; \cancel{\text{mol ZnS}}} \times \frac{32.0 \; \cancel{\text{g O}_2}}{1 \; \cancel{\text{mol O}_2}} \times \frac{1 \; \text{kg O}_2}{10^3 \; \cancel{\text{g O}_2}}$$

$$= 2.46 \; \text{kg O}_2$$

Exercise 4.15   The British chemist Joseph Priestley prepared oxygen in 1774 by heating mercury(II) oxide, HgO. Mercury metal is the other product. If 6.47 g of oxygen is collected, how many grams of mercury metal are also produced?

Solution: The equation is

$$2HgO \longrightarrow O_2 + 2Hg$$

The calculation is

$$6.47 \; \cancel{\text{g O}_2} \times \frac{1 \; \cancel{\text{mol O}_2}}{32.0 \; \cancel{\text{g O}_2}} \times \frac{2 \; \cancel{\text{mol Hg}}}{1 \; \cancel{\text{mol O}_2}} \times \frac{200.6 \; \cancel{\text{g Hg}}}{1 \; \cancel{\text{mol Hg}}} = 81.1 \; \text{g Hg}$$

4.8   LIMITING REACTANT; THEORETICAL AND PERCENTAGE YIELDS

Operational Skill

11.   Calculating with a limiting reactant.   Given the amounts of reactants and the chemical equation, find the limiting reactant; then calculate the amount of a product (Examples 4.14 and 4.15).

   You will need to solve for the limiting reactant whenever amounts of more than one reactant are given in a reaction stoichiometry problem. To determine this limiting reactant, use the amounts of each reactant to find the moles of (any) product that would be produced if that reactant were the limiting reagent. (If the problem asks for a specific product, use that one.) The reactant giving the smaller (smallest) amount of product is the limiting reactant.

   Exercise 4.16   Aluminum chloride, $AlCl_3$, is used as a catalyst in various industrial reactions. It is prepared from hydrogen chloride gas and aluminum metal shavings.

$$2Al(s) + 6HCl(g) \longrightarrow 2AlCl_3(s) + 3H_2(g)$$

Suppose a reaction vessel contains 0.15 mol Al and 0.35 mol HCl. How many moles of $AlCl_3$ can be prepared from this mixture?

Wanted: moles $AlCl_3$.

Given: the equation; 0.15 mol Al; 0.35 mol HCl.

Known: Because two reactant amounts are given, we must first find the limiting reactant by calculating the moles of product each reactant could produce if it were used completely.

Solution: Determine the limiting reactant by calculating the moles of $AlCl_3$ from each reactant.

$$0.15 \text{ mol Al} \times \frac{2 \text{ mol AlCl}_3}{2 \text{ mol Al}} = 0.15 \text{ mol AlCl}_3$$

$$0.35 \text{ mol HCl} \times \frac{2 \text{ mol AlCl}_3}{6 \text{ mol HCl}} = 0.12 \text{ mol AlCl}_3$$

Because less product is formed from the HCl, it is the limiting reactant, and the amount of $AlCl_3$ that can be prepared is 0.12 mol.

Exercise 4.17   In an experiment, 7.36 g of zinc was heated with 6.45 g of sulfur (Figure 4.16). Assume that these substances react according to the equation

$$8Zn + S_8 \longrightarrow 8ZnS$$

What amount of zinc sulfide will be produced?

Wanted: amount of ZnS (assume grams).

Given: 7.36 g Zn, 6.45 g S_8, the equation.

Known: 1 mol Zn = 65.4 g; 1 mol S_8 = 256.8 g; 1 mol ZnS = 97.5 g.

Solution: Determine the limiting reactant:

$$7.36 \text{ g Zn} \times \frac{1 \text{ mol Zn}}{65.4 \text{ g Zn}} \times \frac{8 \text{ mol ZnS}}{8 \text{ mol Zn}} = 0.11\underline{2}5 \text{ mol ZnS}$$

$$6.45 \text{ g } S_8 \times \frac{1 \text{ mol } S_8}{256.8 \text{ g } S_8} \times \frac{8 \text{ mol ZnS}}{1 \text{ mol } S_8} = 0.20\underline{0}9 \text{ mol ZnS}$$

Since Zn gives the smaller amount of product, it is the limiting reactant. Convert moles ZnS (from the Zn calculation) to grams ZnS:

$$0.11\underline{2}5 \text{ mol ZnS} \times \frac{97.5 \text{ g ZnS}}{\text{mol ZnS}} = 11.0 \text{ g ZnS}$$

Exercise 4.18   New industrial plants for acetic acid react liquid methanol with carbon monoxide in the presence of a catalyst.

$$CH_3OH(l) + CO(g) \longrightarrow HC_2H_3O_2(l)$$

In an experiment, 15.0 g of methanol and 10.0 g of carbon monoxide were placed in a reaction vessel. What is the theoretical yield of acetic acid? If the actual yield is 19.1 g, what is the percentage yield?

Solution: First calculate the formula weight of each substance:

FW of CH_3OH = $(1 \times 12.0) + (4 \times 1.01) + (1 \times 16.0)$ = 32.$\underline{0}$4 amu

FW of CO   = $(1 \times 12.0) + (1 \times 16.0)$ = 28.0 amu

FW of $HC_2H_3O_2$ = $(4 \times 1.01) + (2 \times 12.0) + (2 \times 16.0)$ = 60.$\underline{0}$4 amu

Determine the limiting reactant:

$$15.0 \text{ g } CH_3OH \times \frac{1 \text{ mol } CH_3OH}{32.\underline{0}4 \text{ g } CH_3OH} \times \frac{1 \text{ mol } HC_2H_3O_2}{1 \text{ mol } CH_3OH} = 0.46\underline{8}2 \text{ mol}$$

$$10.0 \ \cancel{g\ CO} \times \frac{1 \ \cancel{mol\ CO}}{28.0 \ \cancel{g\ CO}} \times \frac{1 \ mol \ HC_2H_3O_2}{1 \ \cancel{mol\ CO}} = 0.357\underline{1} \ mol$$

The limiting reactant is CO.

$$Theoretical \ yield \ = \ 0.357\underline{1} \ \cancel{mol\ HC_2H_3O_2} \times \frac{60.0\underline{4} \ g \ HC_2H_3O_2}{\cancel{mol\ HC_2H_3O_2}} = 21.4 \ g$$

$$Percentage \ yield \ = \ \frac{actual \ yield}{theoretical \ yield} \times 100\% = \frac{19.1 \ \cancel{g}}{21.4 \ \cancel{g}} \times 100\% = 89.3\%$$

4.9   MOLAR CONCENTRATION

Operational Skills

12.   Calculating molarity from mass and volume.   Given the mass of the solute and the volume of the solution, calculate the molarity   (Example 4.16).

13.   Using molarity as a conversion factor.   Given the volume and molarity of a solution, calculate the amount of solute. Or, given the amount of solute and the molarity of a solution, calculate the volume   (Example 4.17).

The problems in this section are solved by using the definition of molarity (M): moles solute/L soln.

Exercise 4.19   A sample of sodium chloride, NaCl, weighing 0.0678 g is placed in a 25.0-mL volumetric flask. Enough water is added to dissolve the NaCl, and then the flask is filled to the mark with water and carefully shaken to mix the contents. What is the molarity of the resulting solution?

Solution:

$$M = \frac{mol \ solute}{L \ soln} = \frac{0.0678 \ \cancel{g\ NaCl}}{25.0 \times 10^{-3} \ L \ soln} \times \frac{1 \ mol \ NaCl}{58.5 \ \cancel{g\ NaCl}} = 0.0464 \ M \ NaCl$$

Exercise 4.20   How many milliliters of 0.163 M NaCl are required to give 0.0958 g of sodium chloride?

Solution:

$$0.0958 \ \cancel{g} \times \frac{1 \ \cancel{mol\ NaCl}}{58.5 \ \cancel{g}} \times \frac{1 \ \cancel{L}}{0.163 \ \cancel{mol\ NaCl}} \times \frac{10^{-3} \ mL}{\cancel{L}} = 10.0 \ mL$$

Exercise 4.21   How many moles of sodium chloride should be put in a 50.0-mL volumetric flask to give a 0.15 *M* NaCl solution when the flask is filled with water? How many grams of NaCl is this?

Solution:

$$M = \frac{\text{mol NaCl}}{\text{L soln}}$$

Solving algebraically gives

$$M \text{ (L soln)} = \text{mol NaCl}$$

$$\text{mol NaCl} = \frac{0.15 \text{ mol}}{\cancel{L}} \times 50.0 \times 10^{-3} \ \cancel{L} = 7.5 \times 10^{-3} \text{ mol}$$

$$\text{g NaCl} = 7.5 \times 10^{-3} \ \cancel{\text{mol}} \times \frac{58.5 \text{ g}}{\cancel{\text{mol}}} = 0.44 \text{ g}$$

4.10   DILUTING SOLUTIONS

Operational Skill

14.   Diluting a solution.   Calculate the volume of solution of known molarity required to make a specified volume of solution with different molarity   (Example 4.18).

Dilution is a very common laboratory procedure, as the common laboratory acids, as well as the aqueous NH_3 mentioned in your text, are supplied in concentrated form. Certain analytical work is done using successive dilutions of a standard solution, and many solutions are prepared by diluting concentrated reagents.

The important thing to remember in making these calculations is that although there is a change in solute concentration, the moles of solute remain constant before and after dilution: moles$_{initial}$ = moles$_{final}$. Note how the units show this (using liters for volume):

$$M_i \times V_i = M_f \times V_f \qquad \frac{\text{mol}_i}{\cancel{L_i}} \times \cancel{L_i} = \frac{\text{mol}_f}{\cancel{L_f}} \times \cancel{L_f} \qquad \text{moles}_i = \text{moles}_f$$

Exercise 4.22   You have a solution that is 1.5 *M* H_2SO_4 (sulfuric acid). How many milliliters of this acid do you need to prepare 100.0 mL of 0.18 *M* H_2SO_4?

Solution: Rearrange

$$M_i V_i = M_f V_f$$

4. Calculations with Chemical Formulas and Equations

to solve for V_i:

$$\frac{M_f V_f}{M_i} = V_i = \frac{0.18 \; M \times 100.0 \; mL}{1.5 \; M} = 12 \; mL$$

4.11   GRAVIMETRIC ANALYSIS

Operational Skills

15.   Determining the amount of a species by gravimetric analysis.   Given the amount of a precipitate in a gravimetric analysis, calculate the amount of a related species (Example 4.19).

> **Exercise 4.23**   You are given a sample of limestone, which is mostly $CaCO_3$, to determine the mass percentage of Ca in the rock. You dissolve the limestone in hydrochloric acid, which gives a solution of calcium chloride. Then you precipitate the calcium ion in solution by adding sodium oxalate, $Na_2C_2O_4$. The precipitate is calcium oxalate, CaC_2O_4. You find that a sample of limestone weighing 128.3 mg gives 140.2 mg of CaC_2O_4. What is the mass percentage of calcium in the limestone?

Solution: All of the Ca in the limestone is precipitated as CaC_2O_4. Thus, the amount of Ca present in the 140.2 mg (0.1402 g) CaC_2O_4 is the amount present in the original 128.3 mg (0.1283 g) of limestone. You can calculate this as follows:

$$? \; g \, Ca = 0.1402 \; g \; CaC_2O_4 \times \frac{1 \; mol \; CaC_2O_4}{128.1 \; g \; CaC_2O_4} \times \frac{1 \; mol \; Ca}{1 \; mol \; CaC_2O_4} \times \frac{40.08 \; g \; Ca}{1 \; mol \; Ca}$$

$$= 4.3866 \times 10^{-2} \, g \; Ca$$

The percentage of Ca present in the 128.3 mg (0.1283 g) of limestone is:

$$\% \; Ca = \frac{4.3866 \times 10^{-2} \, g}{0.1283 \; g} \times 100\% = 34.19\% \; Ca$$

4.12   VOLUMETRIC ANALYSIS

Operational Skills

16.   Calculating the volume of reactant solution needed.   Given the chemical equation, calculate the volume of solution of known molarity of one substance that just reacts with a given volume of solution of another substance   (Example 4.20).

17.   Calculating the quantity of substance in a titrated solution.   Calculate the mass of one substance that reacts with a given volume of known molarity of solution of another substance   (Example 4.21).

Remember, as in the previous stoichiometry problems, the first step is to write the equation for the reaction. Also be sure to write down the entire unit for each quantity   (i.e., "grams of NaCl," "L KOH soln"). You may think this is a waste of time but it is not. Confusion quickly results if you don't do it.

Exercise 4.24   Nickel sulfate, $NiSO_4$, reacts with trisodium phosphate, Na_3PO_4, to give a pale yellow-green precipitate of nickel phosphate, $Ni_3(PO_4)_2$, and a solution of sodium sulfate, Na_2SO_4.

$$3NiSO_4(aq) + 2Na_3PO_4(aq) \longrightarrow Ni_3(PO_4)_2(s) + 3Na_2SO_4(aq)$$

How many milliliters of 0.375 *M* $NiSO_4$ will react with 45.7 mL of 0.265 *M* Na_3PO_4?

Solution:

$$45.7 \times 10^{-3} \; \text{L Na}_3\text{PO}_4 \text{ soln} \times \frac{0.265 \; \text{mol Na}_3\text{PO}_4}{1 \; \text{L Na}_3\text{PO}_4 \text{ soln}} \times \frac{3 \; \text{mol NiSO}_4}{2 \; \text{mol Na}_3\text{PO}_4}$$

$$\times \frac{1 \; \text{L NiSO}_4 \text{ soln}}{0.375 \; \text{mol NiSO}_4} = 4.84 \times 10^{-2} \text{ L} = 48.4 \text{ mL NiSO}_4 \text{ soln}$$

Exercise 4.25   A 5.00-g sample of vinegar is titrated with 0.108 *M* NaOH. If the vinegar requires 39.1 mL of the NaOH solution for complete reaction, what is the mass percentage of acetic acid, $HC_2H_3O_2$, in the vinegar? The reaction is

$$HC_2H_3O_2(aq) + NaOH(aq) \longrightarrow NaC_2H_3O_2(aq) + H_2O(l)$$

Solution: First find the mass of $HC_2H_3O_2$ in the vinegar sample using titration information:

$$39.1 \times 10^{-3} \text{ L NaOH soln} \times \frac{0.108 \text{ mol NaOH}}{1 \text{ L NaOH soln}}$$

$$\times \frac{1 \text{ mol } HC_2H_3O_2}{1 \text{ mol NaOH}} \times \frac{60.0 \text{ g } HC_2H_3O_2}{1 \text{ mol } HC_2H_3O_2} = 0.25\underline{3}4 \text{ g } HC_2H_3O_2$$

Then solve for mass %:

$$\text{Mass \%} = \frac{0.25\underline{3}4 \text{ g } HC_2H_3O_2}{5.00 \text{ g vinegar}} \times 100\% = 5.07\% \ HC_2H_3O_2$$

An Acid That Matters: ACETIC ACID (a Weak Acid)

Questions for Study

1.   Why is pure acetic acid frequently called glacial acetic acid?

2.   Why is it that pure acetic acid is a poor conductor of electricity whereas solutions of acetic acid are conductors?

3.   Describe some tests that show that acetic acid solutions have acid properties.

4.   Explain why 1 *M* HCl is a better electric conductor than 1 *M* $HC_2H_3O_2$.

5.   New industrial plants produce acetic acid from carbon monoxide and hydrogen. What is the reaction by which CO and H_2 mixtures are prepared from coal?

6.   Why is carbon monoxide being studied as a starting material for the production of organic substances?

7.   Name some of the final products produced from acetic acid.

Answers to Questions for Study

1.   Acetic acid is frequently called glacial acetic acid because years ago in poorly heated laboratories the pure acid, which freezes at 17°C (60°F), was frequently found frozen in the bottle.

2.   Solutions of acetic acid, but not the pure acid, are good conductors, because in solution the uncharged molecules of acetic acid, which do not conduct electricity, combine with water molecules to form ions, which do.

3.   Acetic acid solutions are shown to have acid properties by the sour taste; the indicator bromthymol blue turns yellow in acetic acid; and reactions of acetic acid solutions with carbonates and sulfites produce gaseous products.

4.   A 1 *M* solution of HCl is a better electric conductor than a 1 *M* solution of $HC_2H_3O_2$ because molecules of the strong acid, HCl, are completely ionized, whereas only a very few of the molecules of the weak acid, $HC_2H_3O_2$, are ionized.

5.   The reaction by which CO and H_2 mixtures are prepared from coal is

$$C(s) + H_2O(g) \longrightarrow CO(g) + H_2(g)$$

6.   Carbon monoxide is being studied as a starting material for the production of organic substances because it would be a source of carbon if petroleum becomes scarce.

7.   Some final products produced from acetic acid are ethyl acetate, used as a solvent in lacquers; cellulose acetate for film and textile fibers; and vinyl acetate, used in the manufacture of polyvinyl acetate for latex paints and in paper and wood glues.

A Base That Matters: AMMONIA (a Weak Base)

Questions for Study

1.   What is the normal form of ammonia? What is household ammonia?

2.   Ammonia is said to be a weak base. What does this mean?

3.   What is the origin of the word *ammonia*?

4.   The name sal ammoniac is still used for a certain nitrogen compound. What is the chemical name of this compound?

5.   Give the equation for the preparation of ammonia from ammonium sulfate using potassium hydroxide.

6.   Give two commercial uses of ammonia.

7.   Describe the Haber process for the synthesis of ammonia, including the conditions necessary to produce ammonia by this process.

Answers to Questions for Study

1.   Ammonia is normally a gas. Household ammonia is an aqueous solution of ammonia gas.

2.   Ammonia is a weak base because when added to water, only a small percentage of NH_3 molecules react to form OH^- ion.

3.   The word *ammonia* was applied to the gas obtained when sal ammoniac (salt of Ammon), originally prepared in Egypt near a temple of the Egyptian god Ammon, was heated with a base.

4.   The chemical name of sal ammoniac is ammonium chloride.

5.   The equation is

$$(NH_4)_2SO_4(s) + 2KOH(aq) \longrightarrow 2NH_3(g) + 2H_2O(l) + K_2SO_4(aq)$$

6.   Two commercial uses of ammonia are in the production of nitrogen fertilizer and in the production of nitric acid for the manufacture of explosives.

7.   In the Haber process, nitrogen and hydrogen are heated at moderately high temperature and high pressure with an iron catalyst to form a moderate yield of ammonia.

ADDITIONAL PROBLEMS

1.   Calculate the mass (in grams) of each of the following:

    (a)   an atom of silver
    (b)   a molecule of sucrose, table sugar, $C_{12}H_{22}O_{11}$
    (c)   a formula unit of $CaCl_2$, used to salt streets in winter
    (d)   2.75 mol of aspirin, $C_9H_8O_4$

2.   Calculate:

    (a)   the moles of substance in 47.6 g of allicin, $C_6H_{10}S_2O$, the odor-producing agent in garlic
    (b)   the number of molecules in 6.42 g of allicin
    (c)   the number of atoms in 6.42 g of allicin

3.   Calculate the mass percentage of boron in $(CH_3)_2NBCl_2$.

4. Sulfur dioxide (an air pollutant), SO_2, reacts with marble, principally $CaCO_3$, to produce $CaSO_4$ and CO_2 according to the following equation:

$$2CaCO_3(s) + 2SO_2(g) + O_2(g) \longrightarrow 2CaSO_4(s) + 2CO_2(g)$$

How many grams of $CaCO_3$ are consumed per kilogram of SO_2?

5. Silver may be recovered from solutions of its salts by reaction with zinc. The reaction of zinc with a solution of silver ion, Ag^+, produces silver and zinc ion, Zn^{2+}. A 50.0-g sample of zinc was added to 50.0 L of a solution containing 3.90 g Ag^+ per liter. The reaction is

$$Zn + 2Ag^+ \longrightarrow Zn^{2+} + 2Ag$$

   (a)   Was the 50.0-g sample enough zinc to recover all of the silver present?

   (b)   If not, how much Ag^+ still remained?

6. Commercially available concentrated sulfuric acid is 95% H_2SO_4 by mass and has a density of 1.84 g/mL. How many milliliters of this acid are needed to give 1.0 L of 0.15 M H_2SO_4?

7. 2.488 g of an impure sample of Na_3PO_4 is dissolved in water, and the PO_4^{3-} ions are precipitated as $Ca_3(PO_4)_2$, which is filtered, dried, and found to weigh 1.796 g. What is the percentage of Na_3PO_4 in the original sample?

8. What volume of 12.4 M HCl would you need to make 500.0 mL of 3.50 M HCl?

9. The ore carnotite, which is 3.5% U_3O_8, is an important source of uranium. After a number of processing steps, uranyl sulfate hydrate, $(UO_2)SO_4 \cdot 3H_2O$, is produced. How many kilograms of the hydrate could theoretically be produced from 1000.0 kg of carnotite ore?

10. What is the molarity of the solution made by diluting 75.5 mL of 0.15 M KOH to exactly 250 mL?

11. What volume of 0.10 M HCl reacts completely with 0.455 g Zn if the reaction products are $ZnCl_2$ and H_2?

12. The percentage composition of vitamin C, ascorbic acid, is 40.9% C, 4.58% H, and 54.5% O. What is the molecular formula of this compound if its molecular weight is 176.1 amu?

ANSWERS TO ADDITIONAL PROBLEMS

If you missed an answer, study the text section and operational skill given in parentheses after the answer.

1. (a) $\dfrac{107.9 \text{ g Ag}}{\text{mol Ag}} \times \dfrac{1 \text{ mol Ag}}{6.02 \times 10^{23} \text{ atoms}} = 1.79 \times 10^{-22} \text{ g/atom}$

   (4.2, Op. Sk. 2)

   (b) $\dfrac{342.0 \text{ g}}{\text{mol sucrose}} \times \dfrac{1 \text{ mol sucrose}}{6.02 \times 10^{23} \text{ molecules}} = 5.68 \times 10^{-22} \text{ g/molecule}$

   (4.2, Op. Sk. 2)

   (c) $\dfrac{111.0 \text{ g CaCl}_2}{\text{mol CaCl}_2} \times \dfrac{\text{mol CaCl}_2}{6.02 \times 10^{23} \text{ formula units}}$

   $= 1.84 \times 10^{-22}$ g/formula unit   (4.2, Op. Sk. 2)

   (d) $2.75 \text{ mol C}_9\text{H}_8\text{O}_4 \times \dfrac{180 \text{ g C}_9\text{H}_8\text{O}_4}{1 \text{ mol C}_9\text{H}_8\text{O}_4} = 495 \text{ g}$ (4.2, Op. Sk. 3)

2. (a) $47.6 \text{ g C}_6\text{H}_{10}\text{S}_2\text{O} \times \dfrac{1 \text{ mol}}{162.2 \text{ g C}_6\text{H}_{10}\text{S}_2\text{O}} = 0.293 \text{ mol}$ (4.2, Op. Sk. 3)

   (b) $6.42 \text{ g allicin} \times \dfrac{1 \text{ mol allicin}}{162.2 \text{ g allicin}} \times \dfrac{6.02 \times 10^{23} \text{ molecules allicin}}{\text{mol allicin}}$

   $= 2.38 \times 10^{22} \text{ molecules}$ (4.2, Op. Sk. 4)

   (c) Use solution from part (b).

   $2.38 \times 10^{22} \text{ molecules allicin} \times \dfrac{19 \text{ atoms}}{\text{molecules allicin}} = 4.52 \times 10^{23} \text{ atoms}$

   (4.2, Op. Sk. 4)

3.   First calculate the FW of $(CH_3)_2NBCl_2$.

$$
\begin{aligned}
2 \times \text{AW of C} &= 2(12.0) = 24.0 \text{ amu} \\
6 \times \text{AW of H} &= 6(1.0) = 6.0 \text{ amu} \\
1 \times \text{AW of N} &= 14.0 \text{ amu} \\
1 \times \text{AW of B} &= 10.8 \text{ amu} \\
2 \times \text{AW of Cl} &= 2(35.5) = \underline{71.0 \text{ amu}} \\
\text{FW of } (CH_3)_2NBCl_2 &= 125.8 \text{ amu}
\end{aligned}
$$

Then determine the mass % of B.

$$
\text{mass \% B} = \frac{10.8 \text{ g}}{125.8 \text{ g}} \times 100\% = 8.59\% \qquad (4.3, \text{ Op. Sk. 5})
$$

4. $1 \text{ kg SO}_2 \times \dfrac{10^3 \text{ g SO}_2}{1 \text{ kg SO}_2} \times \dfrac{1 \text{ mol SO}_2}{64.1 \text{ g SO}_2} \times \dfrac{2 \text{ mol CaCO}_3}{2 \text{ mol SO}_2} \times \dfrac{100.1 \text{ g CaCO}_3}{\text{mol CaCO}_3}$

$$
= 1.56 \times 10^3 \text{ g CaCO}_3 \qquad (4.7, \text{ Op. Sk. 9})
$$

5.   (a)   First calculate the mass of Ag^+ present.

$$
\frac{3.90 \text{ g Ag}^+}{1 \text{ L}} \times 50.0 \text{ L} = 195 \text{ g Ag}^+
$$

Then calculate the mass of Ag^+ that would react with 50.0 g Zn.

$$
Zn(s) + 2Ag^+(aq) \longrightarrow Zn^{2+}(aq) + 2Ag(s)
$$

$$
50.0 \text{ g Zn} \times \frac{1 \text{ mol Zn}}{65.4 \text{ g Zn}} \times \frac{2 \text{ mol Ag}^+}{1 \text{ mol Zn}} \times \frac{107.9 \text{ g Ag}^+}{\text{mol Ag}^+} = 165 \text{ g Ag}^+
$$

This is less than the silver present. All of the silver would not be recovered.

(b)   The silver remaining is $195 \text{ g} - 165 \text{ g} = 3.0 \times 10^1 \text{ g}$.   (4.8, Op. Sk. 10)

6.   First find the molarity of concentrated H_2SO_4:

$$M = \frac{\text{mol } H_2SO_4}{\text{L conc. } H_2SO_4}$$

$$= \frac{95 \text{ g } H_2SO_4}{100 \text{ g conc. } H_2SO_4} \times \frac{1 \text{ mol } H_2SO_4}{98.1 \text{ g } H_2SO_4} \times \frac{1.84 \text{ g conc. } H_2SO_4}{10^{-3} \text{ L conc. } H_2SO_4}$$

$$\underbrace{\phantom{\frac{95 \text{ g } H_2SO_4}{100 \text{ g conc. } H_2SO_4}}}_{\substack{\text{\% composition} \\ \text{of conc. } H_2SO_4}} \qquad\qquad \underbrace{\phantom{\frac{1.84 \text{ g conc. } H_2SO_4}{10^{-3} \text{ L conc. } H_2SO_4}}}_{\substack{\text{density of} \\ \text{conc. } H_2SO_4}}$$

$$= 17.8 \ M$$

Rearrange $M_i V_i = M_f V_f$ to solve for V_i:

$$V_i = \frac{M_f V_f}{M_i} = \frac{0.15 \ M \times 1.0 \text{ L}}{17.8 \ M} = 8.4 \times 10^{-3} \text{ L} = 8.4 \text{ mL}$$

This problem can also be done entirely by dimensional analysis:

$$1.0 \text{ L acid} \times \frac{0.15 \text{ mol } H_2SO_4}{1 \text{ L acid}} \times \frac{98.1 \text{ g } H_2SO_4}{1 \text{ mol } H_2SO_4} \times \frac{100 \text{ g conc. } H_2SO_4}{95 \text{ g } H_2SO_4}$$

$$\times \frac{1 \text{ ml conc. } H_2SO_4}{1.84 \text{ g conc. } H_2SO_4} = 8.4 \text{ mL}$$

$$(4.10, \text{ Op. Sk. } 14)$$

7.   Calculate the mol of $Ca_3(PO_4)_2$ present in 1.796 g.

$$1.796 \text{ g } Ca_3(PO_4)_2 \times \frac{1 \text{ mol } Ca_3(PO_4)_2}{310.18 \text{ g } Ca_3(PO_4)_2} = 5.7902 \times 10^{-3} \text{ mol } Ca_3(PO_4)_2$$

Because each mol of $Ca_3(PO_4)_2$ requires 2 mol of Na_3PO_4, you can determine the mass of Na_3PO_4 present in the impure sample.

$$5.7902 \times 10^{-3} \text{ mol } Ca_3(PO_4)_2 \times \frac{2 \text{ mol } Na_3PO_4}{1 \text{ mol } Ca_3(PO_4)_2} \times \frac{163.94 \text{ g } Na_3PO_4}{\text{mol } Na_3PO_4}$$

$$= 1.8985 \text{ g } Na_3PO_4$$

The percentage of Na_3PO_4 in the impure sample is:

$$\frac{1.89\underline{8}5\ g}{2.488\ g} \times 100\% = 76.31\%$$ (4.11, Op. Sk. 15)

8. $V_iM_i\ =\ V_fM_f$

   $V_i\ =\ \dfrac{V_fM_f}{M_i} = \dfrac{500.0\ mL \times 3.50\ \cancel{M}}{12.4\ \cancel{M}} = 141\ mL$ (4.10, Op. Sk. 14)

9.   Find the kg U_3O_8 in the ore.

   $0.035 \times 1000.0\ kg = 35\ kg\ U_3O_8$

Calculate the % U in carnotite.

$$\frac{238\ g/mol\ U \times 3}{(238\ g/mol\ U \times 3 + 16.0\ g/mol\ O \times 8)} = 84.\underline{8}0\%\ U$$

Now calculate the mass of uranium. This determines the amount of hydrate that could be formed.

   $0.84\underline{8}0 \times 35\ kg\ U_3O_8 = 2\underline{9}.7\ kg\ U$

Determine the mass of hydrate from the mass of U per formula weight of the hydrate.

   $2\underline{9}.7\ \cancel{kg\ U} \times \dfrac{420.1\ kg\ hydrate}{238\ \cancel{kg\ U}} = 52\ kg\ hydrate$ (4.1, Op. Sk. 5)

10.   The calculation is

   $M_f = \dfrac{M_iV_i}{V_f} = \dfrac{0.15\ M \times 0.0755\ \cancel{L}}{0.250\ \cancel{L}} = 4.5 \times 10^{-2}\ M\ KOH$

                                                      (4.10, Op. Sk. 14)

11.   The reaction is $2HCl + Zn \longrightarrow H_2 + ZnCl_2$. Thus, 2 mol of HCl molecules react with 1 mol of Zn atoms. The calculation is

   $0.455\ \underline{\cancel{g\ Zn}} \times \dfrac{1\ \cancel{mol\ Zn}}{65.4\ \cancel{g\ Zn}} \times \dfrac{2\ \cancel{mol\ HCl}}{1\ \cancel{mol\ Zn}} \times \dfrac{1\ \cancel{L\ soln}}{0.10\ \underline{\cancel{mol\ HCl}}} \times \dfrac{1\ mL\ soln}{10^{-3}\ \cancel{L\ soln}}$

   $= 1.4 \times 10^2\ mL\ soln$

                                                      (4.12, Op. Sk. 16)

12.   Find moles H:

$$4.58 \; \cancel{g\,H} \times \frac{1 \; mol \; H}{1.01 \; \cancel{g\,H}} = 4.5\underline{3}5 \; mol \; H$$

Find moles C:

$$40.9 \; \cancel{g\,C} \times \frac{1 \; mol \; C}{12.0 \; \cancel{g\,C}} = 3.4\underline{0}8 \; mol \; C$$

Find moles O:

$$54.5 \; \cancel{g\,O} \times \frac{1 \; mol \; O}{16.0 \; \cancel{g\,O}} = 3.4\underline{0}6 \; mol \; O$$

Change to integers by dividing by 3.406 mol (the smallest).

For H: $\dfrac{4.5\underline{3}5 \; \cancel{mol \; H}}{3.4\underline{0}6 \; \cancel{mol}} = 1.33$

For C: $\dfrac{3.4\underline{0}8 \; \cancel{mol \; C}}{3.4\underline{0}6 \; \cancel{mol}} = 1.00$

For O: $\dfrac{3.4\underline{0}6 \; \cancel{mol \; O}}{3.4\underline{0}6 \; \cancel{mol}} = 1.00$

Multiply each by 3 to get 4.0, 3.0, and 3.0; the empirical formula is $C_3H_4O_3$.

Empirical FW $= (3 \times 12.0) + (4 \times 1.01) + (3 \times 16.0) = 88.0 \; amu$

$$n = \frac{176.1}{88.0} = 2.00$$

The molecular formula is $(C_3H_4O_3)_2 = C_6H_8O_6$.   (4.5, Op. Sk. 8, 9)

CHAPTER POST-TEST

1. Calculate the mass percentage of nitrogen in $(NH_4)_3PO_4$.

2. Determine the percentage of Cu in a 1.215-g sample of brass if all the copper in the sample reacted to give 1.743 g of copper phosphate, $Cu_3(PO_4)_2$.

3. Quantitative stoichiometric calculations may be carried out in the absence of a balanced chemical equation if _____
    _____.

4.   A 5.60-g sample of a compound was analyzed to be 2.62 g of oxygen, 2.90 g of chlorine, and 0.0840 g of hydrogen. Determine the empirical formula.

5.   Answer the questions below using the following information. $Ni(CO)_4$ is a volatile, poisonous compound that can be prepared from the reaction of Ni with CO.

$$Ni + 4CO \longrightarrow Ni(CO)_4$$

(a)   Calculate the theoretical yield of $Ni(CO)_4$ obtainable from 54.1 g Ni and 37.6 g CO.

(b)   If only 48.6 g $Ni(CO)_4$ is obtained, what is the percentage yield of this compound?

(c)   How much of the reagent in excess is left at the end of the reaction?

6.   Calculate the mass in grams of H_3PO_4 obtained from 1.91×10^{21} atoms of P in a commercial process that gives only a 26.2% yield of H_3PO_4 from P.

7.   One possible reaction between two common air pollutants is

$$2NO(g) + 2CO(g) \longrightarrow N_2(g) + 2CO_2(g)$$

If 975 g NO is reacted, what would be the mass percentage of CO_2 in the mixture of products?

8.   If 15.6 g NaCl is dissolved in water to make 275 mL of solution, what is the molarity of the solution?

9.   How many milliliters of 5.00 M NaOH would you need to dilute to make 85.0 mL of 0.10 M NaOH?

10. What volume of 0.250 M HCl reacts with 37.4 mL of 0.150 M $Ca(OH)_2$? The reaction products are $CaCl_2$ and water.

ANSWERS TO CHAPTER POST-TEST

If you missed an answer, study the text section and operational skill given in parentheses after the answer.

1.   28.2%   (4.3, Op. Sk. 1, 5)   2.   71.82%   (4.3, Op. Sk. 1, 5, 15)

3.   there is an element in the reaction and it all ends up in a single product. The original material containing the element and this product can be related by stoichiometry. (4.7)

4. $HClO_2$ (4.5, Op. Sk. 8)

5.   (a)   57.3 g   (b)   84.8%   (c)   34.4 g   (4.7, 4.8, Op. Sk. 10, 11)

6.   0.0815 g   (4.7, 4.8, Op. Sk. 10, 11)   7.   75.9%   (4.7, Op. Sk. 10)

8.   0.971 M (4.9, Op. Sk. 12)   9.   1.7 mL   (4.10, Op. Sk. 14)

10.   44.9 mL   (4.12, Op. Sk. 16)

CHAPTER 5   THE GASEOUS STATE

CHAPTER TERMS AND DEFINITIONS

Numbers in parentheses after definitions give the text sections in which the terms are explained. Starred terms are italicized in the text. Where a term does not fall directly under a text section heading, additional information is given for you to locate it.

pressure   force exerted per unit area of surface   (5.1)
acceleration*   change of speed per unit time   (5.1)
pascal (Pa)   SI unit of pressure; 1 Pa = 1 kg/(m • s^2)   (5.1)
barometer   device for measuring the pressure of the atmosphere   (5.1)
manometer   device that measures the pressure of a gas or liquid in a sealed vessel   (5.1)
millimeters of mercury (mmHg or torr)   traditional unit of pressure equal to that exerted
       by a 1-mm column of mercury at 0.00°C in a barometer or manometer (5.1)
atmosphere (atm)   traditional unit of pressure equal to exactly 760 mmHg;
       1 atm = 101.325 kPa, exact   (5.1)
compressibility*   ability to be squeezed into a smaller volume by the application of
       pressure   (5.2)
Boyle's law   the volume of a sample of gas at a given temperature varies inversely with
       the applied pressure   (5.2)
linearly*   term describing how one variable changes with the change in another variable if
       a plot of the two variables gives a straight line   (5.2)
extrapolate*   to extend a line beyond the plotted data points   (5.2)
Kelvin scale*   absolute temperature scale on which the units (kelvins, K) are given by
       K = °C + 273.15   (5.2)
Charles's law   the volume occupied by any sample of gas at a constant pressure is
       directly proportional to the absolute temperature   (5.2)
law of combining volumes*   the volumes of reactant gases at a given pressure and tem-
       perature are in ratios of small whole numbers   (5.2)
Avogadro's law   equal volumes of any two gases at the same temperature and pressure
       contain the same number of molecules   (5.2)
molar gas volume (V_m)   volume occupied by one mole of any gas at a given temperature
       and pressure   (5.2)

standard temperature and pressure (STP) reference conditions for gases chosen by convention to be 0°C and 1 atm pressure (5.2)

molar gas constant (R) constant of proportionality relating the molar volume of a gas to T/P (5.3)

ideal gas law mathematical expression combining all of the gas laws and relating the volume (V), pressure (P), Kelvin temperature (T), and moles (n) of a gas to the molar gas constant R; $PV = nRT$ (5.3)

Amontons's law* the pressure of a given amount of gas at a fixed volume is proportional to the absolute temperature (Example 5.5)

partial pressure pressure exerted by a particular gas in a gas mixture (5.5)

Dalton's law of partial pressures the sum of the partial pressures of all the different gases in a mixture is equal to the total pressure of the mixture (5.5)

mole fraction fraction of moles of a component gas in the total moles of a gas mixture (5.5)

vapor pressure* partial pressure of the molecules of a substance in the gaseous state in the presence of the liquid (or solid) substance (5.5)

kinetic-molecular theory of gases (kinetic theory) idea that a gas consists of molecules in constant random motion (5.6, introductory section)

postulates* basic statements from which all conclusions or predictions of a theory are deduced (5.6)

ideal gas* gas that follows the ideal gas law; its molecules have essentially no volume of their own, and no attraction for each other (5.6)

intermolecular forces* forces of attraction or repulsion between molecules (5.6)

root-mean-square (rms) molecular speed (u) type of average molecular speed, or the speed of a molecule that has the average molecular kinetic energy; can be shown to equal

$$u = \sqrt{\frac{3RT}{M_m}}$$

where R is the molar gas constant, T is the kelvin temperature, and M_m is the molar mass for the gas (5.7)

gaseous diffusion process whereby a gas spreads out through another gas to occupy the space uniformly (5.7)

effusion escape of a gas through a small hole into a vacuum at the same velocity it had in the container (5.7)

Graham's law of effusion the rate of effusion of gas molecules from a particular hole is inversely proportional to the square root of the molecular weight of the gas at constant T and P (5.7)

enrichment* process used to increase the percentage of one isotope in a sample (5.7)

van der Waals equation   equation similar to the ideal gas law, but includes two
   constants, *a* and *b,* to account for deviations from ideal behavior   (5.8)
liquid-air machine*   commercial equipment for liquefying air   (A Gas That Matters:
   Oxygen [a Component of Air])

CHAPTER DIAGNOSTIC TEST

1. A closed-end manometer was constructed using dodecane as the liquid. If the density of dodecane is 0.7487 g/cm^3, or 7.487×10^2 kg/m^3, at 20.0°C, what will be the difference in heights of the liquid levels when the measured pressure of the gas is 745 mmHg? The density of mercury at 20.0°C is 13.546 g/cm^3.

2. Calculate the final pressure of a gas when 15.0 L of the gas at 743 mmHg is transferred to a 39.2-L container at the same temperature.

3. At 1.50 atm and 23°C a gas occupies 13.5 L. What is its volume in liters at 1.04 atm and -5°C?

4. If 2.35 L CO_2 at some pressure and temperature contains 0.0648 mol CO_2, how many moles of helium atoms are there in 2.25 L of helium at the same pressure and temperature?

5. Starting with the ideal gas law, derive the relationship between the pressure and temperature of a gas.

6. What is the temperature of an ideal gas at 531 mmHg if the density is 4.60 g/L and its molecular weight is 63.8 g/mol?

   (a)   155°C   (c)   6.88 K   (e)   none of the above
   (b)   243 K   (d)   118 K

7. What volume of CO_2 can be prepared from the reaction of 152 g $CaCO_3$ with 1.15 mol HCl at 15.0°C and 765 mmHg?

8. A 5.000-L sample of gas at 125°C has the following composition: 0.765 g N_2, 0.843 g O_2, and 0.684 g H_2O.

   (a)   What is the mole fraction of each gas in the mixture?
   (b)   What is the partial pressure of each component gas in the mixture?

9. The total pressure of a mixture of gases at 301 K was 973 mmHg. Analysis of this mixture showed the composition to be 3.33 mol H_2, 1.69 mol NO, and 0.488 mol O_2. Calculate the partial pressure of each gas.

10. A sample of N_2 was collected over water at 26.1°C and a pressure of 1.07 atm. The observed volume of gas was 0.783 L. Calculate the mass of the dry N_2. (Vapor pressure of water at 26.1°C is 25.4 mmHg.)

11. Which of the following statements about the kinetic theory of gases is (are) correct?
    (a) Molecules are viewed as point masses and their masses can be neglected.
    (b) Weak van der Waals forces are responsible for the collisions between gas molecules.
    (c) An increase in temperature causes an increase in the number of gas molecules in a sample.
    (d) The pressure of a gas decreases as the volume of individual gas molecules increases.
    (e) The average kinetic energy of gas molecules is proportional to the absolute temperature of the gas.

12. Which of the following statements about the following figure is (are) correct?

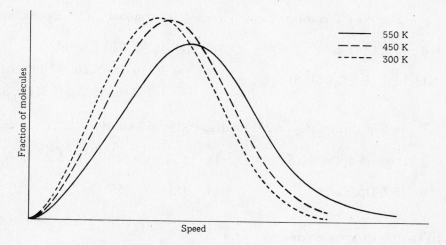

    (a) At a higher temperature there are more gas molecules.
    (b) A greater fraction of higher-speed molecules are observed at a lower temperature.
    (c) A higher temperature is associated with molecules having a higher speed.

(d)   A temperature increase causes an increase in the more probable speeds and in the average speeds of molecules.

(e)   None of the above are correct.

13.   The rate of effusion of Kr compared with that of N_2 $\left(r_{Kr}/r_{N_2}\right)$ is

(a)   1.73   (c)   2.99

(b)   0.578   (d)   0.334   (e)   none of the above.

14.   Calculate the rms molecular speed of CO_2 molecules at room temperature, 25°C.

ANSWERS TO CHAPTER DIAGNOSTIC TEST

If you missed an answer, study the text section and operational skill given in parentheses after the answer.

1. 1.35×10^4 mm   (5.1, Op. Sk. 1)   2.   284 mmHg   (5.2, Op. Sk. 2)

3.   17.6 L   (5.2, Op. Sk. 2)   4.   0.0620 mol   (5.2)

5. $P \propto T$ or $P = CT$, constant n and V, where C is a constant.   (5.3, Op. Sk. 3)

6.   d   (5.3, Op. Sk. 5)   8.   (a)   0.298, 0.287, 0.415

7.   13.5 L   (5.4, Op. Sk. 6)   (b)   0.179 atm N_2, 0.172 atm O_2,
                                                            0.248 atm H_2O (5.5, Op. Sk. 7)

9. P_{H_2} = 588 mmHg, P_{NO} = 299 mmHg, P_{O_2} = 86.2 mmHg   (5.5, Op. Sk. 7)

10.   0.925 g   (5.5, Op. Sk. 8)   11.   e   (5.6)   12.   d   (5.7)

13.   b   (5.7, Op. Sk. 10)   14.   411 m/s   (5.7, Op. Sk. 9)

SUMMARY OF CHAPTER TOPICS

To be able to solve the problems dealing with gases, you will have to know the gas laws presented in the chapter. Table 5.1 is presented to help you learn them.

Table 5.1   Gas Laws

Name of Law	Equation	Conditions
Boyle's law	$PV = $ constant, or $P_f V_f = P_i V_i$	Constant T, n
Charles's law	$\dfrac{V}{T} = $ constant, or $\dfrac{V_f}{T_f} = \dfrac{V_i}{T_i}$	Constant P, n
(Amontons's law)	$\dfrac{P}{T} = $ constant, or $\dfrac{P_f}{T_f} = \dfrac{P_i}{T_i}$	Constant V, n
Combined gas law	$\dfrac{P_f V_f}{T_f} = \dfrac{P_i V_i}{T_i}$	For fixed amount of gas
Avogadro's law	$V_m = $ specific constant	Depending on T, P; independent of gas
Ideal gas law	$PV = nRT$	
Dalton's law of partial pressures	$P = P_A + P_B + P_C + \cdots$	
Graham's law	Rate of effusion of gas $\;$ Rate $\propto 1/\sqrt{M_m}$	Same container, constant T and P
van der Waals equation	$\left(P + \dfrac{n^2 a}{V^2}\right)(V - nb) = nRT$	Moderate pressures

5.1   GAS PRESSURE AND ITS MEASUREMENT

In this section you are introduced to two instruments, the barometer and the manometer. Note how each instrument is constructed and exactly what each measures.

Operational Skill

1.   Relating liquid height and pressure.   Given the density of a liquid used in a barometer or manometer and the height of the column of liquid, obtain the pressure reading in mmHg   (Example 5.1).

> **Exercise 5.1**   An oil whose density is 0.775 g/mL was used in a closed-tube manometer to measure the pressure of a gas in a flask, as shown in Figure 5.4. If the height of the oil column was 7.68 cm, what was the pressure of the gas in the flask in mmHg?
>
> *Wanted:*   pressure in mmHg.
>
> *Given:*   closed-end manometer (Figure 5.3 in text); oil density (d) = 0.775 g/mL; column height (h) = 7.68 cm.
>
> *Known:* $P = gdh$; $g = 9.807$ m/s^2; $P_{gas} = P_{oil} = P_{Hg}$;
>
> $$d_{Hg} = 13.559 \; \frac{g}{\cancel{cm}^3} \times \frac{1 \; \cancel{cm}^3}{1 \; mL} = 13.559 \; g/mL$$
>
> *Solution:* $\cancel{g} \, d_{oil} h_{oil} = \cancel{g} \, d_{Hg} h_{Hg}$
>
> $$h_{Hg} = \frac{d_{oil} h_{oil}}{d_{Hg}} = \frac{(0.775 \; \cancel{g/mL})(7.68 \; cm)}{(13.559 \; \cancel{g/mL})}$$
>
> $$= 0.43\underline{8}9 \; \cancel{cm} \times \frac{10 \; mm}{1 \; \cancel{cm}} = 4.39 \; mm$$

Because pressure is measured in units of mmHg, $P = 4.39$ mmHg.

5.2   EMPIRICAL GAS LAWS

Operational Skill

2.   Using the empirical gas laws.   Given an initial volume occupied by a gas, calculate the final volume when the pressure changes at fixed temperature   (Example 5.2); when the temperature changes at fixed pressure   (Example 5.3); and when both pressure and temperature change   (Example 5.4).

It is important to note that volume, V, is the volume occupied by the gas, *not* the volume of the gas molecules themselves. This will be discussed at greater length in Section 5.6, on kinetic theory.

You can nicely illustrate the inverse relationship between pressure and volume if you blow up a balloon, tie it, and then rather abruptly sit on it! The weight of your body decreases the volume but the resulting increase in the gas pressure inside bursts the balloon.

With eight gas laws to memorize, it is easy to get them mixed up at exam time. The following information may help you to get the right answers when using Boyle's law. In performing the calculations, you will always multiply one quantity, say P, by a ratio of the two values of the other quantity, in this case a ratio of V_i and V_f. Observe whether the volume increases or decreases. If V_f is greater than V_i, then, according to Boyle's law, the pressure must decrease. Therefore, you must put the smaller volume on top when you write the ratio. (The ratio will be less than 1.) See the note just after the solution to Exercise 5.2.

Exercise 5.2   A volume of carbon dioxide gas, CO_2, equal to 20.0 L was collected at 23°C and 1.00 atm pressure. What would be the volume of carbon dioxide if it were collected at 23°C and 0.83 atm?

Given: V_i = 20.0 L P_i = 1.00 atm

         V_f =   ? P_f = 0.830 atm

         (*T* and *n* remain constant.)

Known: Boyle's law states that $V_iP_i = V_fP_f$.

Solution: $V_f = V_i \times \dfrac{P_i}{P_f} = 20.0 \text{ L} \times \left(\dfrac{1.00 \text{ atm}}{0.830 \text{ atm}} \right) = 24 \text{ L}$

Pressure decreases, so volume increases. Thus, the ratio by which the initial volume is multiplied must be greater than 1. (The larger value of P goes on top.)

Note that, beginning in the section on Charles's law, we use lower-case *t* for reporting temperatures in degrees Celsius and capital *T* for reporting temperatures in kelvins. Be sure to use kelvins whenever you work problems involving gases.

Using both Charles's law and the combined gas law (combining Boyle's and Charles's laws), it is again useful to think about forming the ratios of P_i and P_f, V_i and V_f, or T_i and T_f so that the answers will be larger or smaller, as the laws state. The notes included with the following two exercise solutions stress this point.

You will need to memorize the values of STP ($0°C = 273.15$ K, and 1 atm P). Also memorize the molar gas volume (of any gas) at STP, 22.41 L/mol.

Exercise 5.3   If we expect a chemical reaction to produce 4.38 dm^3 of oxygen, O_2, at 19°C and 101 kPa, what would be the volume at 25°C and 101 kPa?

Given: V_i = 4.38 dm^3 T_i = 19°C + 273 = 292 K

$\quad\quad\quad V_f$ = ? $\quad\quad T_f$ = 25°C + 273 = 298 K

$\quad\quad\quad$ (*P* and *n* remain constant.)

Known: Charles's law states that $\dfrac{V_i}{T_i} = \dfrac{V_f}{T_f}$.

Solution: $V_f = V_i \times \dfrac{T_f}{T_i} = 4.38 \text{ dm}^3 \times \left(\dfrac{298\,\textbf{K}}{292\,\textbf{K}}\right) = 4.47 \text{ dm}^3$

Note that *T* increases, so *V* increases. The ratio of temperatures must be greater than 1. (The larger value of *T* goes on top.)

Exercise 5.4   A balloon contains 5.41 dm^3 of helium, He, at 24°C and 101.5 kPa. Suppose the gas in the balloon is heated to 35°C. If the helium pressure is now 102.8 kPa, what is its volume?

Given: V_i = 5.41 dm^3 P_i = 101.5 kPa T_i = 24°C (297 K)

$\quad\quad\quad V_f$ = ? $\quad\quad P_f$ = 102.8 kPa T_f = 35°C (308 K)

$\quad\quad\quad$ (*n* remains constant.)

Known: The combined gas laws give $\dfrac{V_i P_i}{T_i} = \dfrac{V_f P_f}{T_f}$.

Solution:

$$V_f = V_i \times \left(\frac{P_i}{P_f}\right) \times \left(\frac{T_f}{T_i}\right) = 5.41 \text{ dm}^3 \times \left(\frac{101.5\,\cancel{\text{kPa}}}{102.8\,\cancel{\text{kPa}}}\right) \times \left(\frac{308\,\textbf{K}}{297\,\textbf{K}}\right)$$

$$= 5.54 \text{ dm}^3$$

Note that the pressure increases, so the pressure ratio must be less than 1 to give a decrease in the volume. Also note that the temperature increases, so the temperature ratio must be greater than 1 to give an increase in volume.

5.3   THE IDEAL GAS LAW

Operational Skills

3.   Deriving empirical gas laws from the ideal gas law.   Starting from the ideal gas law, derive the relationship between any two variables   (Example 5.5).

4.   Using the ideal gas law.   Given any three of the variables P, V, T, and n for a gas, calculate the fourth from the ideal gas law   (Example 5.6).

5.   Relating gas density and molecular weight.   Given the molecular weight, calculate the density of a gas for a particular temperature and pressure   (Example 5.7); or, given the gas density, calculate the molecular weight   (Example 5.8).

In solving problems using the ideal gas law, you must have your values in the units in which the gas constant, R, is reported. The usual value of R is 0.0821 L • atm/(K • mol). Thus, V must be in liters, P in atmospheres, T in kelvins (as the capital T indicates), and the amount of gas in moles.

Exercise 5.5   Show that the moles of gas are proportional to the pressure for constant volume and temperature.

Known: The ideal gas law includes variables P, V, n, and T.

Solution: Using $PV = nRT$, solve for n:

$$n = P \left(\frac{V}{RT} \right)$$

Since R, T, and V are constant, we can write

$$n = P \cdot C \text{ (where } C \text{ is a constant)} \quad \text{or} \quad n \propto P$$

Exercise 5.6   What is the pressure in a 50.0-L tank that contains 3.03 kg of oxygen, O_2, at 23°C?

Wanted: P.

Given: $V = 50.0$ L; $n = \dfrac{3.03 \times 10^3 \ \cancel{g}}{32.0 \ \cancel{g}/\text{mol}} = 94.\underline{6}9$ mol O_2;

$T = 23°C + 273 = 296$ K

Known: These variables are related by the ideal gas law, which we solve for *P*; $R = 0.0821$ L • atm/(K • mol).

Solution:

$$P = \frac{nRT}{V} = \frac{(94.\underline{69}\ \cancel{mol})(0.0821\ \cancel{L} \bullet atm)(296\ \cancel{K})}{(50.0\ \cancel{L})(\cancel{K} \bullet \cancel{mol})} = 46.0\ atm$$

Exercise 5.7   Calculate the density of helium, He, in grams per liter at 21°C and 752 mmHg. The density of air under these conditions is 1.188 g/L. What is the difference in mass between 1 liter of air and 1 liter of helium? (This mass difference is equivalent to the buoyant, or lifting, force of helium per liter.)

Wanted: density of He (g/L); mass difference of 1 L air and 1 L He.

Given: $t = 21$°C, $P = 752$ mmHg, air density = 1.188 g/L.

Known: Since density is mass per unit volume, calculating moles then mass of He in 1 L, using $PV = nRT$ and the atomic weight of He, gives the density. Then subtract that from the value of the density of air.

Solution: Solve for moles He in 1 L:

$$n = ?$$

$$P = 752\ \cancel{mmHg} \times \frac{1\ atm}{760\ \cancel{mmHg}} = 0.98\underline{9}5\ atm$$

$$T = (21°C + 273)\ K = 294\ K$$

$$V = 1\ L\ (exact)$$

$$n = \frac{PV}{RT} = \frac{(0.98\underline{9}5\ \cancel{atm})(1\ \cancel{L})(\cancel{K} \bullet mol)}{(294\ \cancel{K})(0.0821\ \cancel{L} \bullet \cancel{atm})} = 0.041\underline{0}0\ mol\ He$$

Then solve for grams He, which is also the value of the density of He:

$$0.041\underline{0}0\ \cancel{mol\ He} \times \frac{4.00\ g\ He}{\cancel{mol\ He}} = 0.16\underline{4}0\ g\ He;\ density\ of\ He = 0.16\underline{4}0\ g/L$$

Subtract densities to find the difference in mass:

$$1.188\ g\ (per\ L\ air) - 0.16\underline{4}0\ g\ (per\ L\ He) = 1.024\ g$$

Exercise 5.8   A sample of a gaseous substance at 25°C and 0.862 atm has a density of 2.26 g/L. What is the molecular weight of the substance?

Wanted: molecular weight.

Given: $T = 25$°C + 273 = 298 K; $P = 0.862$ atm; density = 2.26 g/L.

Known: Molecular weight (in amu's) is the same number as molar mass (g/mol). We can use $V = 1$ L (exact). We can use the ideal gas law to solve for moles, then form a ratio of 2.26 g over the calculated number of moles to get g/mol.

Solution: Find moles of vapor:

$$n = \frac{PV}{RT} = \frac{(0.862 \ \cancel{atm})(1 \ \cancel{L})(\cancel{K} \cdot mol)}{(298 \ \cancel{K})(0.0821 \ \cancel{L} \cdot \cancel{atm})} = 0.035\underline{2}3 \ mol$$

Find M_m:

$$M_m = \frac{grams \ \cancel{vapor}}{moles \ \cancel{vapor}} = \frac{2.26 \ g}{0.035\underline{2}3 \ mol} = 64.1 \ g/mol$$

Thus, the molecular weight is 64.1 amu.

5.4 STOICHIOMETRY PROBLEMS INVOLVING GAS VOLUMES

Operational Skill

6. Solving stoichiometry problems involving gas volumes. Given the volume (or mass) of one substance in a reaction, calculate the mass (or volume) of another produced or used up   (Example 5.9).

Now that you are familiar with the properties and behavior of gases, you can use the stoichiometry you learned in Chapter 4 to calculate volumes of gaseous reactants and products in chemical equations. Stoichiometry problems involving gases have one step that causes problems for most students. This step is the conversion between moles of gas and the volume occupied by 1 mole of gas. For any gas at STP, the molar volume is always 22.41 L/mol. In most cases, however, the gas is not at STP; you must then use the ideal gas law, either to find the moles of gas when volume is given or to find the volume once the moles of gas are known.

Exercise 5.9 How many liters of chlorine gas, Cl_2, can be obtained at $40°C$ and 787 mmHg from 9.41 g of hydrogen chloride, HCl, according to the following equation?

$$2KMnO_4(s) + 16HCl(aq) \longrightarrow 8H_2O(l) + 2KCl(aq) + 2MnCl_2(aq) + 5Cl_2(g)$$

Wanted: V_{Cl_2}.

Given: $P = 787 \text{ mmHg} \times \dfrac{1 \text{ atm}}{760 \text{ mmHg}} = 1.0\underline{3}6 \text{ atm};$

$T = 40°C + 273 = 313 \text{ K}$ (Assume 40°C is exact.)

Known: We can get moles Cl_2 from the reaction stoichiometry, then use the ideal gas law to get the volume.

Solution: Find moles Cl_2:

$$n = 9.41 \text{ g HCl} \times \frac{1 \text{ mol HCl}}{36.5 \text{ g HCl}} \times \frac{5 \text{ mol } Cl_2}{16 \text{ mol HCl}} = 0.080\underline{5}7 \text{ mol } Cl_2$$

Find V_{Cl_2}:

$$V = \frac{nRT}{P} = \frac{(0.080\underline{5}7 \text{ mol})(0.0821 \text{ L} \cdot \text{atm})(313 \text{ K})}{(1.0\underline{3}6 \text{ atm})(\text{K} \cdot \text{mol})} = 2.00 \text{ L}$$

5.5   GAS MIXTURES; LAW OF PARTIAL PRESSURES

Operational Skills

   7.   Calculating partial pressures and mole fractions of a gas in a mixture.   Given the masses of gases in a mixture, calculate the partial pressures and mole fractions (Example 5.10).

   8.   Calculating the amount of gas collected over water.   Given the volume, total pressure, and temperature of gas collected over water, calculate the mass of the dry gas (Example 5.11).

   As Dalton observed, each gas in a mixture of gases that don't react with each other exerts a pressure independent of any or all other gases present. So, the pressure of a gas does not change if another gas is added or removed, by chemical reaction, for instance. Neither does the volume change, as "volume" refers to the space the gas occupies — the volume of the container. The total pressure does change, though, since it is the sum of all the partial pressures.

   Exercise 5.10   A 10.0-L flask contains 1.031 g O_2 and 0.572 g CO_2 at 18°C. What are the partial pressures of oxygen and carbon dioxide? What is the total pressure? What is the mole fraction of oxygen in the mixture?

Known: We can find the moles of each gas using the molar mass, and then find the pressure of each using the ideal gas law; mole fraction $O_2 = \dfrac{n_{O_2}}{n_{total}} = \dfrac{P_{O_2}}{P_{total}}$.

Solution: Moles O_2:

$$1.031 \; g \, O_2 \times \frac{1 \; mol \; O_2}{32.0 \; g \, O_2} = 0.03222 \; mol \; O_2$$

$$P_{O_2} = \frac{n_{O_2} RT}{V} = \frac{0.03222 \; mol \times 0.0821 \; L \cdot atm \times (18 + 273) \; K}{(10.0 \; L)(K \cdot mol)}$$

$$= 0.07698 \; atm \; O_2$$

Moles CO_2:

$$0.572 \; g \, CO_2 \times \frac{1 \; mol \; CO_2}{44.0 \; g \, CO_2} = 0.01300 \; mol \; CO_2$$

$$P_{CO_2} = \frac{n_{CO_2} RT}{V} = \frac{0.01300 \; mol \times 0.0821 \; L \cdot atm \times (18 + 273) \; K}{(10.0 \; L)(K \cdot mol)}$$

$$= 0.03105 \; atm \; CO_2$$

Total pressure:

$$P = P_{O_2} + P_{CO_2} = 0.07698 \; atm + 0.03105 \; atm = 0.1080 \; atm$$

Mole fraction O_2:

$$\frac{n_{O_2}}{n_{total}} = \frac{0.03222 \; mol \; O_2}{0.03222 \; mol \; O_2 + 0.01300 \; mol \; CO_2} = 0.713$$

In working problems with gases collected over water, remember that it is the pressure, *not* the volume, that must be adjusted for the presence of the water vapor.

Exercise 5.11   Oxygen can be prepared by heating potassium chlorate, $KClO_3$, with manganese dioxide as a catalyst. The reaction is

$$2KClO_3(s) \xrightarrow{\Delta} 2KCl(s) + 3O_2(g)$$

How many moles of O_2 would be obtained from 1.300 g $KClO_3$? If this amount of O_2 were collected over water at 23°C and at a total pressure of 745 mmHg, what volume would it occupy?

Wanted: moles O_2, V of O_2 collected.

Given: balanced equation; 1.300 g $KClO_3$; $t = 23$°C; $P = 745$ mmHg.

Known: We can find moles from the equation and molar masses;

$$P = P_{H_2O} + P_{O_2}, \quad \text{so} \quad P_{O_2} = P - P_{H_2O};$$

we can find the volume using the ideal gas law.

Solution: Find moles O_2:

$$1.300 \; \cancel{\text{g KClO}_3} \times \frac{1 \; \cancel{\text{mol KClO}_3}}{122.6 \; \cancel{\text{g KClO}_3}} \times \frac{3 \; \text{mol } O_2}{2 \; \cancel{\text{mol KClO}_3}} = 0.015905 = 0.01591 \; \text{mol } O_2$$

Find P_{O_2} in atm (the P_{H_2O} at 23°C from text Table 5.6 is 21.1 mmHg):

$$P_{O_2} = 745 \; \text{mmHg} - 21.1 \; \text{mmHg} \quad = 723.9 \; \text{mmHg}$$

$$723.9 \; \cancel{\text{mmHg}} \times \frac{1 \; \text{atm}}{760 \; \cancel{\text{mmHg}}} \quad = 0.9525 \; \text{atm}$$

Find V_{O_2}:

$$V_{O_2} = \frac{nRT}{P} = \frac{0.01591 \; \cancel{\text{mol } O_2} \times 0.0821 \; \text{L} \cdot \cancel{\text{atm}} \times (23 + 273)\cancel{K}}{(0.9525 \; \cancel{\text{atm}})(K \cdot \cancel{\text{mol}})}$$

$$= 0.406 \; \text{L}$$

5.6   KINETIC THEORY OF AN IDEAL GAS

You might ask why we develop a model for an ideal gas, when no such gas exists. In fact, most gases encountered under normal laboratory conditions, at 1 atm P and at temperatures of 0°C – 100°C, are very well described by this model. Most gases act ideally under these conditions. In Section 5.8, we will see how to describe a gas that does not behave ideally.

5.7 MOLECULAR SPEEDS; DIFFUSION AND EFFUSION

Operational Skills

9. Calculating the rms speed of gas molecules. Given the molecular weight and temperature of a gas, calculate the rms molecular speed (Example 5.12).

10. Calculating the ratio of effusion rates of gases. Given the molecular weights of two gases, calculate the ratio of rates of effusion (Example 5.13); or, given the relative effusion rates of a known and an unknown gas, obtain the molecular weight of the unknown gas (as in Exercise 5.15).

Exercise 5.12 What is the rms speed (in m/s) of a carbon tetrachloride molecule at 22°C?

Known: $u = \sqrt{\dfrac{3RT}{M_m}}$ (We must use the value for R with SI units, and molar mass in kg.)

Solution:

$$u = \left(\frac{3 \times 8.31 \text{ kg} \cdot \text{m}^2 \times (22 + 273) \text{ K}}{(\text{s}^2 \cdot \text{K} \cdot \text{mol}) \, (154.0 \times 10^{-3} \text{ kg} / \text{mol})} \right)^{\frac{1}{2}} = 219 \text{ m/s}$$

Exercise 5.13 At what temperature do hydrogen molecules, H_2, have the same rms speed as nitrogen molecules, N_2, at 455°C? At what temperature do hydrogen molecules have the same average kinetic energy?

Solution: To find T for N_2, set speeds equal:

$$u_{H_2} = u_{N_2}$$

$$\sqrt{\frac{3RT}{M_{mH_2}}} = \sqrt{\frac{3R \, (455 + 273) \text{ K}}{M_{mN_2}}}$$

To solve for T, square both sides:

$$\frac{3RT}{2.02 \times 10^{-3}} = \frac{3R \, (728 \text{ K})}{28.0 \times 10^{-3}}$$

$$T = \frac{728 \text{ K} \times 2.02}{28.0} = 52.5 \text{ K}$$

Because kinetic energy is proportional to T, all molecules would have the same average kinetic energy at the same temperature, whether at 728 K or 52.5 K.

Exercise 5.14   If it takes 3.52 s for 10.0 mL of helium to effuse through a hole in a container at a particular temperature and pressure, how long would it take for 10.0 mL of oxygen, O_2, to effuse from the same container at the same temperature and pressure? (Note that the rate of effusion can be given in terms of volume of gas effused per second.)

Solution:

$$\frac{\text{Rate He}}{\text{Rate O}_2} = \sqrt{\frac{M_{m\,O_2}}{M_{m\,He}}} = \sqrt{\frac{32.0 \text{ g/mol}}{4.00 \text{ g/mol}}} = \sqrt{8.00} = 2.83$$

Setting the first term equal to 2.83 and solving for rate O_2, then substituting the given values for the rate of He, gives

$$\text{Rate O}_2 = \frac{\text{rate He}}{2.83} = \frac{10.0 \text{ mL}/3.52 \text{ s}}{2.83} = \frac{2.84 \text{ mL/s}}{2.83} = 1.00 \text{ mL/s}$$

$$\text{Time O}_2 = 10.0 \text{ mL} \times \frac{\text{s}}{1.00 \text{ mL}} = 10.0 \text{ s}$$

Exercise 5.15   If it takes 4.67 times as long for a particular gas to effuse as it takes hydrogen under the same conditions, what is the molecular weight of the gas? (Note that the rate of effusion is inversely proportional to the time it takes for a gas to effuse.)

Known: The time it takes a gas to effuse is inversely proportional to the rate; Graham's law.

Solution:

$$\overbrace{\frac{\text{Time unknown gas}}{\text{Time H}_2} = \frac{\text{rate H}_2}{\text{rate unknown gas}} = \sqrt{\frac{M_{m\,\text{unknown gas}}}{M_{m\,H_2}}}}^{\text{a statement of Graham's law}} = 4.67$$

Squaring the last two terms gives

$$\frac{M_{m\,\text{unknown gas}}}{2.02 \text{ g/mol}} = 21.\underline{8}1$$

$$M_{m\,\text{unknown gas}} = 44.1 \text{ amu}$$

5.8   REAL GASES

Operational Skill

11.   Using the van der Waals equation.   Given n, T, V, and the van der Waals constants a and b for a gas, calculate the pressure from the van der Waals equation (Example 5.14).

Remember that V in the ideal gas law is the volume of the container. If we do not subtract the correction factor from the V term, the V we use will be too large, since the molecular volume actually takes up some of the space a gas occupies. This molecular volume is negligible when the pressure is low, but it becomes significant at high pressures.

It is easier to see why we *add* the correction for pressure, rather than subtract it, if we rearrange the equation to solve for P:

$$P = \frac{nRT}{(V - nb)} - \frac{n^2 a}{V^2}$$

This shows that pressure is one term minus another (minus the correction factor). The correction factor thus gives an equation that describes reality: the actual pressure is less than ideal. This correction is unnecessary when the temperature is high. When the temperature is low, however, the molecules do not move as quickly and the molecular attraction becomes significant.

Exercise 5.16   Use the van der Waals equation to calculate the pressure of 1.000 mol of ethane, C_2H_6, that has a volume of 22.41 L at 0.0°C. Compare the result with the value predicted by the ideal gas law.

Wanted: van der Waals pressure; compare with ideal pressure.

Given: 1.000 mol C_2H_6, $V = 22.41$ L, $t = 0.0°C$.

Known: van der Waals equation; $R = 0.08206$ L • atm/(K • mol); Table 5.7 gives $a = 5.570$ L^2 • atm/mol^2; $b = 0.06499$ L/mol; $T = (0.0°C + 273.2) = 273.2$ K; P_{ideal} of 1.000 mol of gas at STP = 1.000 atm.

Solution: Rearrange the van der Waals equation to solve for pressure, and substitute in values.

$$P = \frac{nRT}{V - nb} - \frac{n^2 a}{V^2}$$

$$= \frac{(1.000 \text{ mol})[0.08206 \text{ L} \cdot \text{atm}/(\text{K} \cdot \text{mol})](273.2 \text{ K})}{22.41 \text{ L} - (1.000 \text{ mol} \times 0.06499 \text{ L}/\text{mol})}$$

$$- \frac{(1.000^2 \text{ mol}^2)(5.570 \text{ L}^2 \cdot \text{atm}/\text{mol}^2)}{22.41^2 \text{ L}^2}$$

$$= 1.00\underline{33} \text{ atm} - 0.01091\underline{0} \text{ atm} = 0.992 \text{ atm}$$

This value is 0.008 atm lower than the ideal pressure of 1.000 atm.

A Gas That Matters: OXYGEN (a Component of Air)

Questions for Study

1.   What is the reaction used in the main rockets of the space shuttle? Write the chemical equation.

2.   What are the two major components of air?

3.   Explain how liquid air is produced.

4.   How is liquid oxygen obtained from liquid air?

5.   What is the principal commercial use of oxygen?

6.   Describe the classic test for pure oxygen gas.

7.   What is the oxyacetylene torch? What is its purpose?

8.   Describe some end uses of compounds whose commercial production uses oxygen in their preparation.

9.   What would be the approximate optimum mole fraction of oxygen in a helium–oxygen diving mixture whose total pressure is 6.0 atm?

Answers to Questions for Study

1.   The main rockets of the space shuttle burn hydrogen with oxygen to form water vapor. The equation is

$$2H_2(l) + O_2(l) \longrightarrow 2H_2O(g)$$

2.   The two major components of air are nitrogen and oxygen.

3. Liquid air is produced in a liquid air machine. Clean air enters, and it is compressed and cooled. It then passes through a valve, where it expands, cooling sufficiently to become liquid.

4. Oxygen is obtained from liquid air distillation. The liquid air is allowed to warm so that it boils. Nitrogen, then argon, boil off, and the oxygen boils off next at $-183°C$.

5. The principal commercial use of oxygen is in steel making to burn the excess carbon out of the iron produced in a blast furnace.

6. The classic test for pure oxygen gas is to thrust a glowing splint into a bottle of gas. If the gas is oxygen, the splint will burst into a bright white flame.

7. The oxyacetylene torch is a device that supplies pure oxygen with the fuel acetylene which produces temperatures higher than $3000°C$. It is used to weld steel.

8. Compounds whose preparation use oxygen are titanium dioxide, TiO_2 (used as white pigment for paints, paper, and plastics), and ethylene glycol (used as antifreeze and for polyester fibers).

9. The mole fraction of oxygen would be

$$\frac{n_{O_2}}{n_{total}} \quad \text{or} \quad \frac{P_{O_2}}{P_{total}} = \frac{0.21 \text{ atm}}{6.0 \text{ atm}} = 0.035$$

ADDITIONAL PROBLEMS

1. Calculate the density of CO_2 at exactly $100°C$ and standard pressure.

2. Use the van der Waals equation,

$$\left(P + \frac{n^2 a}{V^2} \right)(V - nb) = nRT$$

to calculate the pressure of 0.750 mol $CO_2(g)$ inside a 1.85-L steel cylinder at 273.2 K. The values for a and b are $3.66 \text{ L}^2 \cdot \text{atm/mol}^2$ and 0.0429 L/mol, respectively.

3. A sample of nitrogen gas occupies 46.8 L at $28°C$ and 745 mmHg. What volume would it occupy at $28°C$ and 855 mmHg?

4. What would be the volume of a 255-L sample of gas if it were heated, at constant pressure, from $28°C$ to $42°C$?

5. At 90.0°C, XeF_4 is a gas. When 0.394 g of this noble-gas compound is introduced into a 0.91-L gas bulb at 90.0°C, the pressure is 47.3 mmHg. What pressure would this same amount of gas exhibit when it is transferred to a 2.13-L bulb at 120.0°C?

6. The first step in the industrial production of NH_3 using natural gas as the source of hydrogen can be represented by the following equation:

$$CH_4(g) + H_2O(g) \longrightarrow CO(g) + 3H_2(g)$$

   Assume that natural gas is pure methane, CH_4. Calculate the kilograms of hydrogen gas produced from $1.200 \times 10^3 \ m^3$ of natural gas at 745 mmHg and 25°C.

7. CoF_3 is often used to fluorinate organic compounds. During the fluorination of $(CH_3)_3N$, one of the gaseous products was isolated and purified. At 90.1°C and 105.7 mmHg, 1.46 g of the compound occupied a volume of 1.83 L. What is the molecular weight of this fluorination product?

8. Calculate:
   (a) the mole fraction of O_2 in a container that contains CO_2 at 285 mmHg, O_2 at 305 mmHg, and N_2 at 245 mmHg.
   (b) the root mean square (rms) molecular speed of N_2 at 25°C.
   (c) the ratio of the rates of effusion of H_2 and Rn through a fine pinhole.

9. How many grams of oxygen gas are contained in a 250.0-mL bottle of the gas collected over water at 25°C and 755 mmHg?

10. The height of the mercury column in a barometer is independent of the diameter or bore of the barometer tube. Explain why this is so, if pressure is defined as force F per unit area A, and area is defined as $(\frac{1}{4})\pi d^2$.

ANSWERS TO ADDITIONAL PROBLEMS

If you missed an answer, study the text section and operational skill given in parentheses after the answer.

1. Since $n = m/M_m$,

$$PV = \frac{m}{M_m} \times RT$$

Rearrange to solve for m/V (density):

$$\frac{m}{V} = \frac{PM_m}{RT} = \frac{(1 \ \text{atm}) \times (44.0 \ \text{g}/\text{mol})}{[0.0821 \ \text{L} \cdot \text{atm}/(\text{K} \cdot \text{mol})] \times 373 \ \text{K}} = 1.44 \ \text{g/L}$$

(5.3, Op. Sk. 5)

2. Rearrange the van der Waals equation to give

$$P = \frac{nRT}{V - nb} - \frac{n^2 a}{V^2}$$

Substituting the respective values for n, R, T, V, a, and b gives

$$P = \frac{(0.750 \ \text{mol}) \left[0.08206 \ \text{L} \cdot \text{atm}/(\text{K} \cdot \text{mol}) \right] (273.2 \ \text{K})}{[1.85 \ \text{L} - 0.750 \ \text{mol} \ (0.0429 \ \text{L}/\text{mol})]}$$

$$- \frac{(0.750 \ \text{mol})^2 \ (3.66 \ \text{L}^2 \cdot \text{atm}/\text{mol}^2)}{(1.85 \ \text{L})^2}$$

$$= 8.65 \ \text{atm} \qquad (5.8, \text{Op. Sk. } 11)$$

3. $46.8 \ \text{L} \times \dfrac{745 \ \text{mmHg}}{855 \ \text{mmHg}} = 40.8 \ \text{L}$ (5.2, Op. Sk. 2)

4. $255 \ \text{L} \times \dfrac{315 \ \text{K}}{301 \ \text{K}} = 267 \ \text{L}$ (5.2, Op. Sk. 2)

5. $47.3 \ \text{mmHg} \times \dfrac{0.91 \ \text{L}}{2.13 \ \text{L}} \times \dfrac{393 \ \text{K}}{363 \ \text{K}} = 22 \ \text{mmHg}$ (5.2)

6. $n_{\text{CH}_4} = \dfrac{PV}{RT} = \dfrac{0.9803 \ \text{atm} \ \text{K} \cdot \text{mol} \times 1.200 \times 10^6 \ \text{L}}{0.0821 \ \text{L} \cdot \text{atm} \times 298 \ \text{K}} = 4.808 \times 10^4 \ \text{mol } \text{CH}_4$

Calculate the kilograms of H_2:

$$4.8\underline{0}8 \times 10^4 \; \text{mol CH}_4 \times \frac{3 \; \text{mol H}_2}{1 \; \text{mol CH}_4} \times \frac{2.02 \; \text{g H}_2}{\text{mol H}_2} \times \frac{\text{kg H}_2}{10^3 \; \text{g H}_2}$$

$$= 2.91 \times 10^2 \; \text{kg H}_2 \quad (5.4, \text{Op. Sk. } 6)$$

7. $n = \dfrac{PV}{RT} = \dfrac{(0.1391 \; \text{atm})(1.83 \; L)}{[0.0821 \; L \cdot \text{atm}/(K \cdot \text{mol})](363 \; K)} = 8.5\underline{4}1 \times 10^{-3} \; \text{mol}$

Find M_m:

$$M_m = \frac{\text{grams}}{\text{mole}} = \frac{1.46 \; \text{g}}{8.5\underline{4}1 \times 10^{-3} \; \text{mol}} = 1.71 \times 10^2 \; \text{g/mol} \quad (5.3)$$

8.   (a) $\dfrac{305 \; \text{mmHg}}{(285 + 305 + 245) \; \text{mmHg}} = 0.365 \quad (5.5, \text{Op. Sk. } 7)$

  (b) $u = \sqrt{\dfrac{3 \times 8.31 \; \text{kg} \cdot \text{m}^2 \times 298 \, K}{\text{s}^2 \cdot K \cdot \text{mol} \times 0.0280 \; \text{kg}/\text{mol}}} = 515 \; \text{m/s} \quad (5.7, \text{Op. Sk. } 9)$

  (c) $\dfrac{r_{H_2}}{r_{Rn}} = \sqrt{\dfrac{222 \; \text{g}/\text{mol}}{2.02 \; \text{g}/\text{mol}}} = 10.5 \quad (5.7, \text{Op. Sk. } 10)$

9. $P_{O_2} = (755 - 23.8) \; \text{mmHg} = 73\underline{1}.2 \; \text{mmHg} \times \dfrac{1 \; \text{atm}}{760 \; \text{mmHg}} = 0.96\underline{2}1 \; \text{atm}$

Solve $PV = (m/M_m) RT$ for m and substitute in values.

$$m = \frac{PVM_m}{RT} = \frac{0.96\underline{2}1 \; \text{atm} \times 0.2500 \; L \times K \cdot \text{mol} \times 32.0 \; \text{g}/\text{mol}}{298 \, K \times (0.0821 \, L \cdot \text{atm})}$$

$$= 0.315 \; \text{g O}_2 \quad (5.5, \text{Op. Sk. } 8)$$

10.   By definition, the pressure P is the force F of the liquid column per unit area A.

$$P = F/A$$

The downward force of the liquid may be expressed as the product of the mass m of the liquid and the acceleration due to gravity g. The pressure expression can then be written

$$P = \frac{F}{A} = \frac{mg}{A}$$

Because $A = (\frac{1}{4})\pi d^2$, it looks as though the pressure does depend on the diameter or bore of the tube. However, the mass of a liquid can be expressed as the product of the density d and liquid volume V, where volume is the product of column height h and cross-sectional area A. This gives

$$P = \frac{mg}{A} = \frac{dVg}{A} = \frac{dh\cancel{A}g}{\cancel{A}} = dhg$$

Because g is a constant, $9.81\ \text{ms}^{-2}$, the indicated liquid pressure depends only on the density of the liquid and on the liquid column height, neither of which depends on the diameter or bore of the tube. (5.1)

CHAPTER POST-TEST

1. Indicate whether each of the following statements is true or false. If the statement is false, change it so it is true.

   (a) The accompanying graph depicts the change in temperature with volume at constant pressure for an ideal gas. True/False:

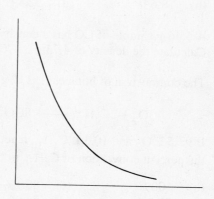

   _____
   _____
   _____
   _____
   _____ .

   (b) At constant pressure, the density of a gas will decrease as the absolute temperature increases. True/False: _____
   _____ .

   (c) At standard temperature and pressure, one mole of H_2 gas will occupy the same volume as one mole of O_2 gas, but the equal volumes will have different masses. True/False: _____
   _____ .

(d)   One assumption of kinetic molecular theory is that gas molecules have virtually no volume and are attracted to each other by van der Waals forces. True/False:

_____

_____ .

(e)   If CO_2 and CO molecules are at the same temperature, the average kinetic energy of each sample of gas molecules would be inversely proportional to the ratio of their molecular weights. True/False: _____

_____ .

(f)   The rate of effusion of NH_3 is about one-half that of HCl. True/False:

_____ .

2.   Calculate the value of the ideal gas constant, R, in units of $(mL \cdot mmHg)/(K \cdot mol)$.

3.   One mole of phosgene, $COCl_2$, at 115°C and 645 mmHg will occupy a volume of

(a)   37.5 L   (c)   27.0 L   (e)   none of the above.
(b)   18.6 L   (d)   22.4 L

4.   A 5.46-g sample of CO has a density of 1.209 g/L at 9°C and 1 atm pressure. Calculate the density of 4.73 g of CO at 28.0°C and 739 mmHg.

5.   The combustion of butane, C_4H_{10}, gives CO_2 and H_2O:

$$13O_2 + 2C_4H_{10} \longrightarrow 8CO_2 + 10H_2O$$

If 93.5 L O_2 and 10.3 L C_4H_{10} react to give 33.0 L CO_2 at 769 mmHg and 554°C, the percent conversion of C_4H_{10} to CO_2 was

(a)   100%   (c)   80.1%   (e)   insufficient data to calculate.
(b)   57.5%   (d)   78.0%

6.   Titanium reacts with chlorine to give a clear, colorless liquid. The elemental analysis of this compound shows a composition of 25.25% Ti and 74.75% Cl. The molecular weight was determined in the gas phase from the following data: mass of sample = 1.63 g, temperature = 171°C, pressure = 736 mmHg, volume = 0.323 L. What is the molecular formula of this titanium–chlorine compound?

7.   Fill in the correct words below. The combustion of propane is represented by the following equation:

$$C_3H_8(g) + 5O_2(g) \longrightarrow 3CO_2(g) + 4H_2O(g)$$

If the reaction is carried out in a closed vessel at a constant temperature and volume:

(a)   the pressure ———————————— as the reaction proceeds to completion.
          (decreases / increases)

(b)   the total number of molecules after the reaction is completed is
          ———————————— the number of molecules before the reaction is initiated.
       (greater than/less than)

8.   A closed container contains an equal number of N_2 and O_2 molecules at 30°C and standard pressure. Assuming ideal-gas behavior, which of the following statements is (are) correct?
(a)   As the temperature increases, the average number of N_2 and O_2 molecules increases.
(b)   The partial pressure of N_2 is the same as that of O_2.
(c)   The N_2 molecules make a greater contribution to the pressure because of their smaller mass.
(d)   If the pressure is decreased, the frequency of impacts of the N_2 and O_2 molecules with the walls of the container increases.
(e)   None of the above are correct statements.

9.   Which of the following statements about Figures A and B (p. 142) is (are) incorrect?
(a)   In each figure a given gas is at a constant temperature with only P and V varying.
(b)   The PV product for one mole of an ideal gas at constant temperature is a constant of value RT.
(c)   The deviations from the dashed horizontal line in Figures A and B for each curve indicate the relative deviation from ideal-gas behavior for the respective gas.
(d)   The minimum in the CO_2 plot results when the volume increases at a faster rate than the increase in pressure.
(e)   Real-gas behavior approaches that of an ideal gas at low P.

10.   Which of the following gases would have the largest value for b in the van der Waals equation for a nonideal gas?

   (a) O_2 (b) CO_2 (c) C_4H_{10} (d)   Ar   (e)   HF

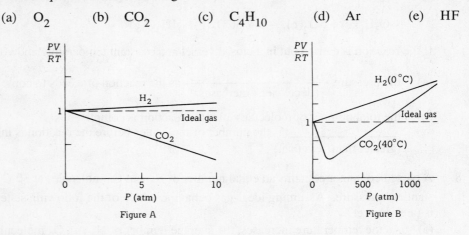

Figure A   Figure B

ANSWERS TO CHAPTER POST-TEST

If you missed an answer, study the text section and operational skill given in parentheses after the answer.

   1.   (a)   False. The graph depicts the change in pressure with volume at constant temperature.   (5.2, Op. Sk. 2)

        (b)   True.   (5.3)   (c)   True.   (5.2)

        (d)   False. One assumption of kinetic theory is that gas molecules have virtually no volume and are not attracted to each other.   (5.6)

        (e)   False. If CO_2 and CO molecules are at the same temperature, the average kinetic energy of each sample of gas molecules would be the same.   (5.6)

        (f)   False. The rate of effusion of NH_3 is about one and one-half times that of HCl.   (5.7, Op. Sk. 10)

   2. 6.24×10^4 (mL • mmHg)/(K • mol)   (5.3)   3.   a   (5.3, Op. Sk. 4)

   4.   1.10 g/L   (5.3, Op. Sk. 5)   5.   c   (5.4, Op. Sk. 6)

   6. $TiCl_4$ (5.3, Op. Sk. 5)   7.   (a)   increases,   (b)   greater than   (5.5)

   8.   b   (5.6)   9.   d   (5.8)   10.   c   (5.8)

CHAPTER 6   THERMOCHEMISTRY

CHAPTER TERMS AND DEFINITIONS

Numbers in parentheses after definitions give the text sections in which the terms are explained. Starred terms are italicized in the text. Where a term does not fall directly under a text section heading, additional information is given for you to locate it.

thermodynamics*   science of the relationships between heat and other forms of energy (6.1, introductory section)

thermochemistry*   one area of thermodynamics; study of the quantity of heat absorbed or evolved by chemical reactions   (6.1, introductory section)

energy   potential or capacity to move matter   (6.1)

kinetic energy (E_k)   energy associated with an object by virtue of its motion   (6.1)

joule (J)   SI unit of energy, $kg \cdot m^2/s^2$ (6.1)

watt*   measure of quantity of energy used per unit time; 1 J/s   (6.1)

calorie (cal)   non-SI unit of energy commonly used by chemists; originally defined as the amount of energy required to raise the temperature of one gram of water by one degree Celsius; 1 cal = 4.184 J (exact definition)   (6.1)

potential energy (E_p)   an object's energy because of its position in a field of force (6.1)

internal energy (U)   sum of kinetic and potential energies of the particles making up a substance   (6.1)

E_{tot}*   total energy of a substance; sum of the kinetic, potential, and internal energies of the substance   (6.1)

law of conservation of energy   energy may be converted from one form to another, but the total quantity of energy remains constant   (6.1)

thermodynamic system (or system)   substance or mixture of substances under study in which a physical or chemical change occurs   (6.2)

surroundings   everything in the vicinity of a thermodynamic system   (6.2)

heat (q)   energy that flows between system and surroundings because of a difference in temperature between the thermodynamic system and its surroundings   (6.2)

144

thermal equilibrium* state in which energy does not flow as heat between system and surroundings; temperature equality   (6.2)

heat of reaction   value of q required, at a given temperature, to return a system to the given temperature at the completion of the reaction   (6.2)

exothermic process   chemical reaction or physical change in which heat is evolved (q is negative)   (6.2)

endothermic process   chemical reaction or physical change in which heat is absorbed (q is positive)   (6.2)

q_p*   heat of reaction at constant pressure   (6.3)

enthalpy (H)   extensive property of a substance used to obtain the heat absorbed or evolved in a chemical reaction   (6.3)

state function   property of a system that depends only on its present state, which is determined by variables such as temperature and pressure, and is independent of any previous history of the system   (6.3)

enthalpy of reaction (ΔH)   change in enthalpy for a reaction at a given temperature and pressure; equals the heat of reaction at constant pressure   (6.3)

enthalpy diagram*   pictorial representation of the enthalpy change for a reaction   (6.3)

pressure–volume work*   energy required by a system to change volume against the constant pressure of the atmosphere   (6.3)

thermochemical equation   chemical equation for a reaction (including phase labels) in which the equation is given a molar interpretation, and the enthalpy of reaction for these molar amounts is written directly after the equation   (6.4)

heat capacity (C)   quantity of heat needed to raise the temperature of the sample of substance one degree Celsius (or one kelvin)   (6.6)

specific heat capacity (specific heat)   quantity of heat required to raise the temperature of one gram of a substance by one degree Celsius (or one kelvin) at constant pressure   (6.6)

calorimeter   device used to measure the heat absorbed or evolved during a physical or chemical change   (6.6)

bomb calorimeter*   calorimeter used for reactions involving gases   (6.6)

Hess's law of heat summation   for a chemical equation that can be written as the sum of two or more steps, the enthalpy change for the overall equation equals the sum of the enthalpy changes for the individual steps   (6.7)

standard state   standard thermodynamic conditions chosen for substances when listing or comparing thermodynamic data: 1 atm pressure and the specified temperature (usually 25°C)   (6.8)

standard enthalpy of reaction ($\Delta H°$)*   enthalpy change for a reaction in which reactants in their standard states yield products in their standard states   (6.8)

allotrope   one of two or more distinct forms of an element in the same physical state   (6.8)

reference form   stablest form (physical state and allotrope) of the element under standard thermodynamic conditions   (6.8)

standard enthalpy of formation (standard heat of formation) (ΔH_f°) enthalpy change for the formation of one mole of a substance in its standard state from its elements in their reference forms and in their standard states   (6.8)

fuel*   any substance that is burned or similarly reacted to provide heat and other forms of energy   (6.9)

*vitriolum**   Latin word meaning "glassy"   (A Chemical That Matters: Sulfuric Acid [an Industrial Acid])

acid mine drainage*   acid drainage from a coal mine   (A Chemical That Matters: Sulfuric Acid [an Industrial Acid])

contact process*   industrial process for the preparation of sulfuric acid from sulfur dioxide   (A Chemical That Matters: Sulfuric Acid [an Industrial Acid])

fuming sulfuric acid*   solution of sulfur trioxide (SO_3) in concentrated sulfuric acid, produced in the contact process   (A Chemical That Matters: Sulfuric Acid [an Industrial Acid])

*aqua fortis**   early name for nitric acid, from the Latin meaning "strong water"   (A Chemical That Matters: Nitric Acid [an Industrial Acid])

Ostwald process*   industrial method of preparing nitric acid by burning ammonia in the presence of a platinum catalyst, reacting the resulting nitric oxide with more oxygen to produce nitrogen dioxide, then reacting this with water   (A Chemical That Matters: Nitric Acid [an Industrial Acid])

CHAPTER DIAGNOSTIC TEST

1.   How much heat is produced when 8.95 g C_2H_5OH is burned in a constant-pressure system? The equation and enthalpy of reaction are

$$C_2H_5OH(l) + 3O_2(g) \longrightarrow 2CO_2(g) + 3H_2O(l); \quad \Delta H = -1.367 \times 10^3 \text{ kJ}$$

2.   Calculate the enthalpy change at 298 K for the reaction

$$Ni(s) + 4CO(g) \longrightarrow Ni(CO)_4(g)$$

Heats of formation at 298 K are: ΔH_f° for CO $= -110.5$ kJ/mol

ΔH_f° for $Ni(CO)_4$ $= -605$ kJ/mol

3. Calculate $\Delta H°$ at 298 K for the reaction C (graphite) + $CO_2(g) \longrightarrow 2CO(g)$ using the following $\Delta H°$ data at 298 K:

$$H_2(g) + CO(g) \longrightarrow C \text{ (graphite)} + H_2O(g); \quad \Delta H° = -131.38 \text{ kJ}$$

$$FeO(s) + H_2(g) \longrightarrow Fe(s) + H_2O(g); \quad \Delta H° = 24.69 \text{ kJ}$$

$$FeO(s) + CO(g) \longrightarrow Fe(s) + CO_2(g); \quad \Delta H° = -16.32 \text{ kJ}$$

4. What is the kinetic energy of an oxygen molecule traveling at a speed of 479 m/s in a tank at 21°C? (*Hint:* Use Avogadro's number and the molar mass of O_2 to get the actual mass of an oxygen molecule.)

5. For the reaction $H_2S(g) + 4H_2O_2(l) \longrightarrow H_2SO_4(l) + 4H_2O(l)$, $\Delta H°$ is -1.186×10^3 kJ. The enthalpy change per mole of H_2O_2 is

   (a) 1.301×10^2 kJ   (c) 2.965×10^2 kJ   (e)   none of the above.

   (b) 4.742×10^3 kJ   (d) -4.741×10^3 kJ

6. Indicate whether each of the following statements is true or false. If a statement is false, change it so it is true.

   (a) The enthalpy of reaction is independent of the exact state of the reactants or products. True/False: _____

   _____ .

   (b) If the enthalpy of reaction for $N_2(g) + O_2(g) \longrightarrow 2NO(g)$ is 180.5 kJ, then the enthalpy of reaction for $(1/2) N_2(g) + (1/2) O_2(g) \longrightarrow NO(g)$ is 90.25 kJ. True/False: _____

   _____ .

   (c) A calorimeter is a useful apparatus for determining heats of reaction. True/False: _____

   _____ .

   (d) For reactions involving gases and carried out at constant pressure, $\Delta H = q_p$. True/False: _____

   _____ .

(e) Hess's law permits the calculation of $\Delta H°$ values for reactions from $\Delta H°_f$ values for reactants and products and the calculation of $\Delta H°$ values for hypothetical reactions. True/False: _____

_____ .

(f) The $\Delta H°_f$ value for an element is always zero. True/False: _____

_____ .

7. Use heat of formation data in Appendix C in the text to calculate the enthalpy of the transition from the liquid to the gaseous state for 1 mole of HCN. Report the answer in kJ and kcal.

8. Determine the enthalpy of ionization, $\Delta H°_I$, for $Cs(g)$ using all of the thermodynamic data below. The equation is $Cs(g) \longrightarrow Cs^+(g) + e^-(g)$.

$$F(g) + e^-(g) \longrightarrow F^-(g); \qquad \Delta H = -336 \text{ kJ}$$

$$F_2(g) \longrightarrow 2F(g); \qquad \Delta H = 158 \text{ kJ}$$

$$Cs(s) \longrightarrow Cs(g); \qquad \Delta H = 78 \text{ kJ}$$

$$Cs(s) + \tfrac{1}{2}F_2(g) \longrightarrow CsF(s); \qquad \Delta H = -555 \text{ kJ}$$

$$CsF(s) \longrightarrow Cs^+(g) + F^-(g); \qquad \Delta H = 757 \text{ kJ}$$

9. When 2.89 g $N_2H_4(g)$ is combusted in a constant-pressure calorimeter containing exactly 1000 g of water, a temperature increase of 6.68°C is observed. The heat capacity of the calorimeter is 1.00 kJ/°C and the specific heat of water is 4.184 J/(g • °C). All products are gaseous. Determine the enthalpy change per mole of N_2H_4 combusted.

10. A 17.9-g sample of an unknown metal was heated to 48.31°C. It was then added to 28.05 g of water in an insulated cup. The water temperature rose from 21.04°C to 23.98°C. What is the specific heat of the metal?

11. Ethane gas (C_2H_6) burns in oxygen to form carbon dioxide gas, CO_2, and gaseous water. For each mole of ethane burned, 1.60 kJ of heat is evolved at constant pressure. Write the thermochemical equation for this reaction, including labels for the states of all reactants and products.

ANSWERS TO CHAPTER DIAGNOSTIC TEST

If you missed an answer, study the text section and operational skill given in parentheses after the answer.

1. -2.66×10^2 kJ   (6.5, Op. Sk. 4)   2. -163 kJ   (6.8, Op. Sk. 9)

3.   172.39 kJ   (6.7, Op. Sk. 7)

4. 6.10×10^{-21} J/molecule   (6.1, Op. Sk. 1)   5.   e   (6.4, Op. Sk. 3)

6.   (a)   False. The enthalpy of reaction is dependent on the exact state of the reactants or products.   (6.4)
     (b)   True.   (6.4, Op. Sk. 3)
     (c)   True.   (6.6)
     (d)   True.   (6.3)
     (e)   True.   (6.7)
     (f)   False. The ΔH_f° value for an element is always zero when the state of the element is the form of the element that exists at standard conditions.   (6.8)

7. 3.0×10^1 kJ; 7.2 kcal   (6.8, Op. Sk. 8)   8.   381 kJ   (6.7, Op. Sk. 7)

9. -383 kJ   (6.6, Op. Sk. 6)   10. 0.792 J/(g • °C)   (6.6, Op. Sk. 5)

11. $C_2H_6(g) + \frac{7}{2}O_2(g) \longrightarrow 2CO_2(g) + 3H_2O(g); \quad \Delta H = -1.60$ kJ

                                                                          (6.4, Op. Sk. 2)

SUMMARY OF CHAPTER TOPICS

Students find thermochemistry one of the more difficult topics in chemistry. This is said not to scare you but to assure you that if you find yourself really scratching your head, you are not alone. It is also said to let you know that this material is going to take a great deal of time and study to master. Your text presents the subject quite well, but you will probably need to read it over several times before the concepts begin to make sense. One of the biggest stumbling blocks for students is the arithmetic signs (+ and –) that go with almost every term. If you memorize the sign conventions stressed in the text, you will find things considerably easier. If you are stuck in your thinking or in an exercise or problem, go back and review the sign conventions to see if your error is there.

6.1 ENERGY AND ITS UNITS

Operational Skill

1. Calculating kinetic energy. Given the mass and speed of an object, calculate the kinetic energy (Example 6.1).

Energy has many forms. Some of these are electromagnetic energy (such as light, heat, and x rays), electric energy, sound energy, gravitational potential energy, elastic potential energy, and chemical potential energy. The forms of energy we will be concerned with in this course are electromagnetic energy, electric energy, and chemical potential energy.

Exercise 6.1 An electron whose mass is 9.11×10^{-31} kg is accelerated by a positive charge to a speed of 5.0×10^6 m/s. What is the kinetic energy of the electron in joules? in calories?

Known: $E_k = \frac{1}{2}mv^2$; 1 J = 1 kg • m^2/s^2; 1 cal = 4.184 J

Solution:

$$E_k = \frac{1}{2} \times 9.11 \times 10^{-31}\ \text{kg} \times (5.0 \times 10^6\ \text{m/s})^2 \times \frac{1\ \text{J}}{1\ \text{kg} \bullet \text{m}^2/\text{s}^2}$$

$$= 1.1 \times 10^{-17}\ \text{J}$$

$$E_k = 1.\underline{1}4 \times 10^{-17}\ \text{J} \times \frac{1\ \text{cal}}{4.184\ \text{J}} = 2.7 \times 10^{-18}\ \text{cal}$$

6.2 HEAT OF REACTION

Heat is a difficult term to understand. The word *heat* makes a good verb but a poor noun. Heat is *energy in transit* from a hotter object to a colder one. Objects do not possess heat. They possess energy that can be transferred as heat. Once the energy arrives at its destination, it is absorbed and is no longer called heat.

In an exothermic reaction, the reactants are always at a higher state of enthalpy than are the products. Heat is given off and the heat of reaction is negative (−). In an endothermic reaction, it is just the reverse. The reactants are at a lower state of enthalpy than are the products. Energy must be added to the reactants to get the products. Thus, the heat of the reaction is positive (+). The following diagrams illustrate these concepts.

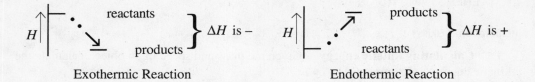

Exothermic Reaction   Endothermic Reaction

Exercise 6.2   Ammonia burns in the presence of a platinum catalyst to give nitric oxide, NO.

$$4NH_3(g) + 5O_2(g) \xrightarrow{\text{Pt}} 4NO(g) + 6H_2O(l)$$

In an experiment, 4 mol NH_3 are burned and evolve 1170 kJ of heat. Is the reaction endothermic or exothermic? What is the value of q?

Wanted: whether reaction is endothermic or exothermic; value of q.

Given: 1170 kJ of heat evolve.

Known: definitions of two terms; sign is + if heat is absorbed.

Solution: Reaction is exothermic; $q = -1170$ kJ.

6.3   ENTHALPY AND ENTHALPY CHANGE

6.4   THERMOCHEMICAL EQUATIONS

Operational Skills

   2.   Writing thermochemical equations.   Given a chemical equation, states of substances, and the quantity of heat absorbed or evolved for molar amounts, write the thermochemical equation   (Example 6.2).

   3.   Manipulating thermochemical equations.   Given a thermochemical equation, write the thermochemical equation for different multiples of the coefficients or for the reverse reaction   (Example 6.3).

   Exercise 6.3   A propellant for rockets is obtained by mixing the liquids hydrazine, N_2H_4, and dinitrogen tetroxide, N_2O_4. These compounds react to give gaseous nitrogen, N_2, and water vapor, evolving 1049 kJ of heat at constant pressure when 1 mol N_2O_4 reacts. Write the thermochemical equation for this reaction.

Solution: The equation and heat of reaction are

$$2N_2H_4(l) + N_2O_4(l) \longrightarrow 3N_2(g) + 4H_2O(g); \quad \Delta H = -1049 \text{ kJ}$$

Exercise 6.4   (a)   Write the thermochemical equation for the reaction described in Exercise 6.3 for the case involving 1 mol N_2H_4.   (b)   Write the thermochemical equation for the reverse of the reaction described in Exercise 6.3.

(a)   *Known:*   Each coefficient would be divided by 2, as would be ΔH_{rxn}.

Solution: $N_2H_4(l) + \frac{1}{2}N_2O_4(l) \longrightarrow \frac{3}{2} N_2 (g) + 2H_2O (g);$

$\Delta H = -5.245 \times 10^2 \text{ kJ}$

(b)   *Solution:* $3N_2(g) + 4H_2O(g) \longrightarrow 2N_2H_4(l) + N_2O_4(l);$

$\Delta H = 1.049 \times 10^3 \text{ kJ}$

Note that ΔH in this case is positive.

6.5   APPLYING STOICHIOMETRY TO HEATS OF REACTION

Operational Skill

4.   Calculating the heat of reaction from the stoichiometry.   Given the value of ΔH for a chemical equation, calculate the heat of reaction for a given mass of reactant or product   (Example 6.4).

Exercise 6.5   How much heat evolves when 10.0 g of hydrazine reacts according to the reaction described in Exercise 6.3?

Wanted: ΔH per 10.0 g hydrazine.

Given: 10.0 g hydrazine; reaction from Exercise 6.3.

Known: $\Delta H_{rxn} = -1.049 \times 10^3$ kJ per 2 mol hydrazine; formula = N_2H_4 = 32.0 g/mol.

Solution: Find the moles N_2H_4, then kJ:

$$10.0 \text{ g } N_2H_4 \times \frac{1 \text{ mol } N_2H_4}{32.0 \text{ g } N_2H_4} \times \frac{-1.049 \times 10^3 \text{ kJ}}{2 \text{ mol } N_2H_4} = -1.64 \times 10^2 \text{ kJ}$$

6.6   MEASURING HEATS OF REACTION

Operational Skills

5.   Relating heat and specific heat.   Given any three of the quantities q, s, m, and Δt, calculate the fourth one   (Example 6.5).

6.   Calculating ΔH from calorimetric data.   Given the amounts of reactants and the temperature change of a calorimeter of specified heat capacity, calculate the heat of reaction (Example 6.6).

In calculations in this section, the temperature used is in °C. Note that whether you use the Celsius or Kelvin scale, the number of units of temperature change is the same, as both scales use the same-sized unit.

When you calculate the temperature difference (Δt), be sure you always take $t_{final} - t_{initial}$ $(t_f - t_i)$. This way the sign of the heat or enthalpy will always come out correctly.

Exercise 6.6   Iron metal has a specific heat of 0.449 J/(g • °C). How much heat is transferred to a 5.00-g piece of iron, initially at 20.0°C, when it is placed in a pot of boiling water? Assume that the temperature of the water is 100.0°C and that the water remains at this temperature, which is the final temperature of the iron.

Wanted: heat transferred (J).

Given: 5.00 g of iron; $t_{initial}$ = 20.0°C; t_{final} = 100.0°C; specific heat of iron is 0.449 J/(g • °C).

Known: Heat transferred = specific heat × mass × temperature change.

Solution: Heat transferred = $\dfrac{0.449\ \text{J}}{g \cdot °\mathcal{C}} \times 5.00\ \mathcal{g} \times 80.0°\mathcal{C} = 1.80 \times 10^2$ J

Exercise 6.7   Suppose 33 mL of 1.20 *M* HCl is added to 42 mL of a solution containing excess sodium hydroxide, NaOH, in a coffee-cup calorimeter. The solution temperature, originally 25.0°C, rises to 31.8°C. Give the enthalpy change, ΔH, for the reaction

$$HCl(aq) + NaOH(aq) \longrightarrow NaCl(aq) + H_2O(l)$$

Express the answer as a thermochemical equation. For simplicity, assume that the heat capacity and the density of the final solution in the cup are those of water. (In more accurate work, these values must be determined.) Also assume that the total volume of the solution equals the sum of the volumes of HCl(*aq*) and NaOH(*aq*).

Wanted: ΔH_{rxn}.

Given: 33 mL 1.20 *M* HCl + 42 mL NaOH = 75 mL; t_i = 25.0°C; t_f = 31.8°C. Use density and *C* of water for calorimeter (cal).

Known: Mass = density × volume; $\Delta H_{rxn} = q_{rxn} = -C_{cal}\,\Delta t$; d_{water} = 1.0 g/mL; sp. ht. water = 4.184 J/(g • °C), so use 4.18 J/(g • °C) for solution.

Solution: First, find C_{cal} as follows, using *C* = sp. ht. × mass:

$$\text{mass soln} = 75 \;\cancel{mL} \times \frac{1.0 \text{ g}}{\cancel{mL}} = 75 \text{ g}$$

$$C = \frac{4.18 \text{ J}}{(\cancel{g} \bullet {}^\circ\text{C})} \times 75\,\cancel{g} = 3.\underline{1}4 \times 10^2 \text{ J/}^\circ\text{C}$$

Then, find ΔH_{rxn} for 33 mL of 1.20 *M* HCl:

$$\Delta H_{rxn} = -C_{cal}\,\Delta t = -3.\underline{1}4 \times 10^2 \frac{\text{J}}{{}^\circ\cancel{C}} \times (31.8 - 25.0)^\circ\cancel{C} = -2.\underline{1}4 \text{ kJ}$$

To find ΔH_{rxn}, determine the moles HCl in 33 mL of 1.20 *M* HCl:

$$0.033\;\cancel{\text{L soln}} \times \frac{1.20 \text{ mol HCl}}{1\;\cancel{\text{L soln}}} = 0.03\underline{9}6 \text{ mol HCl}$$

$$\Delta H_{rxn} = \frac{-2.\underline{1}4 \text{ kJ}}{0.03\underline{9}6 \text{ mol HCl}} = -54 \text{ kJ/mol HCl}$$

The thermochemical equation is

$$\text{HCl}(aq) + \text{NaOH}(aq) \longrightarrow \text{NaCl}(aq) + \text{H}_2\text{O}(l); \quad \Delta H = -54 \text{ kJ}$$

6.7   HESS'S LAW

Operational Skill

7.   Applying Hess's law.   Given a set of reactions with enthalpy changes, calculate ΔH for a reaction obtained from these other reactions by using Hess's law   (Example 6.7).

It is important to understand that the enthalpies (*H*) of the reactants or products describe the state of a given reaction system, just as volumes (*V*) and temperatures (*T*) do. This means that the values of *H* depend only on the state of the chemical system, not on the history of how it got to that state. Thus, we can use Hess's law, combining as many reactions as needed, to come out with the specific ΔH of the reaction wanted. The physical states of the chemical species in the reaction are important to the energy of reaction. Be sure that you use the correct symbols and cancel them properly. Remember that the value given for an enthalpy change (ΔH) of a reaction is specific for *that* written reaction: for the physical states of each species and for the amounts of materials specified by the reaction.

Exercise 6.8   Manganese metal can be obtained by reaction of manganese dioxide with aluminum.

$$4Al(s) + 3MnO_2(s) \longrightarrow 2Al_2O_3(s) + 3Mn(s)$$

What is ΔH for this reaction? Use the following data:

$$2Al(s) + \tfrac{3}{2}O_2(g) \longrightarrow Al_2O_3(s); \qquad \Delta H = -1676 \text{ kJ}$$

$$Mn(s) + O_2(g) \longrightarrow MnO_2(s); \qquad \Delta H = -521 \text{ kJ}$$

Wanted: ΔH_{rxn}.

Given: reactions and enthalpies.

Known: Use Hess's law. Reversing an equation changes the sign of ΔH; if you multiply coefficients by a factor, ΔH_{rxn} must be multiplied by the same factor.

Solution: Write equation (1), multiplying coefficients and ΔH by 2; reverse equation (2), changing the sign of ΔH, and multiply the coefficients and ΔH by 3. Then add the equations and ΔH.

(1) $4Al(s) + \cancel{3O_2(g)} \longrightarrow 2Al_2O_3(s)$ $\Delta H = -3352 \text{ kJ}$

(2) $3MnO_2(s) \longrightarrow 3Mn(s) + \cancel{3O_2(g)}$ $\Delta H = 1563 \text{ kJ}$

         $4Al(s) + 3MnO_2(s) \longrightarrow 2Al_2O_3(s) + 3Mn(s)$ $\Delta H = -1789 \text{ kJ}$

6.8   STANDARD ENTHALPIES OF FORMATION

Operational Skills

8.   Calculating the heat of phase transition from standard enthalpies of formation.   Given a table of standard enthalpies of formation, calculate the heat of phase transition   (Example 6.8).

9.   Calculating the enthalpy of reaction from standard enthalpies of formation. Given a table of standard enthalpies of formation, calculate the enthalpy of reaction (Example 6.9).

The summarizing statement to remember for calculating enthalpy changes for reactions is "the enthalpy change for a reaction at standard conditions is the sum of the enthalpies of formation of the products minus the sum of the enthalpies of formation of the reactants." It is written

$$\Delta H° = \sum n\Delta H_f° \text{ (products)} - \sum m\Delta H_f° \text{ (reactants)}$$

where the Greek letter $\sum$ indicates summation. Memorize this equation. Later you will see other summation statements in similar form. As your text cautions, pay particular attention to signs. This will be your greatest source of error.

Exercise 6.9   Calculate the heat of vaporization, $\Delta H_{vap}°$, of water, using standard enthalpies of formation (Table 6.2).

Known: $\Delta H° = \sum n\Delta H_f° \text{ (products)} - \sum m\Delta H_f° \text{ (reactants)}$.

Solution: Write the reaction. Then write the $\Delta H_f°$ values under each formula and subtract as indicated above.

$$H_2O(l) \longrightarrow H_2O(g)$$

$$-285.8 \qquad\qquad -241.8 \qquad \text{(kJ)}$$

$$\Delta H° = \Delta H_{vap}° \quad = \Delta H_f°\,[H_2O(g)] - \Delta H_f°\,[H_2O(l)]$$

$$= -241.8 \text{ kJ} - (-285.8 \text{ kJ})$$

$$= 44.0 \text{ kJ per mole of water vaporized}$$

Exercise 6.10   Calculate the enthalpy change for the following reaction:

$$3NO_2(g) + H_2O(l) \longrightarrow 2HNO_3(aq) + NO(g)$$

Use standard enthalpies of formation.

Known: $\Delta H = \sum n\Delta H_f^\circ \text{ (products)} - \sum m\Delta H_f^\circ \text{ (reactants)}.$

Solution: Write ΔH_f° under each species in the equation.

$$3NO_2(g) + H_2O(l) \longrightarrow 2HNO_3(aq) + NO(g)$$

$$3 \times 33.2 \quad -285.8 \qquad 2 \times -206.6 \quad 90.3 \qquad \text{(kJ)}$$

Then use the ΔH equation:

$$\Delta H^\circ = \sum n\Delta H_f^\circ \text{ (products)} - \sum m\Delta H_f^\circ \text{ (reactants)}$$

$$= [2(-206.6) + 90.3] - [3(33.2) + (-285.8)] \text{ kJ}$$

$$= -136.7 \text{ kJ}$$

You have now seen two methods for calculating the enthalpy change of a reaction: (1) by using Hess's law and enthalpies of other reactions, and (2) by using heats of formation of reactants and products.

Exercise 6.11   Calculate the standard enthalpy change for the reaction of an aqueous solution of barium hydroxide, $Ba(OH)_2$, with an aqueous solution of ammonium nitrate, NH_4NO_3, at 25°C. (Figure 6.1 illustrated this reaction using solids instead of solutions.) The complete ionic equation is

$$[Ba^{2+}(aq) + 2OH^-(aq)] + 2[NH_4^+(aq) + NO_3^-(aq)] \longrightarrow$$

$$2NH_3(g) + 2H_2O(l) + [Ba^{2+}(aq) + 2NO_3^-(aq)]$$

Known: $\Delta H^\circ = \sum n\Delta H_f^\circ \text{ (products)} - \sum m\Delta H_f^\circ \text{ (reactants)}.$

Solution: Write ΔH_f° under each species in the net ionic equation:

$$2OH^-(aq) + 2NH_4^+(aq) \longrightarrow 2NH_3(g) + 2H_2O(l)$$

$$2 \times -229.9 \quad 2 \times -132.8 \qquad 2 \times -45.9 \quad 2 \times -285.8 \quad \text{(kJ)}$$

Then use the $\Delta H°$ equation:

$$\Delta H° = \left(2\left\{\Delta H_f°\left[NH_3(g)\right]\right\} + 2\left\{\Delta H_f°\left[H_2O(l)\right]\right\}\right)$$

$$- \left(2\left\{\Delta H_f°\left[OH^-(aq)\right]\right\} + 2\left\{\Delta H_f°\left[NH_4^+(aq)\right]\right\}\right)$$

$$= \left[2(-45.9) + 2(-285.8)\right] - \left[2(-229.9) + 2(-132.8)\right]\ kJ$$

$$= \left[(-663.4) - (-725.4)\right]\ kJ = 62.0\ kJ$$

6.9   FUELS—FOODS, COMMERCIAL FUELS, AND ROCKET FUELS

A Chemical That Matters: SULFURIC ACID (an Industrial Acid)

Questions for Study

1. What is the old name for sulfuric acid? What is the origin of this name?

2. Concentrated sulfuric acid has a strong affinity for water. What happens when the concentrated acid is poured on sucrose? Write the equation for the reaction.

3. Why should concentrated sulfuric acid be added to water and not water to the acid?

4. Write the equation for the reaction of zinc metal with sulfuric acid solution.

5. What is acid mine drainage? What mineral is responsible for this acidic solution?

6. What is the ranking of sulfuric acid by pounds produced annually?

7. Write the equations for the preparation of sulfuric acid from sulfur by the contact process.

8. What is fuming sulfuric acid?

9. What is a major use of sulfuric acid?

Answers to Questions for Study

1. The old name for sulfuric acid is oil of vitriol. It was so named because the pure acid is a colorless, oily liquid which was originally prepared by heating green vitriol, iron(II) sulfate heptahydrate.

2. When concentrated sulfuric acid is poured on sucrose, it removes water from the sugar, generating steam and causing the charred mass to froth upward. The equation is

$$C_{12}H_{22}O_{11}(s) \longrightarrow 12C(s) + 11H_2O(g)$$

3. Concentrated sulfuric acid should be added to water to dissipate the heat of reaction. If water is added to concentrated acid, the water will boil violently and spatter the acid.

4. The equation for the reaction of zinc metal with sulfuric acid solution is

$$Zn(s) + H_2SO_4(aq) \longrightarrow ZnSO_4(aq) + H_2(g)$$

5. Acid mine drainage is acidic water draining from a coal mine. The responsible mineral is pyrite, occurring along with coal, which when exposed to weathering with moist oxygen forms iron(III) sulfate and sulfuric acid.

6. Sulfuric acid is the leading industrial chemical in terms of pounds produced.

7. The equations for the contact process are

$$S_8(s) + 8O_2(g) \longrightarrow 8SO_2(g); \quad \Delta H^\circ = -2374 \text{ kJ}$$

$$2H_2S(g) + 3O_2(g) \longrightarrow 2SO_2(g) + 2H_2O(g); \quad \Delta H^\circ = -1037 \text{ kJ}$$

$$2SO_2(g) + O_2(g) \xrightarrow[V_2O_5]{\Delta} 2SO_3(g); \quad \Delta H^\circ = -198 \text{ kJ}$$

$$SO_3(g) + H_2O(l) \longrightarrow H_2SO_4(aq); \quad \Delta H^\circ = -226 \text{ kJ}$$

8. Fuming sulfuric acid is the solution resulting in the contact process from the dissolving of sulfur trioxide in concentrated sulfuric acid.

9. A major use of sulfuric acid is in the fertilizer industry to convert insoluble phosphate rock to a soluble phosphate.

A Chemical That Matters: NITRIC ACID (an Industrial Acid)

Questions for Study

1. What was the old name for nitric acid? What is the origin of that name?

2. Write the chemical equation for the original preparation of nitric acid. How is nitric acid recovered from the reaction mixture?

3. Write the chemical equations for the preparation of nitric acid from ammonia by the Ostwald process.

4. What is the largest use of nitric acid? Write the chemical equation for the reaction involved.

5. What are some other uses of nitric acid?

6. What is dynamite? Who discovered it? What was the significance of the discovery?

Answers to Questions for Study

1. The old name for nitric acid was *aqua fortis*, Latin for "strong water." It was called that presumably because nitric acid reacts with most metals, including copper and silver, which are not attacked by other acids.

2. The equation for the original preparation of nitric acid is

$$KNO_3(s) + H_2SO_4(l) \longrightarrow KHSO_4(s) + HNO_3(g)$$

    The nitric acid is distilled from the reaction vessel.

3. The equations for the Ostwald process are

$$4NH_3(g) + 5O_2(g) \xrightarrow{\text{Pt}} 4NO(g) + 6H_2O(g); \quad \Delta H° = -906 \text{ kJ}$$

$$2NO(g) + O_2(g) \longrightarrow 2NO_2(g); \quad \Delta H° = -114 \text{ kJ}$$

$$3NO_2(g) + H_2O(l) \longrightarrow 2HNO_3(aq) + NO(g); \quad \Delta H° = -137 \text{ kJ}$$

4. The largest use of nitric acid is in the production of ammonium nitrate, NH_4NO_3, used as a nitrogen fertilizer and in the manufacture of explosives. The equation is

$$HNO_3(aq) + NH_3(aq) \longrightarrow NH_4NO_3(aq)$$

5. Other uses of nitric acid are in the manufacture of plastics (including nylon and polyurethane) and of explosive compounds such as nitroglycerin.

6. Dynamite is a mixture of nitroglycerin with clay and similar materials. It was discovered by Swedish chemist and industrialist Alfred Nobel. The significance of the discovery was that nitroglycerin is much less shock-sensitive in this form and therefore safer to use.

ADDITIONAL PROBLEMS

1. When butane gas, C_4H_{10}, burns in oxygen gas, O_2, to form carbon dioxide gas, CO_2, and gaseous water, 2845 kJ of thermal energy is evolved per mole of butane burned.
    (a) Write the thermochemical equation for the reaction, using whole-number coefficients.
    (b) Calculate the thermal energy evolved when 25.0 g of butane is burned.

2. A 100.0-g sample of water at 25.30°C was placed in an insulated cup. Then 45.00 g of lead pellets at 100.00°C was added. The final temperature of the water was 34.34°C. What is the specific heat of lead?

3. A 23.3-g sample of copper at 75.7°C is added to 100.0 mL of benzene at 20.0°C. Assuming that no thermal energy is used in evaporation, calculate the final temperature of the benzene. [The specific heat of copper is 0.389 J/(g • °C), the specific heat of benzene is 1.70 J/(g • °C), and the density of benzene is 0.879 g/mL.] How would this value change if we took into account the energy that would actually go toward the increased evaporation?

4. A 0.875-g sample of anthracite coal was burned in a bomb calorimeter. The temperature rose from 22.50°C to 23.80°C. The heat capacity of the calorimeter was found in another experiment to be 20.5 kJ/°C.
    (a) What was the heat evolved by the reaction?
    (b) What is the energy released on burning one metric ton (exactly 1000 kg) of this type of coal?

5. $POCl_3$ is formed according to the following reaction:

    $$PCl_5(g) + H_2O(g) \longrightarrow POCl_3(g) + 2HCl(g)$$

    The enthalpy of reaction is – 126 kJ per mole of $POCl_3$ formed. Write the thermochemical equation for the reverse reaction, doubling the coefficients.

6. What is the ratio of speeds of H_2 and N_2 molecules at 125°C?

7. Write the equation that represents the heat of formation of each of the following species. Include states of matter.
    (a) $Sb_2O_3(s)$ (c) $N_2H_4(l)$ (e) $Ca(NO_3)_2(s)$
    (b) $C_6H_{12}O_6(s)$ (d) $SF_6(g)$

8. Calculate $\Delta H°$ for the following reaction

$$2NO(g) + 2CO(g) \longrightarrow N_2(g) + 2CO_2(g)$$

given that $\Delta H_f°[NO(g)] = 90.3$ kJ/mol, $\Delta H_f°[CO(g)] = -110.5$ kJ/mol, and $\Delta H_f°$ $[CO_2(g)] = -393.5$ kJ/mol.

9. Using $\Delta H_f°$ data given in Table 6.2 of the text, calculate the amount of heat released from the combustion of a mixture of 15.0 g $C_5H_{12}(l)$, *n*-pentane, and 66.6 g $O_2(g)$.

$$\Delta H_f°[C_5H_{12}(l)] = -173.2 \text{ kJ/mol.}$$

10. Calculate the number of grams of natural gas at 25.0°C that are required to heat 1.00 kg H_2O from 25.0°C to 100.0°C. Assume the natural gas is 80.0% CH_4 and 20.0% C_2H_6 by mass.

$$\Delta H_f°[CH_4(g)] = -74.9 \text{ kJ/mol,} \quad \Delta H_f°[C_2H_6(g)] = -84.7 \text{ kJ/mol,}$$

and the specific heat of liquid water is 4.184 J/ (g • °C).

ANSWERS TO ADDITIONAL PROBLEMS

If you missed an answer, study the text section and operational skill given in parentheses after the answer.

1. (a) $2C_4H_{10}(g) + 13O_2(g) \longrightarrow 8CO_2(g) + 10H_2O(g)$;
   $\Delta H = -5.690 \times 10^3$ kJ   (6.4, Op. Sk. 2)

   (b) $25.0 \text{ g C}_4\text{H}_{10} \times \dfrac{1 \text{ mol C}_4\text{H}_{10}}{58.1 \text{ g C}_4\text{H}_{10}} \times \dfrac{-2845 \text{ kJ}}{\text{mol C}_4\text{H}_{10}} = -1.22 \times 10^3$ kJ

   (6.5, Op. Sk. 4)

2. $-\Delta H_{\text{lead}} = \Delta H_{\text{water}}$

Final temperature of both is the same.

$$-(45.0 \text{ g}) \text{ (sp. ht.) } (34.34 - 100.00)°C$$

$$= (100.0 \text{ g}) [4.184 \text{ J}/(\text{g} \cdot °C)] (34.34 - 25.30)°C$$

$$\text{sp. ht.} = \frac{(100.0) (4.184 \text{ J}) (9.04)}{(45.0 \text{ g}) (65.66°C)} = 1.28 \text{ J}/(\text{g} \cdot °C) \quad (6.6, \text{ Op. Sk. } 5)$$

3. $-\Delta H_{\text{copper}} = \Delta H_{\text{benzene}}$

$$-(23.3 \text{ g}) [0.389 \text{ J}/(\text{g} \cdot °C)] (t_f - 75.7)°C$$

$$= (100.0 \text{ mL} \times 0.879 \text{ g}/\text{mL}) [1.70 \text{ J}/(\text{g} \cdot °C)] (t_f - 20.0)°C$$

$$-9.064 t_f + 686.1°C \quad = 149.4 t_f - 2989°C$$

$$t_f = 23.2°C$$

The final temperature would be lower because not all the energy would go to raising the temperature of the benzene.   (6.6)

4.   (a) $-\Delta H_{\text{rxn}} = \Delta H_{\text{cal}}$

$$\Delta H_{\text{cal}} = 20.5 \text{ kJ}/°C \times (23.80 - 22.50)°C = 26.6 \text{ kJ}$$

$$\Delta H_{\text{rxn}} = -26.6 \text{ kJ}$$

(b) $10^6 \text{ g} \times \dfrac{-26.6 \text{ kJ}}{0.875 \text{ g}} = -3.04 \times 10^7 \text{ kJ} \quad (6.6, \text{ Op. Sk. } 6)$

5. $4HCl(g) + 2POCl_3(g) \longrightarrow 2H_2O(g) + 2PCl_5(g); \quad \Delta H = 252 \text{ kJ}$

$$(6.4, \text{ Op. Sk. } 3)$$

6.   The kinetic energies of gases are the same at the same temperature, so

$$E_k(N_2) = E_k(H_2)$$

$$\tfrac{1}{2} m_{N_2} v_{N_2}^2 = \tfrac{1}{2} m_{H_2} v_{H_2}^2$$

$$\frac{m_{N_2}}{m_{H_2}} = \frac{v_{H_2}^2}{v_{N_2}^2}$$

and as we are interested in the mass ratio, we can use molar masses. Thus,

$$\sqrt{\frac{M_m(N_2)}{M_m(H_2)}} = \frac{v_{H_2}}{v_{N_2}} = \sqrt{\frac{28.0 \text{ g/mol}}{2.02 \text{ g/mol}}} = 3.72 \qquad (6.1)$$

7. (a) $2Sb(s) + \frac{3}{2}O_2(g) \longrightarrow Sb_2O_3(s)$

   (b) $6C(\text{graphite}) + 6H_2(g) + 3O_2(g) \longrightarrow C_6H_{12}O_6(s)$

   (c) $N_2(g) + 2H_2(g) \longrightarrow N_2H_4(l)$

   (d) $S(s) + 3F_2(g) \longrightarrow SF_6(g)$

   (e) $Ca(s) + N_2(g) + 3O_2(g) \longrightarrow Ca(NO_3)_2(s) \qquad (6.8)$

8. $\Delta H° = \sum n\Delta H_f° \text{ (products)} - \sum m\Delta H_f° \text{ (reactants)}$

   $= [0 + 2(-393.5)] - [2(90.3) + 2(-110.5)]$

   $= -787.0 - (180.6 - 221.0) = -746.6 \text{ kJ} \qquad (6.8, \text{ Op. Sk. } 9)$

9. The balanced equation for the combustion of *n*-pentane is

   $$C_5H_{12}(l) + 8O_2(g) \longrightarrow 5CO_2(g) + 6H_2O(l)$$

   The calculation of $\Delta H_{rxn}°$ is

   $\Delta H° = 5(-393.5 \text{ kJ}) + 6(-285.8 \text{ kJ}) - (-173.2 \text{ kJ}) - 8(0 \text{ kJ}) = -3509.1 \text{ kJ}$

   Find the limiting reagent:

   $$15.0 \text{ g } C_5H_{12} \times \frac{1 \text{ mol } C_5H_{12}}{72.1 \text{ g } C_5H_{12}} \times \frac{5 \text{ mol } CO_2}{1 \text{ mol } C_5H_{12}} = 1.040 \text{ mol } CO_2$$

   $$66.6 \text{ g } O_2 \times \frac{1 \text{ mol } O_2}{32.0 \text{ g } O_2} \times \frac{5 \text{ mol } CO_2}{8 \text{ mol } O_2} = 1.301 \text{ mol } CO_2$$

   C_5H_{12} is the limiting reagent. The amount of heat released in the combustion of $0.208 \text{ mol } C_5H_{12}$ is

$$15.0 \; \cancel{g \, C_5H_{12}} \times \frac{1 \; \cancel{mol \, C_5H_{12}}}{72.1 \; \cancel{g \, C_5H_{12}}} \times \frac{-3509.1 \; kJ}{1 \; \cancel{mol \, C_5H_{12}}} = -7.30 \times 10^2 \; kJ$$

$$(6.8, \text{Op. Sk. } 4, 9)$$

10.   Calculate the heats of combustion of CH_4 and C_2H_6.

$$2O_2(g) + CH_4(g) \longrightarrow CO_2(g) + 2H_2O(l)$$

$$\Delta H° = -393.5 \; kJ + 2\,(-285.8 \; kJ) - (-74.9 \; kJ) = -890.2 \; kJ$$

$$\tfrac{7}{2}O_2(g) + C_2H_6(g) \longrightarrow 2CO_2(g) + 3H_2O(l)$$

$$\Delta H° = 2\,(-393.5 \; kJ) + 3\,(-285.8 \; kJ) - (-84.7 \; kJ) = -1559.7 \; kJ$$

The amount of heat required to raise the temperature of 1.00 kg of liquid water $(100.0 - 25.0)°C = 75.0°C$ is:

$$75.0°\cancel{C} \times 1.00 \; \cancel{kg} \times 4.184 \; \cancel{J}/(\cancel{g} \bullet °\cancel{C}) \times 1000 \; \cancel{g}/\cancel{kg}$$

$$\times 1 \; kJ/1000 \; \cancel{J} = 3.1\underline{3}8 \times 10^2 \; kJ$$

The heat released from the combustion of 1 g of natural gas (0.800 g CH_4 and 0.200 g C_2H_6) is:

$$\frac{-890.2 \; kJ}{1 \; \cancel{mol \, CH_4}} \times \frac{1 \; \cancel{mol \, CH_4}}{16.0 \; \cancel{g \, CH_4}} \times \frac{0.800 \; \cancel{g \, CH_4}}{1 \; g \; gas}$$

$$+ \frac{-1559.7 \; kJ}{1 \; \cancel{mol \, C_2H_6}} \times \frac{1 \; \cancel{mol \, C_2H_6}}{30.1 \; \cancel{g \, C_2H_6}} \times \frac{0.200 \; \cancel{g \, C_2H_6}}{1 \; g \; gas} = -54.\underline{8}7 \; kJ/g \; gas$$

The amount of natural gas needed to provide 313.8 kJ of heat is

$$31\underline{3}.8 \; \cancel{kJ} \times \frac{g \; gas}{54.\underline{8}7 \; \cancel{kJ}} = 5.72 \; g \; gas \quad (6.6, 6.8, \text{Op. Sk. } 5, 6, 9)$$

CHAPTER POST-TEST

1.   Calculate the heat produced when 3.76 g CO at 298 K react with an excess of Ni to give $Ni(CO)_4$ in a constant-pressure system.

$$Ni(s) + 4CO(g) \longrightarrow Ni(CO)_4(g); \quad \Delta H° = -163 \; kJ$$

2.   Calculate the enthalpy change at 298 K for the reaction

$$CaO(s) + 2HCl(g) \longrightarrow CaCl_2(s) + H_2O(l)$$

from the following heats of formation at 298 K:

$$\Delta H_f^\circ$$

CaO -635.1 kJ/mol $CaCl_2$ -795.0 kJ/mol

HCl -92.31 kJ/mol H_2O -258.84 kJ/mol

3.   Determine ΔH_f° for $CH_4(g)$ from the following ΔH° data and corresponding reactions at 298 K:

$$C\,(graphite) + O_2(g) \longrightarrow CO_2(g) \qquad \Delta H^\circ = -393.55 \text{ kJ}$$

$$CH_4(g) + 2O_2(g) \longrightarrow CO_2(g) + 2H_2O(l) \qquad \Delta H^\circ = -890.36 \text{ kJ}$$

$$H_2(g) + \tfrac{1}{2}O_2(g) \longrightarrow H_2O(l) \qquad \Delta H^\circ = -285.85 \text{ kJ}$$

4.   If we consider a hot kitchen stove as a system, then the transfer of heat from the stove to the kitchen is taken to be $\dfrac{\rule{3cm}{0.4pt}}{\text{(positive / negative)}}$.

5.   Which of the following is false? For the false statement, change it so it is true.
   (a)   If q_p for a reaction is negative, the reaction is endothermic. _____
   _____

   (b) ΔH is the amount of heat released or absorbed by a system at constant pressure. _____
   _____

   (c)   If heat is evolved in a reaction, the reaction is exothermic. _____
   _____

   (d) $\Delta H = H$ (products) $- H$ (reactants). _____
   (e)   Enthalpy is a state function. _____
   _____

6.   The enthalpy change for a reaction system open to the atmosphere is
   (a)   dependent on the identity of the reactants and products, and/or
   (b)   zero, and/or
   (c)   negative, and/or

(d) q_p, and/or

(e)   positive.

7.   The following thermochemical equation for the decomposition of gaseous ammonia, NH_3,

$$NH_3(g) \longrightarrow \tfrac{1}{2}N_2(g) + \tfrac{3}{2}H_2(g); \quad \Delta H = 45.9 \text{ kJ}$$

indicates that the formation of gaseous ammonia

(a)   evolves 45.9 kJ for each mole of ammonia formed, and/or

(b)   evolves 23 kJ for each mole of nitrogen used, and/or

(c)   absorbs 45.9 kJ for each mole of ammonia formed, and/or

(d)   absorbs 23 kJ for each mole of nitrogen used, and/or

(e)   is an exothermic process.

8.   The following reaction is carried out in a bomb calorimeter:

$$CaO(s) + SO_3(g) \longrightarrow CaSO_4(s)$$

If 1.00 mol SO_3 is reacted with an excess of CaO, 2.000×10^3 g of water increases in temperature from 17.5°C to 55.5°C. The calorimeter has a measured heat capacity of 0.126 kJ/°C. Calculate the heat released in this reaction. [Specific heat of water is 4.184 J/(g • °C).]

9.   What is the speed of a CCl_4 molecule at 22°C when its kinetic energy is 6.13×10^{-21} J/molecule?

10. $SiH_4(g)$ reacts with gaseous oxygen to form $SiO_2(s)$ and liquid water. How much heat is liberated when 5.75 g SiH_4 burns in an excess of oxygen at 298 K? $\Delta H_f^\circ [SiH_4(g)] = +34.3$ kJ/mol.

11. In a blast furnace, solid iron(III) oxide, Fe_2O_3, is reduced by gaseous carbon monoxide, CO, to liquid iron and gaseous carbon dioxide, CO_2. For each mole of iron oxide reduced, 27.6 kJ of heat is evolved. Write the complete equation for this reaction, including the heat of reaction and labels for the states of all reactants and products.

12. A 48.0-g sample of copper was heated to 98.5°C. It was then added to 105 g of water in an insulated container. The temperature of the water rose from 25.6°C to 28.8°C. What is the specific heat of copper?

ANSWERS TO CHAPTER POST-TEST

If you missed an answer, study the text section and operational skill given in parentheses after the answer.

1. 5.47 kJ   (6.5, Op. Sk. 4)

2. 234.1 kJ   (6.8, Op. Sk. 9)

3. −74.89 kJ/mol   (6.7, 6.8, Op. Sk. 7)

4. negative   (6.2)

5. a. If q_p for a reaction is negative, the reaction is exothermic.   (6.2)

6. a and d   (6.3)

7. a and e   (6.4, Op. Sk. 3)

8. 323 kJ released   (6.6, Op. Sk. 6)

9. 219 m/s   (6.1, Op. Sk. 1)

10. −272 kJ   (6.5, 6.8, Op. Sk. 4, 9)

11. $Fe_2O_3(s) + 3CO(g) \longrightarrow 2Fe(l) + 3CO_2(g); \quad \Delta H = -27.6 \text{ kJ}$

    (6.4, Op. Sk. 2)

12. 0.42 J/(g • °C)   (6.6, Op. Sk. 5, 6)

UNIT EXAM 2

1. Match each term in the left-hand column with an appropriate expression in the right-hand column.

   _____ (1) Avogadro's number
   _____ (2) molecular formula
   _____ (3) mass % H in H_2O
   _____ (4) molar mass of H_2O
   _____ (5) empirical formula
   _____ (6) atomic weight of Zn
   _____ (7) mass of one atom

   (a) C_6H_6
   (b) 65.4 amu
   (c) 11.1%
   (d) CH
   (e) 18.02 g/mol
   (f) 6.02×10^{23}
   (g) 63.5 amu
   (h) 1.99×10^{-23} g
   (i) 12 g

2. In three to five sentences, discuss the utility of the mole concept in dealing with atomic and molecular substances.

3. The female silkworm moth produces a sex attractant called bombykol, which is composed of 80.61% carbon, 12.68% hydrogen, and 6.71% oxygen.
   (a) What is the empirical formula of bombykol?
   (b) If the molecular weight of bombykol is 238.4 amu, what is the molecular formula?
   (c) If 5.698 g of bombykol is burned in excess oxygen to form 16.83 g CO_2 and 6.452 g H_2O, how many grams of carbon were in the original sample?

4. Answer the two questions below, given the equation

   $$KClO_3 + 6HCl \longrightarrow KCl + 3H_2O + 3Cl_2$$

   (a) Calculate the theoretical yield of Cl_2 in grams when 41.4 g HCl is reacted with 24.3 g $KClO_3$.

(b)   If 35.6 g Cl_2 is recovered when the reaction is run in the laboratory, what is the percentage yield of product?

5.   A solution of H_2SO_4 is 0.3200 *M*. What volume of solution must be taken to get 4.134×10^{26} atoms of oxygen? Report your answer in liters.

6.   You are to prepare 245 mL of a 0.105 *M* solution of Na_2SO_4.
(a)   How many grams of Na_2SO_4 will you need?
(b)   If you begin with a 0.682 *M* solution, what volume, in mL, of this more concentrated solution must you dilute?
(c)   What is the mass percentage of the prepared solution if the density is 1.19 g/mL?

7.   At 1.00 atm and 30.0°C, the volume of a gas is 50.0 mL. What would be its volume at exactly 770 mmHg and the same temperature?

8.   The density of a gas is 2.194 g/L at exactly 850 mmHg and 298 K. Find its molecular weight.

9.   Calculate the relative rates of effusion of gaseous water and bromine gas.

10.   An ideal gas occupies a volume of 245 mL at 271 mmHg and exactly 50°C. What volume will it occupy at 784 mmHg and 45.0°C?

11.   Starting from the ideal gas law, derive the relationship between pressure and volume.

12.   Potassium chlorate decomposes when heated, according to the following equation:

$$2KClO_3(s) \longrightarrow 2KCl(s) + 3O_2(g)$$

If 855 mL of O_2 is produced at 25°C and 754 mmHg, how many grams of $KClO_3$ will decompose?

13.   The specific heats of Ni and Cu are 0.444 J $\cdot$ g^{-1} $\cdot$ K^{-1} and 0.385 J $\cdot$ g^{-1} $\cdot$ K^{-1}, respectively. When 21.2 J of thermal energy was added to samples of Ni and Cu, their temperatures increased by 5.63 K and 6.22 K, respectively. What is the mass of each sample?

14. The flame of an oxyacetylene torch produces very high temperatures because the enthalpy change in the combustion is very large:

$$C_2H_2(g) + \tfrac{5}{2}O_2(g) \longrightarrow 2CO_2(g) + H_2O(l); \quad \Delta H° = -1255 \text{ kJ}$$

Calculate the thermal energy produced per gram of C_2H_2 at 25°C and 1 atm.

15. Using the thermochemical equation

$$\tfrac{1}{2}N_2(g) + 2H_2(g) + \tfrac{1}{2}Cl_2(g) \longrightarrow NH_4Cl(s); \quad \Delta H_f° = -314.4 \text{ kJ/mol}$$

and $\Delta H_f°$ values from Table 6.2 in the text for $NH_3(g)$ and $HCl(g)$, determine $\Delta H°$ for the synthesis of 1 mol $NH_4Cl(s)$ from the reaction of $NH_3(g)$ and $HCl(g)$.

16. Use data from text Appendix C to calculate the energy released when 1 mol $CS_2(g)$ condenses to the liquid state.

ANSWERS TO UNIT EXAM 2

If you missed an answer, study the text section and operational skill given in parentheses after the answer.

1. (1) f (4.2)   (4) e (4.1, Op. Sk. 1)   (7) h (4.2)
   (2) a (4.5)   (5) d (4.5)
   (3) c (4.3, Op. Sk. 5)   (6) b (4.1, 2.4)

2. The mole concept relates numbers of particles to mass through the unit called the mole and lets us use mass to measure the number of particles in a substance. Since the mole is equal to the number of particles in exactly 12 g of carbon-12, or 6.02×10^{23} particles (Avogadro's number), there is a direct relationship between the mass of an object and the number of particles making up that object. Thus, although all atomic and molecular interactions involve particles, we need measure only mass rather than the number of particles.   (4.2)

3. (a) $C_{16}H_{30}O$ (4.5, Op. Sk. 8)   4. (a) 40.3 g

   (b) $C_{16}H_{30}O$ (4.5, Op. Sk. 9)   (b) 88.3% (4.8, Op. Sk. 11)

   (c) 4.593 g of carbon (4.4, Op. Sk. 7)

5.   536 L   (4.2, 4.9, Op. Sk. 12)

7.   49.4 mL   (5.2, Op. Sk. 2)

8.   48.0 amu   (5.3, Op. Sk. 5)

10.   83.4 mL   (5.2, Op. Sk. 2)

11. $P = k(1/V)$ (5.3, Op. Sk. 3)

12.   2.83 g   (5.4, Op. Sk. 6)

13.   Ni sample: 8.48 g;   Cu sample: 8.85 g   (6.6, Op. Sk. 5)

14. -48.20 kJ/g C_2H_2 (6.5, Op. Sk. 4)

15. -176.2 kJ/mol $NH_4Cl(s)$ (6.8, Op. Sk. 9)

16.   29 kJ   (6.8, Op. Sk. 8)

6.   (a)   3.65 g   (4.9, Op. Sk. 13)
     (b)   37.7 mL   (4.10, Op. Sk. 14)
     (c)   1.25%   (4.13)

9.   2.98 (water:Br_2)   (5.7, Op. Sk. 10)

CHAPTER 7   QUANTUM THEORY OF THE ATOM

CHAPTER TERMS AND DEFINITIONS

Numbers in parentheses after definitions give the text sections in which the terms are explained. Starred terms are italicized in the text. Where a term does not fall directly under a text section heading, additional information is given for you to locate it.

wave* continuously repeating change (oscillation) in matter or in a physical field   (7.1)

electromagnetic radiation* energy in the form of oscillating electric and magnetic fields that can travel through space   (7.1)

wavelength (λ) distance between any two adjacent identical points of a wave   (7.1)

frequency (ν) number of wavelengths of a wave that pass a fixed point in one unit of time (usually one second)   (7.1)

hertz (Hz)* unit of frequency; 1 Hz = /s, or s^{-1} (7.1)

electromagnetic spectrum range of frequencies or wavelengths of electromagnetic radiation   (7.1)

diffraction* spreading out of waves when they encounter an obstruction or opening the size of the wavelength   (7.2)

Planck's constant (h) constant of proportionality relating energy and frequency of vibration or oscillation; $h = 6.63 \times 10^{-34}$ J • s   (7.2)

quantum numbers* integers specifying energy states of quantized particles   (7.2)

quantized* limited to certain values   (7.2)

photons particles of electromagnetic energy, with energy E proportional to the observed frequency of the light   (7.2)

photoelectric effect ejection of electrons from the surface of a metal or from another material when light of the proper frequency shines on it   (7.2)

threshold value* minimum frequency of light that will produce the photoelectric effect for a given metal   (7.2)

absorbed* taken in wholly, as a photon and its energy taken up by an electron   (7.2)

wave–particle duality* idea that wave and particle models are complementary views of light (and all electromagnetic radiation)   (7.2)

continuous spectrum spectrum containing light of all wavelengths   (7.3)

line spectrum   spectrum showing only certain colors or specific wavelengths of light; each element gives a distinctive spectrum   (7.3)

energy levels   specific energy values that an electron can have in an atom   (7.3)

R_H*   constant used to derive the energy levels of the electron in the hydrogen atom; $R_H = 2.179 \times 10^{-18}$ J   (7.3)

n*   symbol for the principal quantum number   (7.3)

transition*   change in energy level of an electron in an atom from E_i to E_f (7.3)

emission*   release of energy from an atom as an electron's energy level changes from an upper to a lower one   (7.3)

excited*   said of an electron gaining energy and boosted to a higher energy level   (7.3)

absorption*   taking in of energy by an atom which can raise an electron to a higher energy level   (7.3)

monochromatic*   of one color; describing light of very narrow wavelength   (A Chemist Looks At: Lasers and Compact Disc Players)

laser*   acronym meaning light amplification by stimulated emission of radiation   (A Chemist Looks At: Lasers and Compact Disc Players)

radiationless transitions*   electron transitions in an atom or ion from a higher to lower energy level that release energy as heat rather than as light   (A Chemist Looks At: Lasers and Compact Disc Players)

spontaneous emission*   release of light energy from an electron transition without external cause   (A Chemist Looks At: Lasers and Compact Disc Players)

stimulated emission*   release of a photon of light from an excited atom or ion that encounters a photon of the same wavelength   (A Chemist Looks At: Lasers and Compact Disc Players)

coherent*   describes light in which all waves forming the beam are in phase; that is, the waves have their maxima and minima at the same points in space and time   (A Chemist Looks At: Lasers and Compact Disc Players)

de Broglie relation   the wavelength associated with a particle of matter given by the equation $\lambda = h/mv$ (7.4)

electron microscope*   microscope in which a beam of electrons, rather than light, is focused on the object to be magnified; electrons passing through or emitted from the object are focused by magnetic lenses on a fluorescent screen   (7.4)

resolving power*   ability of a microscope to distinguish detail   (7.4)

quantum (wave) mechanics   branch of physics that mathematically describes the wave properties of submicroscopic particles   (7.4)

uncertainty principle   the product of the uncertainty in position and the uncertainty in momentum of a particle can be no smaller than Planck's constant (h) divided by 4π:

$$(\Delta x)\,(\Delta p_x) \geq \frac{h}{4\pi} \qquad (7.4)$$

statistical*   of or pertaining to numerical values   (7.4)

probability*   likelihood of finding an electron at a certain point in an atom at a specified time   (7.4)

wave function (ψ)*   mathematical expression providing information about a particle associated with a given energy level (such as an electron in an atom)   (7.4)

ψ^2*   square of the wave function; gives the probability of finding the associated particle within a region of space   (7.4)

scanning tunneling microscope*   microscope in which a very fine-pointed tungsten needle is moved close to a sample so that electrons tunnel to the sample and produce an electric current, the voltage variations of which give rise to an image   (Instrumental Methods: Scanning Tunneling Microscopy)

tunneling*   movement of an electron from one atom to another without extra energy being supplied, due to quantum mechanical effects   (Instrumental Methods: Scanning Tunneling Microscopy)

piezoelectric*   describes the conversion of the variable compression or expansion of a material into a variable voltage   (Instrumental Methods: Scanning Tunneling Microscopy)

atomic orbital   wave function for an electron in an atom, pictured as the region of space about the nucleus in which there is high probability of finding an electron   (7.5)

spin*   magnetic property of electrons   (7.5)

principal quantum number (n)   any positive integer (1, 2, 3, and so on) specifying the major energy level of an electron in an atom   (7.5)

shell*   energy level of an atom   (7.5)

angular momentum (azimuthal) quantum number (l)   integer from 0 to $n - 1$ denoting the shape of an orbital   (7.5)

subshells*   sublevels of an energy level of an atom   (7.5)

magnetic quantum number (m_l)   integer from $-l$ to $+l$ designating the orientation in space of a specific orbital within a subshell   (7.5)

orientation*   position   (7.5)

spin quantum number (m_s)   fraction, either $+\frac{1}{2}$ or $-\frac{1}{2}$, designating one of the two possible orientations of the spin axis of an electron   (7.5)

99% contour*   region of space about an atomic nucleus in which an electron of a given energy is expected to be found 99% of the time   (7.5)

acid rain*   rainfall with unusually high acidity, the major component of which is sulfuric acid formed in moist air from sulfur dioxide produced by burning fuels containing sulfur compounds   (An Element That Matters: Hydrogen: Fuel of the Twenty-first Century?)

greenhouse effect*   absorption and reradiation of infrared rays back to the surface of the earth by increased atmospheric CO_2 (An Element That Matters: Hydrogen: Fuel of the Twenty-first Century?)

hydrogen energy economy* economy in which hydrogen would be a major energy carrier (An Element That Matters: Hydrogen: Fuel of the Twenty-first Century?)

steam-reforming process* process for producing hydrogen from steam and hydrocarbons at high temperature and pressure in the presence of a nickel catalyst (An Element That Matters: Hydrogen: Fuel of the Twenty-first Century?)

water-gas reaction* reaction in which steam is passed over red-hot coke or coal to produce hydrogen (and carbon monoxide) (An Element That Matters: Hydrogen: Fuel of the Twenty-first Century?)

CHAPTER DIAGNOSTIC TEST

Previously, we arranged diagnostic test questions in the same sequence as the presentation of material in the chapter. Now that you have had some experience with problem solving, the diagnostic test and post-test questions will be given in no specific order.

1. Calculate the frequency of electromagnetic radiation having the wavelength of 684.9 nm.

2. When an electron moves from level $n = 4$ to level $n = 5$ in an excited hydrogen atom, what amount of energy is required for this electronic transition?

3. Indicate whether each of the following statements is true or false. If a statement is false, change it so it is true.

   (a) The azimuthal quantum number describes the main energy level of an electron in an atom. True/False: _____

   _____ .

   (b) In an excited hydrogen atom, an electron described by the quantum numbers $n = 4$, $l = 1$, $m_l = 1$ is at higher energy than an electron described by the quantum numbers $n = 4$, $l = 1$, $m_l = -1$. True/False: _____

   _____ .

   (c) The square of the wave function for an electron is related to the charge density of that electron at a point in space. True/False: _____

   _____ .

   (d) Bohr's theory accounted for the line spectra of excited atoms in terms of the electromagnetic radiation absorbed or released from electrons moving in circular orbits about the nucleus. True/False: _____

   _____ .

(e)   Assuming that nature was symmetrical, de Broglie postulated that a wavelength could be associated with a moving particle and calculated from the relationship $\lambda = h/mv$. True/False: _____

_____ .

4.   Which of the following five sets of quantum numbers is (are) not allowed?

(a) $n = 3$ $l = 1$ $m_l = 0$ $m_s = 1/2$
(b) $n = 1$ $l = 0$ $m_l = 0$ $m_s = 3/2$
(c) $n = 4$ $l = 3$ $m_l = -1$ $m_s = 1/2$
(d) $n = 2$ $l = 2$ $m_l = -2$ $m_s = 1/2$
(e) $n = 3$ $l = 2$ $m_l = -2$ $m_s = -1/2$

5.   Describe the types of orbitals you are likely to find in the $n = 3$ shell.

6.   Radiation remains visible to the eye down to a frequency of $4.29 \times 10^{14}/s$. What are the wavelength and the energy of one of these photons?

7.   Which of the following statements concerning the second main energy level is (are) incorrect?
(a)   Its principal quantum number is 2.
(b)   It contains s, p, and d orbitals.
(c)   It can contain an electron having the quantum numbers $n = 2$, $l = 1$, $m_l = 1$, $m_s = 1/2$.
(d)   It has a bilobed (dumbbell) shape.
(e)   It cannot contain any f orbitals.

8.   Match each term in the left-hand column with the appropriate expression in the right-hand column.

_____   (1)   Schrödinger   (a)   wave nature of particles

_____   (2)   main energy level   (b)   wave equation

_____   (3) $h\nu$ (c)   stability of nuclear atom

_____   (4)   de Broglie   (d)   principal quantum number

_____   (5)   Bohr   (e)   energy of a photon

_____   (6)   orbitals   (f)   azimuthal quantum number

9.   A typical wavelength for light in the visible region of the electromagnetic spectrum is 4.8×10^3 pm. If a particle weighs 1.5×10^{-25} g, at what velocity would it have to be moving to exhibit this wavelength?

ANSWERS TO CHAPTER DIAGNOSTIC TEST

If you missed an answer, study the text section and operational skill given in parentheses after the answer.

1. $4.38 \times 10^{14}/s$ (7.1, Op. Sk. 1)

2. 4.905×10^{-20} J   (7.3, Op. Sk. 3)

3.   (a)   False. The principal quantum number describes the main energy level of an electron in an atom.   (7.5)
   (b)   False. In an excited hydrogen atom, an electron described by the quantum numbers $n = 4$, $l = 1$, $m_l = 1$ is of the same energy as an electron described by the quantum numbers $n = 4$, $l = 1$, $m_l = -1$.   (7.5)
   (c)   True.   (7.4)
   (d)   False. Bohr's theory accounted for the line spectra of excited atoms in terms of the electromagnetic radiation absorbed or released when electrons move from one energy level to another in an atom.   (7.3)
   (e)   True.   (7.4)

4.   b, d   (7.5, Op. Sk. 5)

5.   At $n = 3$, the l quantum number has values of 0, 1, and 2. Thus, we find s, p, and d orbitals in this shell.   (7.5)

6. $\lambda = 6.99 \times 10^{-7}$ m   (699 nm)   (7.1, Op. Sk. 1);
   $E = 2.84 \times 10^{-19}$ J   (7.2, Op. Sk. 2)

7.   b, d   (7.5)

8.   (1)   b   (7.4)   (3)   e   (7.2)   (5)   c   (7.3)
   (2)   d   (7.3, 7.5)   (4)   a   (7.4)   (6)   f   (7.5)

9.   9.2 m/s   (7.2, 7.4)

SUMMARY OF CHAPTER TOPICS

7.1   THE WAVE NATURE OF LIGHT

Operational Skill

   1.   Relating wavelength and frequency of light.   Given the frequency of light, calculate the wavelength, or vice versa   (Examples 7.1 and 7.2).

      The chart on page 179 presents information about the electromagnetic spectrum that you may find useful in this and in later courses. (More information is included than you now need, but we will refer to it later.) For now, you should become familiar with the magnitude of the wavelengths for the various "boundaries" in the electromagnetic spectrum.

      You should memorize the formula for the speed of light ($c = \lambda \nu$) and the value of c (3.00×10^8 m/s). If the units of frequency, /s or s^{-1}, are confusing, you can think of this as "waves" per second, although we do not write it that way. The units of wavelength are meters (m) or nanometers (nm), although other units, such as picometers (pm), are sometimes used.

      Exercise 7.1   The frequency of the strong red line in the spectrum of potassium is 3.91×10^{14}/s. What is the wavelength of this light in nanometers?

Wanted: λ (nm).

Given: $\nu = 3.91 \times 10^{14}$/s.

Known: $c = \lambda \nu$; $c = 3.00 \times 10^8$ m/s; 10^9 nm = 1 m.

Solution: Rearrange $c = \lambda \nu$ to get

$$\lambda = \frac{c}{\nu} = \frac{3.00 \times 10^8 \; \text{m/s}}{3.91 \times 10^{14} \; /\text{s}} \times \frac{10^9 \; \text{nm}}{\text{m}} = 767 \; \text{nm}$$

      Note that high energy correlates with high frequency but low wavelength. Even though this may seem easy to grasp, it is a concept that many students stumble over when the equations are not in front of them. You may see a large value for wavelength and automatically think "high energy," which is incorrect. Practice writing wavelength values and putting them in order of increasing energy to emphasize this for yourself.

The Electromagnetic Spectrum

Type of Transition	Spin orientation (NMR)	Molecular rotations	Molecular vibrations			Valence electrons			Inner electrons	Nuclear transitions
Spectral Region	Radiowave	Microwave	Infrared			Visible	Ultraviolet		X Ray	Gamma Ray
			Far	Middle	Near		Near	Far		
λ, cm	100 1		0.01 0.001							
λ, μm		10^{4} 100	10	2	0.8	0.4	0.2	0.01		
λ, nm		10^{7} 10^{5}	10^{4}	2000	800	400	200	10	10^{-1}	10^{-3}
λ, Å			20000	8000	4000	2000		100	1	0.01
ν, /s		3×10^{12}						3×10^{16}	3×10^{18}	
$\bar{\nu}$ / cm	10^{-2}	1 100	1000							
ΔE, eV		10^{-4} 10^{-2}	10^{-1}					10^{2}	10^{4}	
ΔE, kJ/mol		9.65×10^{-3}	9.65					9.65×10^{3}		

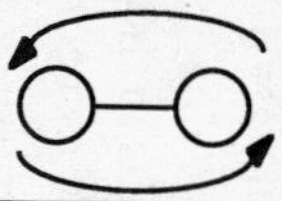

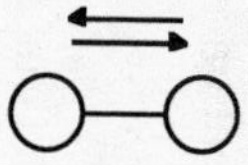

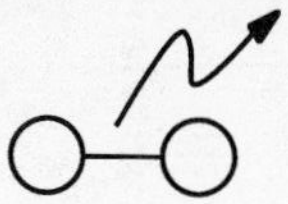

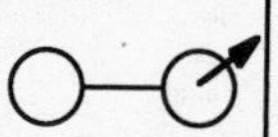

Useful Conversion Units

1 pm = 10^{-8} cm = 10^{-10} m
1 nm = 10^{-9} m = 10 Å
1 eV/atom = 96.5 kJ/mol

Useful Equations

$$\bar{\nu} = 1/\lambda$$
$$\lambda\nu = c$$
$$\Delta E = h\nu$$

Definitions

h = 6.626×10^{-34} J $\bullet$ s
λ = wavelength
ν = frequency
$\bar{\nu}$ = wavenumber

Exercise 7.2   The element cesium was discovered in 1860 by Robert Bunsen and Gustav Kirchhoff, who found two bright blue lines in the spectrum of a substance isolated from a mineral water. One of these spectral lines of cesium has a wavelength of 456 nm. What is its frequency?

Solution: Rearrange $c = \lambda \nu$ to get

$$\nu = \frac{c}{\lambda} = \frac{3.00 \times 10^8 \, \cancel{m}/s}{456 \, \cancel{nm}} \times \frac{10^9 \, \cancel{nm}}{\cancel{m}} = 6.58 \times 10^{14} \, /s$$

7.2   QUANTUM EFFECTS AND PHOTONS

Operational Skill

  2.   Calculating the energy of a photon.   Given the frequency or wavelength of light, calculate the energy associated with one photon   (Example 7.3).

        Diffraction is an interesting phenomenon. To observe it, hold a mesh curtain very close to your eye and look at a street light through the space between the threads. If you are really observant, you will "see" black lines. These appear because the light is being spread as it goes through the holes and some portions of your retina have no light falling on them.
        A good example of a hot solid glowing red is the heating element on your electric stove.

        It is interesting that our concept of the particle nature of light and our understanding of the electronic structure of the atom are interdependent and were worked out together.

        You should memorize the formula for the energy of a photon, $E = h\nu$.

Exercise 7.3   The following are representative wavelengths in the infrared, ultraviolet, and x-ray regions of the electromagnetic spectrum, respectively: 1.0×10^{-6} m, 1.0×10^{-8} m, and 1.0×10^{-10} m. What is the energy of a photon of each radiation? Which has the greatest amount of energy per photon? Which has the least?

Wanted: energy of each photon, and most energetic and least energetic photons.

Given: $\lambda_1 = 1.0 \times 10^{-6}$ m, $\lambda_2 = 1.0 \times 10^{-8}$ m, $\lambda_3 = 1.0 \times 10^{-10}$ m.

Known: $E = h\nu$; $c = \lambda \nu = 3.00 \times 10^8$ m/s;   *h* is given in text.

Solution: Since the energy calculation is the same for each, we will show only the first one. Rearrange $c = \lambda \nu$ to solve for ν:

$$\frac{c}{\lambda} = \nu$$

and substitute this value for ν in the expression for energy:

$$E = h\nu = h\,\frac{c}{\lambda}$$

$$E_1 = h\,\frac{c}{\lambda_1} = \frac{6.63 \times 10^{-34}\ \text{J} \cdot \text{s} \times 3.00 \times 10^8\ \text{m/s}}{1.0 \times 10^{-6}\ \text{m}}$$

$$= 2.0 \times 10^{-19}\ \text{J for the infrared photon}$$
$$E_2 = 2.0 \times 10^{-17}\ \text{J for the ultraviolet photon}$$
$$E_3 = 2.0 \times 10^{-15}\ \text{J for the x-ray photon}$$

The energy values increase, so the x-ray photon has the greatest amount of energy, and the infrared photon the least.

7.3 THE BOHR THEORY OF THE HYDROGEN ATOM

Operational Skill

3. Determining the wavelength or frequency of a hydrogen atom transition. Given the initial and final principal quantum numbers for an electron transition in the hydrogen atom, calculate the frequency or wavelength of light emitted (Example 7.4). You need the value of R_H.

When a hydrogen-gas discharge tube is connected to a source of electric current, several forms of electromagnetic radiation are emitted. Besides those in the visible portion, emissions are also in the infrared and ultraviolet portions of the electromagnetic spectrum. Emissions in the visible region are of greatest interest to us in this course.

Today we accept Bohr's idea about light being emitted by an atom or ion due to its (electron) transitions between allowable energy states. According to his theory, the allowable energies of the hydrogen atom are the allowable energies the electron can have in the atom. A given energy value that corresponds to an energy state (level) of a hydrogen atom is also the energy an electron has when at the specific energy level.

Likewise, we speak of energy transitions in reference to the hydrogen atom. But the atom changes in energy because the electron can absorb energy under certain conditions and thus undergo transitions to higher-energy states. When the electron changes to a lower-energy state, we say the atom has undergone a transition to a lower level. Because the energies of atoms other than hydrogen are so complex, we make reference only to the electron transitions, although, of course, the atom is changing to different energy levels as its electrons are.

The combinations of visible frequencies emitted by excited atoms are the colors we see in a fireworks display. The atoms are excited by the energy of the explosion. Their electrons continually undergo transitions to higher-energy levels, then fall back, emitting specific frequencies that add together to give a characteristic color. For example, strontium produces a red color, and barium a green.

Exercise 7.4   Calculate the wavelength of light emitted from the hydrogen atom when the electron undergoes a transition from level $n = 3$ to level $n = 1$.

Wanted: λ.

Given: Transition is from $n = 3$ to $n = 1$.

Known: The energy of any level is given by

$$E = -\frac{R_H}{n^2}; \quad R = 2.180 \times 10^{-18} \text{ J}.$$

The energy of a photon is

$$E_i - E_f = \Delta E = h\nu$$

$$c = \lambda\nu = 3.00 \times 10^8 \text{ m/s}$$

$$h = 6.63 \times 10^{-34} \text{ J} \cdot \text{s}.$$

Solution: First find the energy of the photon:

$$\Delta E = E_{n=3} - E_{n=1} = \left(\frac{-R_H}{3^2}\right) - \left(\frac{-R_H}{1^2}\right) = \frac{-R_H - (-9R_H)}{9}$$

$$= \frac{-R_H + 9R_H}{9} = \frac{8R_H}{9} = \frac{8 \times 2.180 \times 10^{-18} \text{ J}}{9} = 1.938 \times 10^{-18} \text{ J}$$

To find λ, rearrange $c = \lambda\nu$ to $\nu = c/\lambda$ and substitute for ν in $\Delta E = h\nu$ to give

$$\Delta E = h \frac{c}{\lambda}$$

Now solve this for λ and put in the known values:

$$\lambda = \frac{hc}{\Delta E} = \frac{6.63 \times 10^{-34} \text{ J} \cdot \text{s} \times 3.00 \times 10^8 \text{ m/s}}{1.938 \times 10^{-18} \text{ J}}$$

$$= 1.03 \times 10^{-7} \text{ m } (103 \text{ nm})$$

Exercise 7.5   What is the difference in energy levels of the sodium atom if emitted light has a wavelength of 589 nm?

Known: $\Delta E = h\nu$; $c = \lambda\nu$;   values of h and c from above;   1 nm = 10^{-9} m.

Solution: Solve $c = \lambda\nu$ for ν and substitute in $\Delta E = h\nu$ to give

$$\Delta E = h\frac{c}{\lambda} = \frac{6.63 \times 10^{-34} \text{ J} \cdot \text{s} \times 3.00 \times 10^8 \text{ m/s}}{589 \text{ nm}} \times \frac{1 \text{ nm}}{10^{-9} \text{ m}}$$

$$= 3.38 \times 10^{-19} \text{ J}$$

7.4   QUANTUM MECHANICS

Operational Skill

4.   Applying the de Broglie relation.   Given the mass and speed of a particle, calculate the wavelength of the associated wave   (Example 7.5).

Quantum mechanics can be a difficult topic to understand. Part of this difficulty is the advanced mathematics it uses. But much of the difficulty is that what it describes, the submicroscopic world, is outside of our experience. What is common sense in our macroscopic world has no meaning in a world where 6.02×10^{23} particles can fit on the point of a pin! As you read this section of your textbook, try to get an overall idea of the subject. You may have to read the section several times to do this.

Although the term *wave function* may be new to you, you may recall the term *function* from your math courses. When we plot values of x and y on a two-dimensional graph, we write $y = f(x)$, meaning the y value is some function of x, say $x + 3$.

Quantum mechanics gives us two ways to describe the electron in motion about the nucleus: (1) We cannot know the path of the electron, but we can calculate its average speed as it moves about the nucleus; (2) we can know the probability of finding an electron at a certain point at a given distance from the atomic nucleus.

Exercise 7.6   Calculate the wavelength (in picometers) associated with an electron traveling at a speed of 2.19×10^6 m/s.

Wanted: wavelength in picometers (pm).

Given: Speed v is 2.19×10^6 m/s.

Known: $\lambda = h/mv$; $h = 6.63 \times 10^{-34}$ J $\bullet$ s (text); 1 J $= \text{kg} \bullet \text{m}^2/\text{s}^2$
(1.8, Op. Sk. 6); electron mass $= 9.10939 \times 10^{-31}$ kg   (Table 2.2);
$1 \text{ m} = 10^{10}$ pm   (1.6).

Solution:

$$\lambda = \frac{h}{mv} = \frac{6.63 \times 10^{-34} \, \cancel{\text{J}} \bullet \cancel{\text{s}}}{9.10939 \times 10^{-31} \, \cancel{\text{kg}} \times 2.19 \times 10^6 \, \cancel{\text{m}}/\cancel{\text{s}}}$$

$$\times \frac{\text{kg} \bullet \text{m}^2/\text{s}^2}{1 \text{ J}} \times \frac{10^{10} \text{pm}}{1 \text{ m}} = 3.32 \text{ Å}$$

7.5   QUANTUM NUMBERS AND ATOMIC ORBITALS

Operational Skill

5.   **Using the rules for quantum numbers.**   Given a set of quantum numbers n, l, m_l, and m_s, state whether that set is permissible for an electron   (Example 7.6).

     If we plot all the points where there is high probability of finding an electron of a given energy, we have a three-dimensional region of space (a volume) about the atomic nuclei. This volume of space is a description of an orbital. It is also referred to as an electron cloud. Be sure you do not confuse this with a Bohr orbit, which is a circular path such as a planetary or satellite orbit.

     The first three of the four quantum numbers describing an electron in an atom specify the orbital for the electron. Thus, n specifies the energy level, l the shape of the orbital — the subshell, and m_l the orientation in space — the specific orbital. Do not be concerned about the meaning or derivation of the names (*azimuthal* or *magnetic*). At some later time you can take a course in quantum mechanics and resolve your unanswered questions. The fourth quantum number, m_s, describes the spin orientation of the electron. You should memorize the rules for the allowable values of these quantum numbers in order to more easily work the problems associated with them.

Over the years a number of terms have been used in reference to the energy states of the atom. For your purposes, *energy level*, *quantum state*, and *shell* mean the same thing.

Exercise 7.7   Explain why each of the following sets of quantum numbers is not permissible for an orbital.

(a) $n = 0, l = 1, m_l = 0, m_s = +\frac{1}{2}$ (c) $n = 3, l = 2, m_l = 3, m_s = +\frac{1}{2}$

(b) $n = 2, l = 3, m_l = 0, m_s = -\frac{1}{2}$ (d) $n = 3, l = 2, m_l = 2, m_s = 0$

Known: The text gives the allowable values for all four quantum numbers.

Solution: (a)   The value of n must be a positive whole number.

(b) l cannot be greater than n.

(c) m_l cannot be greater than l.

(d) m_s can equal only $+\frac{1}{2}$ or $-\frac{1}{2}$.

An Element That Matters: HYDROGEN: Fuel of the Twenty-first Century?

Questions for Study

1. Why is hydrogen useful as a rocket fuel? What environmental advantage does hydrogen have as a fuel?

2. What is acid rain? How can the burning of fossil fuels cause acid rain?

3. What is the greenhouse effect? Why do some climatologists feel that the burning of fossil fuels could cause a greenhouse effect?

4. What do we mean by a hydrogen economy?

5. The British chemist Henry Cavendish is credited with discovering hydrogen in 1766. He prepared it by reacting various metals with acids. Write the chemical equation for the preparation of hydrogen from zinc metal and hydrochloric acid.

6. Hydrogen is prepared from natural gas by reacting it with steam (steam-reforming). Natural gas is mainly methane, CH_4. Write the chemical equation for the preparation of hydrogen by steam-reforming of methane.

7. Describe how pure hydrogen could be prepared from wood charcoal (carbon). Use chemical equations where appropriate, accompanied by a verbal description.

8. How might solar energy be converted to hydrogen fuel?

Answers to Questions for Study

1. Hydrogen is useful as a rocket fuel because it produces more heat per gram than any other fuel. Its environmental advantage as a fuel is that when it is burned, the only product is water.

2. Acid rain is rain whose major component is sulfuric acid. Fossil fuels contain sulfur compounds that when burned give sulfur dioxide, which reacts in moist air to produce sulfuric acid.

3. The greenhouse effect is the absorption and reradiation of infrared rays back to the surface of the earth by the increasing concentration of atmospheric CO_2. Some climatologists feel that the burning of fossil fuels could cause a greenhouse effect because CO_2 is formed when they burn.

4. A hydrogen economy is one in which hydrogen would become a major energy carrier.

5. $Zn(s) + 2HCl(aq) \longrightarrow ZnCl_2(aq) + H_2(g)$

6. $CH_4(g) + H_2O(g) \xrightarrow{\text{Ni}} CO(g) + 3H_2(g)$

7. Charcoal could be heated till red-hot, then steam passed over it. The equation is

$$C(s) + H_2O(g) \longrightarrow CO(g) + H_2(g)$$

The carbon monoxide is removed by reaction with steam in the presence of a catalyst to give CO_2 and more hydrogen. The equation is

$$CO(g) + H_2O(g) \xrightarrow{\text{catalyst}} CO_2(g) + H_2(g)$$

The gases could be passed through a basic aqueous solution to dissolve the CO_2, leaving pure hydrogen. The equation for the reaction is

$$CO_2(g) + 2NaOH(aq) \longrightarrow Na_2CO_3(aq) + H_2O(l)$$

8. Researchers are looking into photochemical reactions that would use the light in solar energy to convert water to hydrogen and oxygen.

ADDITIONAL PROBLEMS

1. Calculate:

   (a) the frequency of an x ray with a wavelength of 4.70×10^{-9} nm.

   (b) the wavelength of the transition of cesium-133, frequency 9.193×10^9/s, the standard for the SI unit of time.

   (c) the energy of a photon of red light, wavelength of 7.00×10^2 nm, absorbed by chlorophyll *a* in the process of photosynthesis.

2. An energy of 1.09×10^3 kJ/mol is required to convert gaseous carbon atoms to gaseous C^+ ions and gaseous electrons. Calculate the maximum wavelength (in units of nm) of electromagnetic radiation that can cause the ionization of one carbon atom.

3. (a) Calculate the wavelength of radiation emitted when the electron in the hydrogen atom undergoes a transition from level $n = 2$ to level $n = 1$. (The value of R_H is 2.180×10^{-18} J.)

   (b) Refer to Figure 7.5 and determine in which portion of the electromagnetic spectrum this energy falls.

4. Calculate the wavelength in nm of a vehicle with a mass of 2.200×10^3 kg that is moving with a velocity of 1.10×10^2 km/h.

5. Explain why each of the following sets of quantum numbers is not permissible for an electron in an atom.

   (a) $n = -1,$ $l = 1,$ $m_l = 0,$ $m_s = +\frac{1}{2}$

   (b) $n = 0,$ $l = 0,$ $m_l = 1,$ $m_s = -\frac{1}{2}$

   (c) $n = 3,$ $l = -1,$ $m_l = 2,$ $m_s = +\frac{1}{2}$

   (d) $n = 2,$ $l = 1,$ $m_l = -1,$ $m_s = 1$

ANSWERS TO ADDITIONAL PROBLEMS

If you missed an answer, study the text section and operational skill given in parentheses after the answer.

1.   (a)   Because $c = \lambda\nu$,

$$\nu = \frac{c}{\lambda} = \frac{3.00 \times 10^8 \text{ m/s}}{4.70 \times 10^{-9} \text{ nm}} \times \frac{10^9 \text{ nm}}{1 \text{ m}} = 6.38 \times 10^{25}/\text{s}$$

(7.1, Op. Sk. 1)

(b)   Because $c = \lambda\nu$,

$$\lambda = \frac{c}{\nu} = \frac{3.00 \times 10^8 \text{ m/s}}{9.193 \times 10^9 \text{/s}} = 3.26 \times 10^{-2} \text{ m} \quad (7.1, \text{ Op. Sk. 1})$$

(c)   Because $E = h\nu$ and $c = \lambda\nu$,

$$E = \frac{hc}{\lambda} = \frac{6.63 \times 10^{-34} \text{ J} \cdot \text{s} \times 3.00 \times 10^8 \text{ m/s}}{7.00 \times 10^2 \text{ nm}} \times \frac{10^9 \text{ nm}}{\text{m}}$$

$$= 2.84 \times 10^{-19} \text{ J} \quad (7.2, \text{ Op. Sk. 2})$$

2.   The energy expressed in kJ/mol must be converted to kJ/atom for the ionization of a single carbon atom.

$$E = \frac{1.09 \times 10^3 \text{ kJ}}{\text{mol}} \times \frac{1 \text{ mol}}{6.023 \times 10^{23} \text{ atom}} = 1.8\underline{1}1 \times 10^{-21} \text{ kJ/atom}$$

Because $\Delta E = h\nu = hc/\lambda$, solving for λ gives

$$\lambda = \frac{hc}{\Delta E} = 6.63 \times 10^{-34} \text{ J} \cdot \text{s} \times 3.00 \times 10^8 \text{ m/s} \times \frac{1 \text{ nm}}{10^{-9} \text{ m}}$$

$$\times \frac{1}{1.8\underline{1}1 \times 10^{-21} \text{ kJ}} \times \frac{1 \text{ kJ}}{10^3 \text{ J}}$$

$$= 1.10 \times 10^2 \text{ nm} \quad (7.2, \text{ Op. Sk. 2})$$

3.   (a)   This problem is easier if done algebraically first. The energy of the emitted photon is

$$E = E_i - E_f = \frac{-R_H}{n_i^2} - \frac{-R_H}{n_f^2} = h\nu = \frac{hc}{\lambda}$$

Putting in the principal quantum numbers gives

$$E = \frac{-R_H}{4} - \frac{-R_H}{1} = \frac{3R_H}{4} = \frac{hc}{\lambda}$$

Solving the last equality for λ and putting in values, we get

$$\lambda = \frac{4hc}{3R_H} = \frac{4 \times 6.63 \times 10^{-34} \, J \cdot s \times 3.00 \times 10^8 \text{ m} / s}{3 \times 2.180 \times 10^{-18} \, J}$$

$$= 1.22 \times 10^{-7} \text{ m} \quad (7.3, \text{ Op. Sk. 3})$$

(b)   This is ultraviolet radiation.   (7.3)

4.   Using the de Broglie relation, we get

$$\lambda = \frac{h}{mv} = 6.63 \times 10^{-34} \, J \cdot s \times \frac{1}{2.200 \times 10^3 \, kg} \times \frac{hr}{1.10 \times 10^2 \, km}$$

$$\times \frac{60 \, min}{1 \, hr} \times \frac{60 \, s}{1 \, min} \times \frac{kg \cdot m^2}{s^2 \cdot J} \times \frac{nm}{10^{-9} \, m} \times \frac{1 \, km}{10^3 \, m}$$

$$= 9.86 \times 10^{-30} \text{ nm}; \text{ this value is so small that the wave properties cannot be observed} \quad (7.4, \text{ Op. Sk. 4})$$

5.   (a)   The value of n must be a positive whole number.
     (b)   The value of n cannot be 0, and that of the magnitude of m_l must not be greater than that of l.
     (c)   The value of l must be 0 or a positive whole number.
     (d)   The value of m_s can be only $+\frac{1}{2}$ or $-\frac{1}{2}$.   (7.5, Op. Sk. 5)

CHAPTER POST-TEST

1.   The numerical value of the _____ determines the orbital shape.

2.   The _____ quantum number is most important in determining the energy of the orbital.

3. The *s* orbital may be described as having the shape of a
   (a) dumbbell.   (c) sphere.   (e) circle.
   (b) four-leafed clover.   (d) pyramid.

4. Briefly discuss the interdependence of the numerical values of the quantum numbers *n*, *l*, and m_l.

5. Which of the following sets of quantum numbers is (are) not permitted?
   (a) $n = 1$ $l = 0$ $m_l = 0$ $m_s = -1/2$
   (b) $n = 4$ $l = 3$ $m_l = -3$ $m_s = -1/2$
   (c) $n = 3$ $l = 2$ $m_l = 3$ $m_s = +1/2$
   (d) $n = 2$ $l = 0$ $m_l = 0$ $m_s = -1/2$
   (e) $n = 2$ $l = 1$ $m_l = 0$ $m_s = +1/2$

6. Find the wavelength associated with an electron moving at 95.0% the speed of light. (Electron mass $= 9.11 \times 10^{-31}$ kg; $h = 6.63 \times 10^{-34}$ J • s.)

7. A 68.0-kg runner runs a marathon at an average speed of 4.00 m/s. What is the de Broglie wavelength of the runner? (See question 6 for *h*.)

8. Green light has a wavelength of 5.3×10^3 pm. What is the frequency of green light?

9. When an electron moves from level $n = 1$ to level $n = 3$ in a hydrogen atom, what amount of energy is required for this electronic transition?

10. Molecules undergo electronic transitions similar to those of atoms. If a molecule absorbs radiation of wavelength 380.0 nm, what is the energy of that transition?

ANSWERS TO CHAPTER POST-TEST

If you missed an answer, study the text section and operational skill given in parentheses after the answer.

1. azimuthal quantum number, *l*   (7.5)   2. principal   (7.3)

3. c   (7.5)

4. See Section 7.5 in your text for a complete discussion. The numerical values of these quantum numbers are related as follows:
   (1) n can take on any integral value from 1 to ∞.
   (2) Once the value of n is determined, l can take on any integral value from 0 up to $(n-l)$.
   (3) Once the values of n and l are determined, for any given value of l, m_l can take on any integral value from $-l$ to $+l$. (7.5)

5. c (7.5, Op. Sk. 5)   6. 2.55×10^{-12} m (2.55×10^{-3} nm) (7.4, Op. Sk. 4)

7. 2.44×10^{-36} m (2.44×10^{-27} nm) (7.4, Op. Sk. 4)

8. 5.7×10^{14}/s (7.1, Op. Sk. 1)   9. 1.938×10^{-18} J (7.3, Op. Sk. 3)

10. 5.23×10^{-19} J (7.1, 7.2, Op. Sk. 2)

CHAPTER 8   ELECTRON CONFIGURATIONS AND PERIODICITY

CHAPTER TERMS AND DEFINITIONS

Numbers in parentheses after definitions give the text sections in which the terms are explained. Starred terms are italicized in the text. Where a term does not fall directly under a text section heading, additional information is given for you to locate it.

periodic table*   arrangement of the elemental symbols in rows and columns   (chapter introduction)

electron configuration   particular distribution of electrons among available subshells   (8.1)

orbital diagram   diagram showing how the orbitals of a subshell are occupied by electrons; orbital can hold at most two electrons and then only if the electrons have opposite spins   (8.1)

Pauli exclusion principle   no two electrons in an atom can have the same four quantum numbers   (8.1)

nuclear magnetic resonance (NMR)*   condition wherein an atomic nucleus with a net spin and magnetism aligned with an external magnetic field is caused to change or "flip" to an alignment against the applied field   (Instrumental Methods: Nuclear Magnetic Resonance [NMR])

magnetic resonance imaging (MRI)*   medical diagnostic tool based on nuclear magnetic resonance   (Instrumental Methods: Nuclear Magnetic Resonance [NMR])

ground state*   quantum-mechanical state in which an atom is at its lowest energy level   (8.2)

excited states*   quantum-mechanical states in which an atom is at energy levels other than the lowest   (8.2)

building-up (Aufbau) principle   scheme used to reproduce the electron configuration of the ground states of atoms by successively filling subshells with electrons in a specific order (the building-up order)   (8.2)

building-up order*   order in which electrons successively fill the subshells: $1s$, $2s$, $2p$, $3s$, $3p$, $4s$, $3d$, $4p$, $5s$, $4d$, $5p$, $6s$, $4f$, $5d$, $6p$, $7s$, $5f$, $6d$ (8.2)

noble gases*   Group VIIIA elements   (8.2)

alkaline earth metals*   Group IIA elements   (8.2)

noble-gas core   inner-shell electron configuration corresponding to one of the noble gases   (8.2)

pseudo-noble-gas core   noble-gas core together with $(n-1)d^{10}$ electrons   (8.2)

valence electron   electron in an atom outside the noble-gas or pseudo-noble-gas core   (8.2)

valence-shell configurations*   arrangements of electrons in the outer ns and np sub-shells   (8.2)

main-group (representative) elements*   elements in the A columns of the periodic table, in which an outer s or p subshell is filling   (8.2)

d-block transition (transition) elements*   ten columns of elements inserted between Groups IIA and IIIA in the periodic table, in which d subshells are filling   (8.2)

f-block transition (inner-transition) elements*   two rows of elements, each with 14 columns, at the bottom of the periodic table, in which f subshells are filling; they fit between columns IIIB and IVB of Periods 6 and 7 of the periodic table   (8.2)

x-ray spectroscopy*   analysis of x rays emitted from a target hit by an electron beam   (Instrumental Methods: X Rays, Atomic Numbers, and Orbital Structure [Photoelectron Spectroscopy])

x-ray photoelectron spectroscopy*   analysis of the kinetic energies of electrons ejected from a target irradiated with x rays   (Instrumental Methods: X Rays, Atomic Numbers, and Orbital Structure [Photoelectron Spectroscopy])

photoelectric effect*   ejection of electrons from the surface of a metal when light of the proper frequency shines on it   (Instrumental Methods: X Rays, Atomic Numbers, and Orbital Structure [Photoelectron Spectroscopy])

ionization energy*   energy necessary to remove an electron from an atom   (Instrumental Methods: X Rays, Atomic Numbers, and Orbital Structure [Photoelectron Spectroscopy])

Hund's rule   the lowest-energy arrangement of electrons in a subshell is obtained by putting electrons into separate orbitals of the subshell with the same spin before pairing electrons   (8.4)

paramagnetic substance   substance that is weakly attracted by a magnetic field; generally due to unpaired electrons   (8.4)

diamagnetic substance   substance that is not attracted by a magnetic field, or is weakly repelled by such a field; generally means the presence of only paired electrons (8.4)

ferromagnetism*   strong, permanent magnetism of iron objects, due to the cooperative alignment of electron spins in many iron atoms   (8.4, marginal note)

eka*   Sanskrit word meaning "first"   (8.5)

melting point*   temperature at which a solid substance changes to a liquid   (8.5)

boiling point*   temperature at which the vapor pressure of a liquid equals the external pressure   (8.5)

periodic law   when the elements are arranged by atomic number, their physical and
        chemical properties vary periodically   (8.6)
covalent radii*   lengths of atomic radii obtained from measurements of distances
        between the nuclei of atoms in the chemical bonds of molecular substances   (8.6)
effective nuclear charge   positive charge that an electron experiences from the nucleus,
        equal to the nuclear charge but reduced by any shielding or screening from any
        intervening electron distribution   (8.6)
first ionization energy (first ionization potential)   minimum energy needed to remove
        the highest-energy (the outermost) electron from a neutral atom in the gaseous state
        (8.6)
alkali metals*   Group IA elements   (8.6)
electron affinity   energy change for the process of adding an electron to a neutral atom in
        the gaseous state to form a negative ion   (8.6)
basic oxide (basic anhydride)   oxide that reacts with acids   (8.7)
acidic oxide (acidic anhydride)   oxide that reacts with bases   (8.7)
amphoteric substance   substance that has both acidic and basic properties   (8.7)
alloys*   metallic mixtures   (8.7, marginal note)
halogens*   reactive nonmetals with the general molecular formula X_2, where X
        symbolizes a halogen   (8.7)
hydrogen halides*   binary compounds of hydrogen with a halogen   (8.7)
hydrohalic acids*   solutions of hydrogen halides (for example, HCl) in water   (8.7)
phosphorus*   name of an element deriving from the Greek and meaning "bearer of
        light"   (A Nonmetal That Matters: Phosphorus [Group VA])
chemiluminescence*   emission of light from excited molecular products of a chemical
        reaction   (A Nonmetal That Matters: Phosphorus [Group VA])

CHAPTER DIAGNOSTIC TEST

1.   Which of the following electron configurations is (are) incorrect? Give the correct
     one (s).

     (a)   Al: [Ne] $3s^2 3p^1$ (c)   Br: [Ar] $4d^{10} 5s^2 5p^5$ (e)   S: [Ne] $3s^2 3p^4$

     (b)   Co: [Ar] $3d^8 4s^1$ (d)   Cs: [Xe] $6s^1$

2.   Atoms in the same group of the periodic table have the same _____
     _____ .

3.   Silicon has the following number of valence electrons:
     (a)   two   (c)   one   (e)   none of the above
     (b)   five   (d)   eight

4. Determine whether each of the following statements is true or false. If a statement is false, change it so it is true.

    (a) The orbital diagram of the nitrogen atom in the ground state is

    [He] (↑↓) (↑↓) (↑) . True/False: _____

    _____ .

    (b) Paramagnetism arises when a species has unpaired electrons and is demonstrated experimentally when a sample of the species is attracted into a magnetic field. True/False: _____

    _____ .

    (c) According to Hund's rule, the lowest energy state of an atom will have the maximum number of parallel spins for a given n and l designation. True/False: _____

    _____ .

    (d) The correct order in increasing size of atomic radii is Na, S, Cs, Ba. True/False: _____

    _____ .

5. Write electron configurations for the following:
    (a) Sr _____
    (b) Ni _____
    (c) Cl^- _____

6. Choose the answer (or answers) that will complete the following phrase. The Pauli exclusion principle
    (a) states that no two electrons in an atom can have the same four quantum numbers.
    (b) states that the ground-state electron configuration of an atom will have the maximum number of parallel spins.
    (c) implies that the energy ordering of orbitals depends on the numerical values of n and l.
    (d) excludes the possibility of having two electrons with opposed spins.
    (e) determines the number of electrons in each orbital.

7.   Following is a labeled drawing of the periodic table:

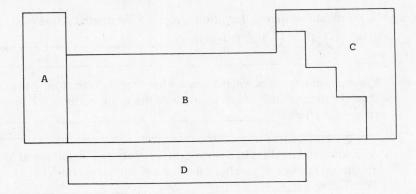

Which of the following sets correctly identifies the labeled parts?

(a)   A – main-group, B – nonmetal, C – transition, D – lanthanide elements

(b)   A – transition, B – main-group, C – inner-transition, D – actinide elements

(c)   A – inner-transition, B – transition, C – main-group, D – lanthanide and
       actinide elements

(d)   A – main-group, B – transition, C – main-group, D – inner-transition elements

(e)   None of the above is correct.

8.   Of the five atoms Rb, O, P, Sr, and Se, which should exhibit the lowest first
      ionization potential?

(a)   Rb   (c)   P   (e)   Se

(b)   O   (d)   Sr

ANSWERS TO CHAPTER DIAGNOSTIC TEST

If you missed an answer, study the text section and operational skill given in parentheses
after the answer.

1.   b   should be [Ar] $3d^7 4s^2$ c   should be [Ar] $3d^{10} 4s^2 4p^5$ (8.3, Op. Sk. 2)

2.   number of valence electrons and similar chemical and physical properties.
      (8.5, 8.6, Op. Sk. 3)

3.   e   (8.3, Op. Sk. 3)

4. (a) False. [He] (↑↓) (↑) (↑) (↑)   (8.1, 8.4, Op. Sk. 1, 4)
                2s   2p

    (b) True. (8.4)
    (c) True. (8.4)
    (d) False. The correct order is S, Na, Ba, Cs. (8.6, Op. Sk. 5)

5. (a) $1s^2 2s^2 2p^6 3s^2 3p^6 3d^{10} 4s^2 4p^6 5s^2$, or [Kr] $5s^2$

    (b) $1s^2 2s^2 2p^6 3s^2 3p^6 3d^8 4s^2$, or [Ar] $3d^8 4s^2$

    (c) $1s^2 2s^2 2p^6 3s^2 3p^6$, or [Ar]   (8.3, Op. Sk. 2)

6. a and e (8.1)   7. e (8.2)   8. a (8.6)

SUMMARY OF CHAPTER TOPICS

8.1 ELECTRON SPIN AND THE PAULI EXCLUSION PRINCIPLE

Operational Skill

1. Applying the Pauli exclusion principle. Given an orbital diagram or electron configuration, decide whether it is possible or not, according to the Pauli exclusion principle (Example 8.1).

You will need to remember the essence of the Pauli exclusion principle to work problems. In short, it is that no two electrons in an orbital can have the same spin.

Exercise 8.1   Look at the following orbital diagrams and electron configurations. Which are possible and which are not, according to the Pauli exclusion principle? Explain.

(a) (↑) (↑) (◯) (◯) (◯)   (d) $1s^2 2s^2 2p^4$
    1s   2s   2p

(b) (↑) (↑) (↑↓) (↑↓) (↑↓)   (e) $1s^2 2s^4 2p^2$
    1s   2s   2p

(c) (↑↓) (↑↓) (↑↑) (↑↓) (↑↓)   (f) $1s^2 2s^2 2p^6 3s^2 3p^{10} 3d^{10}$
    1s   2s   2p

Known: Pauli exclusion principle (from text).

Solution: (a)   Possible.

                (b)   Possible.

                (c)   Impossible; two electrons in a $2p$ orbital have the same spin.

                (d)   Possible.

                (e)   Impossible; there are four electrons in the $2s$ subshell.

                (f)   Impossible; there are ten electrons in the $3p$ subshell.

8.2   BUILDING-UP PRINCIPLE AND THE PERIODIC TABLE

A memory device often called "the diagonal rule" is shown below. It is a very useful aid for writing the arrangement of electrons in atoms.

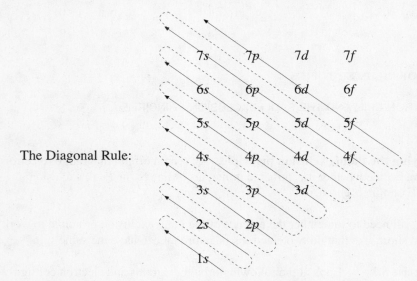

The Diagonal Rule:

To obtain the order in which orbitals are filled, start at the right of the bottom arrow, and follow it to its point. Then begin at the top right of the next arrow and follow it to its point, and so forth.

It is a good idea to memorize the following exceptions to the building-up principle and their configurations: Cr and Cu in Period 4 and Ag in Period 5. There are others, as well. See Appendix D.

8.3 WRITING ELECTRON CONFIGURATIONS USING THE PERIODIC TABLE

Operational Skills

2. Determining the configuration of an atom using the building-up principle.
Given the atomic number of an atom, write the complete electron configuration for the ground state, according to the building-up principle (Example 8.2).

3. Determining the configuration of an atom using the period and group numbers. Given the period and group for an element, write the configuration of the outer electrons (Example 8.3).

Mendeleev and Meyer found that when the elements were ordered by atomic weight, properties of the main-group elements recurred every eighth element. (The noble gases were not discovered until around 1900.) They arranged the elements in horizontal rows with like elements under one another. The reason for the recurring similarities, we believe, is the recurrence of a similar electron configuration. In our periodic table, arranged by atomic number, all elements in Group (column) IA have one electron in the s subshell of the highest occupied quantum level, n. All elements in Group IIA have two electrons in the s subshell of the highest occupied quantum level, etc. Thus, the group A number is the number of valence electrons. Moreover, each period of the table corresponds to a major quantum level. In Period 1, for example, quantum level 1 is the quantum level that fills.

Thus, it is the electrons with their particular energies that are the stars of the chemical drama. In Groups IA and IIA, the s subshell (and orbitals) are filling. In Groups IIIA to VIIIA, the p subshell (and orbitals) are filling, except for helium, in which the $1s$ sublevel is filled. In the transition metals, the d subshell (and orbitals) are filling; and in the inner-transition metals, the f subshell (and orbitals) are filling.

Once you are comfortable with this quantum level, subshell, and orbital correlation with periodic-table groups and sections, you can read the building-up order directly from the table. Begin at the upper left with hydrogen, and read across to helium: $1s^1$ and $1s^2$. Then begin at the left again with lithium, $2s^1$, and continue across to neon, $2p^6$. Skipping down to potassium, which begins Period 4, you have $4s^1$, $4s^2$, then $3d^1$, etc. (remember the exceptions, chromium and copper). Practice doing this until you can write the electron configurations of main-group elements and at least the $3d$ transition elements directly from the table.

If you look into the subject, you will find that there are several forms of the periodic table. The one in your text is called the long form. The short form takes less space but is much more difficult to read.

In some texts you will see the lanthanides called lanth<u>anoids</u> and the actinides called actin<u>oids</u>. According to the currently accepted rules in naming compounds, an ionic compound of two elements is a salt and ends in -<u>ide</u>. Recall that NaCl is sodium chloride. Some authors feel that the group names should reflect these rules accurately, and since these substances are elements they should not be named as salts. Other authors, including your textbook author, believe history should be preserved, and call them by their original names.

Exercise 8.2   Use the building-up principle to obtain the electron configuration for the ground state of the manganese atom ($Z = 25$).

Known: the diagonal-rule memory aid; Z, the atomic number = number of protons; number of electrons in the neutral atom equals the number of protons = 25; 2 electrons fill an s subshell, 6 fill a p subshell, 10 fill a d subshell.

Solution: $1s^2 2s^2 2p^6 3s^2 3p^6 4s^2 3d^5$. To be consistent with the text, we order the shells with $3d$ before $4s$: $1s^2 2s^2 2p^6 3s^2 3p^6 3d^5 4s^2$, which could also be written [Ar] $3d^5 4s^2$.

Exercise 8.3   Using the periodic table on the inside front cover, write the valence-shell configuration of arsenic (As).

Known: definition of valence-shell configuration; the period gives the number of the highest occupied quantum level; the Group A number gives the number of valence electrons.

Solution: Arsenic is in Period 4 and Group VA. Its valence-shell configuration is $4s^2 4p^3$.

Exercise 8.4   The lead atom has the ground-state configuration [Xe] $4f^{14} 5d^{10} 6s^2 6p^2$. Find the period and group for this element. From its position in the periodic table, would you classify lead as a main-group, a transition, or an inner-transition element?

Known: Valence electrons are in the highest occupied quantum level (shell); the quantum number of this level is the period number; the number of valence electrons gives the column (group) A number.

Solution: The highest occupied quantum level is 6, so lead is in Period 6. There are 4 valence electrons, 2 in the $6s$ subshell and 2 in the $6p$ subshell, indicating Group IVA. In this position, lead is a main-group element.

8.4   ORBITAL DIAGRAMS OF ATOMS; HUND'S RULE

Operational Skill

4.   Applying Hund's rule.   Given the electron configuration for the ground state of an atom, write the orbital diagram   (Example 8.4).

You will need to remember the essence of Hund's rule to work problems: one electron goes into each orbital of a subshell before two electrons occupy any of them, and all electrons in the singly occupied orbitals have the same spin.

> **Exercise 8.5**   Write an orbital diagram for the ground state of the phosphorus atom ($Z = 15$).   Write all orbitals.
>
> *Known:*   There are 15 electrons to place in the filling order; Hund's rule; Pauli exclusion principle (no more than 2 electrons per orbital; spins must be opposite).
>
> *Solution:*   First write the electron configuration:
>
> $$1s^2 2s^2 2p^6 3s^2 3p^3$$
>
> Then write the orbital diagram:

8.5   MENDELEEV'S PREDICTIONS FROM THE PERIODIC TABLE

8.6   SOME PERIODIC PROPERTIES

Operational Skill

5.   Applying periodic trends.   Using the known trends and referring to a periodic table, arrange a series of elements in order by atomic radius   (Example 8.5) or ionization energy   (Example 8.6).

> **Exercise 8.6**   Using a periodic table, arrange the following in order of increasing atomic radius: Na, Be, Mg.
>
> *Known:*   Size increases down a group (column), decreases across a period; we can get the order from the periodic table.
>
> *Solution:*   Be < Mg < Na.

Exercise 8.7   The first ionization energy of the chlorine atom is 1251 kJ/mol. Without looking at Figure 8.17, state which of the following values would be the more likely ionization energy for the iodine atom. Explain.

(a)   1000 kJ/mol   (b)   1400 kJ/mol

Known: Ionization energies decrease going down a column; the periodic table shows that iodine is below chlorine.

Solution: Iodine should have a lower ionization energy than 1251 kJ/mol, so (a) is the more likely value.

Electron affinity is the energy change when an electron is added to a neutral atom in the gaseous state to form a negative ion. Note that the associated sign indicates whether the energy is lost (–) or gained (+). The more stable the ion, the more energy is released on its formation and the higher the electron affinity. The less stable the ion, the more positive is the electron affinity. Across a period, electron affinity generally increases. This trend is the same as the trend for ionization energy. (Recall, however, that ionization energy is the energy *needed to remove* an electron from a neutral atom in the gaseous state.)

Exercise 8.8   Without looking at Table 8.4 but using the general comments in this section, decide which has the larger negative electron affinity, C or F.

Known: definition of electron affinity; the more stable the anion formed by the addition of an electron, the greater electron affinity; anions with half-filled p subshells are quite stable; anions with filled p subshells, and thus noble-gas configurations, are more stable; the valence-electron configuration gives the number and arrangement of electrons in the subshells.

Solution: The valence-electron configuration of C^- would be $2s^2 2p^3$. The valence-electron configuration of F^- would be $2s^2 2p^6$. Because F^- has the filled p sublevel and noble-gas configuration, it will be more stable and thus have the larger electron affinity.

8.7   BRIEF DESCRIPTIONS OF THE MAIN-GROUP ELEMENTS

Exercise 8.9   (a)   Write the formula for a compound of hydrogen and selenium.   (b)   The formula of calcium sulfate is $CaSO_4$. Predict the formula of the similar compound of selenium, calcium selenate.

Known: Selenium is in the same column as oxygen and sulfur and thus has the same valence-electron configuration. We would expect it to form compounds with the same ratios as they do. Because selenium is below sulfur, it is more like sulfur than it is like oxygen.

Solution: (a)   Since water is H_2O, the compound would be H_2Se.

              (b) $CaSeO_4$.

A Nonmetal That Matters: PHOSPHORUS (Group VA)

Questions for Study

1. What causes the glow of white phosphorus in air?

2. What are the starting materials for the commercial production of white phosphorus?

3. Explain why white phosphorus should not be picked up with bare fingers.

4. What are the two common allotropic forms of phosphorus? List several properties of each allotrope.

5. What is the principal compound into which commercial white phosphorus is converted? Give chemical equations for the conversion.

Answers to Questions for Study

1. White phosphorus glows in air due to chemiluminescence, the emission of light from excited molecular products of the reaction between white phosphorus and oxygen.

2. Starting materials for the commercial production of white phosphorus are calcium phosphate, $Ca_3(PO_4)_2$; coke (carbon), C; and sand, SiO_2.

3. White phosphorus should not be picked up with bare fingers because the heat from your fingers will ignite it.

4. The two common allotropic forms of phosphorus are white phosphorus, P_4, and red phosphorus, $P(s)$. White phosphorus is a soft, waxy solid, initially pure white but becoming yellowish on exposure to light. It burns spontaneously in air, and the solid and vapor are extremely poisonous. Red phosphorus reacts much less readily with oxygen than does white phosphorus, does not have the blue-green chemiluminescence, and is relatively nontoxic.

5. Most of the commercially produced white phosphorus is converted into phosphoric acid. The equations are

$$P_4(s) + 5O_2(g) \longrightarrow P_4O_{10}(s)$$

$$P_4O_{10}(s) + 6H_2O(l) \longrightarrow 4H_3PO_4(aq)$$

ADDITIONAL PROBLEMS

1. Briefly explain how the quantum-mechanical model of the atom helps us to understand the groupings of the elements in the periodic table.

2. Write the ground-state electron configuration for each of the following species:

    (a) Br   (b) Al   (c) Ca^{2+} (d) N^{3-} (e) V   (f) S^{2-}

3. Arrange the following atomic orbitals in order of increasing energy: $5p$, $4p$, $3p$, $4s$, and $3d$. Explain your ordering.

4. Decide whether each of the following orbital diagrams or electron configurations is possible according to the quantum-mechanical view of the atom, the Pauli exclusion principle, and Hund's rule. Explain your decisions.

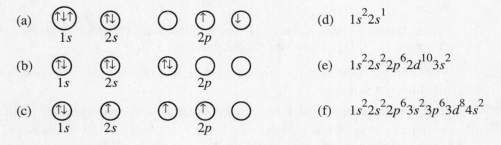

    (d) $1s^2 2s^1$

    (e) $1s^2 2s^2 2p^6 2d^{10} 3s^2$

    (f) $1s^2 2s^2 2p^6 3s^2 3p^6 3d^8 4s^2$

5. Write an orbital diagram for each of the following electron configurations. Indicate whether each atom is diamagnetic or paramagnetic, and explain why.

    (a) $1s^2 2s^2 2p^6 3s^2 3p^5$ (b) $1s^2 2s^2 2p^3$ (c) [Ne] $3s^2$

6. Write the valence-electron configuration for atoms of the element in each of the following positions in the periodic table.

    (a) Period 4, Group VIIA
    (b) Period 6, Group VA
    (c) Period 3, Group IA
    (d) Period 2, Group VIA

7. Indicate the expected trends (increase or decrease) in atomic radii, cation radii, ionization energy, and electron affinity for the elements as we move from left to right across the periodic table, then as we move down the periodic table.

8. The observed trends in atomic and ionic radii for the halogens and halide ions are as follows, in picometers:

F	Cl	Br	I		F^-	Cl^-	Br^-	I^-
64	99	114	133		119	167	182	206

    Discuss these data in terms of the valence-shell electron configuration, nuclear charge, and ionic charge of these species.

9. Arrange the elements in each of the following groups in order of decreasing ionization energy. Use only the periodic table.

    (a) Zn, K, Br   (b) Ca, Be, Ba   (c) Sr, Cs, Rb   (d) Al, Ca, Cs

10. Show, in a drawing, your idea of what the first four periods of the periodic table would look like if each orbital could have three electrons.

ANSWERS TO ADDITIONAL PROBLEMS

If you missed an answer, study the text section and operational skill given in parentheses after the answer.

1. The valence-shell electron configurations of elements appearing in the same column of the periodic table are the same, except that the principal quantum number, n, increases by one as you go down a given column.   (8.5)

2. (a) Br: [Ar] $4s^2 3d^{10} 4p^5$

    (b) Al: [Ne] $3s^2 3p^1$

    (c) Ca^{2+}: [Ar]

    (d) N^{3-}: [Ne]

    (e) V: [Ar] $4s^2 3d^3$

    (f) S^{2-}: [Ar]   (8.1, Op. Sk. 2)

3.   The relative energy ordering of these atomic orbitals is as follows:

$$3p < 4s < 3d < 4p < 5p$$

From the building-up principle, we know that the lower the value of the principal quantum number, n, the lower the energy of the orbital. For orbitals with the same principal quantum number, the energies of orbitals increase with the l quantum number. The $3d$ energy is quite close to that of the $4s$ orbital, and the energy of the atom is lowest when the $4s$ orbital is filled before the $3d$ orbitals.   (8.2)

4.   (a)   Not possible; three electrons cannot occupy an orbital ($1s$), and single electrons in orbitals in the same subshell must have like spins ($2p$).
   (b)   Not possible; two electrons cannot occupy one orbital when other orbitals in the same subshell are empty ($2p$).
   (c)   Possible; this atom is in an excited state, because the $2s$ orbital contains only one electron.
   (d)   Possible; the s subshell can contain two electrons.
   (e)   Not possible; there is no $2d$ subshell.
   (f)   Possible. No violations.   (8.1, Op. Sk. 1)

5.   (a)   (↑↓)   (↑↓)   (↑↓) (↑↓) (↑↓)   (↑↓)   (↑↓) (↑↓) (↑)
             $1s$ $2s$ $2p$ $3s$ $3p$

   The spin of the one unpaired electron makes the atom paramagnetic.

   (b)   (↑↓)   (↑↓)   (↑) (↑) (↑)
             $1s$ $2s$ $2p$

   Three unpaired electrons (with like spins) make the atom paramagnetic.

   (c)   (↑↓)   (↑↓)   (↑↓) (↑↓) (↑↓)   (↑↓)
             $1s$ $2s$ $2p$ $3s$

   The atom is diamagnetic, because all electron spins are paired.
                                             (8.4, Op. Sk. 4)

6.   (a) $4s^2 4p^5$ (c) $3s^1$
   (b) $6s^2 6p^3$ (d) $2s^2 2p^4$ (8.3, Op. Sk. 3)

7.   The trends in the observed properties of the elements are as follows:

Observed Property	Trend from Left to Right in Periodic Table	Trend from Top to Bottom in Periodic Table
Atomic radii	decrease	increase
Cation radii	decrease	increase
Ionization energy	increase	decrease
Electron affinity	increase	decrease (8.6)

8.   Going down the group from fluorine to iodine, the atoms, as well as their ions, increase in size. This is due to the increased value of the principal quantum number, n, causing the average distance of the valence electrons from the nucleus to increase.

The valence-shell electron configuration of each of the elements is ns^2np^5, and that for each of their ions is ns^2np^6 (the configuration of the nearest noble gas in the periodic table).

The nuclear charge of F and that of F^- are the same, and similarly for Cl and Cl^-, Br and Br^-, and I and I^-. The negative ion, in each case, has a larger radius than the neutral atom, because the electrons (one more than in the neutral atom) of the ion are held less tightly by the nucleus than those of the neutral atom.   (8.6)

9.   (a)   Br > Zn > K   (c)   Sr > Rb > Cs

(b)   Be > Ca > Ba   (d)   Al > Ca > Cs   (8.6, Op. Sk. 5)

10.

1	2																3
H	He																Li
4	5	6								7	8	9	10	11	12	13	14	15
Be	B	C								N	O	F	Ne	Na	Mg	Al	Si	P
16	17	18								19	20	21	22	23	24	25	26	27
S	Cl	Ar								K	Ca	Sc	Ti	V	Cr	Mn	Fe	Co
28	29	30	31	32	33	34	35	36	37	38	39	40	41	42	43	44	45	46	47	48	49	50	51	52	53	54
Ni	Cu	Zn	Ga	Ge	As	Se	Br	Kr	Rb	Sr	Y	Zr	Nb	Mo	Tc	Ru	Rh	Pd	Ag	Cd	In	Sn	Sb	Te	I	Xe

(8.2)

CHAPTER POST-TEST

1. For each of the following pairs, indicate which atom or ion should have the larger radius.

   (a) Au^{3+} or Au^+ (c) Br or Br^- (e) O or B
   (b) Ar or Xe   (d) F or I

2. _____ electrons are found in the outermost energy level.

3. Which of the following statements about periodic properties of the elements is (are) incorrect?
   (a) The ionization energies of the elements in a given period generally increase from left to right.
   (b) The electron affinities of the elements in a given period generally increase from left to right.
   (c) Chemical properties of the elements are periodic functions of the atomic number.
   (d) Atomic radii increase across a period and down a group.
   (e) The ionization energies of the elements generally increase in going down a given group in the periodic table.

4. The element with atomic number 53 has how many electrons in its valence shell?

   (a) 7   (c) 8   (e) 126
   (b) 53   (d) 2

5. Write the abbreviated electron configuration for each of the following in the ground state.

   (a) potassium, $Z = 19$ (c) aluminum, $Z = 13$
   (b) titanium, $Z = 22$ (d) Ag, $Z = 47$

6. The *d* subshell can accommodate, for any given principal quantum number, the following number of electrons:

   (a) 6   (c) 2   (e) none of the above
   (b) 14   (d) 10

7. Which of the following orbital diagrams is (are) incorrect for the respective ground-state electron configuration?

    (a) $4s^2 3d^5$ = (↑↓)   (↑) (↑) (↑) (↑) (↑)

    (b) $4s^2 3d^{10} 4p^4$ = (↑↓)   (↑↓) (↑↓) (↑↓) (↑↓) (↑↓)   (↑↓) (↑↓)

    (c) $3s^2 3p^3$ = (↑↓)   (↑) (↑) (↑)

    (d) $5s^1$ = (↑)

    (e) $4f^6$ = (↑) (↑) (↑) (↑) (↑) (↑) ( )

8. Which of the following is (are) diamagnetic?

    (a) Sn^{2+} (c) H   (e) K^+
    (b) F   (d) Si

9. Arrange the following sets of atoms in order of increasing first ionization potential.
    (a) K   Cs   Rb   Li   (c) O   N   Se   Te
    (b) Ga   As   Br   Ca   (d) B   He   Li   Ne

10. Given the general trend in electron affinities across any period in the periodic table, explain why the electron affinities of C and N are -122 and 0 kJ/mol, respectively.

11. Explain in simple terms how (or why) the Pauli exclusion principle works.

ANSWERS TO CHAPTER POST-TEST

If you missed an answer, study the text section and operational skill given in parentheses after the answer.

1. (a) Au^+ (b) Xe   (c) Br^- (d) I   (e) B   (8.6, Op. Sk 5)

2. Valence (8.2)   3. d and e (8.6)   4. a (8.3, Op. Sk. 3)

5. (a) [Ar] $4s^1$ (c) [Ne] $3s^2 3p^1$

    (b) [Ar] $3d^2 4s^2$ (d) [Kr] $4d^{10} 5s^1$ (An exception)   (8.3, Op. Sk. 2)

6. d (8.1)   7. b (8.1, 8.4, Op. Sk. 1, 4)   8. a and e (8.4)

9.   (a)   Cs < Rb < K < Li   (c)   Te < Se < O < N
     (b)   Ga < Ca < As < Br   (d)   Li < B < Ne < He   (8.6, Op. Sk. 5)

10.   Electron affinity generally increases across a given period. Nitrogen is an exception, as are all Group VA elements. They have the valence electron configuration $ns^2 np^3$ with the half-filled p subshell. Adding an additional electron sacrifices the stability attributed to the half-filled subshell and produces an unstable negative ion.   (8.6)

11.   Electrons are negatively charged particles that repel each other. When degenerate orbitals exist (separate orbitals with the same energy), it is energetically favorable for the electrons to fill the orbitals singly until all available degenerate orbitals are used. At that point, it becomes energetically favorable for electrons to pair rather than to move to a new subshell or to a new quantum level.   (8.1)

CHAPTER 9   IONIC AND COVALENT BONDING

CHAPTER TERMS AND DEFINITIONS

Numbers in parentheses after definitions give the text sections in which the terms are explained. Starred terms are italicized in the text. Where a term does not fall directly under a text section heading, additional information is given for you to locate it.

chemical bond* strong attractive force that exists between certain atoms in a substance (chapter introduction)

metallic bonding* type of bonding exhibited in metals; all valence electrons of these atoms move in a crystal, attracted to the positive cores of all the metal ions (chapter introduction)

salts* compounds composed of ions (9.1, introductory section)

ionic bond chemical bond formed by the electrostatic attraction between positive and negative ions (9.1)

cation* positive ion (9.1)

anion* negative ion (9.1)

Lewis electron-dot symbol symbol in which the electrons in the valence shell of an atom or ion are represented as dots placed around the letter symbol of the element (9.1)

Coulomb's law* the energy E obtained in bringing two ions with electric charges Q_1 and Q_2 to a distance r apart is $E = kQ_1Q_2/r$ (9.1)

lattice energy change in energy that occurs when an ionic solid is separated into isolated ions in the gas phase (9.1)

Born–Haber cycle* use of Hess's law to relate lattice energy to other thermochemical quantities that have been measured experimentally (9.1)

sublimation* transformation of a solid to a gas (9.1)

ionic radius measure of the size of the spherical region around the nucleus of an ion within which the electrons are most likely to be found (9.3)

isoelectronic different species having the same number and configuration of electrons (9.3)

covalent bond chemical bond formed by the sharing of a pair of electrons between atoms (9.4, introductory section)

bond dissociation energy*   energy that must be added to separate bonded atoms   (9.4)

Lewis electron-dot formula   formula using dots to represent valence electrons   (9.4)

bonding pair   electron pair shared between two atoms   (9.4)

lone (nonbonding) pair   electron pair that remains on one atom and is not shared   (9.4)

coordinate covalent bond   covalent bond in which both electrons are donated by one atom   (9.4)

octet rule   tendency of atoms in molecules to have eight electrons in their valence shells, except for hydrogen, which tends to have two   (9.4)

single bond   covalent bond in which a single pair of electrons is shared by two atoms (9.4)

double bond   covalent bond in which two pairs of electrons are shared by two atoms (9.4)

triple bond   covalent bond in which three pairs of electrons are shared by two atoms (9.4)

polar covalent bond   covalent bond in which the bonding electrons are not shared equally and spend more time near one atom than the other   (9.5)

nonpolar*   bond or molecule in which the centers of positive and negative charge coincide   (9.5)

electronegativity   measure of the ability of an atom in a molecule to draw bonding electrons to itself   (9.5)

electropositive*   descriptive of metals, which have small electronegativity values   (9.5)

polar molecule*   molecule in which the centers of positive and negative charge are separated   (9.5)

skeleton structure*   information about a molecule that indicates which atoms are bonded to one another (without regard to whether the bonds are single or not)   (9.6)

delocalized bonding   model of bonding in which a bonding electron pair is spread over a number of atoms rather than localized between two atoms   (9.7)

resonance description   representation of a molecule having delocalized bonding and showing all possible electron-dot formulas   (9.7)

resonance formulas*   all possible electron-dot formulas for a molecule described by localized bonding   (9.7)

bridge*   position of an atom that usually forms one bond when that atom is bonded covalently between two atoms; for example, the Br atom in the Al_2Br_6 molecule (9.8)

formal charge   hypothetical charge you obtain by assuming that bonding electrons are equally shared between bonded atoms and that the electrons of each lone pair belong completely to one atom   (9.9)

bond length (bond distance)   distance between the nuclei in a bond   (9.10)

covalent radii   radius values assigned to atoms in such a way that the sum of covalent radii of atoms A and B predicts an approximate A—B bond length   (9.10)

bond order   number of electron pairs in a bond in a Lewis electron-dot formula   (9.10)

bond energy (*BE*)   average enthalpy change for the breaking of a bond between two specific atoms in a molecule in the gas phase   (9.11)

wavenumbers*   frequency units obtained by dividing the frequency by the speed of light expressed in cm/s   (Instrumental Methods: Infrared Spectroscopy and Vibrations of Chemical Bonds)

percent transmittance*   percent of radiation that passes through a sample   (Instrumental Methods: Infrared Spectroscopy and Vibrations of Chemical Bonds)

Claus process*   method of producing sulfur by burning hydrogen sulfide recovered from natural gas and petroleum and reacting the resulting sulfur dioxide with additional hydrogen sulfide   (A Molecule That Matters: Sulfur [a Group VIA Nonmetal])

viscose*   syrupy solution formed when cellulose is dissolved in sodium hydroxide and carbon disulfide in the manufacture of rayon or cellophane   (A Molecule That Matters: Sulfur [a Group VIA Nonmetal])

CHAPTER DIAGNOSTIC TEST

1.   In one or two sentences, distinguish between molecular and ionic compounds.

2.   Write Lewis formulas for the following:

   (a)   NaCl   (c) CCl_4 (e) XeO_3

   (b) CaF_2 (d) NCl_3 (f)   HOCl

3.   _____ results from a transfer of electron(s) between atoms in a chemical bond.

4.   Using the following enthalpy data and a Born–Haber cycle calculation, determine $\Delta H°$ for $Ca(s) + \frac{1}{2}O_2(g) \longrightarrow CaO(s)$. Enthalpy change for the sublimation of calcium is 192.4 kJ/mol; for the bond dissociation of O_2, it is 495.0 kJ/mol of O_2; for the first and second ionization energies of $Ca(g)$, it is 595.8 and 1151 kJ/mol, respectively; for the first and second electron affinities for $O(g)$, it is -138.1 and 790.8 kJ/mol, respectively; and for the lattice energy for the formation of CaO, it is -3556 kJ/mol.

5.   Write the abbreviated electron configuration for each of the following:

   (a) N^{3-} (b) Ca^{2+} (c) Fe^{3+}

6. Determine whether each of the following statements is true or false. If a statement is false, change it so it is true.

(a) The octet rule accurately gives Lewis formulas for the compounds NH_2Cl, PCl_3, $SOCl_2$, and $CaCl_2$, since no *d* orbitals are involved in the bonding.
True/False: _____

(b) For a given species, the relative sizes of the neutral atom, positive ion, and negative ion would be predicted as

neutral atom < positive ion < negative ion

True/False: _____

(c) The bond length for the CN bond increases in the order

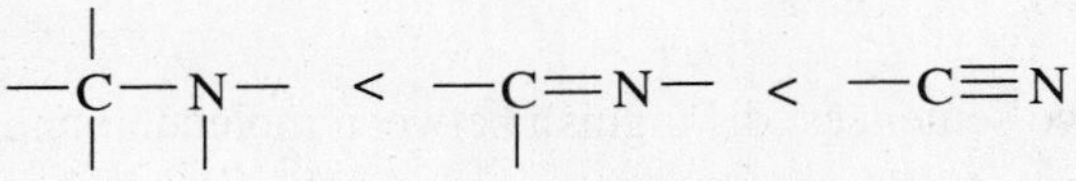

$$-\overset{|}{\underset{|}{C}}-\overset{|}{\underset{|}{N}}- \quad < \quad -C=\overset{|}{N}- \quad < \quad -C\equiv N$$

True/False: _____

(d) In a polar covalent bond, the unequal attraction for the shared electron pair creates an unsymmetrical distribution of electron density in the covalent bond.
True/False: _____

(e) In the resonance description of a molecule, the double-headed arrows indicate that the molecule flips between the various dot structures.
True/False: _____

7. For each of the following four pairs, pick the bond you expect to be the more polar. (Use a table of electronegativities. See text Figure 9.11.)

(a) N — O and N — F   (c) C — F and C — Br
(b) H — O and H — S   (d) C — H and C — F

8. Given the following enthalpy data, calculate the average P — Cl bond energy, BE (P — Cl).

$$\Delta H_f^\circ$$

P(g)	333.9	kJ/mol
Cl(g)	121.0	kJ/mol
$PCl_3(g)$	-271	kJ/mol

(a) 242 kJ/mol   (c) 968 kJ/mol   (e) none of the above
(b) 323 kJ/mol   (d) 142 kJ/mol

9. List the following in order of increasing radii: N^{3-}, Na^+, O^{2-}, Mg^{2+}.

10. Given the following enthalpy data, determine the average H — O bond energy, BE (H — O), in H_2O.

$$2H_2(g) + O_2(g) \longrightarrow 2H_2O(g) \qquad \Delta H^\circ = -493.7 \text{ kJ}$$

BE (H — H) = 435.1 kJ/mol, BE (O = O) = 493.7 kJ/mol

11. Describe the bonding in the ozone molecule, O_3, using resonance formulas.

12. What is the observed effect that resonance has upon the bond length between two atoms?

ANSWERS TO CHAPTER DIAGNOSTIC TEST

If you missed an answer, study the text section and operational skill given in parentheses after the answer.

1. Molecular compounds are composed of molecules having no net electric charge, whereas ionic compounds are composed of ions of positive and negative electric charge. (9.1, 9.4)

---

2. 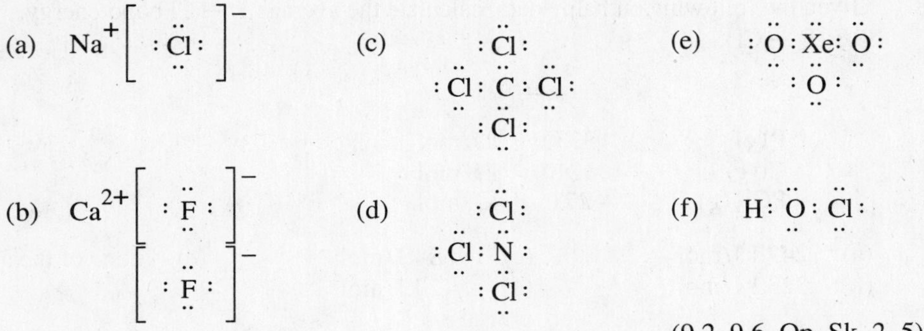 (a) Na^+ $[:\ddot{Cl}:]^-$ (c) $:\ddot{Cl}:$; $:\ddot{Cl}:\ddot{C}:\ddot{Cl}:$; $:\ddot{Cl}:$ (e) $:\ddot{O}:Xe:\ddot{O}:$; $:\ddot{O}:$

(b) Ca^{2+} $[:\ddot{F}:]^-$ $[:\ddot{F}:]^-$ (d) $:\ddot{Cl}:$; $:\ddot{Cl}:\ddot{N}:$; $:\ddot{Cl}:$ (f) $H:\ddot{O}:\ddot{Cl}:$

(9.2, 9.6, Op. Sk. 2, 5)

3. An ionic bond   (9.1)

4. -717 kJ   (9.1, 9.11, Op. Sk. 8)

5. (a) [Ne]   (b) [Ar]   (c) [Ar] $3d^5$ (9.2, Op. Sk. 2)

6. (a) True.   (9.2, 9.6, Op. Sk. 2, 5)
    (b) False. The relative order of size would be predicted as

    positive ion < neutral atom < negative ion   (9.3)

    (c) False. The bond length for the CN bond increases in the order

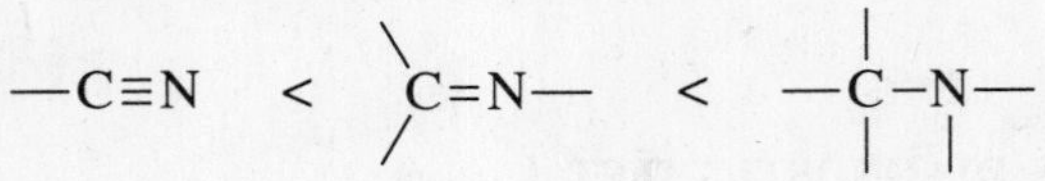

$$-C\equiv N \quad < \quad \overset{\backslash}{\underset{/}{C}}=N- \quad < \quad -\overset{|}{\underset{|}{C}}-\overset{|}{N}- \quad (9.10, \text{Op. Sk. } 8)$$

(d) True.   (9.5)
    (e) False. The arrows indicate that the molecule is a composite of all the dot
    structures; we cannot draw its electronic structure.   (9.7)

7. (a) $N-F$,   (b) $H-O$,   (c) $C-F$,   (d) $C-F$ (9.5, Op. Sk. 4)

8. b   (9.11, Op. Sk. 9)

9. $Mg^{2+}, Na^+, O^{2-}, N^{3-}$ (9.3, Op. Sk. 3)

10. 464.4 kJ/mol   (9.11, Op. Sk. 9)

11. 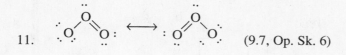 (9.7, Op. Sk. 6)

12. Resonance is a type of averaging process. The net result is to average the bond lengths involved in the resonance structures.   (9.7)

SUMMARY OF CHAPTER TOPICS

9.1   DESCRIBING IONIC BONDS

Operational Skill

1.   Using Lewis symbols to represent ionic bond formation.   Given a metallic and a nonmetallic main-group element, use Lewis symbols to represent the transfer of electrons to form ions of noble-gas configurations   (Example 9.1).

Use the periodic table to your advantage when writing the Lewis symbols for the main-group elements. Recall that for the main-group elements the group number is the number of valence electrons. Orbital drawings of the outer shells indicate the electron pairing for Groups VA through VIIIA. Note, however, that for Groups IIA through IVA the electrons are placed singly (see text Table 9.1), which better reflects the chemistry of these elements. For example, the orbital notations for aluminum (Group IIIA) and oxygen (Group VIA) are

The Lewis symbol for aluminum is $\cdot \text{Al} \cdot$, showing all electrons unpaired. The symbol for

oxygen is $: \overset{\cdot}{\underset{\cdot \cdot}{\text{O}}} \cdot$, showing electrons paired as they are in the orbital diagram.

Exercise 9.1   Represent the transfer of electrons from magnesium to oxygen atoms to assume noble-gas configurations. Use Lewis electron-dot symbols.

Known: Magnesium (Group IIA) has 2 valence electrons; oxygen (Group VIA) has 6 valence electrons, 2 of them unpaired; 8 valence electrons form the noble-gas configuration.

Solution:

$$\cdot Mg \cdot \ + \ :\!\overset{\displaystyle\cdot}{\underset{\displaystyle\cdot\cdot}{O}}: \ \longrightarrow \ Mg^{2+} + \left[:\overset{\displaystyle\cdot\cdot}{\underset{\displaystyle\cdot\cdot}{O}}: \right]^{2-}$$

9.2 ELECTRON CONFIGURATIONS OF IONS

Operational Skill

2.   Writing electron configurations of ions.   Given an ion, write the electron configuration. For an ion of a main-group element, give the Lewis symbol   (Examples 9.2 and 9.3).

It cannot be stressed enough that you must know the element names and symbols and the ion names, formulas, and charges. Consult your instructor to be sure of the particular ones you should have on the tip of your tongue. You will soon be bogged down in your study if you have not learned the great majority of them.

Exercise 9.2   Write the electron configuration and the Lewis symbol for Ca^{2+} and for S^{2-}.

Known: Ca^{2+} is Group IIA, and $Z = 20$, so there are 20 electrons in the neutral atom and 18 in the ion. S^{2-} is Group VIA, and $Z = 16$, so there are 16 electrons in the neutral atom and 18 in the ion.

Solution: The electron configuration for Ca^{2+} is [Ar]; the Lewis symbol is Ca^{2+}. The electron configuration for S^{2-} is [Ne] $3s^2 3p^6$ (which is also [Ar]); the Lewis

symbol is $\left[:\overset{\displaystyle\cdot\cdot}{\underset{\displaystyle\cdot\cdot}{S}}: \right]^{2-}$.

Exercise 9.3   Write the electron configurations of Pb and Pb^{2+}.

Known: Pb is Group IVA, and $Z = 82$, so there are 82 electrons in Pb and 80 in Pb^{2+}.

Solution: Pb is [Xe] $4f^{14} 5d^{10} 6s^2 6p^2$. Pb^{2+} is [Xe] $4f^{14} 5d^{10} 6s^2$.

Exercise 9.4   Write the electron configuration of Mn^{2+}.

Known: $Z = 25$, so Mn^{2+} would have 23 electrons. Transition metals lose the *ns* electrons first.

Solution: $[Ar]\, 3d^5$.

9.3   IONIC RADII

Operational Skill

3.   Using periodic trends to obtain relative ionic radii.   Given a series of ions, arrange them in order of increasing ionic radius   (Example 9.4).

Exercise 9.5   Which has the larger radius, S or S^{2-}? Explain.

Known: Radius depends on *n* and effective nuclear charge (from Section 8.6).

Solution: Since there are more electrons for S^{2-}, the effective nuclear charge experienced by the valence electrons is less. Thus, S^{2-} is larger.

Exercise 9.6   Without looking at Table 9.3, arrange the following ions in order of increasing ionic radius: $Sr^{2+}, Mg^{2+}, Ca^{2+}$. (You may use a periodic table.)

Known: All are Group IIA elemental ions; ionic radius increases down a group.

Solution: $Mg^{2+} < Ca^{2+} < Sr^{2+}$.

Exercise 9.7   Without looking at Table 9.3, arrange the following ions in order of increasing ionic radius: Cl^-, Ca^{2+}, P^{3-}. (You may use a periodic table.)

Known: All are isoelectronic with the [Ar] configuration; in any isoelectronic sequence, radius decreases with increasing *Z*.

Solution: $Ca^{2+} < Cl^- < P^{3-}$.

9.4   DESCRIBING COVALENT BONDS

The octet rule is a great aid in determining the bonding in a molecule. Use it to your advantage. It is not ideal (there are exceptions to any "rule" in chemistry), but it is a good first step in solving problems.

9.5   POLAR COVALENT BONDS; ELECTRONEGATIVITY

Operational Skill

4.   Using electronegativities to obtain relative bond polarities.   Given the electronegativities of the atoms, arrange a series of bonds in order by polarity   (Example 9.5).

Electronegativity is an important concept. You will use it to understand bond characteristics, dipole moment, intermolecular bonding, and many other properties of molecules that you will study later on. Electronegativity is the ability of an atom to attract electrons *in a chemical bond*. The final phrase here distinguishes electronegativity from electron affinity, which relates to the ability of an atom to attract a *free* electron.

Learn the trend in the periodic table that is associated with electronegativity. Electronegativity increases moving to the right in the periodic table and decreases down a group. Since noble gases are often exceptions to periodic trends, fluorine is the most electronegative element, and francium the least electronegative, or most electropositive. In some texts you will find cesium listed as the most electropositive element, as it is more common than francium.

When a chemical bond is formed, electrons are either transferred from one atom to another or shared between the two atoms, depending upon the attraction of an atomic nucleus for valence electrons. It is convenient to think of bonding as a horizontal line with electron transfer at one end and equal sharing at the other. The difference in electronegativities of the two atoms gives a rough guide to the type of bond to expect. When the difference is 0.5 or less, the bond is essentially covalent. When the difference is greater than 1.8, the bond is considered ionic. The polar covalent bond would thus lie within electronegativity differences of 0.6 and 1.8. Generally speaking, elements in Groups IA and IIA form ionic bonds with elements in Groups VIA and VIIA, although there are exceptions. The nonmetals, then, with similar electronegativities, form covalent bonds.

Exercise 9.8   Using electronegativities, decide which of the following bonds is most polar: C — O, C — S, H — Br.

Known: The greater the electronegativity difference, the more polar the bond. The elements have the following electronegativities: H = 2.1, C = 2.5, S = 2.5, Br = 2.8, O = 3.5 (from text Figure 9.11).

Solution: C — O is 3.5 – 2.5 = 1.0 (most polar bond)
C — S is 2.5 – 2.5 = 0
H — Br is 2.8 – 2.1 = 0.7

9.6 WRITING LEWIS ELECTRON-DOT FORMULAS

Operational Skill

5. Writing Lewis formulas. Given the molecular formula of a simple compound or ion, write the Lewis electron-dot formula (Examples 9.6, 9.7, 9.8, and 9.10).

Exercise 9.9 Dichlorodifluoromethane, CCl_2F_2, is a gas used as a refrigerant and aerosol propellant. Write the Lewis formula for CCl_2F_2.

Known: rules and steps given in text; valence electrons = group number (from periodic table); central atom is C (least electronegative). (See text Figure 9.11.)

Solution: Number of valence electrons is

4 (from C) + (4 × 7) (from 2Cl and 2F) = 32

Skeleton structure with outer-atom electrons uses all 32 electrons, and each atom has an octet of electrons.

Lewis formula is

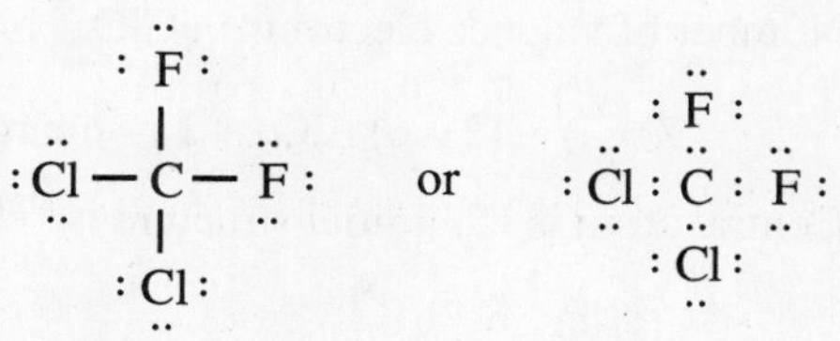

$$
\begin{array}{ccc}
\ddot{\text{F}} \colon & & \ddot{\text{F}} \colon \\
| & & \\
\colon\!\ddot{\text{Cl}} - \text{C} - \ddot{\text{F}} \colon & \text{or} & \colon\!\ddot{\text{Cl}} \colon \text{C} \colon \ddot{\text{F}} \colon \\
| & & \\
\colon\!\ddot{\text{Cl}} \colon & & \colon\!\ddot{\text{Cl}} \colon
\end{array}
$$

(Dash represents an electron pair.)

Exercise 9.10 Write the electron-dot formula of carbon dioxide, CO_2.

Known: Central atom is C, from number considerations only.

Solution: Number of valence electrons is

4 (from C) + (2 × 6) (from O) = 16

Initial structure is $\colon\!\ddot{\text{O}} - \text{C} - \ddot{\text{O}}\colon$

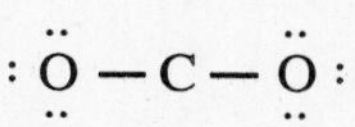

This accounts for the 16 valence electrons, but carbon needs four more electrons for an octet. This suggests two double bonds. Thus, the electron-dot formula is

$$\ddot{O} = C = \ddot{O} \quad \text{or} \quad \ddot{O}::C::\ddot{O}$$

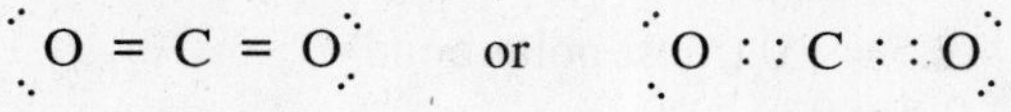

Exercise 9.11   Write the electron-dot formula of (a) the hydronium ion, H_3O^+; (b) the chlorite ion, ClO_2^-.

Solution:

(a)   Number of valence electrons in H_3O^+ is

$$(3 \times 1)\,(3H) + 6\,(O) - 1\,(\text{for + charge}) = 8$$

Central atom is O (H can't be). Initial structure is

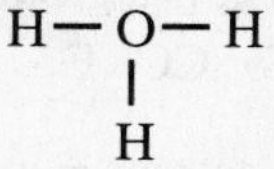

$$H - O - H$$
$$|$$
$$H$$

leaving $8 - 6 = 2$ electrons to give O an octet. Electron-dot formula is

$$\left[H - \ddot{O} - H \atop \quad | \atop \quad H \right]^+ \quad \text{or} \quad \left[H : \ddot{O} : H \atop \quad H \right]^+$$

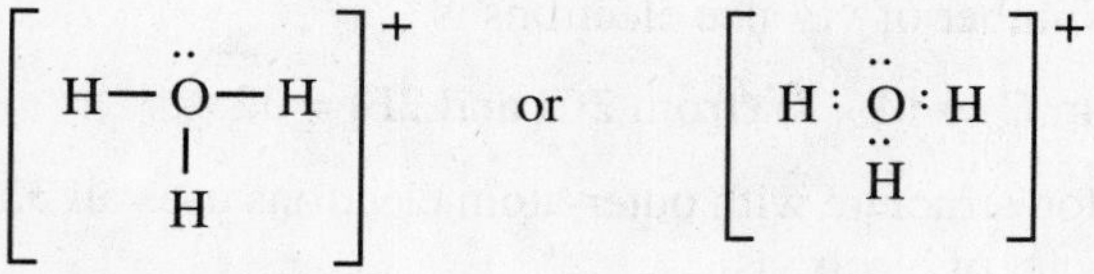

(b)   Number of valence electrons in ClO_2^- is

$$7\,(Cl) + (2 \times 6)\,(2O) + 1\,(-\text{charge}) = 20$$

Central atom is Cl. Initial structure is

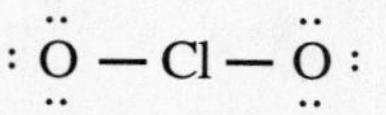

$$: \ddot{O} - Cl - \ddot{O} :$$

leaving $20 - 16 = 4$ electrons to give Cl an octet. Electron-dot formula is

$$\left[: \ddot{O} - \ddot{Cl} - \ddot{O} : \right]^- \quad \text{or} \quad \left[: \ddot{O} : \ddot{Cl} : \ddot{O} : \right]^-$$

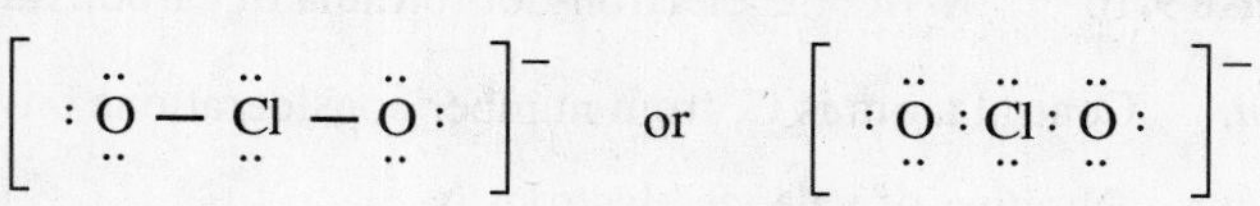

9.7   DELOCALIZED BONDING; RESONANCE

Operational Skill

6.   Writing resonance formulas.   Given a simple molecule with delocalized bonding, write the resonance description   (Example 9.9).

The bonding theory you have been using is valence bond theory, which will be discussed at length in Chapter 10 of the text. In this model, bonding occurs by overlap of the valence orbitals. The electrons in the bond spend most of their time in the space between the two nuclei. However, this model and its Lewis structures cannot account for bond-length data taken on molecules such as SO_2, O_3, and HNO_3. For these molecules we would expect bonds of different lengths on the basis of the dot formulas, but this is not observed. The resonance model is added to valence bond theory to make the theory agree with experimental data. Another model, the molecular orbital theory, will be discussed in Chapter 10.

When drawing resonance formulas, make sure you understand that the molecule never exists as any of the dot structures we draw. We cannot draw the electron arrangement. The double-headed arrows indicate that we must mentally combine all of the structures to describe the actual molecule.

Exercise 9.12   Describe the bonding in NO_3^- using resonance formulas.

Known: Resonance formula is an electron-dot formula.

Solution: Valence electrons = $5 + (3 \times 6) + 1 = 24$; N is the central atom. One resonance formula is

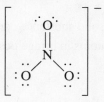

The resonance description is

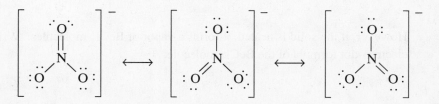

(A dash represents a bonding electron pair; two dashes represent two bonding pairs, or a double bond.)

9.8   EXCEPTIONS TO THE OCTET RULE

Exercise 9.13   Sulfur tetrafluoride, SF_4, is a colorless gas. Write the electron-dot formula of the SF_4 molecule.

Solution: Number of valence electrons is

$$6 \text{ (from S)} + (4 \times 7) \text{ (4F)} = 34$$

Central atom is S. Initial structure is

leaving $34 - 32 = 2$ electrons to place on S. Electron-dot formula is

Exercise 9.14   Beryllium chloride, $BeCl_2$, is a solid substance consisting of long (essentially infinite) chains of atoms with Cl atoms in bridge positions.

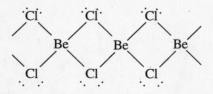

However, if the solid is heated, it forms a vapor of $BeCl_2$ molecules. Write the electron-dot formula of the $BeCl_2$ molecule.

Solution: Number of valence electrons is

$$2 \text{ (from Be)} + (2 \times 7) \text{ (2Cl)} = 16$$

Central atom is Be. Skeleton structure uses all the valence electrons, so the electron-dot formula is

$$:\overset{..}{\text{Cl}} - \text{Be} - \overset{..}{\underset{..}{\text{Cl}}}: \qquad \text{or} \qquad :\overset{..}{\underset{..}{\text{Cl}}} : \text{Be} : \overset{..}{\underset{..}{\text{Cl}}} :$$

9.9 FORMAL CHARGE AND LEWIS FORMULAS

Operational Skill

7. Using formal charges to determine the best Lewis formula. Given two or more Lewis formulas, use formal charges to determine which formula best describes the electron distribution or gives the most plausible molecular structure   (Example 9.11).

Exercise 9.15   Write the Lewis formula that best describes the phosphoric acid molecule, H_3PO_4.

Known: When you assign formal charges, you must know the number of valence electrons on an atom. H_3PO_4 contains H, O, and P atoms, which have 1, 6, and 5 valence electrons, respectively. The number of electrons assigned to each atom will depend upon the possible Lewis formulas.

Solution: Two probable Lewis formulas can be written for H_3PO_4:

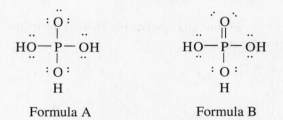

Formula A   Formula B

The formal charges in Formula A are as follows. To the top O, you assign 6 electrons from the lone pairs and 1 from the bond. The formal charge for the top O is $6 - 7 = -1$. To the other oxygens, you assign 4 electrons from the lone pairs and 2 from the bonds. The formal charges on these oxygens are $6 - 6 = 0$. To a H you assign one electron from the bond. The formal charge on each H is $1 - 1 = 0$. To P, you assign 4 electrons from the bonds. Thus, its formal charge is $5 - 4 = 1$.

The formal charges in Formula B are as follows. For the top O, you assign 4
electrons from the lone pairs and 2 from the bonds. This gives a formal charge of
$6 - 6 = 0$. The formal charges for the other oxygens and the hydrogens have the
same values as they did in Formula A. To P you assign 5 electrons from the
bonds. This gives P a formal charge of $5 - 5 = 0$.

Because the atoms in Formula B have the lowest magnitude of formal charges, you
would predict that this is the best description of H_3PO_4. In a resonance description
of H_3PO_4, which involves both formulas, you would predict that Formula B is the
greater contributor.

9.10   BOND LENGTH AND BOND ORDER

Operational Skill

 8.   Relating bond order and bond length.   Know the relationship between bond
order and bond length   (Example 9.12).

> **Exercise 9.16**   Estimate the O — H bond length in H_2O from the covalent radii
> listed in Table 9.4.
>
> *Known:*   Bond length is the sum of the covalent radii.
>
> *Solution:*   Radii (Table 9.4) are
>
> $$\begin{array}{rcl} O & = & 66 \text{ pm} \\ H & = & \underline{37 \text{ pm}} \\ \text{Bond length} & = & 103 \text{ pm} \end{array}$$

> **Exercise 9.17**   Formic acid, isolated in 1670, is the irritant in ant bites. The
> structure of formic acid is
>
> $$\begin{array}{l} H - C = O \\ \quad\;\; | \\ \quad\;\; O \\ \quad\;\; | \\ \quad\;\; H \end{array}$$
>
> One of the carbon–oxygen bonds has a length of 136 pm; the other is 123 pm long.
> What is the length of the $C = O$ bond in formic acid?
>
> *Wanted:*   length of $C = O$ bond.
>
> *Given:*   the two carbon–oxygen bond lengths: 136 pm, 123 pm.

Known: Double bonds are shorter than single bonds.

Solution: The C=O bond is the shorter one, 123 pm.

9.11   BOND ENERGY

Operational Skill

9.   Estimating ΔH from bond energies.   Given a table of bond energies, estimate the heat of reaction   (Example 9.13).

This section introduces you to the estimation of the enthalpy change of a reaction from estimated bond energies. Be sure to understand that this method is only approximate. The other two methods of calculating the enthalpy change for a reaction — using Hess's law to add experimentally determined enthalpy changes, and using enthalpies of formation of reactants and products — are based on exact experimental measurement.

In determining ΔH using bond-energy calculations, you will be breaking all the bonds of the reactants and forming all the bonds of the products. Make sure that you add the bond energies when bonds are broken and subtract the bond energies when bonds are formed. Also, you will find it very helpful to write out the structural formula of each species to determine the bonding. Remember that the energies given are per mole of bonds made or broken.

Exercise 9.18   Use bond energies to estimate the enthalpy change for the combustion of ethylene, C_2H_4, according to the equation

$$C_2H_4(g) + 3O_2(g) \longrightarrow 2CO_2(g) + 2H_2O(g)$$

Known: We must break all bonds in the reactants and form all bonds in the products. For bond breaking, ΔH is +; for bond forming, ΔH is –. Use bond energies from Table 9.5 in the text. The overall equation for ΔH is:

ΔH = Σ bond energies for bonds broken – Σ bond energies for bonds formed

Solution: First write the reaction using the structural formula of each species:

Write the bonds under each species, as follows:

$$1C = C \qquad\qquad 3O = O \qquad 4C = O \qquad 4H - O$$
$$4C - H$$

Then, substitute into the ΔH equation:

$$\Delta H \simeq BE\,(C = C) + 4BE\,(C - H) + 3BE\,(O = O) - 4BE\,(C = O)$$
$$- 4BE\,(H - O)$$
$$\simeq 602\ kJ + 4\,(411\ kJ) + 3\,(494\ kJ) - 4\,(799\ kJ) - 4\,(459\ kJ)$$
$$\Delta H \simeq (602 + 1644 + 1482 - 3196 - 1836)\ kJ = -1304\ kJ$$

Round to two significant figures as an estimate: -1.3×10^3 kJ.

A Molecule That Matters: SULFUR (a Group VIA Nonmetal)

Questions for Study

1. What substances are believed to be responsible for the different colors of the surface of Io? Which substance causes the patches of white color?

2. Describe the structure of the sulfur molecule.

3. Sulfur forms in volcanic gases from SO_2 and H_2S. Give the chemical equation for the reaction.

4. Describe the Claus process for the production of sulfur from natural gas and petroleum.

5. What is the principal end product obtained from sulfur? Give the chemical equations for its production from sulfur.

6. Why is sulfur used in the production of rubber?

7. List formulas for the binary compounds of sulfur with oxygen and with carbon. If any of these compounds has a commercial use, describe it.

Answers to Questions for Study

1. Sulfur and sulfur dioxide spewing from active volcanos are believed to cause the colors on the surface of Io. The patches of white color are believed to be caused by frozen sulfur dioxide, SO_2.

2. The sulfur molecule is a ring of eight sulfur atoms in the shape of a crown.

3. The equation for the formation of sulfur from volcanic gases is

$$16H_2S(g) + 8SO_2(g) \longrightarrow 16H_2O(g) + 3S_8(s)$$

4. In the Claus process, hydrogen sulfide, separated out during the purification of natural gas and petroleum, is burned in air to form sulfur dioxide, which is reacted with additional hydrogen sulfide to give sulfur and water vapor.

5. Sulfuric acid, H_2SO_4, is the principal end product obtained from sulfur. The equations for its production are

$$S_8(s) + 8O_2(g) \longrightarrow 8SO_2(g)$$

$$2SO_2(g) + O_2(g) \xrightarrow[V_2O_5]{\Delta} 2SO_3(g)$$

$$SO_3(g) + H_2O(l) \longrightarrow H_2SO_4(aq)$$

6. Sulfur is used to vulcanize rubber to remove its tackiness and give it greater elasticity.

7. Binary compounds of sulfur with oxygen include sulfur dioxide, SO_2 (used to produce paper pulp from wood and to bleach textile fibers), and sulfur trioxide, SO_3 (used to make sulfuric acid). A binary compound of sulfur with carbon is carbon disulfide, CS_2 (used to manufacture rayon and cellophane).

ADDITIONAL PROBLEMS

1. In which of the following pairs of elements are the members most likely to form an ionic compound? Explain.

    (a)   N and O   (c)   C and F   (e)   P and Br
    (b)   K and Cl   (d)   Ca and O

2. Write the abbreviated electron configuration and Lewis symbol for these ions:

    (a) Ni^{2+} (b) Ba^{2+} (c) I^- (d) Li^+

3. Referring only to the periodic table, arrange the ions of each of the following groups in order of increasing radius. Explain your arrangements.

    (a) Cl^-, K^+, S^{2-} (b) N^{3-}, F^-, O^{2-}

4. Using the electronegativity values given in Figure 9.11 of the text, arrange the following bonds in order of increasing polarity: O — H, N — Cl, P — Cl, Ge — S, and N — H.

5. Which of the following dot formulas is (are) incorrect for the species indicated? For this (these), write the correct formula(s).

    (a) H_2CO_3 : O :   (c) CS_2 : S̈ :: C :: S̈ :

                                 H : O :: C : O : H

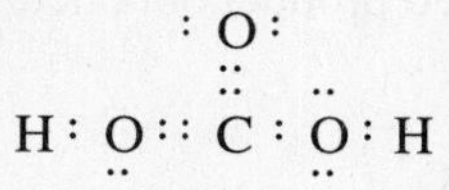

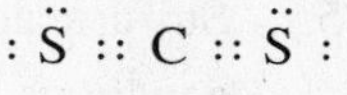

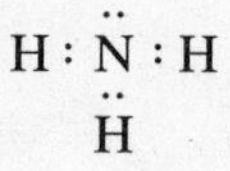

    (b) NH_3 H : N̈ : H   (d)   HOCl   H : Ö : C̈l :

                                 H

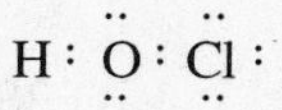

6. Write the electron-dot formula for each of the following:

    (a) H_2Se (b) BrO_3^- (c)   CO   (d) HPO_3^{2-}

7. Use the table of covalent radii in the text (Table 9.4) to estimate the bond length for each of the following. Tell which of the three bonds is the longest and which is the shortest.

    (a)   the C — Cl bond in methyl chloride, CH_3 — Cl.

    (b)   the O — O bond in hydrogen peroxide, H — O — O — H.

    (c)   the N — O bond in hydroxyl amine, NH_2 — OH.

8. Write the dot formula for the nitrite ion, NO_2^-, a molecule with equivalent NO bonds.

9. (a)   Use bond energies from Table 9.5 in the text to estimate the heat of formation of water from its elements in the gas phase.

    (b)   Compare this value with the measured heat of formation of water given in Appendix C of the text, and calculate the percent error in your estimated value.

10. Using the following enthalpy data, calculate $\Delta H°$ for the reaction

$$Na(s) + \frac{1}{2}Cl_2(g) \longrightarrow NaCl(s)$$

First ionization enthalpy for Na is 496 kJ/mol. Electron affinity for Cl is − 349 kJ/mol. Bond energy for Cl_2 is 240 kJ/mol (assume exact). Sublimation enthalpy for Na is 108 kJ/mol. Lattice enthalpy for the formation of $NaCl(s)$ from ions is -786 kJ/mol.

11.   Two likely Lewis formulas for $OXeF_3$, which has Xe as the central atom, differ in that one shows an O — Xe single bond and the other an O = Xe double bond. Which is the more likely structure?

ANSWERS TO ADDITIONAL PROBLEMS

If you missed an answer, study the text section and operational skill given in parentheses after the answer.

1.   b and d. In both cases, the elemental pairs consist of a highly electropositive and a highly electronegative element.   (9.5)

2.   (a)   [Ar] $3d^8$, Ni^{2+} (c)   [Xe], $\left[\; : \overset{..}{\underset{..}{I}} : \; \right]^-$

       (b)   [Xe], Ba^{2+} (d)   [He], Li^+ (9.2, Op. Sk. 2)

3.   (a) $K^+ < Cl^- < S^{2-}$

              In a new period, the column I ion, with larger nuclear charge, will be smaller than anions of the previous period. Sulfide ion and chloride ion are isoelectronic, but the larger nuclear charge in chloride pulls the electrons in more tightly, making it smaller than sulfide.

       (b) $F^- < O^{2-} < N^{3-}$

              These are in the same period and are isoelectronic. As we move from left to right in the period, the effective nuclear charge gets larger and the valence electrons are pulled in more tightly. The sizes thus become progressively smaller.   (9.3, Op. Sk. 3)

4.   The relative bond polarities are N — Cl < Ge — S < P — Cl = N — H < O — H. (9.5, Op. Sk. 4)

5.   (a) The correct formula is $: \overset{..}{O} :$ (9.4, 9.6, Op. Sk. 5)

                                          $H : \overset{..}{\underset{..}{O}} : \overset{::}{C} : \overset{..}{\underset{..}{O}} : H$

6.   (a)   H : S̈e :   (c)   : C ⋮⋮ O :
               Ḧ

     (b) $\left[\; : \overset{..}{\underset{..}{O}} : \overset{..}{\underset{..}{Br}} : \overset{..}{\underset{..}{O}} : \;\right]^{-}$ (d) $\left[\; \overset{H}{\underset{}{: \overset{..}{\underset{..}{O}} : P : \overset{..}{\underset{..}{O}} :}} \;\right]^{2-}$ (9.6, Op. Sk. 5)
                     $: \overset{..}{\underset{..}{O}} :$ $: \overset{..}{\underset{..}{O}} :$

7.   (a)   77 pm + 99 pm = 176 pm
     (b)   66 pm + 66 pm = 132 pm
     (c)   70 pm + 66 pm = 136 pm

     The C — Cl bond is the longest, the O — O bond the shortest.
                                                                     (9.10, Op. Sk. 8)

8.   Two resonance formulas can be written.

     $$\left[\; \overset{..}{\underset{}{N}} \; \right]^{-} \longleftrightarrow \left[\; \overset{..}{\underset{}{N}} \;\right]^{-}$$
     $$: \overset{}{O} \diagup \diagdown O :$$

                                                                     (9.7, Op. Sk. 6)

9.   (a)   The reaction is

           $$2H_2(g) + O_2(g) \longrightarrow 2H_2O(g)$$

           $\Delta H \simeq 2 \; \text{mol} \; (432 \; kJ/ \text{mol}) + 1 \; \text{mol} \; (494 \; kJ/ \text{mol})$
                                                   $- 4 \; \text{mol} \; (459 \; kJ/ \text{mol})$

           $\simeq -478 \; kJ$ (9.11, Op. Sk. 9)

     (b)   Dividing ΔH for the reaction by 2 gives – 239 kJ as the estimated enthalpy of
           formation of one mole of water. The measured value from the appendix is
           – 241.826 kJ/mol.

           $$\text{Percent error} = \frac{|242 - 239|}{242} \times 100 = 1\%$$

10. – 411 kJ   (9.2, 9.11)

11. The structure with the O = Xe double bond. Each atom has a formal charge of 0.
                                                                     (9.9, Op. Sk. 7)

CHAPTER POST-TEST

1. Write Lewis formulas for the following:

    (a) NH_4^+ (c) PCl_5 (e) H_2CO_3 (g) CH_4O

    (b) H_3O^+ (d) XeF_4 (f) N_2H_4

2. Which of the following ions are isoelectronic?

    $$Na^+, S^{2-}, Mg^{2+}, O^{2-}, Li^+, F^-$$

3. Briefly describe how electronegativity values may be used to predict the relative polarity of a chemical bond between two atoms.

4. Using only a periodic table, arrange the following elements in order of increasing electronegativity: P, Na, Cs, O, Al, Mg.

5. The NO bond in NO_2^- should be (shorter/longer) than the NO bond in H_2NOH, because _____

    _____ .

6. Write the abbreviated electron configuration for each of the following:

    (a) Ti^{3+} (b) Br^- (c) Sn^{2+}

7. List the following in order of increasing radii: $O^{2-}, N^{3-}, Mg^{2+}, Ne, Na^+$.

8. The molecule S_2Cl_2 contains one S — S bond and two S — Cl bonds. The average bond energy for S — Cl bonds is 255.2 kJ/mol. From the following additional thermodynamic data, estimate the bond energy for the S — S bond in S_2Cl_2:

    $$BE \ (Cl - Cl) = 242.1 \ kJ/mol$$

    $$S(s) \longrightarrow S(g); \qquad \Delta H° = 276.1 \ kJ$$

    $$2S(s) + Cl_2(g) \longrightarrow S_2Cl_2(g); \qquad \Delta H° = 22.93 \ kJ$$

9. Draw the resonance structures for OCN^-.

10. Calculate the enthalpy associated with the deprotonation of NH_4^+,

$$NH_4^+(g) \longrightarrow H^+(g) + NH_3(g)$$

from the following thermodynamic data:

$$BE\ (N_2) = 944.7 \qquad BE\ (Cl_2) = 242.1 \qquad BE\ (H_2) = 435.9\ (kJ/mol)$$

$$\Delta H_f^\circ(NH_4Cl) = -314.2 \qquad \Delta H_f^\circ(NH_3) = -45.6\ (kJ/mol)$$

and (in kJ):

$$Cl(g) + 1e^- \longrightarrow Cl^-(g); \qquad\qquad \Delta H = -387.0$$

$$H(g) \longrightarrow H^+(g) + 1e^-; \qquad\qquad \Delta H = 1305.5$$

$$NH_4Cl(s) \longrightarrow NH_4^+(g) + Cl^-(g); \qquad \Delta H = 640.2$$

11. An inorganic chemistry student drew the following four unique resonance structures for SO_3Cl^-. On her examination, she selected structure C as the most likely one. Do you agree? State your reasons for your answer.

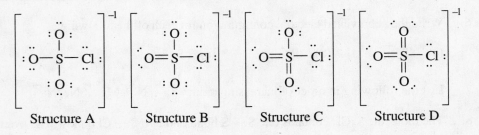

Structure A   Structure B   Structure C   Structure D

ANSWERS TO CHAPTER POST-TEST

If you missed an answer, study the text section and operational skill given in parentheses after the answer.

1. (a), (b), (c), (d), (e), (f), (g)

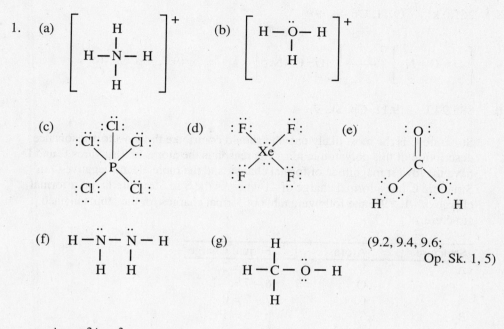

(9.2, 9.4, 9.6; Op. Sk. 1, 5)

2. Na^+, Mg^{2+}, O^{2-}, F^- (9.3)

3. Electronegativity values reflect the relative degree of attraction for electrons by atoms in a chemical bond. In a given chemical bond, the atom with the higher electronegativity value will have a greater attraction for electrons than the other atom. Therefore, the electron density will be greater in the direction of the atom with the higher electronegativity value, giving a polar chemical bond.   (9.5, Op. Sk. 4)

4. $Cs < Na < Mg < Al < P < O$ (9.5, Op. Sk. 4)

5. The NO bond in NO_2^- should be shorter because it has multiple-bond character in NO_2^- but single-bond character in H_2NOH.   (9.10, Op. Sk. 8)

6. (a) [Ar] $3d^1$ (b) [Kr]   (c) [Kr] $4d^{10}5s^2$ (9.2, Op. Sk. 2)

7. Because all have the same number of electrons, the smallest would be the one with the most protons. The order is thus

$$Mg^{2+} < Na^+ < Ne < O^{2-} < N^{3-} \quad \text{(9.3, Op. Sk. 3)}$$

8. 261.0 kJ   (9.11, Op. Sk. 9)

9. $$\left[\ddot{O} = C = \ddot{N} \right]^- \longleftrightarrow \left[:\ddot{O}-C\equiv N: \right]^- \qquad \text{(9.7, Op. Sk. 6)}$$

10. 885.9 kJ   (9.11, Op. Sk. 9)

11. Structure C is the most likely one and should contribute the most to a resonance description of this polyatomic ion. The reason is the atoms in Structures C and D have the lowest magnitude of formal charge, and the more electronegative O in Structure C has a formal charge of –1 whereas the S in Structure D has a formal charge of –1. (See the following table of formal charges for the atoms in each structure.)

Structure	Atoms	Formal Charge
A	S	$6 - 4 = 2$
	O	$6 - 7 = -1$
	Cl	$7 - 7 = 0$
B	S	$6 - 5 = 1$
	O=	$6 - 6 = 0$
	O	$6 - 7 = -1$
	Cl	$7 - 7 = 0$
C	S	$6 - 6 = 0$
	O=	$6 - 6 = 0$
	O	$6 - 7 = -1$
	Cl	$7 - 7 = 0$
D	S	$6 - 7 = -1$
	O=	$6 - 6 = 0$
	Cl	$7 - 7 = 0$

(9.9, Op. Sk. 7)

CHAPTER 10   MOLECULAR GEOMETRY AND CHEMICAL BONDING THEORY

CHAPTER TERMS AND DEFINITIONS

Numbers in parentheses after definitions give the text sections in which the terms are explained. Starred terms are italicized in the text. Where a term does not fall directly under a text section heading, additional information is given for you to locate it.

cis*   double-bonded compound with distinctive groups on the same side of the double bond   (chapter introduction)

trans*   double-bonded compound with distinctive groups on opposite sides of the double bond   (chapter introduction)

molecular geometry   general shape of a molecule, as determined by the relative positions of the atomic nuclei   (10.1, introductory section)

valence-shell electron-pair repulsion (VSEPR) model   predicts the shapes of molecules and ions in which valence-shell electron pairs are arranged about each atom so that electron pairs are kept as far away from one another as possible, thus minimizing electron-pair repulsions   (10.1)

linear*   two points at an angle of 180° to one another   (10.1)

trigonal planar*   three points directed at 120° angles to one another   (10.1)

tetrahedral*   four points directed at approximately 109.5° angles to one another   (10.1)

bent (angular)*   two points directed to one another at an angle of less than 180°   (10.1)

trigonal pyramidal*   four points describing a pyramid with a triangular base   (10.1)

trigonal bipyramidal arrangement*   five points at the apexes of the six-sided figure formed by placing the face of one tetrahedron onto the face of another   (10.1)

octahedral arrangement*   six points directed to the apexes of a figure with eight triangular faces and six vertexes   (10.1)

axial directions*   directions 180° apart forming an axis through the central atom, in a trigonal bipyramidal arrangement   (10.1)

equatorial directions*   directions 120° apart pointing toward the corners of the equilateral triangle lying on the plane through the central atom and perpendicular to the axial directions, in a trigonal bipyramidal arrangement   (10.1)

seesaw (distorted tetrahedral) geometry*   four atoms arranged about a fifth atom in a tetrahedral arrangement but with three different bond angles (10.1)

T-shaped geometry*   four atoms lying in one plane with one atom at the intersection of the "T"   (10.1)

square pyramidal geometry*   five atoms arranged about a sixth atom so that all bonding angles are 90°   (10.1)

square planar geometry*   four atoms arranged about a fifth atom so that all atoms lie in a plane and bonding angles are 90°   (10.1)

dipole moment   quantitative measure of the degree of charge separation in a molecule   (10.2)

debyes (D)*   units of dipole moment; $1 \text{ D} = 3.34 \times 10^{-30} \text{ C} \cdot \text{m}$ (10.2)

capacitance*   capacity of charged plates to hold a charge   (10.2)

vector*   quantity having both magnitude and direction   (10.2)

nonpolar*   molecules in which bonds are directed symmetrically about the central atom and, therefore, have a zero dipole moment   (10.2)

polar*   molecules in which bonds are not symmetrical about the central atom and, therefore, have a nonzero dipole moment   (10.2)

valence bond theory   approximate theory to explain the electron pair or covalent bond by quantum mechanics   (10.3)

overlap*   occupation of the same region of space by two orbitals   (10.3)

promoted*   said of an electron that is raised to a higher orbital level   (10.3)

hybrid orbitals   orbitals used to describe the bonding in some atoms; obtained by taking combinations of atomic orbitals of the isolated atoms   (10.3)

sp^3*   set of four hybrid orbitals constructed from one s orbital and three p orbitals   (10.3)

sp^2*   set of three hybrid orbitals constructed from one s orbital and two p orbitals   (10.3)

sp*   set of two hybrid orbitals constructed from one s orbital and one p orbital   (10.3)

σ **(sigma) bond**   chemical bond with cylindrical shape about the bond axis; formed when two s orbitals overlap or by overlap along the axis of an orbital with directional character along the bond axis   (10.4)

π **(pi) bond**   chemical bond with electron distribution above and below the bond axis; formed by sideways overlap of two parallel p orbitals   (10.4)

geometric (*cis–trans*) isomers*   compounds with the same chemical formula and the same bonding but with different arrangements as about a double bond   (10.4)

molecular orbital theory   explanation of the electron structure of molecules in terms of molecular orbitals, which may spread over several atoms or the entire molecule   (10.5, introductory section)

bonding orbitals   molecular orbitals that are concentrated in regions between nuclei   (10.5)

antibonding orbitals   molecular orbitals that are concentrated in regions other than between nuclei   (10.5)

bond order* in molecular orbital theory, for a diatomic molecule, one-half the difference between the number of electrons in bonding orbitals, n_b, and the number in antibonding orbitals, n_a:

$$\frac{1}{2}(n_b - n_a) \qquad (10.5)$$

homonuclear diatomic molecules molecules composed of two like nuclei (10.6)
heteronuclear diatomic molecules molecules composed of two different nuclei (10.6)
nonbonding* describes orbitals, atomic or molecular, which are neither bonding nor antibonding (10.7)
ozone common name for trioxygen, O_3 (A Gas That Matters: Ozone [An Absorber of Ultraviolet Radiation in the Stratosphere])
ozein* Greek word meaning "to smell" (A Gas That Matters: Ozone [An Absorber of Ultraviolet Radiation in the Stratosphere])
photochemical smog* blend of smoke and fog produced when sunlight interacts with a mixture of nitrogen oxides and hydrocarbons in polluted air (A Gas That Matters: Ozone [An Absorber of Ultraviolet Radiation in the Stratosphere])
stratosphere* region of the atmosphere beginning at about 15 km (9 miles) above the surface of the earth (A Gas That Matters: Ozone [An Absorber of Ultraviolet Radiation in the Stratosphere])
troposphere* lower portion of the atmosphere near the surface of the earth (A Gas That Matters: Ozone [An Absorber of Ultraviolet Radiation in the Stratosphere])

CHAPTER DIAGNOSTIC TEST

1. Match each of the following molecules with its corresponding geometry:

    _____ (1) $SeCl_2$ (a) square planar

    _____ (2) CCl_4 (b) angular

    _____ (3) $(CH_3)_3$ As   (c) linear (possesses zero dipole moment)

    _____ (4) XeF_4 (d) tetrahedral (possesses zero dipole moment)

    _____ (5) CO_2 (e) pyramidal

2. Determine whether each of the following statements is true or false. If a statement is false, change it so it is true.

    (a) In valence bond theory, chemical bonding is the result of the overlap of outer-shell atomic orbitals. True/False: _____

    _____

    _____ .

(b)   In sp^2 hybridization, the third p orbital disappears. True/False: _____

_____

_____ .

(c)   When the twisting motion of atoms about a bond is restricted, geometrical
(*cis–trans*) isomers become possible for a molecule. True/False: _____

_____

_____ .

(d)   The C_2 molecule should be stable and diamagnetic, whereas the Mg_2 molecule
should be unstable. True/False: _____

_____

_____ .

3.   In molecular orbital theory, each molecular orbital

(a)   can accommodate a maximum of two electrons.
(b)   can accommodate the number of electrons in the valence shell of the atoms.
(c)   is formed from the atomic orbitals on the less electronegative atom.
(d)   is concentrated in the region of a specific chemical bond.
(e)   defines the spectroscopic characteristics of the atoms forming a chemical bond.

4.   In the compound C_2H_2, the expected value for the H — C — C bond angle is

(a)   90°   (b)   120°   (c)   109.5°   (d)   180°

(e)   none of the above, since the molecule should not exist.

5.   Draw and name the molecular geometry for each of the following species:

(a) ICl_2^- (c) SbF_5^{2-} (e) NH_2^-
(b) GeF_4 (d) SF_6 (f) BCl_3

6.   Write the structural formulas showing the geometric isomers possible for the compound with the condensed formula $CH_3CH = CHCH_2CH_3$. (The carbons are
joined to each other, and there is a double bond between the second and third carbons.) Also, name the prefix for each isomer.

7.   Answer the three questions below given the following information: The electron
configuration of sulfur is [Ne] $3s^2 3p^4$.

(a)   What type of hybridization or bonding exists in the sulfur hexafluoride
molecule?

(b) What type of hybridization or bonding exists in the sulfur tetrafluoride molecule?

(c) What type of hybridization or bonding would be expected in a SF_2 molecule?

8. Describe in your own words just what orbital hybridization is.

ANSWERS TO CHAPTER DIAGNOSTIC TEST

If you missed an answer, study the text section and operational skill given in parentheses after the answer.

1. (1) b,   (2) d,   (3) e,   (4) a,   (5) c   (10.1, 10.2; Op. Sk. 1, 2)

2. (a) True.   (10.3)
    (b) False. The third *p* orbital remains in position and can contain electrons. It is also available for bonding with other nonhybridized *p* orbitals.   (10.4)
    (c) True.   (10.4)
    (d) True.   (10.5, 10.6, Op. Sk. 4)

3. a   (10.5)   4. d   (10.3, 10.4, Op. Sk. 3)

5. (a)

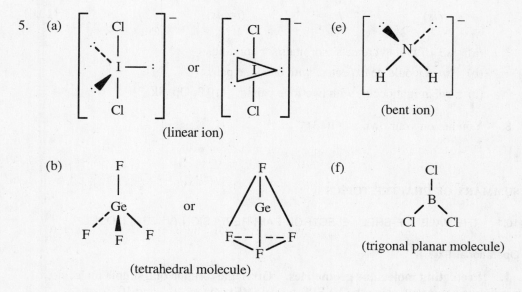

(linear ion)

(bent ion)

(tetrahedral molecule)

(trigonal planar molecule)

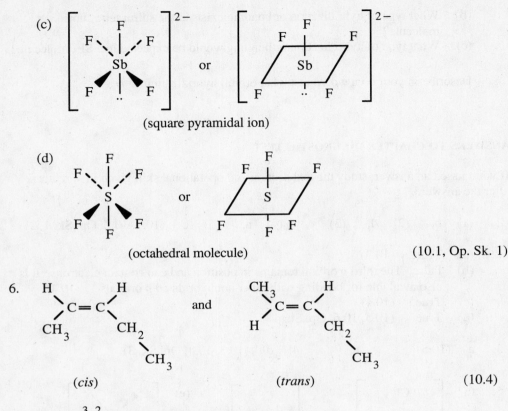

(square pyramidal ion)

(octahedral molecule)   (10.1, Op. Sk. 1)

6. (*cis*)   and   (*trans*)   (10.4)

7. (a) sp^3d^2 would create six orbitals for bonding.

   (b) sp^3d would be expected, with one lone pair.

   (c) sp^3 might occur, with two lone pairs.   (10.3, Op. Sk. 3)

8. You are on your own.   (10.3)

SUMMARY OF CHAPTER TOPICS

10.1   THE VALENCE-SHELL ELECTRON-PAIR REPULSION (VSEPR) MODEL

Operational Skill

   1.   Predicting molecular geometries.   Given the formula of a simple molecule, predict its geometry, using the VSEPR model   (Examples 10.1 and 10.2).

Molecular geometry and directional bonding are crucial topics because these phenomena are major factors in why molecules react as they do. The VSEPR model gives us a very good way to approach these topics. You must remember, however, that it is just a model. Whether a predicted geometry agrees with experiment depends upon many factors. The excellent discussion of molecular-geometry models by R. S. Drago ["A Criticism of the Valence-Shell Electron-Pair Repulsion Model as a Teaching Device," *J. Chem. Educ.*, 50 (1973), p. 244] points out the utility and limitations of models. A review of this article is very appropriate for Chapter 10.

To be able to predict molecular geometry, you must first write the Lewis structure to determine where the valence electrons are. You must then determine the electron arrangement (geometry). To do this you must memorize the correlations between number of electron pairs (or groups) around the central atom and the geometry (2 pairs indicate linear; 3 pairs, trigonal planar; etc.). You will find this information in text Figure 10.3. Now you are ready to place atoms with the electron pairs (or groups) and to determine the atomic arrangement (molecular geometry), shown in text Figure 10.4. Note that a tetrahedral electron arrangement can lead to three different molecular geometries, depending on the number of lone pairs.

You should learn to draw the molecular geometries. There are two basic ways. One is to use different types of lines (simple, wedged, and dotted) for bonds, to indicate three-dimensional structure. In this method, the wedged line, ◢, indicates that the atom is coming out of the plane of the paper. A dotted line means the atom is going behind the plane of the paper. A simple straight line indicates the atom is in the plane of the paper.

The other method is to draw an outline of the geometric figure representing the electron geometry, put the central atom in the center of the figure, and place the outer atoms or electron pairs at the vertexes of the figure. Common figures are

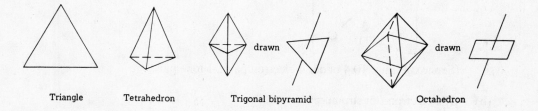

Triangle   Tetrahedron   Trigonal bipyramid   drawn   Octahedron   drawn

Both methods of drawing molecular geometries are shown in the solutions to Chapter 10 exercises. Always place the lone pairs on the central atom in the drawing.

You should also memorize the angles (in degrees) that go with the particular electron geometries. Linear is 180°; trigonal planar, 120°; tetrahedral, 109.5°; trigonal bipyramidal, 120° and 90°; and octahedral, 90°. This, then, gives you the expected bond angles.

If you have difficulty visualizing molecular geometry in three dimensions, the stereoscopic plots of molecules found in L. H. Hall's book should be very helpful (*Group Theory and Symmetry in Chemistry* [New York: McGraw-Hill Book Co., 1969], pp. 15–31). The textual material in this book is advanced, so you may ignore it. The stereographic drawings, however, are well worth your effort in trying to locate a copy of this text in your library.

Exercise 10.1   Use the VSEPR method to predict the geometry of the following ion and molecules: (a) ClO_3^-;   (b) OF_2;   (c) SiF_4.

Solution:

(a) ClO_3^- electron-dot structure:

$$\left[\begin{array}{c} \ddot{\underset{\cdot\cdot}{O}} : \ddot{Cl} : \ddot{\underset{\cdot\cdot}{O}} : \\ : \ddot{\underset{\cdot\cdot}{O}} : \end{array} \right]^-$$

Electron arrangement is tetrahedral (text Figure 10.3). Molecular geometry is trigonal pyramidal:

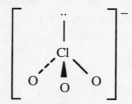

(See text Figure 10.4 under 4 electron pairs, 1 lone pair.)

(b) OF_2 electron-dot structure:

$$\begin{array}{c} : \ddot{\underset{\cdot\cdot}{O}} : \ddot{\underset{\cdot\cdot}{F}} : \\ : \ddot{\underset{\cdot\cdot}{F}} : \end{array}$$

Electron arrangement is tetrahedral. Molecular geometry is bent:

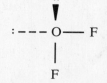

(See text Figure 10.4 under 4 electron pairs, 2 lone pairs.)

(c) SiF_4 electron-dot structure:

$$
\begin{array}{c}
:\ddot{F}: \\
:\ddot{F}:Si:\ddot{F}: \\
:\ddot{F}:
\end{array}
$$

Electron arrangement is tetrahedral. Molecular geometry is tetrahedral:

(See text Figure 10.4 under 4 electron pairs, 0 lone pairs.)

Exercise 10.2   According to the VSEPR model, what molecular geometry would you predict for iodine trichloride, ICl_3?

Solution: Electron-dot structure is as follows:

$$
\begin{array}{c}
:\ddot{Cl}: \\
:\ddot{Cl}:\overset{..}{I}\overset{..}{} \\
:\ddot{Cl}:
\end{array}
$$

Electron arrangement is trigonal bipyramidal. Molecular geometry is T-shaped:

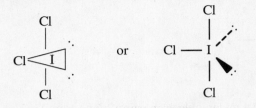

(See text Figure 10.9 under 5 electron pairs, 2 lone pairs.)

10.2   DIPOLE MOMENT AND MOLECULAR GEOMETRY

Operational Skill

2.   Relating dipole moment and molecular geometry.   State what geometries of a molecule AX_n are consistent with the information that the molecule has a nonzero dipole moment   (Example 10.3).

Exercise 10.3   Bromine trifluoride, BrF_3, has a nonzero dipole moment. Indicate which of the following geometries are consistent with this information:
(a)   trigonal planar;   (b)   trigonal pyramidal;   (c)   T-shaped.

Known: Dipole moment is nonzero in molecules where there is separation of the centers of + and – charge. This happens when bond polarities do not cancel, or where lone electron pairs exist and their effect is not cancelled by bond polarity in the opposite direction.

Solution: In a trigonal planar arrangement, (a), bond polarities cancel and the dipole moment would be zero. In (b) and (c), bond polarities would not cancel and, depending on the bond polarities in relation to the lone pair or pairs, the molecule could have a dipole moment. Therefore, both (b) and (c) are consistent molecular geometries to give a dipole moment. (Note that the electron geometry of BrF_3 is trigonal bipyramidal; therefore, only (c) is a possibility for the molecular geometry of BrF_3.)

Exercise 10.4   Which of the following would be expected to have a dipole moment of zero on the basis of symmetry? Explain.

(a) $SOCl_2$ (b) SiF_4 (c) OF_2

Known: the relationship between molecular geometries in an AX_n species and dipole moment (from the text); bond polarities (Section 9.5).

Solution:

(a)   The geometry of $SOCl_2$ is pyramidal:

On the basis of symmetry, this molecule would be expected to have a nonzero dipole moment.

(b)   The geometry of SiF_4 is tetrahedral:

This geometry gives a dipole moment of zero as long as bonds are of equal polarity, as they are. Therefore, this molecule is expected to have a dipole moment of zero.

(c)   The geometry of OF_2 is bent:

This geometry indicates a nonzero dipole moment because bond polarities do not cancel. However, one must consider bond polarities versus the lone pairs on oxygen and thus must look to laboratory measurements. One would predict a nonzero dipole moment for this molecule, believing the effect of the lone pairs to be greater than the opposing electronegativity of the fluorine atoms.

10.3   VALENCE BOND THEORY

Operational Skill

3.   Applying valence bond theory.   Given the formula of a simple molecule, describe its bonding, using valence bond theory   (Examples 10.4, 10.5, and 10.6).

The essence of valence bond theory is that electrons in atoms making up molecules remain in orbitals on the individual atoms, that only the valence electrons in a particular atomic orbital are involved in bonding, and that the bonding electrons spend most of their time between the two nuclei, in the space of the orbital overlap.

Many students have difficulty comprehending orbital hybridization. It will be helpful to remember that an orbital represents the volume of space in which there is a high probability of finding an electron. In some molecules, however, spectroscopic studies show that the electrons are not where we would expect them. In short, they are in different volumes of space. We call these new regions hybrid orbitals because we describe them by mathematically combining the valence orbitals of an atom to give new orbitals. (Recall hybrid plants from your biology class.)

Note that, according to theory, all valence orbitals do not have to be involved in the hybridization process. When one s and two p orbitals are hybridized to form three sp^2 hybrid orbitals, one valence p orbital remains unhybridized. This volume of space will remain perpendicular to the plane of the three sp^2 orbitals in their trigonal planar arrangement. (See Figure A.) When one s and one p orbital combine to form two sp hybrid orbitals, two valence p orbitals are left unhybridized. They are at right angles to each other and to the plane of the linear sp hybrid orbitals. (See Figure B.)

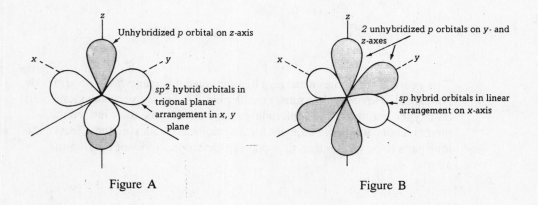

Figure A   Figure B

A few words need to be said about electron promotion as part of hybridization. As you will note from the text discussion of the bonding in carbon, boron, oxygen, and xenon, promotion is used in the case of carbon, boron, and xenon, but not in oxygen. The concept of promotion is used when both filled and unoccupied orbitals are mathematically combined. In the case of oxygen, all orbitals involved in the hybridization have electrons in them, and promotion is not possible.

The following steps summarize how to describe bonding in a given molecule according to the valence bond theory. Do enough exercises and problems so you can perform the steps in order without looking at the list.

1. Write the Lewis formula for the molecule.
2. Determine the electron arrangement about the central atom (text Figure 10.3).
3. Determine the hybrid orbitals used. Refer to text Table 10.2. It is useful to memorize the orbitals that go along with each geometry so you will not be dependent on the table at test time.
4. Draw the orbital diagram of the hybridized central atom. It is important to do this when you begin to work these problems. Later you will probably be able to omit this step on paper, although you must certainly do it in your head.
5. Describe the bonding as overlap of appropriate valence orbitals.
6. Write the orbital diagram of the bonded central atom and place valence electrons therein. Indicate the extra ones, from the bonded atoms, with dotted arrows.

Exercise 10.5 Using hybrid orbitals, describe the bonding in NH_3 according to valence bond theory.

Known: the steps to solving the problem (listed in the previous paragraph).

Solution: The Lewis formula is

$$H : \overset{\displaystyle ..}{\underset{\displaystyle ..}{N}} : H$$
$$H$$

Since there are four electron pairs, the electron arrangement is tetrahedral; thus, four sp^3 hybrid orbitals are used. The orbital diagram of the hybridized nitrogen atom is

[He] (↑↓) (↑) (↑) (↑)

sp^3

Each $N - H$ bond is formed by overlap of the $1s$ orbital of a hydrogen atom with one singly occupied sp^3 hybrid orbital of the nitrogen atom. The expected bond angle would be 109.5°. The lone pair of electrons occupies the remaining sp^3 hybrid orbital. The orbital diagram of the bonded nitrogen atom is

lone
pair   N — H bonds

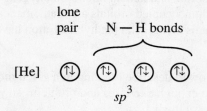

sp^3

Exercise 10.6   Describe the bonding in PCl_5 using hybrid orbitals.

Solution: The Lewis formula is

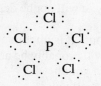

Since there are five electron pairs, the electron arrangement is trigonal bipyramidal; thus, five sp^3d hybrid orbitals are used. The orbital diagram of the hybridized phosphorus atom is

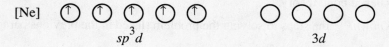

sp^3d $3d$

Each P — Cl bond is formed by overlap of the singly filled $3p$ orbital of a chlorine atom with one singly occupied sp^3d hybrid orbital of the phosphorus atom. The orbital diagram of the bonded phosphorus atom is

P — Cl bonds

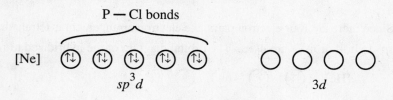

sp^3d $3d$

10.4   DESCRIPTION OF MULTIPLE BONDING

Exercise 10.7   Describe the bonding on the carbon atom in carbon dioxide, CO_2, using valence bond theory.

Solution: The electron-dot formula of CO_2 is

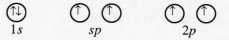

Since there are two groups of electrons, the electron arrangement is linear, and two *sp* hybrid orbitals are used. The orbital diagram of the hybridized carbon is

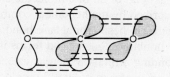

The σ (sigma) bond of each double bond is formed from a singly occupied *sp* hybrid orbital on carbon. The unhybridized *2p* orbitals form two π (pi) bonds in planes perpendicular to one another:

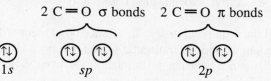

The orbital diagram of the bonded carbon atom is

2 C=O σ bonds   2 C=O π bonds

Exercise 10.8   Dinitrogen difluoride (see Example 10.6) exists as *cis* and *trans* isomers. Write structural formulas for these isomers and explain (in terms of the valence bond theory of the double bond) why they exist.

Known: the electron-dot structure is given in Example 10.6; definition of structural formula (Section 2.6). Double bonds are rigid, nonrotating bonds. The electron arrangement about a double-bonded atom is trigonal planar.

Solution:

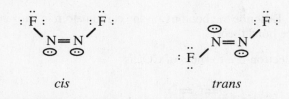

     cis   *trans*

These molecules exist because it is difficult to break the π bond to interconvert them.

10.5   PRINCIPLES OF MOLECULAR ORBITAL THEORY

As its name implies, the molecular orbital theory of the electronic structure of molecules, and thus of their bonding, holds that when bonding occurs, the electrons from each of the atoms occupy completely different volumes of space than they do on the free atoms. These regions encompass part or all of a molecule and are called molecular orbitals. These regions are described by mathematical combinations of like atomic orbitals from the individual atoms. Some of these orbitals are bonding orbitals, others are antibonding, and a few can be nonbonding. A bond forms if there are more electrons in bonding orbitals than in antibonding orbitals. Whereas valence bond theory and the accompanying VSEPR model are useful in predicting molecular geometries, molecular orbital theory is useful in explaining observed magnetic properties of molecules and explaining molecules where electron delocalization comes into play.

Here we again find the term *bond order*. In valence bond theory, this is defined as the number of electron pairs in a bond. In molecular orbital theory, it is defined mathematically as one-half the difference between the number of electrons in bonding and anti-bonding orbitals, or $\frac{1}{2}(n_b - n_a)$. Sometimes this is the same number as the number of electron pairs in a bond, and sometimes it isn't. This molecular orbital definition is more elegant and more useful than valence bond theory in describing the nature of bonding.

10.6   ELECTRON CONFIGURATIONS OF DIATOMIC MOLECULES
   OF THE SECOND-PERIOD ELEMENTS

Operational Skill

4.   Describing molecular orbital configurations.   Given the formula of a diatomic molecule obtained from first- or second-period elements, deduce the molecular orbital configuration, the bond order, and whether the molecular substance is diamagnetic or para-magnetic   (Examples 10.7 and 10.8).

We discuss diatomic molecules, as they are the simplest molecules and thus the easiest to work with. The following steps summarize how to describe bonding in a given molecule according to molecular orbital theory:

1. Count the number of valence electrons on both atoms.
2. Write the molecular orbital diagram.
3. Write the electron configuration of the molecule.
4. Calculate the bond order.

Exercise 10.9 The C_2 molecule exists in the vapor phase over carbon at high temperature. Describe the molecular orbital structure of this molecule; that is, give the orbital diagram and electron configuration. Would you expect the molecular substance to be diamagnetic or paramagnetic? What is the bond order for C_2?

Wanted: molecular orbital description of bonding in C_2; magnetic character; bond order.

Given: C_2.

Known: steps to follow in molecular orbital description (given in previous paragraph); molecular orbital filling order (Section 10.5); definitions of paramagnetism and diamagnetism.

Solution: The number of valence electrons is $2 \times 4 = 8$. The molecular orbital diagram is

$$\sigma_{2s} \qquad \sigma_{2s}^* \qquad \pi_{2p}$$

The electron configuration is

$$KK \, (\sigma_{2s})^2 \, (\sigma_{2s}^*)^2 \, (\pi_{2p})^4$$

The molecule should be diamagnetic, since all electrons are paired. There are six bonding electrons and two antibonding electrons. Therefore, bond order = $\frac{1}{2}(6-2) = 2$.

Exercise 10.10 Give the orbital diagram and electron configuration for the carbon monoxide molecule, CO. What is the bond order of CO? Is the molecule diamagnetic or paramagnetic?

Solution: The number of valence electrons is $4 + 6 = 10$. The orbital diagram is

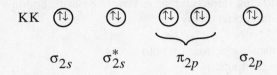

$$\text{KK}\,(\sigma_{2s})^2\,(\sigma_{2s}^*)^2\,(\pi_{2p})^4\,(\sigma_{2p})^2$$

The electron configuration is

$$\text{KK}\,(\sigma_{2s})^2\,(\sigma_{2s}^*)^2\,(\pi_{2p})^4\,(\sigma_{2p})^2$$

The molecule is diamagnetic, since all electrons are paired. There are eight bonding electrons and two antibonding electrons; thus, bond order $= \frac{1}{2}(8 - 2) = 3$.

10.7   MOLECULAR ORBITALS AND DELOCALIZED BONDING

A Gas That Matters: OZONE (an Absorber of Ultraviolet Radiation in the Stratosphere)

Questions for Study

1.   List some physical properties of ozone.

2.   What are the commercial uses of ozone?

3.   Describe the bonding in ozone. What is the geometry?

4.   Describe the formation of ozone in the stratosphere.

5.   Why is the presence of ozone of vital importance to us?

6.   Biologists believe that life appeared on land no earlier than 600 million years ago. Why do they believe that land plants and other organisms could not have appeared before this?

7.   What are the reactions responsible for the catalysis of ozone to dioxygen?

8.   In 1987, airplanes flew through the Antarctic stratosphere measuring the quantity of ClO. What was the result of these measurements and what is their significance?

Answers to Questions for Study

1.   Ozone is a faintly blue gas that has a slightly disagreeable odor.

2.   Ozone is used as a bleaching agent and a water disinfectant.

3. The electronic structure of ozone can be described by two resonance formulas:

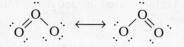

The O — O bond distances in the molecule are consistent with partial double-bond character. From the VSEPR model, we predict angular or V-shaped geometry with a bond angle of near 120°, which is consistent with sp^2 hybridization. The actual O — O — O bond angle is 117°.

4. In the stratosphere, O_3 is formed by the reaction of O_2 and oxygen atoms, which result from the cleavage of the O_2 molecule when it absorbs ultraviolet radiation of less than 200 nm. The equation that describes the formation of O_3 is

$$O_2(g) + O(g) \longrightarrow O_3(g)$$

5. O_3 plays a very important role in protecting biological organisms from the lower-energy ultraviolet radiation from the sun that reaches the earth's atmosphere. This longer-wavelength ultraviolet radiation is absorbed by O_3 in the stratosphere.

6. Biologists believe that prior to 600 million years ago there was insufficient ozone in the atmosphere to absorb the harmful ultraviolet radiation, which prevented the development of living organisms.

7. The following equations describe the catalytic decomposition of O_3 to O_2 by chlorine atoms:

$$Cl(g) + O_3(g) \longrightarrow ClO(g) + O_2(g)$$

$$ClO(g) + O(g) \longrightarrow Cl(g) + O_2(g)$$

8. The measurements indicated that wherever depletion of ozone occurs, ClO appears. This suggests that chlorine atoms, arising from chlorofluorocarbons that have absorbed ultraviolet radiation and then decomposed in the stratosphere to chlorine atoms, are catalyzing the ozone depletion.

ADDITIONAL PROBLEMS

1. Give the molecular geometry and hybrid orbitals used in each of the following species.

    (a) BF_3 (c) SeF_4 (e) NO_2^-

    (b) XeF_2 (d) SF_6 (f) BrO_4^-

2. Indicate which of the following species have polar bonds but a dipole moment of zero.

    (a) NH_3 (c) H_2O (e) BCl_3

    (b) PF_5 (d) N_2 (f) CO_2

3. Each of the following molecules has a nonzero dipole moment. Select the molecular geometry that is consistent with this fact. Explain your choices.

    (a) ClO_2 linear or bent?

    (b) BrF_3 trigonal pyramidal or T-shaped?

    (c) SF_4 tetrahedral, seesaw, or square planar?

4. Use valence bond theory to explain the bonding about boron in boric acid, $B(OH)_3$, a crystalline substance used as an antiseptic and in many other ways.

5. Indicate the approximate $\angle BAB$ bond angle in each of the following AB_x species.

    (a) SiH_4 (c) $BeCl_2$ (e) BrF_4^-

    (b) NCl_3 (d) PF_5 (f) $SnCl_2$

6. Use molecular orbital theory to explain the bonding and magnetic character of molecular nitrogen, N_2.

7. C_2 is diamagnetic with a bond order of 2, but O_2 is paramagnetic with a bond order of 2. Using molecular orbital theory, account for these facts.

8. Determine the bond order of O_2, O_2^-, O_2^+, and O_2^{2-}, and arrange these species in order of increasing bond length.

ANSWERS TO ADDITIONAL PROBLEMS

If you missed an answer, study the text section and operational skill given in parentheses after the answer.

1. (a)   trigonal planar, sp^2

   (b)   linear, sp^3d

   (c)   seesaw, sp^3d

   (d)   octahedral, sp^3d^2

   (e)   bent (or angular), sp^2

   (f)   tetrahedral, sp^3

                                    (10.1, 10.3; Op. Sk. 1, 3)

2. b, e, f   (10.1, 10.2, Op. Sk. 2)

3. (a)   Bent. Linear geometry gives a symmetrical molecule with a dipole moment of zero.

   (b)   T-shaped. A trigonal pyramidal geometry gives a symmetrical molecule with a dipole moment of zero.

   (c)   Seesaw. Square planar and tetrahedral geometries give symmetrical molecules with dipole moments of zero.   (10.2, Op. Sk. 2)

4.   The Lewis structure is

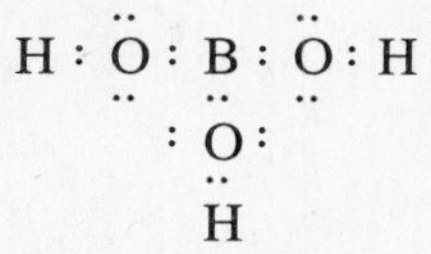

$$H : \overset{..}{\underset{..}{O}} : \overset{..}{\underset{..}{B}} : \overset{..}{\underset{..}{O}} : H$$
$$: \overset{}{\underset{..}{O}} :$$
$$H$$

showing three groups of electrons about boron, all single bonds. The hybridization would be sp^2.

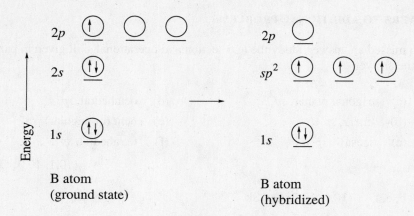

B atom
(ground state)

B atom
(hybridized)

We will assume that each oxygen atom is sp^3 hybridized, as in water. Thus, bonds will form as the singly occupied sp^2 hybrid orbitals on boron overlap with the singly occupied sp^3 hybrid orbitals on oxygen. The bonding to the boron can be represented as follows:

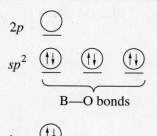

B—O bonds

B atom
[in $B(OH)_3$]   (10.3, Op. Sk.3)

5.   (a)   109°   (d)   90°, 120°, and 180°
     (b)   < 109°   (e)   90°
     (c)   180°   (f)   < 120°   (10.1, 10.3)

6. There are five valence electrons on each nitrogen atom, which occupy the molecular orbitals

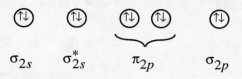

$$\sigma_{2s} \qquad \sigma_{2s}^* \qquad \pi_{2p} \qquad \sigma_{2p}$$

The electron configuration is

$$KK\,(\sigma_{2s})^2\,(\sigma_{2s}^*)^2\,(\pi_{2p})^4\,(\sigma_{2p})^2$$

There are eight bonding electrons and two antibonding electrons. Bond order $= \frac{1}{2}(8-2) = 3$. Because all electrons are paired, the molecule is diamagnetic.

(10.5, 10.6, Op. Sk. 4)

7. The valence electron configuration for C_2 gives 8 valence electrons in 4 molecular orbitals: $(\sigma_{2s})^2\,(\sigma_{2s}^*)^2\,(\pi_{2p})^4$. The calculated bond order is $(6-2)/2 = 2$. Because all electrons are paired in the molecular orbitals, the molecule is diamagnetic.

The molecular orbital description for O_2 gives 12 valence electrons in 7 molecular orbitals: $(\sigma_{2s})^2\,(\sigma_{2s}^*)^2\,(\pi_{2p})^4\,(\sigma_{2p})^2\,(\pi_{2p}^*)^2$. Because there are two unpaired electrons in the (π_{2p}^*) molecular orbitals, the molecule is paramagnetic. The calculated bond order is $(8-4)/2 = 2$.   (10.5, 10.6, Op. Sk. 4)

8. O_2, bond order = 2; O_2^-, bond order = 1.5; O_2^+, bond order = 2.5; O_2^{2-}, bond order = 1. The predicted order of increasing bond length is

$$\xrightarrow{\text{increasing bond length}}$$
$$O_2^+ < O_2 < O_2^- < O_2^{2-}$$ (10.5, 10.6, Op. Sk. 4)

CHAPTER POST-TEST

1. Using the VSEPR model, give the molecular geometry and bond angle (s) predicted for each of the following species ($\angle$ denotes "angle of"):

		Geometry		Bond angle (s)
(a)	ClF_5	_____	$\angle FClF \simeq$	_____
(b)	$PbBr_2$	_____	$\angle BrPbBr \simeq$	_____
(c)	SbF_6^{-}	_____	$\angle FSbF \simeq$	_____
(d)	$As(CH_3)_3Cl_2$	_____	$\angle CAsC \simeq$	_____
			$\angle ClAsC \simeq$	_____
(e)	$Au(CN)_2^{-}$	_____	$\angle CAuC \simeq$	_____
(f)	Cl_3PO	_____	$\angle ClPCl \simeq$	_____
			$\angle ClPO \simeq$	_____

2. For many years noble gases were thought never to form compounds. In 1962 xenon tetrafluoride was made. Predict the orbital hybridization for xenon.

3. Predict the geometry and orbital hybridization for each of the following molecules:

    (a) BI_3, with zero dipole moment

    (b) NCl_3, a polar molecule

    (c) C_2H_2, with zero dipole moment

4. Answer the two questions below, given the following information: Carbon tetra-chloride, CCl_4, has a tetrahedral structure with bond angles of 109.5°.

    (a) Are the bonds polar?
    (b) Does the molecule possess a dipole moment?

5. Suppose you replace a chlorine in the molecule in question 4 with a fluorine. Would the molecule now be likely to have a dipole moment?

6. A _____ pair of electrons gives a larger electron density on a central atom than a _____ pair of electrons.

7. Is rotation about a multiple bond in a species such as

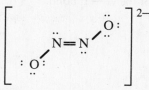

restricted or unrestricted?

8. Discuss in qualitative terms how molecular orbital theory explains the formation of a covalent bond.

9. Explain the structure of F_2 using (a) valence bond theory, (b) molecular orbital theory.

10. Show, by molecular orbital theory, why neon does not form a diatomic molecule.

ANSWERS TO CHAPTER POST-TEST

If you missed an answer, study the text section and operational skill given in parentheses after the answer.

1. (a) square pyramidal, 90°   (b) bent, $\angle 120°$ (c) octahedral, 90°
    (d) trigonal bipyramidal;   (e) linear, 180°   (f) tetrahedral;
    $\angle CAsC \simeq 120°$ $\angle ClPCl \simeq 109.5°$
    $\angle ClAsC \simeq 90°$ $\angle ClPO \simeq 109.5°$

    (10.1, Op. Sk. 1)

2. Xenon would have sp^3d^2 hybridization.   (10.1, 10.3, Op. Sk. 3)

3. (a) trigonal planar, sp^2
    (b) pyramidal, sp^3 with one lone pair of electrons
    (c) linear, sp with $C \equiv C$ (10.1, 10.2, 10.3; Op. Sk. 1, 2, 3)

4. (a) Yes, the bonds are polar, since carbon and chlorine have different electro-negativities.
    (b) The molecule is nonpolar, as the tetrahedral geometry allows for a perfect cancellation of bond dipoles.   (10.2, Op. Sk. 2)

5.   Yes. Since the C — F bond is more polar than the C — Cl bond, a resultant dipole
     moment will exist, pointing in the direction of the C — F bond.   (10.2, Op. Sk. 2)

6.   lone, bonded   (10.1)

7.   Rotation is restricted.   (10.4)

8.   Molecular orbital theory explains covalent bonding as arising from the combination
     of atomic orbitals on the bonding atoms to give molecular orbitals that belong to the
     entire molecule. The molecular orbitals are analogous to atomic orbitals but are not
     centered on any one atom in a molecule. Each molecular orbital can have a maxi-
     mum of two electrons with opposed spins.   (10.5)

9.   (a)   From the electron-dot structure $\ddot{:}\ddot{F}:\ddot{F}:$ we see that each fluorine shares one

           electron with the other fluorine. This completes each octet and leads to a linear
           nonpolar bond. Under valence bond theory, the bonding is explained as the
           overlap of the singly occupied $2p$ orbital on each atom.   (10.3, Op. Sk. 3)

     (b)   The molecular orbital description gives 14 valence electrons in 7 molecular
           orbitals:

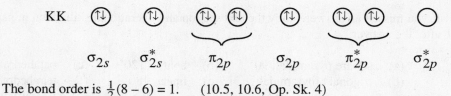

           The bond order is $\frac{1}{2}(8-6) = 1$.   (10.5, 10.6, Op. Sk. 4)

10. Two neon atoms would have a total of 16 valence electrons. Putting these into the
     molecular orbital diagram gives

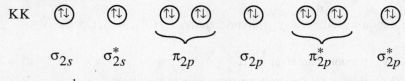

     The bond order is $\frac{1}{2}(8-8) = 0$; thus, no bond forms.   (10.6, Op. Sk. 4)

UNIT EXAM 3

Use the following information in your calculations:

$$h = 6.63 \times 10^{-34} \text{ J} \cdot \text{s}$$
$$c = 3.00 \times 10^{8} \text{ m/s}$$
$$R_{\text{H}} = 2.180 \times 10^{-18} \text{ J}$$

1. A strong line in the spectrum of atomic mercury has a wavelength of 5461 pm. What is the energy of a photon with this wavelength?

2. Calculate the energy (in joules) emitted by an electron in a hydrogen atom that falls from the $n = 4$ to the $n = 2$ energy level.

3. The ground state of a boron atom has the electron configuration $1s^2 2s^2 2p^1$. Which of the following is *not* an acceptable set of quantum numbers for the highest-energy electron in this configuration?
   (a) 2, 1, -1, $-1/2$ (c) 2, 0, 0, 1/2   (e) 2, 1, 1, 1/2
   (b) 2, 1, 0, $-1/2$ (d) 2, 1, -1, 1/2

4. Which of the following sets of quantum numbers is (are) not permitted?
   (a) $n = 3, l = 2, m_l = -2, m_s = -1/2$
   (b) $n = 1, l = 0, m_l = -1, m_s = +1/2$
   (c) $n = 2, l = 1, m_l = 0, m_s = +1/2$
   (d) $n = 0, l = 0, m_l = 0, m_s = -1/2$

5. Consider the following possible orbital diagrams for a nitrogen atom or the hypothetical singly charged positive ion N^+.

(a) ⬤ 1s   ⬤ 2s   ⬤ ⬤ ⬤ 2p   (d) ⬤ 1s   ⬤ 2s   ⬤ ⬤ ⬤ 2p

(b) ⬤ 1s   ⬤ 2s   ⬤ ⬤ ⬤ 2p   (e) ⬤ 1s   ⬤ 2s   ⬤ ⬤ ⬤ 2p

(c) ⬤ 1s   ⬤ 2s   ⬤ ⬤ ⬤ 2p

(a) Which of these orbital diagrams represents the ground state of a nitrogen atom?

(b) Which of the diagrams represents an excited state of N^+ (the nitrogen positive ion)?

6. Write the electron configuration for each of the following:

(a) $_{29}Cu$ (b) $_{16}S^{2-}$

7. Write the orbital diagram for the ground-state oxygen atom whose electron configuration is $1s^2 2s^2 2p^4$.

8. Write the electron configuration for the outer electrons of the elements in each of the following groups: (a) VIIA   (b) IIIB

9. Write equations using Lewis symbols to show the transfer of electrons to form ions of noble-gas configuration for calcium and phosphorus.

10. Arrange the following in order of

(a) increasing ionic radius: $Al^{3+}, F^-, Mg^{2+}, Na^+, O^{2-}$

(b) increasing bond polarity: Pb — C, C — Br, Be — F, In — O, B — H, B — N

11. Write the Lewis formula for each of the following:

(a) SO_2 (b) NaCl

12. $\Delta H°$ for the following reaction is 1564.8 kJ.

$$CH_3Cl(g) \longrightarrow C(g) + Cl(g) + 3H(g)$$

If the C — H bond energy is 411 kJ/mol, what is the C — Cl bond energy in CH_3Cl?

13. Predict and draw the geometry for each of the following, using the VSEPR model:
    (a) OF_2 (b)   HCN

14. What geometry (or geometries) of a molecule AX_4 is (are) consistent with the information that the molecule has a nonzero dipole moment? Draw the geometry.

15. Deduce the molecular orbital configuration and bond order for the CO molecule. Determine whether the molecule is diamagnetic or paramagnetic.

16. Describe the bonding in SCl_2 using hybrid orbitals.

ANSWERS TO UNIT EXAM 3

If you missed an answer, study the text section and operational skill given in parentheses after the answer.

1. 3.64×10^{-19} J   (7.2, Op. Sk. 2)   2. 4.088×10^{-19} J   (7.2, 7.3; Op. Sk. 2, 3)

3.   c   (7.5, Op. Sk. 4)   4.   b, d   (7.5, Op. Sk. 5)

5.   (a)   e   (b)   d   (8.1, 8.2, 8.3; Op. Sk. 1, 2)

6.   (a) $1s^2 2s^2 2p^6 3s^2 3p^6 4s^1 3d^{10}$ (b) $1s^2 2s^2 2p^6 3s^2 3p^6$ (8.2, 8.3, Op. Sk. 2)

7.   ⓝ   ⓝ   ⓝ ⓣ ⓣ   (8.4, Op. Sk. 4)
     $1s$ $2s$ $2p$

8.   (a) $ns^2 np^5$ (b) $ns^2 (n-1) d^1$ (8.3, 8.4, Op. Sk. 3)

9.   Ca : $\longrightarrow$ Ca^{2+} + 2e⁻;   · $\overset{.}{P}$: + 3e⁻ $\longrightarrow$: $\overset{..}{\underset{..}{P}}$: $^{3-}$ (9.1, Op. Sk. 1)

10. (a) Al^{3+}, Mg^{2+}, Na^+, F^-, O^{2-} (9.3, Op. Sk. 3)

(b)

$$\xrightarrow{\text{Increasing bond polarity}}$$

B — H < C — Br < Pb — C < B — N < In — O < Be — F

(9.5, Op. Sk. 4)

11. (a) 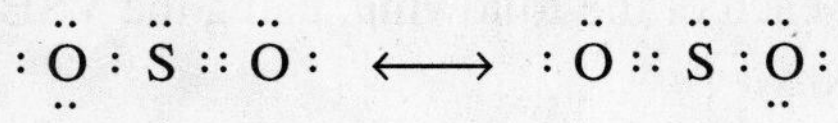 $:\overset{..}{\underset{..}{O}}:S::\overset{..}{\underset{..}{O}}: \longleftrightarrow :\overset{..}{\underset{..}{O}}::S:\overset{..}{\underset{..}{O}}:$

(b) 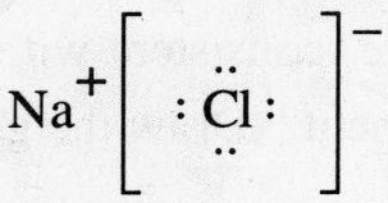 $Na^+\left[:\overset{..}{\underset{..}{Cl}}:\right]^-$ (9.6, 9.7, Op. Sk. 5, 6)

12. 332 kJ/mol   (9.2, 9.11, Op. Sk. 9)

13. (a) bent

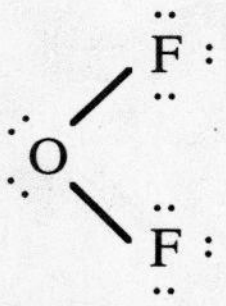

$:\overset{}{O}$ with two F atoms $\overset{..}{\underset{..}{F}}:$

(b) linear H — C $\equiv$ N :   (10.1, Op. Sk. 1)

14. seesaw (or distorted tetrahedral)

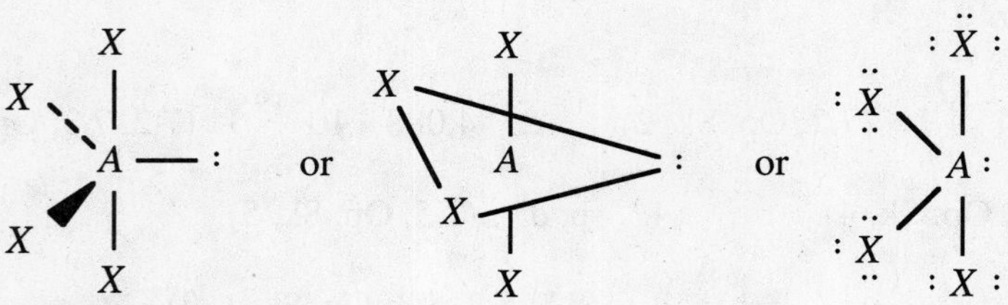

or

or

(like SF_4)

(10.2, Op. Sk. 2)

15. $KK\,(\sigma_{2s})^2\,(\sigma_{2s}^*)^2\,(\pi_{2p})^4\,(\sigma_{2p})^2$; bond order = 3; diamagnetic.

(10.5, 10.6, Op. Sk. 4)

16. Each S — Cl bond can be described as overlap of a singly occupied sp^3 hybrid orbital on sulfur with a singly occupied p orbital on chlorine.   (10.3, Op. Sk. 3)

CHAPTER 11   STATES OF MATTER; LIQUIDS AND SOLIDS

CHAPTER TERMS AND DEFINITIONS

Numbers in parentheses after definitions give the text sections in which the terms are explained. Starred terms are italicized in the text. Where a term does not fall directly under a text section heading, additional information is given for you to locate it.

change of state (phase transition)   change of a substance from one state to another; transition from the solid to the liquid form, from the liquid to the gaseous form, from the solid to the vapor form, or the reverse   (11.2, introductory section)

phase*   homogeneous portion of a system; a given state of a substance or solution (11.2, introductory section)

melting   change of a solid to the liquid state; fusion   (11.2)

freezing   change of a liquid to the solid state   (11.2)

vaporization   change of a solid or a liquid to the vapor   (11.2)

sublimation   change of a solid directly to the vapor   (11.2)

condensation   change of a gas to either the liquid or solid state   (11.2)

deposition*   change of a vapor to the solid   (11.2)

liquefaction*   change of a substance that is normally a gas to the liquid state   (11.2)

vapor pressure   partial pressure of the vapor over a liquid (or solid), measured at equilibrium   (11.2)

dynamic equilibrium*   state of a system in which the rates of two opposing molecular processes (e.g., vaporization and condensation) are equal and are continuously occurring   (11.2)

volatile*   describes liquids and solids with relatively high vapor pressure at normal temperatures   (11.2)

boiling point   temperature at which the vapor pressure of a liquid equals the pressure exerted on the liquid (atmospheric pressure, unless the vessel containing the liquid is closed)   (11.2)

normal boiling point*   boiling point of a liquid at one atmosphere   (11.2)

freezing point; melting point   temperature at which a pure liquid changes to a crystalline solid and at which the solid changes to a liquid; thus, the temperature at which the solid and liquid are in dynamic equilibrium   (11.2)

heat of phase transition* heat energy added to a substance during a phase change without any change in temperature of the substance (11.2)

heat (or enthalpy) of fusion (ΔH_{fus}) enthalpy change for the melting of a solid (11.2)

heat (or enthalpy) of vaporization (ΔH_{vap}) enthalpy change for the vaporization of a liquid (11.2)

Clausius–Clapeyron equation* $\log \dfrac{P_2}{P_1} = \dfrac{\Delta H_{vap}}{2.303R} \left(\dfrac{1}{T_1} - \dfrac{1}{T_2} \right)$ or

$\ln \dfrac{P_2}{P_1} = \dfrac{\Delta H_{vap}}{R} \left(\dfrac{1}{T_1} - \dfrac{1}{T_2} \right)$ (11.2)

phase diagram graph that summarizes the conditions under which the different states of a substance are stable (11.3)

triple point point on a phase diagram representing the temperature and pressure at which three phases of a substance coexist in equilibrium (11.3)

critical temperature temperature above which the liquid state of a substance no longer exists regardless of the pressure (11.3)

critical pressure vapor pressure at the critical temperature (11.3)

critical point* on a phase diagram, the point where the vapor-pressure curve ends and at which the temperature and pressure have their critical values (11.3)

surface tension energy required to increase the surface area of a liquid by a unit amount (11.4)

capillary rise* rise of a column of liquid in an upright, small-diameter tube because of surface tension (11.4)

meniscus* curved upper surface of a liquid in a container (11.4)

viscosity resistance to flow that is exhibited by all liquids and gases (11.4)

intermolecular forces interactive forces between molecules (11.5)

molecular beam* group of molecules caused to travel together in the same direction; when two beams collide, the resulting data can be used to calculate the energy of molecular interactions (11.5, marginal note)

van der Waals forces weak attractive forces between molecules, including dipole–dipole and London forces (11.5)

dipole–dipole force attractive intermolecular force resulting from the tendency of polar molecules to align themselves such that the positive end of one molecule is near the negative end of another (11.5)

instantaneous dipoles* small partial charges on molecules due to momentary shifts in the distributions of electrons as they move about atomic nuclei (11.5)

London (dispersion) forces weak attractive forces between molecules resulting from the small, instantaneous dipoles that occur because of the varying positions of the electrons during their motion about nuclei (11.5)

polarizable* describes molecules that are easily distorted to give instantaneous dipoles (11.5)

hydrogen bonding weak to moderate attractive force between a hydrogen atom covalently bonded to a very electronegative atom, particularly nitrogen, oxygen, or fluorine, and the lone pair of electrons of another small, electronegative atom (usually on another molecule)   (11.5)

molecular solid   solid that consists of atoms or molecules held together by intermolecular forces   (11.6)

metallic solid   solid that consists of positive cores of atoms held together by the surrounding "sea" of electrons (metallic bonding)   (11.6)

ionic solid   solid that consists of cations and anions held together by electrical attraction of opposite charges (ionic bonds)   (11.6)

covalent network solid   solid that consists of atoms held together in large networks or chains by covalent bonds   (11.6)

malleable*   able to be shaped by hammering, as are metallic crystals   (11.6)

crystalline solid   solid composed of one or more crystals in which each crystal has a well-defined, ordered structure in three dimensions   (11.7)

amorphous solid   solid that has a disordered structure; it lacks the well-defined arrangement of basic units (atoms, molecules, or ions) found in a crystal   (11.7)

crystal lattice   geometric arrangement of lattice points of a crystal, in which there is one lattice point at the same location within each of the basic units of the crystal   (11.7)

unit cell   smallest boxlike unit (each box having faces that are parallelograms) from which you can imagine constructing a crystal by stacking the units in three dimensions   (11.7)

crystal systems*   seven basic shapes possible for unit cells; a classification of crystals   (11.7)

simple (primitive) lattice*   crystalline lattice in which the unit cell has lattice points only at the corners of the unit cell   (11.7)

simple cubic unit cell   cubic unit cell in which lattice points are situated only at the corners   (11.7)

body-centered cubic unit cell   cubic unit cell in which there is a lattice point at the center of the cubic cell as well as at each of the corners   (11.7)

face-centered cubic unit cell   cubic unit cell in which there are lattice points at the centers of each face of the unit cell, in addition to those at the corners   (11.7)

nonstoichiometric*   describes a compound whose composition varies slightly from the idealized formula   (11.7)

nematic phase*   liquid-crystal phase in which the molecules are aligned parallel to each other in one direction, but otherwise have random positions (A Chemist Looks At: Liquid-Crystal Displays)

polarized light*   light rays with waves vibrating in a given plane (A Chemist Looks At: Liquid-Crystal Displays)

hexagonal close-packed structure (hcp)   crystal structure composed of close-packed atoms (or other units) with the stacking ABABABA ...; structure has a hexagonal unit cell   (11.8)

cubic close-packed structure (ccp)   crystal structure composed of close-packed atoms (or other units) with the stacking ABCABCABCA ...   (11.8)

coordination number   number of nearest-neighbor atoms of an atom   (11.8)

polymorphic*   describes a substance that can crystallize in more than one crystal structure   (11.8)

diffraction pattern*   series of spots on a photographic plate, produced by the reflection of x rays by the atoms in a crystal   (11.10)

in phase*   said of two waves of the same wavelength that come together so that their peaks (maxima) and troughs (minima) match   (11.10)

constructive interference*   result of combining two waves that are of the same wavelength and in phase such that the intensity of the resultant ray is increased (11.10)

amplitude*   height of a wave   (11.10)

out of phase*   said of two waves of the same wavelength that come together with their peaks at opposite points and their troughs at opposite points   (11.10)

destructive interference*   result of combining two waves that are of the same wavelength and out of phase such that the intensity of the resultant ray is decreased or reduced to zero   (11.10)

Bragg equation*   equation relating the wavelength of x rays, λ, to the distance between atomic planes, *d*, and the angle of reflection, θ; $n\lambda = 2d \sin \theta$, $n = 1, 2, 3 \ldots$
(Instrumental Methods: Automated X-Ray Diffractometry)

CHAPTER DIAGNOSTIC TEST

1.   Match each term in the left-hand column with its description in the right-hand column.

E	(1)	Mn	(a)	ionic compound
D	(2)	SiC	(b)	expected to exhibit hydrogen bonding
A	(3)	$CaCl_2$	(c)	London forces are major attractive force
C	(4)	CH_4 and SiH_4	(d)	network solid
f	(5)	RbCl and H_2O	(e)	expected to exhibit electrical conductivity in the solid state
B	(6)	H_2O	(f)	expected to participate in ion-dipole intermolecular forces

2.   On the phase diagram in Figure A, locate and label the following:

(a) melting point
at 1.5 atm

(b) pressure at
which the boiling
point is 55°C

(c) effect of
changing the pressure
from 1 atm to 0.4
atm, if the temperature
is held constant at 25°C

(d) triple point

(e) solid vapor-
pressure curve

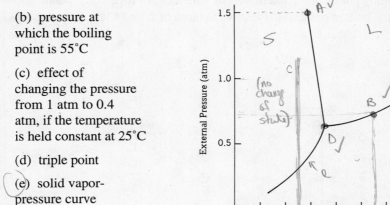

Figure A

3.   Arrange the following compounds in order of increasing vapor pressure: propane,
C_3H_8; propanol, C_3H_7OH; methane, CH_4.

4.   How much heat will be required to melt a tray of ice cubes? The mass of the ice is
575 g, and the heat of fusion of ice is 6.01 kJ/mol. $575g\left(\frac{mol}{18g}\right)\left(6.01\left(\frac{kJ}{mol}\right)\right)=192kJ$.

5.   On a phase diagram, the intersection of the solid vapor-pressure, liquid vapor-
pressure, and melting-point curves is known as the ___triple point___ .

6.   Titanium metal (atomic weight 48 amu) crystallizes in a body-centered cubic structure with a unit-cell volume of 3.3×10^{-2} nm^3. If the density of Ti is 4.8 g/cm^3,
calculate Avogadro's number from these data.

7. On the graph in Figure B, the boiling points of the halogens and some interhalogen compounds are plotted relative to their respective molecular weights. Discuss the trends in boiling points of (a) the halogens and (b) the interhalogens. (c) Compare the boiling points of the interhalogens with those of their corresponding halogens (i.e., the boiling point of ICl relative to those of I_2 and Cl_2).

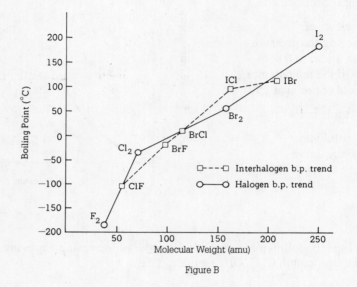

Figure B

8. If Ti crystallizes in a body-centered cubic structure, the number of Ti atoms per unit cell would be
    (a)   9   (b)   3   (c)   5   (d)   2   (e)   none of the above.

9. The boiling point of a liquid is defined as _____ .

10. From a molecular viewpoint, molecules evaporate from the surface of a liquid if they
    (a)   have sufficient kinetic energy to overcome the attractive forces in the liquid.
    (b)   are attracted to other molecules in the vapor phase.
    (c)   are in a state of dynamic equilibrium with the water molecules in the atmosphere.
    (d)   are repelled by other molecules in the liquid phase.
    (e)   none of the above.

11. Which of the following represent(s) the *incorrect* trend in the indicated phenomenon for the given species?

    (a) Melting point: $NaF > GeF_4 > Cl_2$

    (b) Boiling point: $H_2O < H_2S < H_2Se < H_2Te$ ✓

    (c) Electrical conductivity of melting: $NaCl \cong NaBr > S$

    (d) Hydrogen bonding: $H_2O > CH_3OH > PH_3 > CH_4$

    (e) London forces: $Ne > S_8 > Br_2 > I_2$ at 25°C

12. Indicate whether each of the following statements is true or false. If a statement is false, change it so it is true.

    (a) Graphite can serve as a good lubricant because the weak dipole–dipole forces *London* between the layers of hexagonal rings of carbon atoms permit the layers to slide over one another. True/False: __F__

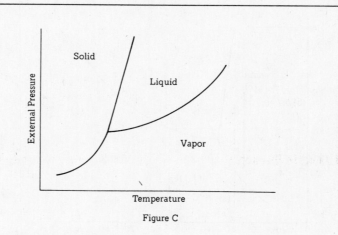

Figure C

    (b) In the phase diagram in Figure C, for a system where the solid is more dense than the liquid, an increase in external pressure will cause the melting point to increase. True/False: __True__

    (c) London forces are very short-range forces and tend to be the greatest for molecules composed of small atoms. True/False: __F__ *large*

(d)   Hydrogen bonding can result when the hydrogen atom is bonded to an atom of low electronegativity. This permits a shift in electron density toward the hydrogen atom. True/False: _____

_____ .

ANSWERS TO CHAPTER DIAGNOSTIC TEST

If you missed an answer, study the text section and operational skill given in parentheses after the answer.

1.   (1)   e   (11.6)   (4)   c   (11.5, Op. Sk. 4)
     (2)   d   (11.6, Op. Sk. 6)   (5)   f   (11.5, Op. Sk. 4)
     (3)   a   (11.6, Op. Sk. 6)   (6)   b   (11.5, Op. Sk. 4)

2.

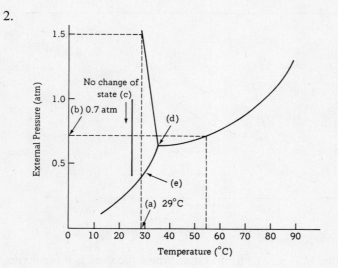

(11.3, Op. Sk. 3)

3. $C_3H_7OH < C_3H_8 < CH_4$ (11.5, Op. Sk. 5)

4.   192 kJ   (11.2, Op. Sk. 1)   5.   triple point   (11.3, Op. Sk. 3)

6. $$\frac{\text{atoms}}{\text{mol}} = \frac{2 \text{ atoms}}{\text{unit cell}} \times \frac{\text{unit cell}}{3.3 \times 10^{-2} \text{ nm}^3} \times \left(\frac{1 \text{ nm}}{10^{-9} \text{ m}}\right)^3$$

$$\times \left(\frac{10^{-2} \text{ m}}{1 \text{ cm}}\right)^3 \times \frac{1 \text{ cm}^3}{4.8 \text{ g}} \times \frac{48 \text{ g}}{\text{mol}}$$

$$= 6.\underline{0}6 \times 10^{23} \text{ atoms/mol} = 6.1 \times 10^{23} \text{ atoms/mol} \quad (11.9, \text{ Op. Sk. 9, 10})$$

7. (a) The general trend in boiling points of the halogens is an increasing boiling point with increasing molecular weight and size of the halogen. More thermal energy is required to raise the kinetic energy of heavier molecules to enable them to escape the liquid phase. Also, van der Waals forces are greater for larger molecules because they are more polarizable.

    (b) The general trend in boiling points of the interhalogens is an increasing boiling point with increasing molecular weight and size of the interhalogen. The same reasons apply here as discussed for the halogens. Interhalogens also have dipole–dipole forces.

    (c) In general, the boiling points of the interhalogens are slightly greater than the average of the corresponding parent halogens (i.e., ICl, b.p. 97°C; Cl_2, b.p. 35°C; I_2, b.p. 180°C). A possible explanation for these higher boiling points is the dipole–dipole force contributions in the liquid phase for the polar interhalogen molecules. The halogens are nonpolar molecular species.
    (11.2, 11.5, Op. Sk. 5)

8. d   (11.7, Op. Sk. 8)

9. the temperature at which the vapor pressure of the liquid equals the pressure exerted on the liquid.   (11.2)

10. a   (11.2, 11.6)

11. b and e are incorrect   (11.5, Op. Sk. 4, 5); section references for other responses: a (11.6, Op. Sk. 7); c   (11.6, Op. Sk. 6); d   (11.5, Op. Sk. 4)

12. (a) False. Graphite can serve as a good lubricant because the weak London forces between the layers of hexagonal rings of carbon atoms permit the layers to slide over one another.   (11.6)
    (b) True.   (11.3)
    (c) False. London forces are very short-range forces and tend to be the greatest for molecules composed of large atoms.   (11.5)

(d)   False. Hydrogen bonding can result when the hydrogen atom is bonded to an atom of high electronegativity. This creates a partial positive charge on the hydrogen atom. The hydrogen atom can then polarize highly electronegative atoms on neighboring unsymmetrical molecules.   (11.5, Op. Sk. 4)

SUMMARY OF CHAPTER TOPICS

11.1   COMPARISON OF GASES, LIQUIDS, AND SOLIDS

11.2   PHASE TRANSITIONS

Operational Skills

1.   Calculating the heat required for a phase change of a given mass of substance. Given the heat of fusion (or vaporization) of a substance, calculate the amount of heat required to melt (or vaporize) a given quantity of that substance   (Example 11.1).

2.   Calculating vapor pressures and heats of vaporization.   Given the vapor pressure of a liquid at one temperature and its heat of vaporization, calculate the vapor pressure at another temperature   (Example 11.2). Given the vapor pressures of a liquid at two temperatures, calculate the heat of vaporization   (Example 11.3).

        Have you ever heated milk in a pan on an electric or gas stove to make a cup of instant cocoa? If not, do so, and watch carefully as the temperature of the milk rises. As the milk warms, small bubbles appear at the interface where the milk touches the pan. What is the composition of these bubbles? As you continue to heat the milk, the bubbling stops. Eventually, if you overheat it (boiled milk makes lousy cocoa), large bubbles roll up to the surface of the liquid and pop. This is the phenomenon called boiling. What is the composition of these second bubbles?

        The first bubbles you observed were air that was dissolved in the liquid. At higher temperatures the solubility of gas in liquid is lowered. But the large bubbles associated with boiling are not air. They are the vapor state of the liquid, in this case the water in the milk. As thermal energy is continually applied, the liquid molecules reach a temperature — and thus an energy — high enough so that the gaseous state forms within the body of the liquid. Because vapor is less dense than the liquid, it quickly moves up to the surface as a bubble and escapes into the atmosphere above the liquid.

        An important concept to understand is that there is no temperature change during a phase change. As thermal energy is continually and evenly added to a boiling liquid, the temperature does not increase. The same is true if you heat an equilibrium mixture of ice and water, which under normal conditions is at 0°C. If you heat this mixture slowly and evenly, the temperature does not begin to rise until all the ice has melted.

Exercise 11.1   The heat of vaporization of ammonia is 23.4 kJ/mol. How much heat is required to vaporize 1.00 kg of ammonia? How many grams of water at 0°C could be frozen to ice at 0°C by the evaporation of this amount of ammonia?

Wanted: q to vaporize NH_3, in kJ; g water.

Given: ΔH_{vap} for NH_3 = 23.4 kJ/mol; 0°C.

Known: ΔH_{fus} for H_2O (ice) = 6.01 kJ/mol; 1.00 kg NH_3 = 1.00×10^3 g NH_3; NH_3 = 17.0 g/mol; H_2O = 18.0 g/mol. Evaporating NH_3 takes heat from the freezing H_2O.

Solution: Find q needed to vaporize the NH_3.

$$1.00 \times 10^3 \text{ g NH}_3 \times \frac{\text{mol NH}_3}{17.0 \text{ g NH}_3} \times \frac{23.4 \text{ kJ}}{\text{mol NH}_3} = 13\underline{7}6 \text{ kJ} = 1.38 \times 10^3 \text{ kJ}$$

Find the g water that would freeze if this amount of heat was transferred away. (In this case both heat values are −.)

$$-13\underline{7}6 \text{ kJ} \times \frac{\text{mol H}_2\text{O}}{-6.01 \text{ kJ}} \times \frac{18.0 \text{ g H}_2\text{O}}{\text{mol H}_2\text{O}} = 41\underline{2}1 \text{ g} = 4.12 \times 10^3 \text{ g} = 4.12 \text{ kg H}_2\text{O}$$

Vapor pressure is a concept that many students find difficult to grasp. The important thing to remember is that at each temperature, water or any liquid has a specific equilibrium vapor pressure, plotted as in text Figure 11.7.

In Exercise 11.2 you will use logarithms to calculate the answer. The mathematical work is greatly simplified with the use of a calculator with log and antilog functions. However, it is a good idea for you to review the subject of logarithms in text Appendix A.2 or in a math book; you will need to know how to manipulate the log expressions you will be using. Specific information about using the log functions on your calculator is given in the text and with exercise solutions in this chapter of the study guide. You will use logarithms frequently as we move on to the study of kinetics, equilibrium, and electrochemistry.

Exercise 11.2   Carbon disulfide, CS_2, has a normal boiling point of 46°C and a heat of vaporization of 26.8 kJ/mol. What is the vapor pressure of carbon disulfide at 35°C?

Solution: The Clausius–Clapeyron equation written in "two-point" form is

$$\log \frac{P_2}{P_1} = \frac{\Delta H_{vap}}{2.303R} \left(\frac{1}{T_1} - \frac{1}{T_2} \right)$$

Substituting appropriate known values and multiplying through gives

$$\log \frac{P_2}{760 \text{ mmHg}} = \frac{26.8 \times 10^3 \, \cancel{J}/\cancel{mol}}{2.303 \times 8.31 \, \cancel{J}/(K \cdot \cancel{mol})} \left(\frac{1}{319 \text{ K}} - \frac{1}{308 \text{ K}} \right)$$

$$\text{(exact)}$$

$$= (1.4\underline{0}0 \times 10^3 \, \cancel{K}) \times (-1.12 \times 10^{-4} \, /\cancel{K})$$

$$= -0.15\underline{68}$$

Now take the antilog of both sides, using a log table or a hand calculator. If you are using a hand calculator look for an *inverse* key, or a y^x key. If the calculator has an inverse key, enter -0.1568, press *inverse*, then *log*. The result will be 0.69\underline{69}, the antilog of $-0.15\underline{68}$. To use the y^x key, enter 10, press y^x, enter -0.1568, and press *equals*. The result of $10^{-0.15\underline{68}}$ = antilog $-0.15\underline{68}$ = 0.69\underline{69}. Thus

$$\frac{P_2}{760 \text{ mmHg}} = 0.69\underline{69}$$

$$P_2 = 0.69\underline{69} \times 760 \text{ mmHg} = 5.30 \times 10^2 \text{ mmHg}$$

Significant figures must be observed in calculations with logarithms. The rule is that the number of decimal places (including zeros) given in a logarithm is the number of figures that are significant in the antilog. Note in the exercise above that the logarithm, 0.15\underline{68}, has three decimal places, and that the antilog, 0.69\underline{69}, has three significant figures. In each case, the fourth digit is kept for accuracy.

Exercise 11.3   Selenium tetrafluoride, SeF_4, is a colorless liquid. It has a vapor pressure of 757 mmHg at 105°C and 522 mmHg at 95°C. What is the heat of vaporization of selenium tetrafluoride?

Solution: Let $P_1 = 757$ mmHg, $T_1 = 105$°C or 378 K, $P_2 = 522$ mmHg, and $T_2 = 95$°C or 368 K. Substituting appropriate known values into the Clausius–Clapeyron equation and multiplying through gives

$$\log \frac{522\ \cancel{mmHg}}{757\ \cancel{mmHg}} = \frac{\Delta H_{vap}}{2.303 \times 8.31\ J/(\cancel{K} \bullet mol)} \left(\frac{1}{378} - \frac{1}{368} \right)\ \text{or}\ K^{-1}\ \cancel{K}$$

$$-0.1614 = \frac{(-7.19 \times 10^{-5})}{19.\underline{1}4\ J/mol} \times \Delta H_{vap}$$

Rearranging to solve for ΔH_{vap} gives $\Delta H_{vap} = 4.2\underline{9}6 \times 10^{4}\ J/mol = 43.0\ kJ/mol.$

11.3   PHASE DIAGRAMS

Operational Skill

3.   Relating the conditions for the liquefaction of gases to the critical temperature.
Given the critical temperature and pressure of a substance, describe the conditions
necessary for liquefying the gaseous substance   (Example 11.4).

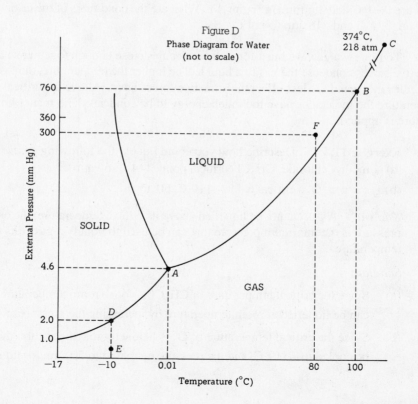

Figure D
Phase Diagram for Water
(not to scale)

A phase diagram is a comprehensive summary of the behavior of a substance with varying temperature and pressure. Figure D is the phase diagram of water (not drawn to scale). Acquaint yourself with the three equilibrium lines along each of which two phases exist. These lines were plotted from measurements of the vapor pressures of the solid and liquid at each temperature and from determinations of the freezing point of water at various pressures. Find the triple point, at which all three phases exist; the critical point, the highest temperature at which you can liquefy the gas; and the pressure necessary at the critical point.

Learn to use the diagram by working in the following way: Think of a temperature and pressure, then find the point on the diagram and identify the state of water at those conditions. For example, at a temperature of $-10°C$ and a pressure of 0.2 mmHg, water is a gas (point E). At 80°C and 300 mmHg, water is a liquid (point F). Ask and answer other questions about the phase diagram. What is the equilibrium vapor pressure of the solid at $-10°C$? Looking at the diagram to see what pressure is at $-10°C$ on the solid–gas equilibrium line, we see it is 2.0 mmHg (point D). What is the normal boiling point of water? It is 100°C at 760 mmHg (point B). What are the conditions of the triple point? They are 0.01°C and 4.6 mmHg (point A). What are the conditions of the critical point? They are 374°C and 218 atm (point C).

To liquefy a gas, we must force the molecules close enough together so that the attractive forces condense the gas to a liquid. The higher the temperature, the faster the molecules move and the less effect the attractive forces have. Above the critical temperature the molecules have too much energy to be condensed, no matter how much pressure is applied.

Exercise 11.4   Describe how you could liquefy the following gases:
(a)   methyl chloride, CH_3Cl (critical point, 144°C, 66 atm);
(b)   oxygen, O_2 (critical point, $-119°C$, 50 atm).

Known: A gas cannot be liquefied above its critical temperature; the critical pressure is the minimum pressure that can be used to liquefy a gas at its critical temperature.

Solution:

(a)   Since the critical temperature of CH_3Cl is above room temperature, the gas
        can be liquefied at room temperature by increasing the pressure.

(b)   Since the critical temperature of O_2 is below room temperature, the gas must
        be cooled to $-119°C$, and the pressure increased to 50 atm, for liquefaction.

11.4   PROPERTIES OF LIQUIDS: SURFACE TENSION AND VISCOSITY

11.5   INTERMOLECULAR FORCES; EXPLAINING LIQUID PROPERTIES

Operational Skills

   4.   Identifying intermolecular forces.   Given the molecular structure, state the kinds of intermolecular forces expected for a substance   (Example 11.5).

   5.   Determining relative vapor pressure on the basis of intermolecular attraction. Given two liquids, decide on the basis of the intermolecular forces which has the higher vapor pressure at a given temperature or which has the lower boiling point   (Example 11.6).

   Understanding the various forces acting between particles of matter will enable you to understand and explain observed trends in physical properties such as melting points, boiling points, solubility, etc. You should memorize the types of forces and their relative strengths, and the types of molecules in which they exist. Remember especially that all of these forces are electrical in nature, the attraction of + and −.

   To determine intermolecular forces, it is helpful to first draw the Lewis structure. You can then envision the molecular geometry in order to find whether the molecule is polar (see text Sections 10.1 and 10.2).

   Exercise 11.5   List the different intermolecular forces you would expect for each of the following compounds: (a) propanol, $CH_3CH_2CH_2OH$; (b) carbon dioxide, CO_2;   (c) sulfur dioxide, SO_2.

   Known: All molecules exhibit London forces; a polar molecule results from unsymmetrically arranged polar bonds; atoms that hydrogen bond are O, N, and F. The electron dot structures and polarities are

(a) $CH_3CH_2CH_2OH$ (b) CO_2 (c) SO_2

nonpolar

polar

polar

Solution: (a) $CH_3CH_2CH_2OH$ has London forces, dipole–dipole forces, H-bonding;   (b) CO_2 has London forces;   (c) SO_2 has London forces, dipole-dipole forces.

Exercise 11.6   Arrange the following hydrocarbons in order of increasing vapor pressure: ethane, C_2H_6; propane, C_3H_8; and butane, C_4H_{10}. Explain your answer.

Given: C_2H_6, C_3H_8, C_4H_{10}.

Known: Molecular attractive forces decrease vapor pressure; the factors to consider are molecular weight and London forces, dipole–dipole forces, and hydrogen bonding. The structural formulas are

$$\begin{array}{ccc}
\begin{array}{c}
\text{H} \quad \text{H} \\
| \quad\; | \\
\text{H}-\text{C}-\text{C}-\text{H} \\
| \quad\; | \\
\text{H} \quad \text{H}
\end{array}
&
\begin{array}{c}
\text{H} \quad \text{H} \quad \text{H} \\
| \quad\; | \quad\; | \\
\text{H}-\text{C}-\text{C}-\text{C}-\text{H} \\
| \quad\; | \quad\; | \\
\text{H} \quad \text{H} \quad \text{H}
\end{array}
&
\begin{array}{c}
\text{H} \quad \text{H} \quad \text{H} \quad \text{H} \\
| \quad\; | \quad\; | \quad\; | \\
\text{H}-\text{C}-\text{C}-\text{C}-\text{C}-\text{H} \\
| \quad\; | \quad\; | \quad\; | \\
\text{H} \quad \text{H} \quad \text{H} \quad \text{H}
\end{array}
\end{array}$$

These molecules are not dipoles, as the bonds are symmetrically arranged. Hydrogen bonding is not possible. London forces increase with molecular weight.

Solution: The order of increasing vapor pressure is

$$C_4H_{10} < C_3H_8 < C_2H_6$$

Exercise 11.7   Methyl chloride, CH_3Cl, has a vapor pressure of 1490 mmHg, and ethanol has a vapor pressure of 42 mmHg. Explain why you might expect methyl chloride to have a higher vapor pressure than ethanol, even though methyl chloride has a somewhat larger molecular weight.

Solution: The structure of ethanol is

$$\begin{array}{c}
\text{H} \quad \text{H} \\
| \quad\; | \\
\text{H}-\text{C}-\text{C}-\overset{..}{\text{O}}: \\
| \quad\; | \quad\; | \\
\text{H} \quad \text{H} \quad \text{H}
\end{array}$$

and the $O-H$ group can participate in hydrogen bonding. This strong intermolecular force would reduce the vapor pressure of ethanol.

11.6 CLASSIFICATION OF SOLIDS BY TYPE OF ATTRACTION OF UNITS

Operational Skills

6. Identifying types of solids. From what you know about the bonding in a solid, classify it as a molecular, metallic, ionic, or covalent network (Example 11.7).

7. Determining relative melting points based on types of solids. Given a list of substances, arrange them in order of increasing melting point from what you know of their structures (Example 11.8).

Again, remember that all of the forces discussed are electrical in nature, whether called intermolecular forces or chemical bonds. We differentiate between them in terms of their differing strengths and the situations in which they exist.

You may need to go back and review material on ionic and covalent bonding (text Sections 9.1 and 9.4) to be able to decide what types of forces of attraction exist in various substances.

Exercise 11.8 Classify each of the following solids according to the forces of attraction that exist between the structural units: (a) zinc; (b) sodium iodide, NaI; (c) silicon carbide, SiC; (d) methane, CH_4.

Known: The periodic table placement indicates the bonding type.

Solution: (a) Zinc is a metal, so we would expect a metallic solid. (b) NaI is ionic and would exist as an ionic solid. (c) SiC is composed of two atoms from the same group or family of nonmetals; these could form a molecular solid or a covalent network solid similar to diamond with Si atoms in half the atomic positions. We would have to know the properties to determine which. (d) CH_4 is a molecular substance, so we would expect a molecular solid.

Exercise 11.9 Decide what type of solid is formed for each of the following substances: C_2H_5OH, CH_4, CH_3Cl, $MgSO_4$. On the basis of the type of solid and the expected magnitude of intermolecular forces (for molecular crystals), arrange these substances in order of increasing melting point. Explain your reasoning.

Wanted: types of solids; increasing order of m.p.

Given: four substance formulas.

Known: Type of solid depends on the unit particles and the forces within them; m.p. depends on attractive force between particles; types and strengths of attractive forces are London < dipole < H-bond <<< ionic; electron-dot structures and forces are

$$
\begin{array}{cccc}
\underset{\displaystyle C_2H_5OH}{H-\overset{\displaystyle \overset{H}{|}}{\underset{\displaystyle \underset{H}{|}}{C}}-\overset{\displaystyle \overset{H}{|}}{\underset{\displaystyle \underset{H}{|}}{C}}-\ddot{O}:} & \underset{\displaystyle CH_4}{H-\overset{\displaystyle \overset{H}{|}}{\underset{\displaystyle \underset{H}{|}}{C}}-H} & \underset{\displaystyle CH_3Cl}{H-\overset{\displaystyle \overset{H}{|}}{\underset{\displaystyle \underset{H}{|}}{C}}-\ddot{Cl}:} & Mg^{2+}\ \left[\ :\!\ddot{O}\!:\ \overset{\displaystyle :\ddot{O}:}{\underset{\displaystyle :\ddot{O}:}{-\ \overset{|}{\underset{|}{S}}\ -}}\ \ddot{O}\!:\ \right]^{2-}
\end{array}
$$

      C$_2$H$_5$OH   CH$_4$ CH$_3$Cl   MgSO$_4$

forces within:

covalent	covalent	covalent	covalent
bonds	bonds	bonds	bonds in SO$_4^{2-}$

forces between:

London, dipole–	London	London,	ionic bonds
dipole, hydrogen		dipole–dipole	between Mg^{2+}
bond			and SO$_4^{2-}$

Solution:

type of solid:

molecular	molecular	molecular	ionic

Order by increasing melting point is

$$CH_4 < CH_3Cl < C_2H_5OH < MgSO_4$$

The first three substances listed are molecular because the forces within the units are covalent bonds. The fourth is ionic because the forces holding the units together are ionic bonds. The order of increasing melting points is as it is because the strength of attraction between the units increases.

The table on the next page summarizes physical properties related to solid structure and attractive forces, and gives examples in each case.

PHYSICAL PROPERTIES RELATED TO SOLID STRUCTURE

	Solid Structure Type				
	Ionic	Polar Molecule	Nonpolar Molecule	Covalent Network	Metallic
Examples	NaCl, KCl, CaBr, $MgCl_2$, CaF_2, Na_2O	HCl, H_2O, NH_3, PCl_3, C_2H_5OH, C_6H_5OH	I_2, CH_4, SF_6, Cl_2, $SiCl_4$, BF_3	SiO_2, C (diamond) SiC	Cu, Fe, Ni, Cr
Particle Type in Solid Crystalline Structure	Positive and negative ions	Polar molecular species	Nonpolar molecular species	Atoms	Positive ions and electrons
Major Type of Attractive Forces Between Particles in Solid Crystal	Electrostatic attractive	Electrostatic attractive (dipole–dipole)	London	Covalent bonds	Metallic bonds
Relative Strength of Attractive Forces	Strong	Strong to weak (depending on the magnitude of the molecular forces)	Wea to very weak (depending on the size of the molecular species)	Strong	Strong to weak
Physical Properties: Melting Points	High	Low to average	Low	Very high	High
Boiling Points	Very high	Low to average	Low	Very high	High to very high
Electrical Conductivity of Solid Phase	Very low	Very low	Very low	Very low	Very high
Electrical Conductivity of Melt	Very high	Very low	Very low	Very low	Very high
Solubility	More soluble in a polar solvent than in a nonpolar solvent	More soluble in a polar solvent than in a nonpolar solvent	More soluble in a nonpolar solvent than in a polar solvent	Insoluble in all ordinary solvents (except those which react with the network solid)	Soluble in other metals. Insoluble in all ordinary solvents (except those which react)
Hardness	High	High to average	Low	Very high	High
Brittleness	High	High to average	Low	Very high	High to average
Malleability	Very low	Low	Low	Very low	High
Thermal Conductivity	Low	Low to average	Low	Low	High

11.7 CRYSTALLINE SOLIDS; CRYSTAL LATTICES AND UNIT CELLS

Operational Skill

8. Determining the number of atoms per unit cell. Given the description of a unit cell, find the number of atoms per cell   (Example 11.9).

> **Exercise 11.10**   Figure 11.31 shows solid dots ("atoms") forming a two-dimensional lattice. A unit cell is marked off by the dashed line. How many "atoms" are there per unit cell in such a two-dimensional lattice?
>
> *Wanted:*   atoms per unit cell.
>
> *Given:*   Figure 11.31 in text.
>
> *Known:*   There are four corners, each part of four cells; there is one atom in the center of the cell.
>
> *Solution:*
>
> $$\frac{4 \text{ corners}}{\text{cell}} \times \frac{1/4 \text{ atom}}{\text{corner}} + \frac{1 \text{ (central) atom}}{\text{cell}} = 2 \text{ atoms/cell}$$

11.8 STRUCTURES OF SOME CRYSTALLINE SOLIDS

11.9 CALCULATIONS INVOLVING UNIT-CELL DIMENSIONS

Operational Skills

9. Calculating atomic mass from unit-cell dimension and density. Given the edge length of the unit cell, the crystal structure, and the density of a metal, calculate the mass of a metal atom   (Example 11.10).

10. Calculating unit-cell dimension from unit-cell type and density. Given the unit-cell structure, the density, and the atomic weight for an element, calculate the edge length of the unit cell   (Example 11.11).

> **Exercise 11.11**   Lithium metal has a body-centered cubic structure with all atoms at the lattice points and a unit-cell length of 350.9 pm. Calculate Avogadro's number. The density of lithium is 0.534 g/cm^3.
>
> *Wanted:*   Avogadro's number (atoms/mol).
>
> *Given:*   Li; bcc unit-cell length 350.9 pm; density of Li = 0.534 g/cm^3.

Known: $V = l^3$; density = mass/V; Li = 6.941 g/mol; 10^{10} pm = 1 cm; atoms per unit cell = $\{ [8 \times (1/8)] + 1 \} = 2$.

Solution:

$$\frac{2 \text{ atoms}}{\cancel{\text{cell}}} \times \frac{\cancel{\text{cell}}}{(350.9 \cancel{\text{pm}})^3} \times \frac{(10^{10} \cancel{\text{pm}})^3}{\cancel{\text{cm}^3}} \times \frac{\cancel{\text{cm}^3}}{0.534 \cancel{\text{g Li}}} \times \frac{6.941 \cancel{\text{g Li}}}{\text{mol}}$$

$$= 6.02 \times 10^{23} \text{ atoms/mol}$$

Exercise 11.12 Potassium metal has a body-centered cubic structure with all atoms at the lattice points. The density of the metal is 0.856 g/cm^3. Calculate the edge length of a unit cell.

Wanted: edge length (in cm or pm).

Given: K; bcc unit cell; density of K = 0.856 g/cm^3.

Known: K = 39.0983 g/mol; bcc unit cell has 2 atoms per cell; 10^{10} pm = 1 cm; $N_A = 6.022 \times 10^{23}$ atoms/mol.

Solution: Find the volume of one unit cell using dimensional analysis, as follows:

$$1 \cancel{\text{unit cell}} \times \frac{2 \cancel{\text{atoms}}}{\cancel{\text{unit cell}}} \times \frac{1 \cancel{\text{mol}}}{6.022 \times 10^{23} \cancel{\text{atoms}}} \times \frac{39.098 \cancel{\text{g}}}{1 \cancel{\text{mol}}} \times \frac{1 \text{ cm}^3}{0.856 \cancel{\text{g}}}$$

$$= 1.5\underline{1}7 \times 10^{-22} \text{ cm}^3$$

Take the cube root to find the edge length:

Edge length = 5.33×10^{-8} cm = 533 pm

11.10 DETERMINING CRYSTAL STRUCTURE BY X-RAY DIFFRACTION

A Substance That Matters: CARBON DIOXIDE (a Commercial Chemical; the Greenhouse Effect)

Questions for Study

1. Describe some of the physical properties of carbon dioxide.

2. Describe the production of carbon dioxide from natural gas.

3. How has the concentration of carbon dioxide in the atmosphere changed in recent years? Explain.

4. Explain what is meant by the greenhouse effect. Of what significance is the greenhouse effect on our present climate? Why is it of concern?

Answers to Questions for Study

1. Carbon dioxide is a colorless gas that is nontoxic at usual concentrations and more dense than air.

2. Natural gas (methane, CH_4) treated with steam yields a mixture of hydrogen and carbon dioxide as a result of the following reactions:

$$CH_4(g) + H_2O(g) \longrightarrow CO(g) + 3H_2(g)$$

$$CO(g) + H_2O(g) \longrightarrow H_2(g) + CO_2(g)$$

3. The concentration of CO_2 in the atmosphere has increased primarily from the burning of fossil fuels, such as natural gas, coal, and oil.

4. Light in the visible region of the electromagnetic spectrum, which is not absorbed by the gases in the upper atmosphere, is converted to heat when it is absorbed on the surface of the earth. This heat causes atoms in materials on the earth's surface to give off infrared radiation. Since CO_2 absorbs infrared radiation, this traps the sun's warmth in the atmosphere close to the earth's surface. Thus, the earth's atmosphere is warmed much the way that a greenhouse is warmed through the absorbing of infrared radiation by glass. As the CO_2 concentration in the atmosphere increases, more of the sun's warmth is trapped. This could potentially alter climatic conditions, which would affect food production and living conditions on our planet. Because the concentration of CO_2 continues to increase and is predicted to rise considerably over the next fifty years, the greenhouse effect is of particular concern.

ADDITIONAL PROBLEMS

1. Calculate the amount of thermal energy necessary to change 10.0 g of ice at $-15.0°C$ to steam at 110°C. The specific heat of water is 4.18 J/(g • °C), that of ice is 2.09 J/(g • °C), and that of steam is 2.03 J/(g • °C). The heat of fusion of ice at 0°C is

333 J/g; the heat of vaporization of liquid water at 100°C is 2.26×10^3 J/g. You may wish to look back at text Section 6.6.

2.   Describe the conditions necessary for liquefying the following gaseous substances: argon (critical point, – 122°C, 48 atm); hydrogen (critical point, –240°C, 12.8 atm); and sulfur dioxide (critical point, 157°C, 77.7 atm).

3.   Describe the types of forces that must be overcome in melting crystals of the following substances:

(a) SiO_2 (b)   Na   (c) I_2 (d) CH_3OH (e) $MgCl_2$

4.   Account for the following:
(a)   The glass in very old windows is sometimes found to be thicker at the bottom than at the top.
(b)   Soap bubbles always appear spherical in shape.
(c)   A sharp 1-to-2-degree range in melting point is usually characteristic of a pure crystalline solid.
(d)   Glass does not give a sharp melting point; instead it softens over a wide temperature range.

5.   Arrange the following compounds in order of increasing boiling point: H_2S, CH_3OH, PH_3, H_2O.

6.   List (by letter) the following substances in order of increasing melting point. Explain your answer.

(a) $AlPO_4$ (b)   LiI   (c) $C_{diamond}$

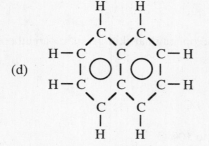

(d)   naphthalene, $C_{10}H_8$

(e) $O = C - C - C - C - C - C - H$ glucose, $C_6H_{12}O_6$

7.   Sodium (density 0.97 g/cm^3) crystallizes in a body-centered cubic structure. The measured length of the edge of a unit cell is 0.428 nm. Use this information to calculate Avogadro's number.

8.   Copper (density 9.0 g/cm^3) crystallizes in a face-centered cubic lattice. Calculate (a) the volume of a unit cell in cubic nanometers, and (b) the radius of the copper atom.

9.   The pressure regulator on a kitchen pressure cooker weighs 2.5 oz. It sits over the steam pipe, the inner diameter of which is 2.00 mm. See Figure E. What must be the temperature of the water inside the cooker to lift and "jiggle" the weight? Why can foods be prepared more quickly in a pressure cooker than in conventional cook-ware? (1 atm = 14.7 lb/in^2; the heat of vaporization of water is 40.7 × 10^3 J/mol.)

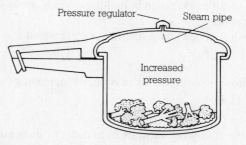

Figure E   A Pressure Cooker

ANSWERS TO ADDITIONAL PROBLEMS

If you missed an answer, study the text section and operational skill given in parentheses after the answer.

1.   30.6 kJ.
     (1)   31<u>3</u>.5 J to raise the ice temperature to 0°C
     (2)   33<u>30</u> J to melt the ice
     (3)   41<u>80</u> J to raise the water temperature to 100°C
     (4)   22,<u>6</u>00 J to vaporize the water to steam
     (5)   203 J to raise the steam temperature to 110°C   (11.2, Op. Sk. 1)

2. Argon can be liquefied by lowering the temperature to $-122°C$ and applying 48 atm pressure. Hydrogen can be liquefied by lowering the temperature to $-240°C$ and applying 12.8 atm pressure. Sulfur dioxide can be liquefied at room temperature by applying 77.7 atm pressure; its critical temperature is above room temperature.
   (11.3, Op. Sk. 3)

3. (a) covalent bonds   (d) hydrogen bonding and London forces
   (b) metallic bonding   (e) ionic bonds   (11.5, 11.6, Op. Sk. 4, 6)
   (c) London forces

4. (a) Glass is an amorphous solid that is essentially a highly viscous liquid. It flows very slowly. (11.6)
   (b) Surface tension causes the surface to assume the shape with the minimum surface area — a spherical surface. (11.4)
   (c) The strength of the forces holding the lattice particles together is the same throughout the entire crystal. (11.7)
   (d) Glass is an amorphous solid. The forces holding the particles together vary throughout the body of the glass. (11.6)

5. $PH_3 < H_2S < CH_3OH < H_2O$ (11.5, Op. Sk. 5)

6. d, e, b, a, c. Molecular substances (d and e) have lower melting points than do ionic substances (a and b). Ionic substances are lower-melting than network covalent ones (c). The reason is increasing bond strengths between the units making up the substances. Glucose (e) melts at a higher temperature than does naphthalene (d), because glucose molecules can hydrogen bond. Aluminum phosphate (a) is higher-melting than lithium iodide (b) because of the greater attraction due to higher ionic charges.   (11.6, Op. Sk. 7)

7. 6.0×10^{23} atoms/mol   (11.9, Op. Sk. 9)

8. (a) 4.7×10^{-2} nm^3 (b) 0.13 nm   (11.9, Op. Sk. 10)

9. 136°C. The area of the steam pipe opening $= \pi r^2 = 0.031\underline{4}2$ cm^2. The pressure from the regulator is

$$\frac{2.50 \text{ lb} / (16 \text{ lb} / \text{lb})}{0.031\underline{4}2 \text{ cm}^2} \times \frac{(2.54)^2 \text{ cm}^2}{\text{in}^2} \times \frac{\text{atm}}{14.7 \text{ lb} / \text{in}^2} = 2.1\underline{8}3 \text{ atm}$$

Adding the regulator pressure to atmospheric pressure gives a water vapor pressure of 2419 mmHg. Use the Clausius–Clapeyron equation to find T.

Foods cook faster in a pressure cooker because the temperature of the boiling water is higher than the temperature of boiling water in conventional cookware.

(11.2, Op. Sk. 2)

CHAPTER POST-TEST

1. On a scale of 1 to 5 (strongest to weakest), rank the following intermolecular forces:

    (a) dipole–dipole ____   (d) ion–dipole ____
    (b) ion-induced dipole ____   (e) dipole-induced dipole ____
    (c) London ____

2. Of the following list of compounds, which would you expect to have the lowest melting point?

    (a) PF_3 (c) GeF_4 (e) CsF

    (b) CF_4 (d) AlF_3

3. Which of the following statements concerning ionic substances is incorrect?
    (a) Typical ionic substances are NaCl, MgF_2, BaO, KI, and CsCl.
    (b) Ionic substances are formed between atoms that have markedly different attractions for valence-shell electrons.
    (c) In ionic substances, the strong attractive forces are electrostatic attractions between ions.
    (d) Electrical conductance of solid ionic substances is high because of the existence of ions in the crystalline lattice.
    (e) Ionic substances are more soluble in polar than in nonpolar solvents.

4. Compound A has a higher vapor pressure than compound B. However, both compounds have the same molecular formula, C_3H_6O. What may this suggest about the structure of A in comparison to B?

5. Copper is known to crystallize in a face-centered cubic lattice. Calculate the percent of the cell volume unoccupied by copper atoms if the unit cell measures 0.362 nm on an edge. (The radius of a copper atom is 1.28×10^{-8} cm; the volume of a sphere is $4/3 \pi r^3$.)

6.   The crystal-packing pattern in Figure F represents a
     (a)   hexagonal closest-packed structure.
     (b)   body-centered cubic structure.
     (c)   halite structure.
     (d)   zinc blende structure.
     (e)   cubic closest-packed structure.

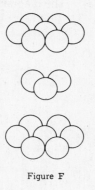

Figure F

7.   In a cubic closest-packed structure, what is the number of atoms per unit cell of crystallized solid?

8.   Indicate whether each of the following statements is true or false. If a statement is false, change it so it is true.
     (a)   Molecules of a nonpolar covalent solid are held together by van der Waals forces. True/False: _____

     _____ .

     (b)   As the molecular weights of similar molecules decrease, the boiling points usually increase. True/False: _____

     _____ .

     (c)   The critical pressure of a gas is the pressure required to liquefy a gas at 0 K. True/False: _____

     _____ .

     (d)   Molecular motion is more restricted in the solid phase than in the liquid phase. True/False: _____

     _____ .

(e)   The diagram in Figure G represents a body-centered cubic crystal.

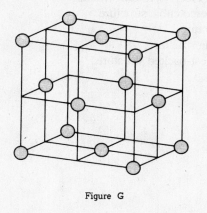

Figure G

True/False: _____ .

9.   Using the phase diagram in Figure H, explain what will happen as the substance illustrated is heated from 10°C to 90°C at a constant pressure of 0.75 atm.

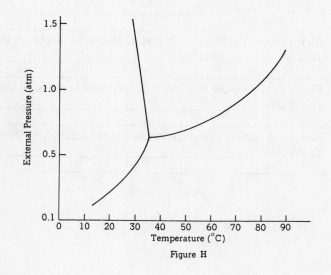

Figure H

10.   In the sublimation process, a substance undergoes a phase transition from a _____ state to a _____ state.

11. Elements A and B have identical densities and unit-cell volumes for their crystalline lattices. Element A crystallizes in a body-centered cubic lattice, and element B crystallizes in a face-centered cubic lattice. Which of the following is a valid conclusion concerning A and B?
    (a) The atomic weight of A is greater than that of B.
    (b) The atomic volume of A is less than that of B.
    (c) Element A has a higher melting point than that of B.
    (d) The atomic weight of A is less than that of B.
    (e) The atomic volume of A is greater than that of B.

12. Which of the following substances requires more heat to vaporize?
    (a) a 10.0-g sample of H_2O (heat of vaporization, 40.7 kJ/mol)
    (b) a 15.0-g sample of diethyl ether, $C_4H_{10}O$ (heat of vaporization, 27.7 kJ/mol)

ANSWERS TO CHAPTER POST-TEST

If you missed an answer, study the text section and operational skill given in parentheses after the answer.

1. (a) 2 (b) 3 (c) 5 (d) 1 (e) 4 (11.5, Op. Sk. 4)

2. b (11.2, 11.5, 11.6, Op. Sk. 6, 7)   3. d (11.6)

4. In compound A, hydrogen is not attached to oxygen. Compound B does contain an H — O bond and can hydrogen bond in the liquid state; thus, it has a lower vapor pressure. (11.5, Op. Sk. 5)

5. 25.9% (11.7, 11.9, Op. Sk. 8, 10)   6. a (11.8)

7. 4 (11.7, 11.8, Op. Sk. 8)

8. (a) True. (11.5, Op. Sk. 4)
   (b) False. As the molecular weights of similar molecules increase, the boiling points usually increase. (11.2, 11.5, Op. Sk. 5)
   (c) False. The critical pressure of a gas is the minimum pressure required to liquefy a gas at its critical temperature. (11.3, Op. Sk. 3)
   (d) True. (11.6)
   (e) False. The diagram in Figure G represents a face-centered cubic crystal structure. (11.7, Op. Sk. 8)

9. At 10°C the substance is in the solid state. Warming at a constant pressure will result in the solid melting at about 33°C. Once it is melted, continued heating will warm the liquid to about 58°C. At this point, the liquid will boil, converting itself to a gas. Any further warming will increase the temperature of the gas. (11.3)

10. solid, gaseous   (11.2)   11.   a   (11.7, 11.9)

12. a   (11.2, Op. Sk. 1)

CHAPTER 12   SOLUTIONS

CHAPTER TERMS AND DEFINITIONS

Numbers in parentheses after definitions give the text sections in which the terms are explained. Starred terms are italicized in the text. Where a term does not fall directly under a text section heading, additional information is given for you to locate it.

solution*   homogeneous mixture of two or more substances, consisting of ions or molecules   (12.1, introductory section)

solute   gas or solid dissolved in a liquid, or the solution component in smaller amount   (12.1)

solvent   liquid that dissolves a gas or solid, or the solution component in greater amount   (12.1)

miscible fluids   fluids that mix with or dissolve in each other in all proportions   (12.1)

immiscible*   said of two fluids that do not mix but form two layers   (12.1)

dynamic equilibrium*   equilibrium in which the forward and reverse processes, such as the dissolving and depositing of crystals, are always occurring and at the same rate   (12.2)

saturated solution   solution that is in equilibrium with respect to a given dissolved substance   (12.2)

solubility   amount that dissolves in a given quantity of liquid at a given temperature to give a saturated solution   (12.2)

unsaturated solution   solution not in equilibrium with respect to a given dissolved substance and in which more of the substance can dissolve   (12.2)

supersaturated solution   solution that contains more dissolved substance than a saturated solution   (12.2)

entropy*   measure of disorder   (12.2, marginal note)

ion–dipole force*   attraction between an ion and a polar molecule   (12.2)

hydration (of ions)   attraction of ions for water molecules   (12.2)

lattice energy*   energy holding ions together in a crystal lattice   (12.2)

heat of solution*   heat that is released or absorbed when a substance dissolves in a solvent   (12.3)

Le Chatelier's principle   when a system in equilibrium is altered by a change of
     temperature, pressure, or concentration variable, the system shifts in equilibrium
     composition in a way that tends to counteract this change of variable   (12.3)
Henry's law   the solubility of a gas is directly proportional to the partial pressure of the
     gas above the solution; $S = k_H P$ (12.3)
colligative properties   properties of solutions that depend on the concentration of solute
     particles (molecules or ions) in solution but not on the chemical identity of the
     solute   (12.4, introductory section)
concentration*   amount of solute dissolved in a given quantity of solvent or solution
     (12.4)
molarity (*M*)*   number of moles of solute per liter of solution   (12.4)
mass percentage of solute   percentage by mass of solute contained in a solution   (12.4)
molality (*m*)   number of moles of solute per kilogram of solvent   (12.4)
mole fraction (*X*)   number of moles of a substance divided by the total number of moles
     of solution   (12.4)
mole percent*   mole fraction times 100   (12.4)
vapor-pressure lowering   colligative property of a solution; equal to the vapor pressure
     of the pure solvent minus the vapor pressure of the solution   (12.5)
Raoult's law   the partial pressure of solvent, P_A, over a solution is equal to the vapor
     pressure of the pure solvent, P_A°, multiplied by the mole fraction of solvent in the
     solution: $P_A = P_A^{\circ} X_A$ (12.5)
ideal solution*   solution in which both substances follow Raoult's law for all values of
     mole fractions   (12.5)
normal (boiling point of a liquid)*   temperature at which the vapor pressure of the
     liquid equals 1 atm   (12.6)
boiling-point elevation (ΔT_b)   colligative property of a solution; equal to the boiling
     point of the solution minus the boiling point of the pure solvent   (12.6)
boiling-point-elevation constant (K_b)*   proportionality constant between the
     boiling-point elevation and the molality of a solution   (12.6)
freezing-point depression (ΔT_f)   colligative property of a solution; equal to the freezing
     point of the pure solvent minus the freezing point of the solution   (12.6)
freezing-point-depression constant (K_f)*   proportionality constant between the
     freezing-point lowering and the molality of a solution   (12.6)
semipermeable*   describes a membrane that allows solvent molecules, but not solute
     molecules, to pass through   (12.7)
osmosis   phenomenon of solvent flow through a semipermeable membrane from lower
     solute concentration to higher concentration to equalize concentrations on both sides
     of the membrane   (12.7)
osmotic pressure   colligative property of a solution; equal to the pressure that, when
     applied to the solution, just stops osmosis   (12.7)

reverse osmosis* reversing the osmosis process by applying greater pressure to the more concentrated solution so that solvent flows from the concentrated solution through a membrane to the more dilute solution (12.7)

desalinate* to remove salts from seawater to make it usable for drinking or industrial uses (12.7)

activities* effective concentrations of ions in solution (12.8)

Debye–Hückel theory* describes the distribution of ions in a salt solution and enables us to calculate ionic activities (12.8)

colloid dispersion of particles of one substance (the dispersed phase) throughout another substance or solution (the continuous phase) (12.9)

Tyndall effect scattering of light by colloidal-size particles (12.9)

aerosols colloidal dispersions of liquid droplets or solid particles throughout a gas (12.9)

emulsion colloidal dispersion of liquid droplets throughout another liquid (12.9)

sol colloidal dispersion of solid particles in a liquid (12.9)

hydrophilic colloid colloid in which there is a strong attraction between the dispersed phase and the continuous phase (water) (12.9)

hydrophobic colloid colloid in which there is a lack of attraction between the dispersed phase and the continuous phase (water) (12.9)

nuclei (in crystallization)* centers about which crystallization occurs (12.9)

coagulation process by which the dispersed phase of a colloid is made to aggregate and thereby separate from the continuous phase (12.9)

micelle colloidal-size particle formed in water by the association of molecules or ions that each have a hydrophobic and a hydrophilic end (12.9)

association colloid colloid in which the dispersed phase consists of micelles (12.9)

hydrologic cycle* natural cycle of water from the oceans to freshwater sources and back to the oceans (A Solvent That Matters: Water [A Special Substance for Planet Earth])

hard water* naturally occurring water containing certain metal ions, such as Ca^{2+} and Mg^{2+} (A Solvent That Matters: Water [A Special Substance for Planet Earth])

water softening* process of removing Ca^{2+} and Mg^{2+} ions from hard water (A Solvent That Matters: Water [A Special Substance for Planet Earth])

ion exchange* method of softening water in which a water solution is passed through a column of a material that replaces one kind of ion in solution with another kind (A Solvent That Matters: Water [A Special Substance for Planet Earth])

CHAPTER DIAGNOSTIC TEST

1. When solid A dissolves in water, there occurs a corresponding decrease in tempera-
    ture. Therefore, heating a mixture of solid A and water should _____
    (decrease/increase) the solubility of A in water.

2. The graph in Figure A plots solubility as a function of temperature.

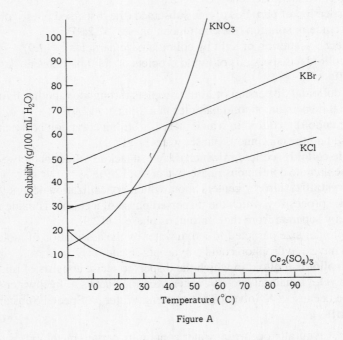

Figure A

Which of the following statements concerning information available from the
solubility curves in Figure A is incorrect?
(a)   At all indicated temperatures, KBr is more soluble in water than is KCl. ✓
(b)   The solubility of KBr can be increased by warming the solution. ✓
(c)   At 35°C the order of solubilities is KBr > KNO_3 > KCl > $Ce_2(SO_4)_3$. ✓
(d)   The solubility of $Ce_2(SO_4)_3$ is an endothermic process.
(e)   The solubility of KNO_3 is more affected by temperature than that of KCl. ✓

3. The energy of hydration increases with decreasing cation size, when cationic charge
    is constant, because
    (a)   a larger cation can coordinate more water molecules about it.
    (b)   cationic charge has no effect on the coordination of water molecules.

(c) electrostatic attractive forces are inversely proportional to cation size.
(d) the cation size has no effect on hydration when the cationic charge changes.
(e) cations are attracted to water molecules and anions are repelled by water molecules.

4. For the following chemical reaction at equilibrium,

$$H_3PO_4 + KOH \rightleftharpoons KH_2PO_4 + H_2O; \quad \Delta H \text{ is negative } (exothermic)$$

the addition of water to the reaction mixture will

(a) produce a larger ΔH per mole of H_3PO_4.
(b) increase the amount of KH_2PO_4 when the new equilibrium is established.
(c) have no effect.
(d) increase the amount of H_3PO_4 and KOH when equilibrium is re-established.
(e) have no effect other than giving a more dilute solution of KH_2PO_4.

5. Which of the following carbonates would you expect to be insoluble in water?

(a) Na_2CO_3 (c) $(NH_4)_2CO_3$ (e) $BaCO_3$

(b) H_2CO_3 (d) Rb_2CO_3

6. A solution of 132.4 g $Cu(NO_3)_2$ per liter has a density of 1.116 g/mL. The weight percent of $Cu(NO_3)_2$ in the solution is

(a) 11.86.   (b) 8.40.   (c) 9.00.   (d) 22.4.
(e) insufficient data given to calculate weight percent.

7. Calculate the mole fraction of each component in a solution that contains 46.5 g of ethylene glycol, CH_2OHCH_2OH, and 236 g of methanol, CH_3OH.

8. Calculate the molecular formula of a species that has the empirical formula CH_3O if, when 25.1 g of the compound was added to 0.150 kg of water, the solution froze at $-5.0°C$. $K_f = 1.86°C/m$.

9. Which of the following should have the highest boiling point?
(a) pure water
(b) a 0.2 m $Ca(NO_3)_2$ aqueous solution
(c) a 0.2 m $(CH_3)_2CO$ aqueous solution
(d) a 0.1 m KBr aqueous solution
(e) a 0.1 m $Ba(NO_3)_2$ aqueous solution

10.   The graph in Figure B shows the effect of the addition of a nonvolatile solute on the vapor pressure of water.

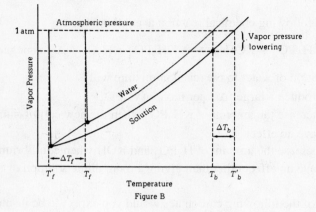

Figure B

    Which of the following statements is a logical deduction from this graph?

✗ (a)   If a nonvolatile solute is added to water, the boiling point of the solution will be T_b.

✗ (b)   The addition of a volatile solute will change the boiling point of the water.

✗ (c)   Atmospheric pressure will affect the composition of the aqueous solution.

(d)   If a nonvolatile solute is added to water, the freezing point of the solution will be lower than that of water.

✗ (e)   If a nonvolatile solute is added to water, the boiling and melting points of the solution will be greater than those of the water by ΔT_b and ΔT_f, respectively.

11.   A liter of water at 25°C and 1.0 atm dissolves 1.45 g of carbon dioxide. If the partial pressure of CO_2 is increased to 15 atm, what is its solubility in water? $\dfrac{S_1}{S_2} = \dfrac{P_1}{P_2}$

12.   An 8.0 M solution of KCl in water has a density of 1.218 g/mL. The density of pure water is 1.000 g/mL. Calculate the mole fraction of KCl in the solution.

$$1 L \left(1218 g/L\right) = 1218 g \left(\frac{mol}{\,\,}\right) \qquad \frac{\downarrow\,moles}{total\,moles} = \frac{8}{8+}$$

13.   Calculate the boiling point of a glucose solution consisting of 9.0 g of glucose dissolved in 100.0 g of water (glucose = $C_6H_{12}O_6$, 180.1 g/mol, K_b for water is 0.512°C/m). $\Delta T_b = (T_b) T_b° = K_b C_m$

                                                       $100.26°C$

14.   A solution consists of 0.100 mole of naphthalene, $C_{10}H_8$, and 9.90 moles of benzene, C_6H_6, at 25°C. Calculate the vapor pressure and vapor-pressure lowering

$$P_A = P°_A X_A \qquad \Delta P = P°_A X_B$$
$$= 267.3 \qquad\qquad = 2.7$$

of the solution. (The vapor pressure of pure benzene at 25°C is 2.70×10^2 mmHg. Assume naphthalene is a nonvolatile, nonelectrolytic solute.)

15. Calculate the osmotic pressure of a solution that consists of 4.68 g of hemoglobin (molar mass = 6.83×10^4 g/mol) in 125 mL of water (molar mass = 18.0 g/mol) at 3.00×10^2 K. $\Pi = MRT$.

16. Classify each of the following colloids as an aerosol, foam, emulsion, sol, or gel:
    (a) cigarette smoke ①   (e) carbon ink ④
    (b) strawberry-flavored jello ⑤   (f) pearl ⑤
    (c) muddy water ④   (g) raised dough ②
    (d) milk ③

ANSWERS TO CHAPTER DIAGNOSTIC TEST

If you missed an answer, study the text section and operational skill given in parentheses after the answer.

1. increase   (12.2, 12.3)   2. d   (12.2, 12.3)   3. c   (12.2)

4. d   (12.3)   5. e   (12.2)   6. a   (12.3)

7. $X_{glycol} = 9.23 \times 10^{-2}$; $X_{CH_3OH} = 9.08 \times 10^{-1}$ (12.4, Op. Sk. 2)

8. molecular formula: $C_2H_6O_2$ (12.6, Op. Sk. 5, 6)

9. b   (12.6, 12.8)   10. d   (12.5, 12.6)

11. 22 g CO_2/L H_2O (12.3, Op. Sk. 1)

12. mole fraction KCl = 0.19   (12.4, Op. Sk. 3)

13. b.p. = 100.26°C   (12.4, 12.6, Op. Sk. 2, 5)

14. vapor pressure = 267 mmHg; vapor-pressure lowering = 2.70 mmHg
(12.5, Op. Sk. 4)

15.   0.0135 atm   (12.7, Op. Sk. 7)

16.   (a)   aerosol   (c)   sol   (e)   sol   (g)   foam   (12.9)
      (b)   gel   (d)   emulsion   (f)   gel

SUMMARY OF CHAPTER TOPICS

12.1   TYPES OF SOLUTIONS

Whether we call a mixture of one or more substances in another a suspension, a colloid, or
a solution depends on the size of the particles, particularly of those we are adding. To
illustrate, assume we have water as the solvent. If the particles we add are large and heavy,
they will sink to the bottom. If we stir the mixture vigorously, we will have a *suspension*,
wherein the added particles are momentarily suspended in the continuous phase, the water.
If the particles are small and light, they may stay suspended; light passing through this
suspension, or *colloid*, will be scattered. If the particles are the size of molecules and ions,
they are so small that we cannot discern them, unless they color the *solution*, as does the
Cu^{2+} ion in $CuSO_4$.

The definitions of the key terms in this chapter are of particular importance.
Memorize them as soon as possible. Your ability to do the exercises and problems will
depend on your knowing them.

Exercise 12.1   Give an example of a solid solution prepared from a liquid and a
solid.

Solution: The dental-filling alloy mentioned in the text has liquid mercury in silver (and other metals), which gives a solid solution.

12.2   SOLUBILITY AND THE SOLUTION PROCESS

The solubility of one substance in another at a given temperature and pressure is a function
of two factors: the attraction the substance has for its own species, and the attraction the
substance has for the other species. When we dissolve a solid in water, we can measure
these attractions in terms of lattice energy and energy of hydration, respectively.

Exercise 12.2   Which of the following compounds is likely to be more soluble
in water: C_4H_9OH or C_4H_9SH? Explain.

Known: The molecules differ only in S and O; oxygen can hydrogen bond, whereas sulfur cannot.

Solution: C_4H_9OH is the more soluble, as it can hydrogen bond with water.

Exercise 12.3   Which ion has the larger hydration energy, Na^+ or K^+?

Known: Hydration energy is inversely proportional to ionic radius and proportional to charge; K^+ has the same charge as Na^+ and is below it on the periodic table, so it is larger.

Solution: Na^+, being smaller, has the larger hydration energy.

12.3   EFFECTS OF TEMPERATURE AND PRESSURE ON SOLUBILITY

Operational Skill

1.   Applying Henry's law.   Given the solubility of a gas at one pressure, find its solubility at another pressure   (Example 12.1).

According to the principle of Le Chatelier (pronounced "la shaught lee ay' "), if a system in a state of equilibrium is disturbed by a change of some variable, the system shifts in equilibrium composition in a way that tends to counter that change of variable. This statement is an observation. It is not an explanation of what happens on a molecular or ionic level. A state of equilibrium exists when the rates of both the forward and reverse reactions are equal. On the molecular level, when a system at equilibrium is disturbed, the rate of one of the reactions is increased. Thus, we say the equilibrium is shifted. The system will again reach equilibrium if left undisturbed, but the composition of the mixture will not be the same. It appears as if the system has tried to negate the change.

Exercise 12.4   A liter of water at 25°C dissolves 0.0404 g O_2 when the partial pressure of the oxygen is 1.00 atm. What is the solubility of oxygen from air, in which the partial pressure of O_2 is 159 mmHg?

Wanted: solubility (S_2) in water of O_2 from air.

Given: $t = 25°C$; $S_1 = 0.0404$ g/L; $P_1 = 1$ atm = 760 mmHg; $P_2 = 159$ mmHg.

Known: Henry's law: the ratio of solubilities is proportional to the ratio of pressures:

$$\frac{S_2}{S_1} = \frac{P_2}{P_1}$$

Solution: Solving for S_2 and substituting, we get

$$S_2 = \frac{P_2 S_1}{P_1} = \frac{159 \ \cancel{mmHg} \times 0.0404 \ g}{760 \ \cancel{mmHg} \ L} = 0.00845 \ g/L$$

12.4   WAYS OF EXPRESSING CONCENTRATION

Operational Skills

2.   Calculating solution concentration.   Given the mass percent of solute, state how to prepare a given mass of solution   (Example 12.2). Given the masses of solute and solvent, find the molality (Example 12.3) and mole fractions   (Example 12.4).

3.   Converting concentration units.   Given the molality of a solution, calculate the mole fractions of solute and solvent; and given the mole fractions, calculate the molality (Examples 12.5 and 12.6). Given the density, calculate the molarity from the molality, and vice versa   (Examples 12.7 and 12.8).

This and the remaining sections of the chapter deal with colligative properties of solutions, which are properties that depend on the number of particles in a given amount of solution or solvent, rather than on the identity of the particles. The word *colligative* comes from the same Latin root that the word *collection* comes from — "to gather together." If you think of *collection* when you see *colligative*, it will help you remember these properties. Memorize the four colligative properties of solutions. They are (1) vapor pressure; the two properties related to vapor pressure, that is, (2) boiling-point elevation and (3) freezing-point depression; and (4) osmotic pressure.

Exercise 12.5   An experiment calls for 35.0 g of hydrochloric acid that is 20.2% HCl by mass. How many grams of HCl is this? How many grams of water?

Wanted: g HCl, g H_2O.

Given: need 35.0 g solution; solution is 20.2% HCl by mass.

Known: 20.2 mass % means

$$\frac{20.2 \text{ g HCl}}{100 \text{ g solution}} \times 100.$$

Solution: Find g HCl by dimensional analysis. Begin with what you can hold in a container — the 35.0 g of solution:

$$35.0 \text{ g } \cancel{\text{solution}} \times \frac{20.2 \text{ g HCl}}{100 \text{ g } \cancel{\text{solution}}} = 7.07 \text{ g HCl}$$

Find g H_2O by subtracting g HCl from 35.0 g solution:

$$35.0 \text{ g solution} - 7.07 \text{ g HCl} = 27.9 \text{ g } H_2O$$

Exercise 12.6   Toluene, $C_6H_5CH_3$, is a liquid compound similar to benzene, C_6H_6. It is the starting material for other substances, including trinitrotoluene (TNT). Find the molality of toluene in a solution that contains 35.6 g of toluene and 125 g of benzene.

Wanted: molality (m) of toluene solution.

Given: Solution contains 35.6 g toluene, 125 g benzene; formulas are given.

Known: Definition of molality $= \dfrac{\text{moles solute}}{\text{kg solvent}}$.

　　　　Toluene, $C_6H_5CH_3 = 92.1$ g/mol.

Solution: Convert g toluene to moles, and g benzene to kg:

$$\frac{35.6 \cancel{\text{ g}}}{92.1 \cancel{\text{ g}} / \text{mol}} = 0.38\underline{6}5 \text{ mol toluene}$$

125 g benzene = 0.125 kg

$$m = \frac{0.38\underline{6}5 \text{ mol}}{0.125 \text{ kg}} = 3.09 \text{ } m \text{ } C_6H_5CH_3$$

Exercise 12.7   Calculate the mole fractions of toluene and benzene in the solution described in Exercise 12.6.

Wanted: X_{toluene}, X_{benzene}.

Given: 0.38\underline{6}5 mol toluene (calculated in Exercise 12.6); 125 g benzene. Formula of benzene is C_6H_6.

Known: Mole fraction $= \dfrac{\text{mol substance}}{\text{mol total}}$; $C_6H_6 = 78.1$ g/mol.

Solution: Calculate moles benzene:

$$\frac{125 \ \cancel{g} \ \text{benzene}}{78.1 \ \cancel{g} \ /\text{mol}} = 1.6\underline{0}1 \ \text{mol benzene}$$

Calculate total moles:

0.38$\underline{65}$ mol toluene + 1.6$\underline{0}$1 mol benzene = 1.9$\underline{88}$ total mol

Then,

$$\text{Mole fraction toluene} \ = \frac{0.38\underline{65} \ \cancel{\text{mol}}}{1.9\underline{88} \ \cancel{\text{mol}}} = 0.194$$

$$\text{Mole fraction benzene} \ = \frac{1.6\underline{0}1 \ \cancel{\text{mol}}}{1.9\underline{88} \ \cancel{\text{mol}}} = 0.805$$

Exercise 12.8 A solution is 0.120 m methanol dissolved in ethanol. Calculate the mole fractions of methanol, CH_3OH, and ethanol, C_2H_5OH, in the solution.

Known: $m = \dfrac{\text{mol methanol}}{\text{kg ethanol}}$, so, since the solution is 0.120 m, moles of methanol in 1 kg solvent = 0.120 mol; ethanol, $C_2H_5OH = 46.1$ g/mol.

Solution: Assume 1 kg solvent and calculate moles ethanol:

$$\frac{1000 \ \cancel{g}}{46.1 \ \cancel{g} \ /\text{mol}} = 21.\underline{69} \ \text{mol ethanol}$$

Calculate total moles:

0.120 mol methanol + 21.$\underline{69}$ mol ethanol = 21.$\underline{81}$ total mol

Then,

$$\text{Mole fraction of methanol} \ = \frac{0.120 \ \cancel{\text{mol}}}{21.\underline{81} \ \cancel{\text{mol}}} = 5.50 \times 10^{-3}$$

$$\text{Mole fraction of ethanol} \ = \frac{21.\underline{69} \ \cancel{\text{mol}}}{21.\underline{81} \ \cancel{\text{mol}}} = 0.994$$

Exercise 12.9 A solution is 0.250 mole fraction methanol, CH_3OH, and 0.750 mole fraction ethanol, C_2H_5OH. What is the molality of methanol in the solution?

Known: $m = \dfrac{\text{mol methanol}}{\text{kg ethanol}}$; $C_2H_5OH = 46.1$ g/mol.

Solution: Assume one mole of solution and find the kg of ethanol. Since the mole fraction of ethanol is 0.750, in one mole of solution there would be 0.750 mole of ethanol. The mass of this ethanol is

$$0.750 \; \text{mol} \times 46.1 \; \text{g}/\text{mol} = 34.\underline{5}8 \; \text{g ethanol} = 0.034\underline{5}8 \; \text{kg}$$

$$m = \frac{0.250 \; \text{mol}}{0.034\underline{5}8 \; \text{kg}} = 7.23 \; m \; CH_3OH$$

Exercise 12.10 Urea, $(NH_2)_2CO$, is used as a fertilizer. What is the molar concentration of an aqueous solution that is 3.42 *m* urea? The density of the solution is 1.045 g/mL.

Wanted: molarity (*M*).

Given: Molality (*m*) is 3.42; solution density is 1.045 g/mL.

Known: definitions; urea = 60.0 g/mol. We must convert

$$m, \frac{3.42 \; \text{mol urea}}{1 \; \text{kg water}}, \text{ to } M, \frac{\text{mol urea}}{\text{L solution}}.$$

Solution: First find the mass of solution that contains 3.42 mol urea and 1 kg water:

The mass of urea is

$$3.42 \; \text{mol} \; \text{urea} \times 60.0 \; \text{g}/\text{mol} = 205.2 \; \text{g urea}$$

The total weight of solution is

$$100\underline{0} \; \text{g} + 20\underline{5}.2 \; \text{g} = 120\underline{5}.2 \; \text{g of solution}$$

The volume of solution is its mass divided by its density

$$\frac{120\underline{5}.2 \; \text{g}}{1.045 \; \text{g}/\text{mL}} = 115\underline{3}.3 \; \text{mL} = 1.15\underline{3}3 \; \text{L}$$

So,

$$M = \frac{3.42 \; \text{mol}}{1.15\underline{3}3 \; \text{L}} = 2.97 \; \text{mol/L} = 2.97 \; M$$

Exercise 12.11 An aqueous solution is 2.00 *M* urea. The density of the solution is 1.029 g/mL. What is the molal concentration of urea in the solution?

Known: We must convert

$$M, \frac{2.00 \; \text{mol urea}}{\text{L solution}} \text{ to } m, \frac{\text{mol urea}}{\text{kg water}}; \text{ urea} = 60.0 \; \text{g/mol}.$$

Solution: Find the mass of water in one liter of a 2.00 M solution:

$$\text{Total mass of solution} = 1000 \, \text{mL} \times 1.029 \, \text{g/mL} = 1029 \, \text{g}$$

$$\text{Mass of solute} = 2.00 \, \text{mol urea} \times 60.0 \, \text{g/mol} = 12\underline{0}.0 \, \text{g urea}$$

$$\text{Mass of solvent} = 1029 \, \text{g soln} - 12\underline{0}.0 \, \text{g urea} = 909 \, \text{g water}$$

$$= 0.909 \, \text{kg water}$$

Thus,

$$m = \frac{2.00 \, \text{mol}}{0.909 \, \text{kg}} = 2.20 \, m$$

12.5   VAPOR PRESSURE OF A SOLUTION

Operational Skill

4.   Calculating vapor-pressure lowering.   Given the mole fraction of solute in a solution of nonvolatile, undissociated solute and the vapor pressure of pure solvent, calculate the vapor-pressure lowering and vapor pressure of the solution   (Example 12.9).

Recall that the vapor pressure of a solution is the pressure of the vapor above the liquid when equilibrium exists — when the rate of vaporization equals the rate of condensation. We consider cases where only the solvent is appreciably volatile, so in the discussion of colligative properties we will be comparing the vapor pressure of the pure solvent with its vapor pressure in solution.

Memorize Raoult's law, $P_A = P_A^\circ X_A$, which states that the vapor pressure of the solvent in solution equals the vapor pressure of the pure solvent times its mole fraction in the solution.

Exercise 12.12   Naphthalene, $C_{10}H_8$, is used to make mothballs. Suppose a solution is made by dissolving 0.515 g of naphthalene in 60.8 g of chloroform, $CHCl_3$. Calculate the vapor-pressure lowering of chloroform at 20°C from the naphthalene. The vapor pressure of chloroform at 20°C is 156 mmHg. Naphthalene can be assumed to be nonvolatile compared to chloroform. What is the vapor pressure of the solution?

Wanted: P_{CHCl_3}

Given: $t = 20°C$; 0.515 g naphthalene, $C_{10}H_8$; 60.8 g chloroform, $CHCl_3$; $P_{CHCl_3}^\circ = 156$ mmHg

Known: $\Delta P_{CHCl_3} = P^\circ_{CHCl_3} X_{C_{10}H_8}$; $\Delta P_{CHCl_3} = P^\circ_{CHCl_3} - P_{CHCl_3}$;
naphthalene = 128.1 g/mol.

Solution: Calculate the mole fraction, X, of naphthalene by first calculating the moles of each compound.

$$\frac{0.515 \ g \ C_{10}H_8}{128.1 \ g/mol} = 0.004020 \ mol \ C_{10}H_8$$

$$\frac{60.8 \ g \ CHCl_3}{119.5 \ g/mol} = 0.5088 \ mol \ CHCl_3$$

$$X_{C_{10}H_8} = \frac{0.004020 \ mol}{(0.004020 + 0.5087) \ mol} = \frac{0.004020}{0.5128} = 0.007839$$

Putting values in the expression for vapor-pressure lowering gives

$$\Delta P_{CHCl_3} = P^\circ_{CHCl_3} X_{C_{10}H_8} = 156 \ mmHg \times 0.007839 = 1.223 \ mmHg$$

Assuming that naphthalene is nonvolatile, the vapor pressure of the solution is the vapor pressure of the chloroform in solution:

$$P_{CHCl_3} = P^\circ_{CHCl_3} - \Delta P_{CHCl_3} = (156 - 1.223) \ mmHg = 155 \ mmHg$$

12.6   BOILING-POINT ELEVATION AND FREEZING-POINT DEPRESSION

Operational Skills

5.   Calculating boiling-point elevation and freezing-point depression.   Given the molality of a solution of nonvolatile, undissociated solute, calculate the boiling-point elevation and freezing-point depression   (Example 12.10).

6.   Calculating molecular weights.   Given the masses of solvent and solute and the molality of the solution, find the molecular weight of the solute   (Example 12.11). Given the masses of solvent and solute, the freezing-point depression, and K_f, find the molecular weight of the solute   (Example 12.12).

Boiling-point elevation is not too difficult a concept to grasp. Since the vapor pressure of a solvent is decreased when a solute is added, we have to heat the solution to a higher temperature to make the vapor pressure equal the atmospheric pressure, and thus to make the solution boil.

Freezing-point depression is a more difficult concept. You may have to read your text and the following explanation a few times. It will also be a good idea to take another look at phase diagrams, text Section 11.3, to be sure you understand all the information there.

The freezing point of a substance is the temperature at which the liquid and solid are in dynamic equilibrium; melting and freezing occur at the same rate. At this temperature, the vapor pressures of the liquid and solid phases are the same. If we decrease the vapor pressure of the liquid by adding solute particles, the liquid in solution will have a lower vapor pressure than the solid at the freezing point of the pure liquid, and freezing will not occur. As we continue to lower the temperature, the vapor pressure of the solid decreases more rapidly than that of the liquid in solution, and we reach a temperature at which both solid and liquid in solution have the same vapor pressure. (See Figure B in the chapter diagnostic test.) At this lower temperature, then, the solution will freeze.

Memorize the formulas for boiling-point elevation and freezing-point depression. Do not worry about memorizing the constants (they are provided in text Table 12.3) unless your instructor asks you to do so.

Exercise 12.13   How many grams of ethylene glycol, CH_2OHCH_2OH, must be added to 37.8 g of water to give a freezing point of $-0.150°C$?

Wanted: g ethylene glycol.

Given: formula; 37.8 g water; $T_f = -0.150°C$.

Known: $\Delta T_f = K_f c_m$; $K_f = 1.858°C/m$ for water; $m = \dfrac{\text{moles solute}}{\text{kg solvent}}$; ethylene glycol = 62.1 g/mol.

Solution: Rearrange $\Delta T_f = K_f c_m$ to solve for molality:

$$m = \frac{\Delta T_f}{K_f} = \frac{0.150°\cancel{C}}{1.858°\cancel{C}/m} = 0.0807\underline{3}\ m$$

Solve for g ethylene glycol (e. gly.):

$$37.8\ \cancel{\text{g H}_2\text{O}} \times \frac{0.08073\ \cancel{\text{mol e. gly.}}}{1000\ \cancel{\text{g H}_2\text{O}}} \times \frac{62.1\ \text{g e. gly.}}{\cancel{\text{mol e. gly.}}} = 0.190\ \text{g e. gly.}$$

Exercise 12.14   A 0.930-g sample of ascorbic acid (vitamin C) was dissolved in 95.0 g of water. The concentration of ascorbic acid, as determined by freezing-point depression, was 0.0555 m. What is the molecular weight of ascorbic acid?

Wanted: molecular weight of ascorbic acid (amu).

Given: 0.930 g ascorbic acid; 95.0 g = 0.0950 kg H_2O; $m = 0.0555\ m$.

Known: $m = \dfrac{\text{moles ascorbic acid}}{\text{kg } H_2O}$; molar mass = g/mol = same number as for molecular weight.

Solution: Find moles of ascorbic acid using a rearranged definition of molality:

$$\text{Moles} = m \times \text{kg } H_2O = \frac{0.0555\ \text{mol}}{\text{kg } H_2O} \times 0.0950\ \text{kg } H_2O = 0.005272\ \text{mol}$$

Then,

$$\text{Molar mass} = \frac{0.930\ \text{g asc. acid}}{0.005272\ \text{mol}} = 176\ \text{g/mol}$$

$$\text{Molecular weight} = 176\ \text{amu}$$

Exercise 12.15 A 0.205-g sample of white phosphorus was dissolved in 25.0 g of carbon disulfide, CS_2. The boiling-point elevation of the carbon disulfide solution was found to be 0.159°C. What is the molecular weight of the phosphorus in solution? What is the formula of molecular phosphorus?

Wanted: MW of P (amu); formula of molecular P.

Given: 0.205 g P; 25.0 g CS_2 (0.0250 kg); $\Delta T_b = 0.159$°C.

Known: $\Delta T_b = K_b c_m$; definition of molality; K_b of $CS_2 = 2.40$°C/m (text Table 12.3). To get formula of elemental molecular substance, divide molecular weight by atomic weight.

Solution: Find molality from ΔT_b formula:

$$m = \frac{\Delta T_b}{K_b} = \frac{0.159°\text{C}}{2.40°\text{C}/m} = 0.06625\ m$$

Find moles P from definition of molality:

$$\text{mol P} = m \times \text{kg solvent} = 0.06625\ \text{mol}/\text{kg } CS_2 \times 0.0250\ \text{kg } CS_2$$

$$= 0.001656\ \text{mol P}$$

Thus,

$$\text{MW} = \frac{0.205\ \text{g P}}{0.001656\ \text{mol P}} = 124\ \text{g/mol}$$

Atomic weight of P is 31 amu. Find molecular formula:

$$\frac{\text{molecular weight}}{\text{atomic weight}} = \frac{124}{31} = 4.0$$

Formula is P_4.

12.7   OSMOSIS

Operational Skill

7.   Calculating osmotic pressure.   Given the molarity and the temperature of a solution, calculate the osmotic pressure   (Example 12.13).

> **Exercise 12.16**   Calculate the osmotic pressure at 20°C of an aqueous solution containing 5.0 g of sucrose, $C_{12}H_{22}O_{11}$, in 100.0 mL of solution.

Wanted: osmotic pressure, π.

Given: 5.0 g $C_{12}H_{22}O_{11}$; 20.0°C (293 K); 100.0 mL solution.

Known: $\pi = MRT$; $M = \dfrac{\text{mol solute}}{\text{L solution}}$; $C_{12}H_{22}O_{11} = 342.3$ g/mol.

Solution: Find moles of sucrose:

$$\frac{5.0 \text{ g}}{342.3 \text{ g /mol}} = 0.014\underline{6} \text{ mol } C_{12}H_{22}O_{11}$$

The solution concentration is

$$M = \frac{0.014\underline{6} \text{ mol}}{0.100 \text{ L}} = 0.14\underline{6} \, M$$

Find osmotic pressure:

$$\pi = MRT = \frac{0.14\underline{6} \text{ mol} \times 0.0821 \text{ L} \cdot \text{atm} \times 293 \text{ K}}{(\text{L}) (\text{K} \cdot \text{mol})}$$

$$= 3.5 \text{ atm}$$

12.8 COLLIGATIVE PROPERTIES OF IONIC SOLUTIONS

Operational Skill

8.   Determining colligative properties of ionic solutions.   Given the concentration of ionic compound in a solution, calculate the magnitude of a colligative property; if i is not given, assume the value based on the formula of the ionic compound . (Example 12.14).

Exercise 12.17   Estimate the boiling point of a 0.050 m aqueous $MgCl_2$ solution. Assume a value of i based on the formula.

Wanted: T_b (°C).

Given: 0.050 m aqueous $MgCl_2$ solution.

Known: $T_b = T_b$ pure solvent $+ \Delta T_b$ solution; $\Delta T_b = iK_b c_m$; K_b for water = 0.521°C/m. $MgCl_2$ is ionic; each formula unit produces 3 ions, so $i = 3$.

$$MgCl_2(s) \longrightarrow Mg^{2+}(aq) + 2Cl^-(aq) \quad \text{(3 ions)}$$

Solution: $\Delta T_b = 3 \times 0.521°C/m \times 0.050\ m = 0.0782°C$

$T_b = T_b$ pure solvent $+ \Delta T_b$ solution

$= 100°C + 0.0782°C = 100.078°C$

12.9 COLLOIDS

Exercise 12.18   Colloidal sulfur particles are negatively charged with thiosulfate ions, $S_2O_3^{2-}$, and other ions on the surface of the sulfur. Indicate which of the following would be most effective in coagulating colloidal sulfur: $NaCl$, $MgCl_2$, or $AlCl_3$.

Known: Coagulation is caused by charge neutralization; the more compact the layer of charge, the more effective the coagulation; the higher the ionic charge, the more compact the layer.

Solution: The positive ions will cause the coagulation as they interact with $S_2O_3^{2-}$. $AlCl_3$ would be most effective because it has the most highly charged cation, Al^{3+}.

A Solvent That Matters: WATER (a Special Substance for Planet Earth)

Questions for Study

1. How does hydrogen bonding in water explain why ice is less dense than liquid water?

2. How does hydrogen bonding affect the heat capacity and heat of vaporization of water?

3. How does the unusually large heat capacity of water affect the earth's weather?

4. Explain why water dissolves so many different substances.

5. What does the term *hard water* mean? Why does hard water reduce the effectiveness of soap?

6. Explain how water is softened by ion exchange.

Answers to Questions for Study

1. The hydrogen-bonded structure (text Figure 12.32) of water has each oxygen atom in the structure of ice surrounded tetrahedrally by four hydrogen atoms. Two of these are close and covalently bonded to the H_2O molecule. Two others are farther away and are held by hydrogen bonds. The tetrahedral angles give rise to a three-dimensional structure that contains more open space and fewer molecules per unit volume than the more compact arrangement of molecules in liquid water. Since density = mass/volume, the density of ice is less than that of liquid water.

2. To increase the temperature of a body of water (heat capacity) or to evaporate it (heat of vaporization), energy must be supplied to increase molecular agitation. Because hydrogen bonding is such a strong intermolecular attraction, the heat capacity and heat of vaporization are much greater than would be expected for such a small molecule.

3. The evaporation of surface water absorbs over 30% of the solar energy that reaches the earth's surface. When the water vapor condenses, this energy is released and thunderstorms and hurricanes can result. Because of its exceptionally large heat capacity, a body of water can absorb or release large quantities of heat with only moderate changes in temperature. Thus large bodies of water (lakes and oceans) affect the weather and moderate the climate of adjacent land masses.

4. Since water is a polar molecule and participates in hydrogen bonding, it can dissolve ionic and polar covalent compounds. These classes of compounds comprise a large number of substances.

5. Because of the ability of water to dissolve ionic and polar compounds, naturally occurring water always contains dissolved ionic substances (salts). Hard water

contains metal ions, such as Ca^{2+} and Mg^{2+}, that react with soaps. These ions displace the sodium ions and precipitate the soaps as calcium and magnesium salts. These precipitates adhere to clothing and make it appear dirty. In the bathtub, they form the "bathtub ring."

6. In the water-softening process, the water is passed through a column that contains cation-exchange resins, which are insoluble macromolecular substances that contain negatively charged groups whose negative charges are counterbalanced by Na^+ ions. When the hard water, which contains Ca^{2+} and Mg^{2+} ions, passes through the resin column, the Na^+ ions in the resin are released into the water and replaced by the Ca^{2+} and Mg^{2+} ions.

ADDITIONAL PROBLEMS

1. List the letters of the following compounds in order of their increasing solubility in water. Explain your answer.

   (a) $C_5H_{11}OH$, 1-pentanol   (b) C_5H_{12}, pentane   (c) $C_5H_{11}SH$, pentanethiol

2. The solubility of carbon dioxide gas, CO_2, in water at 20°C and 5.00 atm pressure is 8.05 g/L. What mass of CO_2 would dissolve in exactly 250 mL of water at 1 atm pressure?

3. A 35.5%-by-mass hydrochloric acid solution has a density of 1.18 g/mL. What are the molality, mole fraction, and molarity of HCl in the solution?

4. State how you would prepare each of the following solutions:

   (a) 255 g of an aqueous solution containing 5.00% by mass of $BaCl_2$, starting with $BaCl_2 \cdot 2H_2O$

   (b) 755 g of a 1.40 *m* solution of naphthalene, $C_{10}H_8$, in benzene, C_6H_6

   (c) 455 g of a solution of glycerin, $C_3H_5(OH)_3$, in water in which the mole fraction of glycerin is 0.250

5. The vapor pressure of benzene at 25°C is 57.4 mmHg. Calculate the vapor-pressure lowering when 160.0 g of naphthalene (MW = 128.19 g/mol) is added to 1.50×10^3 g of benzene (MW = 78.12 g/mol) at 25°C.

6. A solution of 2.24×10^{-1} g of an organic compound of unknown molecular weight in 1.664 g of camphor gave a freezing point of 176.44°C. The freezing point of pure camphor is 179.80°C, and the freezing-point-depression constant, K_f, is 39.7°C/m. Calculate the molecular weight of the unknown compound.

7. Methanol and ethanol form a nearly ideal solution over a wide concentration range. What is the vapor pressure of a 33.4%-by-weight solution of methanol in ethanol at 40.0°C? The vapor pressures of the pure alcohols at 40.0°C are:

    $P^o_{\text{methanol}} = 236$ mmHg; $P^o_{\text{ethanol}} = 119$ mmHg.

8. The freezing point of a 0.0766 m H_3PO_4 solution is observed to be -0.218°C. Does the H_3PO_4 dissociate primarily to $H_2PO_4^-$, HPO_4^{2-}, or PO_4^{3-}? The freezing point of pure water is 0°C, and the freezing-point-depression constant, K_f, is 1.86°C/m.

9. An aqueous solution of urea, a molecular compound, has a freezing point of -0.0631°C. Calculate the osmotic pressure of the solution at body temperature (37.0°C).

10. Calcium chloride is used to salt icy roads in many communities in winter. Sodium chloride is also used and is much cheaper. Compare the freezing-point lowering of a ton of each of the salts. If you were the city manager, what other factors would you consider in choosing which to use?

ANSWERS TO ADDITIONAL PROBLEMS

If you missed an answer, study the text section and operational skill given in parentheses after the answer.

1. b, c, a. The more similar the attractions between the molecules of a substance are to the attractions between water molecules, the more they can interact with water, and the more soluble the compound is. Water molecules are bound by strong hydrogen bonds. Pentane is bound by only London forces, and pentanethiol by dipole–dipole interactions, but 1-pentanol forms hydrogen bonds.   (12.2)

2. 0.402 g   (12.3, Op. Sk. 1)

3. molality: 15.1 m; mole fraction HCl: 0.214; molarity: 11.5 M (12.4, Op. Sk. 3)

4. (a) Dissolve 15.0 g $BaCl_2 \cdot 2H_2O$ in 240.0 g of water.
    (b) Dissolve 115 g of naphthalene in exactly 640 g of benzene.
    (c) Add 287 g of glycerin to 168 g of water.   (12.4, Op. Sk. 2)

5. 3.50 mmHg   (12.5, Op. Sk. 4)   6.   1590 g/mol   (12.6, Op. Sk. 6)

7. 168 mmHg   (12.5)

8. $i = 1.53$; dissociates primarily to $H_2PO_4^-$ (12.6, 12.8)

9. 0.864 atm   (12.7, Op. Sk. 7)

10. A ton of calcium chloride, $CaCl_2$, is theoretically only ~ 79% as effective as a ton of sodium chloride, NaCl. (The theoretical value of i for $CaCl_2$ is 3 and for NaCl is 2, but the formula weight of NaCl is only 53% of that of $CaCl_2$.) Other factors would include supply, storage, and transportation costs, as well as effect on the road surface and on the environment.   (12.8)

CHAPTER POST-TEST

1. From a molecular viewpoint, briefly discuss the meaning of the phrase "like dissolves like."

2. Which of the following statements is incorrect with respect to the reaction

    $$Sn(s) + 2Cl_2(g) \rightleftharpoons SnCl_4(l); \quad \Delta H = -130 \text{ kcal}$$

    (a) Heating the reaction will decrease the yield of $SnCl_4$.
    (b) Increasing the pressure will decrease the yield of $SnCl_4$.
    (c) Lowering the temperature of the reaction will shift the equilibrium to the right.
    (d) Light should have little effect on the overall K_{eq}.

3. For each pair of ions below, choose the ion that should have the larger hydration energy.

    (a) Ca^{2+}, Ba^{2+} (c) Al^{3+}, Mg^{2+}

    (b) Na^+, Sr^{2+} (d) Fe^{3+}, Fe^{2+}

4. List the following compounds in order of decreasing solubility in C_2H_5OH:

    $ClCH_2CH_2CH_2Cl$, H_2O, $(CH_3)_3CCH_2Cl$, $[CH_3(CH_2)_{16}CH_2]_3N$, Cs_2SO_4.

5. Calculate the minimum mass of NaCl that would have to be added to 1.200×10^3 g H_2O so the resulting solution would not freeze outside on a cold day ($-10.0°F$). ($K_f = 1.858°C/m$)

6. The solubility of $K_2Pt(NO_2)_4$ is endothermic and is measured to be 38.0 g per liter of water at 15°C. If 14.6 g of the compound is dissolved in 412 mL of water after considerable stirring and heating and remains dissolved after the solution is cooled to 15°C, which of the following statements is a logical conclusion?
    (a) $K_2Pt(NO_2)_4$ is relatively insoluble in water.
    (b)   The final solution is supersaturated.
    (c)   Heating the $K_2Pt(NO_2)_4$ solution decreases the solute solubility.
    (d)   The final solution is not saturated.
    (e)   A change in temperature has no effect on solute solubility.

7. Which of the following is generally true for solutions that obey Raoult's law?
    (a)   The vapor phase is richer in the more volatile component than is the liquid phase.
    (b)   The component richer in the liquid phase will be the first to distill from the solution.
    (c)   The mole fraction of the most volatile component will be less than that of the less volatile components.
    (d)   The components in the solutions are usually not chemically similar.
    (e)   All of the above are generally correct.

8. What is the molality of a sulfuric acid solution that contains 30.0% by weight H_2SO_4 and has a density of 1.21 g/mL?

9. Calculate the quantities of water and acetone, C_3H_6O, necessary for the preparation of 150.0 g of solution that is 2.00 *m* in acetone.

10. Indicate whether each of the following statements is true or false. If a statement is false, change it so it is true.
    (a)   The spontaneous flow of solute from high to low concentration is known as osmosis. True/False: _____
    _____

(b)   The addition of a nonvolatile solute to a solvent produces a vapor-pressure lowering of the solvent. True/False: _____

_____ .

✗ (c)   The molality of a solution can be calculated if only the percentage composition by weight and the identity of the solute are known. True/False: _____

_____ .

(d)   The specific gravity of a solution is a colligative property of the solution. True/False: _____

_____ .

✗ (e)   When NaCl is sprinkled on grass, the grass withers and dies. This occurs because of the poisonous nature of NaCl. True/False: _____

_____ phenomenon y osmotic pressure _____ .

✗ (f)   The freezing-point depression of a 0.25 *m* aqueous solution of Na_2CO_3 will be three times that of a 0.25 *m* CH_3OH aqueous solution. True/False: _____

_____ .

(g)   Raoult's law relates the vapor pressure of solvent to the mole fraction of solvent in the solution. True/False: _____

_____ .

11.   A solution is made of two volatile components, toluene and benzene. At 60°C, the vapor pressure of benzene, C_6H_6, is 396 mmHg and the vapor pressure of toluene, C_7H_8, is 1.40×10^2 mmHg. The solution is made of equimolar amounts of these two substances. Calculate the total vapor pressure above the solution. (Assume Raoult's law is valid for volatile solutes.)

12.   Calculate the osmotic pressure of a solution that consists of 43.5 g of an enzyme with a molar mass of 9.80×10^4 g/mol dissolved in benzene to make 1.500 L of solution at 30.0°C. The density of benzene is 0.879 g/mL, and $K_b = 2.61°C/m$. Pure benzene boils at 80.2°C.

13.   A liter of water dissolves 0.0045 g of oxygen at 25°C and 1.0 atm. If the partial pressure of O_2 is increased to 5.6 atm, what is its solubility in water?

14.   A 4.28-g sample of a non-ionizing solute is dissolved in exactly 250 mL of water (molar mass = 18.02 g/mol) at 25°C. The molality of the solution is 0.121. The vapor pressure of water at 25°C is 23.8 mmHg. Calculate (a) the molecular weight of the solute, (b) the mole fraction of the solute, (c) the vapor-pressure lowering due to the solute, and (d) the vapor pressure of the solution.

ANSWERS TO CHAPTER POST-TEST

If you missed an answer, study the text section and operational skill given in parentheses after the answer.

1.   Solutes and solvents that have similar forces of interaction, functional groups, and/or structural features tend to be completely miscible.   (12.2)

2.   b   (12.3)

3.   (a) Ca^{2+},   (b) Sr^{2+},   (c) Al^{3+},   (d) Fe^{3+} (12.2)

4. $H_2O > ClCH_2CH_2CH_2Cl > (CH_3)_3CCH_2Cl > [CH_3(CH_2)_{16}CH_2]_3N > Cs_2SO_4$
     (12.2)

5.   440 g NaCl   (12.4, 12.6, 12.8; Op. Sk. 2, 5)

6.   d   (12.2, 12.3)   7.   a   (12.5)

8.   4.37 moles H_2SO_4/1 kg solvent, or 4.37 m (12.4, Op. Sk. 2)

9.   134.4 g of water, 15.6 g of acetone   (12.4, Op. Sk. 2)

10. (a) False. The spontaneous flow of solvent from low to high concentration of solute is known as osmosis.   (12.7)
     (b) True.   (12.5)
     (c) True.   (12.4)
     (d) False. The specific gravity of a solution is not a colligative property of the solution, since it is not a property that depends on the number of particles of substance.   (12.4)
     (e) False. This occurs because of the phenomenon of osmotic pressure.   (12.7)
     (f) True.   (12.6, 12.8)
     (g) True.   (12.5)

11. 268 mmHg   (12.5)   12. $\pi = 7.36 \times 10^{-3}$ atm   (12.7, Op. Sk. 7)

13. 0.025 g O_2/L   (12.3, Op. Sk. 1)

14. (a) 141 g/mol (12.4, 12.6, Op. Sk. 6)   (c) 0.0519 mmHg (12.5, Op. Sk. 4)
     (b) 0.00218 (12.4, Op. Sk. 3)   (d) 23.7 mmHg (12.5, Op. Sk. 4)

CHAPTER 13   RATES OF REACTION

CHAPTER TERMS AND DEFINITIONS

Numbers in parentheses after definitions give the text sections in which the terms are explained. Starred terms are italicized in the text. Where a term does not fall directly under a text section heading, additional information is given for you to locate it.

chemical kinetics*   study of reaction rates, how they change under varying reaction conditions, and what molecular events occur during the overall reaction   (13.1, introductory section)

catalyst   substance that increases the rate of a reaction without being consumed in the overall reaction   (13.1, introductory section)

reaction rate   increase in molar concentration of product of a reaction per unit time, or decrease in molar concentration of reactant per unit time   (13.1)

average (rate)*   rate of reaction over a time interval, Δt (13.1)

instantaneous (rate)*   rate of reaction at a particular instant of time; also the value of $\Delta [x]/\Delta t$ for the tangent to the concentration-versus-time curve at a given instant   (13.1)

rate law   equation that relates the rate of a reaction to the concentrations of reactants (and catalyst) raised to various powers   (13.3)

rate constant (k)   proportionality constant in the relationship between rate and reactant concentrations   (13.3)

reaction order   experimentally determined exponent of the concentration of a species in a rate law   (13.3)

overall order (of a reaction)*   sum of the exponents in a rate law   (13.3)

isomerization*   reaction in which one geometric isomer of a compound is converted to another   (13.3)

initial-rate method*   process of determining reaction orders by varying the starting concentrations of reactants   (13.3)

half-life ($t_{1/2}$)   time required for the reactant concentration to decrease to one-half of its initial value   (13.4)

collision theory   model that assumes that, for reaction to occur, reactant molecules must collide with an energy greater than some minimum value and with the proper orientation   (13.5)

activation energy (E_a)   minimum energy that two molecules must possess in order to react upon collision   (13.5)

transition-state theory   model that considers the collision of two reactant molecules in terms of an unstable grouping (activated complex) that can break up to form products   (13.5)

activated complex (transition state)   in transition-state theory, an unstable grouping of atoms that can break up to form products   (13.5)

Arrhenius equation $k = Ae^{-E_a/RT}$; mathematical equation that expresses the dependence of the rate constant on temperature   (13.6)

frequency factor   symbol A, a constant in the Arrhenius equation related to the frequency of collisions with proper orientation   (13.6)

elementary reaction   single molecular event, such as a collision of molecules, resulting in a reaction   (13.7)

reaction mechanism   set of elementary reactions whose overall effect is given by the net chemical equation   (13.7)

reaction intermediate   species produced during a reaction that does not appear in the net equation because it reacts in a subsequent step in the mechanism   (13.7)

molecularity   number of molecules on the reactant side of an elementary reaction   (13.7)

unimolecular reaction   elementary reaction involving one reactant molecule   (13.7)

bimolecular reaction   elementary reaction involving two reactant molecules   (13.7)

termolecular reaction   elementary reaction involving three reactant molecules   (13.7)

rate-determining step   slowest step of a reaction mechanism, the rate of which governs the overall rate of reaction   (13.8)

dynamic equilibrium*   state in which the rates of the forward and reverse reactions are equal   (13.8)

enzymes*   proteins that catalyze biochemical processes   (13.9)

homogeneous catalysis   use of a catalyst that is in the same phase as the reacting species   (13.9)

heterogeneous catalysis   use of a catalyst (usually a solid) that is in a different phase from the reacting species   (13.9)

adsorption*   attraction of molecules to a surface   (13.9)

physical adsorption*   attraction of molecules to a surface through weak intermolecular forces   (13.9)

chemisorption   binding of one substance to the surface of another due to chemical bonding forces between the surfaces of both substances   (13.9)

catalytic hydrogenation*   addition of hydrogen to carbon–carbon multiple bonds of an organic compound in the presence of a catalyst   (13.9)

contact process*   industrial method of preparing sulfuric acid by the catalytic oxidation of sulfur dioxide   (13.9)

substrate* substance whose reaction an enzyme catalyzes (13.9)

active site* in an enzyme-catalyzed reaction, the particular place on the enzyme where the substrate binds and the catalysis takes place (13.9)

*$\overline{iodes}$*** Greek word meaning "violet-colored" (An Element That Matters: Iodine [The Clock Reaction and the Oxidation States of Iodine])

CHAPTER DIAGNOSTIC TEST

1. Determine whether each of the following statements is true or false. If the statement is false, change it so it is true.

    (a) An isotope with a decay constant of 7.32×10^{-4}/s has a half-life of 5.07×10^{-4} s. True/False: _____ $9.47 \times 10^2 s$ _____ .

    (b) The exponents of concentrations in the rate-law expression are determined by the stoichiometry of the chemical reaction. True/False: ___ by exp't ___ .

    (c) A catalyst operates by reducing the value of ΔH of a reaction. True/False: __
    _____ Activ. Energy _____ .

    (d) In gas-phase kinetics, the molecularity is limited to bimolecular reactions. True/False: _____ intur _____ .

    (e) The rate-law expression of rate = $0.91\,[SO_3]\,[F_2]^2$ indicates the rate law is second order with respect to F_2 and third order overall. True/False: _____
    _____ .

    (f) The existence of an activated complex can be substantiated by isolating the species at extremely low temperature. True/False: _____
    _____ .

    (g) When a reaction is influenced by a homogeneous catalyst, the reaction-rate expression is a function of the catalyst concentration. True/False: _____
    _____ .

2. For the following reactions and data, provide the missing information.

    (a) Reaction: $2NO(g) + H_2(g) \longrightarrow N_2O(g) + H_2O(g)$

    Rate law: rate = $k\,[H_2]\,[NO]^2$

Order with respect to H_2:   1

N_2O: 0

NO: 2

H_2O: 0

Overall order: 3

(b)   Reaction: $2H_2 + O_2 \longrightarrow 2H_2O$

Rate law: rate = $K[H_2]^2[O_2]$

Order with respect to H_2:   second

O_2:   first

H_2O: zero

Overall order: 3

(c)   Reaction: $2NO_2(g) + F_2(g) \longrightarrow 2NO_2F(g)$

Rate law: rate = $k[NO_2][F_2]$

Order with respect to NO_2:   1

F_2:   1

NO_2F: 0

Overall order: second

3.   The decomposition of acetaldehyde, CH_3CHO, has an activation energy of 188.3 kJ/mol and occurs by the mechanism

$$CH_3CHO \longrightarrow CH_3CHO^{2+} \longrightarrow CH_4 + CO$$

When this reaction is carried out in the presence of iodine, the activation energy is 138.1 kJ/mol. Sketch the potential-energy diagram for these two cases, on the same graph, as the reaction progresses. Identify the following on the curves: reactant, product, activation energy, activated complex.

4.   Determine which of the following is a correct expression for the average rate of the reaction

$$aA + bB \longrightarrow eE + fF$$

(a) $\quad \text{rate} = \dfrac{1}{a} \dfrac{\Delta[A]}{\Delta t}$

(c) $\quad \text{rate} = -\dfrac{\Delta[B]^{b}}{\Delta t}$

(e) $\quad$ none of the above

(b) $\quad \text{rate} = -f \dfrac{\Delta[F]}{\Delta t}$

(d) $\quad \text{rate} = \dfrac{1}{e} \dfrac{\Delta[E]}{\Delta t}$

5. The debromination of stilbene dibromide by $SnCl_2$ occurs according to the following overall stoichiometry:

$$(C_6H_5CHBr)_2 + SnCl_2 \longrightarrow (C_6H_5CH)_2 + SnCl_2Br_2$$

Kinetic studies indicate the reaction mechanism entails the following elementary reactions:

$$(C_6H_5CHBr)_2 + SnCl_2 \longrightarrow \left[\begin{array}{c} H \\ | \\ C_6H_5C-C-C_6H_5 \\ | \quad | \\ Br \quad H \end{array} \right]^{-} + SnCl_2Br^{+} \quad \text{(slow)}$$

$$\left[\begin{array}{c} H \\ | \\ C_6H_5C-C-C_6H_5 \\ | \quad | \\ Br \quad H \end{array} \right]^{-} \longrightarrow \begin{array}{c} H \\ | \\ C_6H_5C-C-C_6H_5 \\ | \\ H \end{array} + Br^{-} \quad \text{(fast)}$$

$$\begin{array}{c} H \\ | \\ C_6H_5C-C-C_6H_5 \\ | \\ H \end{array} \longrightarrow \begin{array}{c} H \\ | \\ C_6H_5C=C-C_6H_5 \\ | \\ H \end{array} \quad \text{(fast)}$$

$$SnCl_2Br^{+} + Br^{-} \longrightarrow SnCl_2Br_2 \quad \text{(fast)}$$

On the basis of this information, predict the rate-law expression for the debromination reaction and give the order of reaction. $\quad$ rate $= [C_6H_5CHBr][SnCl_2]$
2nd order.

6. The following concentration data were collected for the reaction

$$F_2 + 2ClO_2 \longrightarrow 2FClO_2$$

at 350°C and various times:

Rate of formation of $FClO_2$ (M/s)	$[F_2]$ (M)	$[ClO_2]$ (M)
1.6×10^{-3}	0.15	0.015
4.8×10^{-3}	0.15	0.045
6.4×10^{-3}	0.15	0.060
9.6×10^{-3}	0.30	0.045
1.9×10^{-2}	0.60	0.045

Calculate the rate constant and order of this reaction.

7. The hydrolysis of $(CH_2)_6CClCH_3$ is first order. Using the following data, calculate the rate constant, k:

t (s)	$[(CH_2)_6CClCH_3]$ (M)
0	0.650
184	0.613
293	0.592
314	0.588
495	0.555

8. Which of the following statements is true?
   (a) Knowledge of a rate law is useful in understanding how a reaction takes place at the molecular level.
   (b) A plot of concentration versus time for a first-order reaction is linear.
   (c) The overall order of a reaction equals the largest exponent in the rate-law expression.
   (d) A catalyst increases the reaction rate by increasing the activation energy of the reaction.
   (e) None of the above.

9. When heated, ammonium cyanate, NH_4CNO, forms urea, $(NH_2)_2CO$, according to the following stoichiometry:

$$NH_4^+ + CNO^- \longrightarrow (NH_2)_2CO$$

The assumed mechanism encompasses the following elementary reactions:

$$NH_4^+ + CNO^- \underset{k_{-1}}{\overset{k_1}{\rightleftharpoons}} NH_3 + HCNO \qquad \text{(fast, equilibrium)}$$

$$NH_3 + HCNO \xrightarrow{k_2} (NH_2)_2CO \qquad \text{(slow)}$$

Verify that the mechanism agrees with the experimentally determined rate law:

rate $= k_{obs} [NH_4^+][CNO^-]$.

10. Acetonecarbonic acid decomposes to dimethylketone and carbon dioxide. The rate constants for the decomposition were measured at 283 K and 323 K and found to be 1.08×10^6/s and 1.85×10^8/s, respectively. Calculate the activation energy for this reaction.

11. A substance decomposes according to the first-order rate law. With a rate constant of 4.80×10^{-4}/s at 45°C, calculate the following:
    (a) half-life of the substance.
    (b) concentration after 725 s if the initial concentration is 0.0824 *M*.
    (c) time required for concentration of the substance to decrease to 20.0%.

12. The rate constant for the decomposition of PH_3 is 0.024 s^{-1}. What is the half-life for the decomposition reaction?

13. The reaction between the hydroxyl radical and carbon monoxide, which is important in atmospheric chemistry, is proposed to occur through the following elementary reactions:

$$2\,\overset{2}{\cancel{3}}OH + \overset{2}{\cancel{3}}CO \longrightarrow 3HOCO^* \quad \rightarrow \quad [OH]^*[CO]^? \quad bimol.$$
$$HOCO^* \longrightarrow OH + CO \quad \rightarrow \quad [HOCO] \quad unimol$$
$$HOCO^* \longrightarrow H + CO_2 \quad \rightarrow \quad [HOCO] \quad unimol.$$
$$HOCO^* + M \longrightarrow HOCO + M \quad \rightarrow \quad [HOCO][M] \quad bimol$$
$$\underline{HOCO \longrightarrow H + CO_2 \quad \rightarrow \quad [HOCO] \quad unimol}$$
$$2\,OH + 2CO \rightarrow 2H + 2CO_2$$

Write the overall chemical equation for the hydroxyl radical/CO reaction.

14. Write the rate equation for each elementary reaction in Problem 13 and indicate the molecularity of each.

ANSWERS TO CHAPTER DIAGNOSTIC TEST

If you missed an answer, study the text section and operational skill given in parentheses after the answer.

1. (a) False. An isotope with a decay constant of 7.32×10^{-4}/s has a half-life of 9.47×10^2 s. (13.4, Op. Sk. 6)

   (b) False. The exponents of concentrations in the rate-law expression are determined experimentally. (13.3)

   (c) False. A catalyst operates by providing a mechanism in which the energy of activation, E_a, is lower than in the uncatalyzed reaction. (13.9)

   (d) False. In gas-phase kinetics, the molecularity is limited to intermolecular reactions. (13.7)

   (e) True. (13.3, Op. Sk. 3)

   (f) False. The activated complex cannot be isolated because it is an unstable species. (13.5)

   (g) True. (13.9)

2. (a) Order with respect to H_2:   first

                                      N_2O: zero

                                      NO: second

                                      H_2O: zero

      Overall order: third

   (b) Rate law: rate $= k [H_2]^2 [O_2]$

      Overall order: third

   (c) Order with respect to NO_2:   first

                                      F_2:   first

                                      NO_2F: zero   (13.3, Op. Sk. 3)

3.

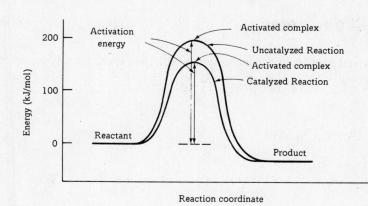

Reaction coordinate   (13.5, 13.9)

4.   d   (14.1, Op. Sk. 1)

5.   rate = $k_{obs} [(C_6H_5CHBr)_2] [SnCl_2]$; reaction is second order

(13.3, 13.8, Op. Sk. 11)

6. $k = 7.1 \times 10^{-1}/(M \cdot s)$; reaction is second order   (13.2, 13.3, Op. Sk. 4)

7. $k = 3.20 \times 10^{-4}/s$ (13.2, 13.4, Op. Sk. 5)

8.   a   (14.8); sections for other responses are b   (13.4),   c   (13.3),   d   (13.9)

9.   According to the mechanism, the rate-determining step is the reaction of NH_3 and $HCNO$. The rate-law expression is

rate = $k_2 [NH_3] [HCNO]$

The concentration of NH_3 and that of $HCNO$ are related to NH_4CNO by the fast, equilibrium reaction with the equilibrium expression

$$\frac{k_1}{k_{-1}} = \frac{[NH_3] [HCNO]}{[NH_4^+] [CNO^-]}$$

In terms of the product, $[NH_3] [HCNO]$,

$$[NH_3] [HCNO] = \frac{k_1}{k_{-1}} [NH_4^+] [CNO^-]$$

Substituting this into the rate-law expression gives

$$\text{rate} = \frac{k_2 k_1}{k_{-1}} [NH_4^+][CNO^-]$$

The rate law derived from the mechanism agrees with the experimentally observed rate law, if $\dfrac{k_2 k_1}{k_{-1}} = k_{obs}$.   (13.8, Op. Sk. 11)

10. 9.77×10^4 J   (13.6, Op. Sk. 7)

11.   (a) 1.44×10^3 s

      (b) $0.0582\ M$

      (c) 3.35×10^3 s   (13.4, Op. Sk. 5, 6)

12.   29 s   (13.4, Op. Sk. 6)

13.   OH + CO $\longrightarrow$ H + CO_2 (13.7, Op. Sk. 8)

14.   rate = k [OH] [CO], bimolecular

      rate = k [HOCO*], unimolecular

      rate = k [HOCO*], unimolecular

      rate = k [HOCO*] [M], bimolecular

      rate = k [HOCO], unimolecular   (13.7, 13.8; Op. Sk. 9, 10)

SUMMARY OF CHAPTER TOPICS

Chemical kinetics requires a good deal of mathematical computation — involving calculus and logarithms — to answer questions related to reaction rates. Coverage of this topic in your text and this study guide, and in most freshman chemistry courses, does not use calculus but does use logarithms. You were introduced to the use of logarithms in Section 11.2 of this study guide. You may wish to look back for a review.

13.1 DEFINITION OF REACTION RATE

Operational Skills

1. **Relating the different ways of expressing reaction rates.** Given the balanced equation for a reaction, relate the different possible ways of defining the rate of the reaction (Example 13.1).

2. **Calculating the average reaction rate.** Given the concentration of reactant or product at two different times, calculate the average rate of reaction over that time interval (Example 13.2).

Exercise 13.1 For the reaction given in Example 13.1, how is the rate of formation of NO_2F related to the rate of reaction of NO_2?

Known: By definition the rate of formation of NO_2F is

$$\frac{\Delta[NO_2F]}{\Delta t}$$

and the rate of reaction of NO_2 is

$$-\frac{\Delta[NO_2]}{\Delta t}$$

Solution: Divide each rate by the coefficient of the appropriate substance in the reaction and set these expressions equal:

$$\frac{\Delta[NO_2F]/\Delta t}{2} = -\frac{\Delta[NO_2]/\Delta t}{2}$$

which can be written as

$$\frac{\Delta[NO_2F]}{\Delta t} = -\frac{\Delta[NO_2]}{\Delta t}$$

In words this reads, "The rate of formation of NO_2F is equal to the rate of reaction of NO_2."

Exercise 13.2 Iodide ion is oxidized by hypochlorite ion in basic solution.

$$I^-(aq) + ClO^-(aq) \longrightarrow Cl^-(aq) + IO^-(aq)$$

In 1.00 M NaOH at 25°C, the iodide-ion concentration (equal to the ClO^- concentration) at different times was as follows:

Time	I$^-$
2.00 s	0.00169 *M*
8.00 s	0.00101 *M*

Calculate the average rate of reaction of I$^-$ during this time interval.

Wanted: average reaction rate of I$^-$.

Given: data table.

Known: rate $= -\dfrac{\Delta[\text{I}^-]}{\Delta t}$

Solution: Calculations are

$$\Delta[\text{I}^-] = [\text{I}^-]_{\text{final}} - [\text{I}^-]_{\text{initial}}$$

$$= (0.00101\ M) - (0.00169\ M) = -0.00068\ M$$

$$\Delta t = t_{\text{final}} - t_{\text{initial}} = 8.00\ \text{s} - 2.00\ \text{s} = 6.00\ \text{s}$$

$$\text{rate} = -\frac{-0.00068\ M}{6.00\ \text{s}} = 1.1 \times 10^{-4}\ M/\text{s}$$

13.2   EXPERIMENTAL DETERMINATION OF RATE

13.3   DEPENDENCE OF RATE ON CONCENTRATION

Operational Skills

3.   Determining the order of reaction from the rate law.   Given an empirical rate law, obtain the orders with respect to each reactant (and catalyst, if any) and the overall order   (Example 13.3).

4.   Determining the rate law from initial rates.   Given initial concentrations and initial-rate data (in which the concentrations of all species are changed one at a time, holding the others constant), find the rate law for the reaction   (Example 13.4).

It is important to remember that the reaction orders with respect to each species in a rate law must be determined experimentally. They may or may not be related to the coefficients in the equation.

Exercise 13.3   What are the reaction orders with respect to each reactant species for the following reaction?

$$NO_2(g) + CO(g) \longrightarrow NO(g) + CO_2(g)$$

Assume the rate law is

$$Rate = k[NO_2]^2$$

What is the overall order?

Wanted: reaction order with respect to each reactant species; overall order.

Given: reaction; rate law.

Known: definition of reaction order.

Solution: The order with respect to NO_2 is 2, as given in the rate law. The order with respect to CO is 0, as CO does not appear in the rate law. The overall order is $2 + 0 = 2$.

In determining reaction orders with respect to reactant species, it certainly saves time if you can look at the data and think out the orders. Sometimes, however, it is not that easy to do. It is helpful to know how to solve the problem algebraically, as Exercise 13.4 is solved below.

Exercise 13.4   The initial-rate method was applied to the decomposition of nitrogen dioxide.

$$2NO_2(g) \longrightarrow 2NO(g) + O_2(g)$$

It yielded the following results:

	Initial NO$_2$ Concentration	Initial Rate of Formation of O$_2$
Experiment 1	0.010 mol/L	7.1×10^{-5} mol/(L • s)
Experiment 2	0.020 mol/L	28×10^{-5} mol/(L • s)

Find the rate law and the value of the rate constant with respect to O_2 formation.

Known: The rate law is of the form

$$rate = k[NO_2]^m$$

The data give rate and concentration; m and k must be calculated.

Solution: Find *m* by making a ratio of the rate law using data from Experiment 2 over data from Experiment 1:

$$\frac{\text{Exp 2}}{\text{Exp 1}} = \frac{\text{rate 2}}{\text{rate 1}} = \frac{28 \times 10^{-5}}{7.1 \times 10^{-5}} = \frac{\cancel{k}\,(0.020)^m}{\cancel{k}\,(0.010)^m}$$

$$= 3.94 \approx 4 = \left[\frac{0.020}{0.010}\right]^m = 2^m$$

$$m = 2$$

Find *k* by solving the rate law using data from one experiment and the calculated value of *m*. In this case, data from Experiment 1 are used. Also note that the units are written so they can be manipulated algebraically to determine the correct units for *k*.

$$7.1 \times 10^{-5} \frac{\text{mol}}{\text{L} \cdot \text{s}} = k\left[0.010 \frac{\text{mol}}{\text{L}}\right]^2$$

$$= k\left[1.0 \times 10^{-4} \frac{\text{mol}^2}{\text{L}^2}\right]$$

$$\frac{7.1 \times 10^{-5} \ \cancel{\text{mol}}}{\cancel{\text{L}} \cdot \text{s}} \times \frac{\text{L}^{\cancel{2}}}{1.0 \times 10^{-4} \ \text{mol}^{\cancel{2}}} = k$$

$$k = 0.71 \frac{\text{L}}{\text{mol} \cdot \text{s}}$$

13.4   CHANGE OF CONCENTRATION WITH TIME

Operational Skills

   5.   Using the concentration–time equation for a first-order reaction.   Given the rate constant and initial reactant concentration for a first-order reaction, calculate the reactant concentration after a definite time, or calculate the time it takes for the concentration to decrease to a prescribed value   (Example 13.5).

   6.   Relating the half-life of a reaction to the rate constant.   Given the rate constant for a reaction, calculate the half-life   (Example 13.6).

   In the solution to Example 13.5, the text author introduces the use of the natural logarithm (base *e* = 2.718). If you have an e^x key on your calculator, it is simpler to use natural logarithms in your calculations, following the directions given in text Appendix A.2. Study guide problems are worked with logarithms to the base 10.

Exercise 13.5   (a)   What would be the concentration of dinitrogen pentoxide in the experiment described in Example 13.5 after 6.00×10^2 s?   (b)   How long would it take for the concentration of N_2O_5 to decrease to 10.0% of its initial value?

Wanted: (a) $[N_2O_5]$ at $t = 6.00 \times 10^2$ s; compare with table value;

  (b) t when $[N_2O_5] = 10.0\%\ [N_2O_5]_0$.

Given: $t = 6.00 \times 10^2$ s, text Example 13.5.

Known: Reaction is first order; $k = 4.80 \times 10^{-4}$/s at 45°C; $[N_2O_5]_0 = 1.65 \times 10^{-2}$ mol/L.

Solution: (a)   The integrated rate law from calculus, using the logarithm to the base 10, is

$$\log \frac{[N_2O_5]_0}{[N_2O_5]_t} = \frac{kt}{2.303}$$

Putting values into the right side gives

$$\log \frac{[N_2O_5]_0}{[N_2O_5]_t} = \frac{4.80 \times 10^{-4} \times 6.00 \times 10^2\ \cancel{s}}{\cancel{s} \times 2.303}$$

$$= 0.125\underline{1}$$

Now take the antilog of both sides. Do this operation on an electronic calculator. If the calculator has an inverse key, enter 0.1251, press *inverse*, then *log*. The result will be 1.334, which is antilog 0.1251. Or, to use the y^x key, enter 10, press y^x, enter 0.1251, and press *equals*. The result of $10^{0.1251}$ = antilog 0.125$\underline{1}$ is 1.3$\underline{3}$4. Note that because there are three decimal places in the logarithm, there are three significant figures in the antilog. (You may wish to review the explanation following the solution to Exercise 11.2 in this study guide.) Thus,

$$\frac{[N_2O_5]_0}{[N_2O_5]_t} = 1.3\underline{3}4$$

Using the known value of $[N_2O_5]_0$, we solve for $[N_2O_5]_t$:

$$[N_2O_5]_t = \frac{1.65 \times 10^{-2} \text{ mol/L}}{1.334}$$

$$= 1.24 \times 10^{-2} \text{ mol/L}$$

(b)   Solve for time. Given that $[N_2O_5]_t = 0.100\,[N_2O_5]_0$, we can write the integrated rate law as

$$\log \frac{\cancel{[N_2O_5]_0}}{0.100\,\cancel{[N_2O_5]_0}} = \frac{kt}{2.303}$$

which becomes

$$-\log 0.100 = \frac{kt}{2.303}$$

Rearranging to solve for t and using the known value of k gives

$$t = \frac{2.303 \text{ s}}{4.80 \times 10^{-4}} \times 1.000$$

$$= 4.80 \times 10^3 \text{ s}$$

Half-life is one of the many fascinating realities of science. It carries with it the mysterious aura of nuclear disintegrations and transmutation of elements. As the rate of a reaction slows down as the reaction proceeds, common sense suggests that the time it takes for half of the substance to react should depend on how much you start with. But this is not the case in a first-order reaction. No matter how much substance you begin with, it always takes the same amount of time for half of that substance to disappear in a given first-order reaction! The half-lives of different reactions are different, but no half-life for a first-order reaction depends on the initial concentration of the reacting substance.

Exercise 13.6   The isomerization of cyclopropane, C_3H_6, to propylene, $CH_2 = CHCH_3$, is first order in cyclopropane and first order overall. At 1000°C, the rate constant is 9.2/s. What is the half-life of cyclopropane at 1000°C? How long would it take for the concentration of cyclopropane to decrease to 50% of its initial value?   to 25% of its initial value?

Wanted: half-life, $t_{1/2}$; time for $[C_3H_6]$ to decrease to 50% and 25% of $[C_3H_6]_{initial}$.

Given: Reaction is first order in cyclopropane; $k = 9.2/s$.

Known: definition of half-life; $t_{1/2} = \dfrac{0.693}{k}$.

Solution: The calculation is

$$t_{1/2} = \frac{0.693 \text{ s}}{9.2} = 0.075\underline{3}$$

$$= 0.075 \text{ s}$$

The time for $[C_3H_6]$ to decrease by 50% is the half-life, which is 0.075 s. The time for concentration to decrease to 25% would be two half-lives, as 25% is half of the second 50% of substance. This would be $2 \times 0.075\underline{3}$ s $= 0.15$ s.

With your introduction to the plotting of the integrated rate law, you have now seen two methods of determining the overall reaction order. Recall that the first method was using the rate law and data on initial rates of reaction.

13.5 TEMPERATURE AND RATE; COLLISION AND TRANSITION-STATE THEORIES

13.6 ARRHENIUS EQUATION

Operational Skill

7. Using the Arrhenius equation. Given the values of the rate constant for two temperatures, find the activation energy and calculate the rate constant at a third temperature (Example 13.7).

Exercise 13.7 Acetaldehyde, CH_3CHO, decomposes when heated.

$$CH_3CHO(g) \longrightarrow CH_4(g) + CO(g)$$

The rate constant for the decomposition is $1.05 \times 10^{-3}/(M^{1/2} \cdot s)$ at 759 K and $2.14 \times 10^{-2}/(M^{1/2} \cdot s)$ at 836 K. What is the activation energy for this decomposition? What is the rate constant at 865 K?

Wanted: E_a; k_3 at 865 K (T_3).

Given: $k_1 = 1.05 \times 10^{-3}/(M^{1/2} \cdot s)$ at 759 K (T_1); $k_2 = 2.14 \times 10^{-2}/(M^{1/2} \cdot s)$ at 836 K (T_2).

Known: Arrhenius equation in logarithmic form for two different temperatures:

$$\log \frac{k_2}{k_1} = \frac{E_a}{2.303R} \left[\frac{1}{T_1} - \frac{1}{T_2} \right]$$

Also, $R = 8.31$ J/(K · mol).

Solution: Substitute values into the Arrhenius equation and solve for E_a.

$$\log \left[\frac{2.14}{0.105} \right] = \frac{E_a \cdot K \cdot \text{mol}}{2.303 \times 8.31 \text{ J}} \left[\frac{1}{759 \ K} - \frac{1}{836 \ K} \right]$$

$$1.3092 = \frac{E_a \cdot \text{mol}}{19.14 \text{ J}} \ (1.214 \times 10^{-4})$$

$$E_a = \frac{1.3092 \ (19.14 \text{ J/mol})}{1.214 \times 10^{-4}} = 2.06 \times 10^5 \text{ J/mol}$$

Find k_3, the rate constant at 865 K (T_3), by writing the Arrhenius equation as follows:

$$\log k_3 - \log k_1 = \frac{E_a}{2.303R} \left[\frac{1}{T_1} - \frac{1}{T_3} \right]$$

Then rearrange, put in values, and solve.

$$\log k_3 = \frac{2.06 \times 10^5 \ J \ (K \cdot \text{mol})}{2.303 \ \text{mol} \times 8.31 \ J} \left[\frac{1}{759 \ K} - \frac{1}{865 \ K} \right] + \log 1.05 \times 10^{-3}$$

$$= \left[1.076 \times 10^4 \ (1.615 \times 10^{-4}) \right] + (-2.9788) = -1.2411$$

$$k_3 = \text{antilog} - 1.2411 = 5.74 \times 10^{-2}/(M^{1/2} \cdot \text{s})$$

13.7   ELEMENTARY REACTIONS

Operational Skills

8.   Writing the overall chemical equation from a mechanism.   Given a mechanism for a reaction, obtain the overall chemical equation   (Example 13.8).

9.   Determining the molecularity of an elementary reaction.   Given an elementary reaction, state the molecularity   (Example 13.9).

10.   Writing the rate equation for an elementary reaction.   Given an elementary reaction, write the rate equation   (Example 13.10).

> **Exercise 13.8**   The iodide ion catalyzes the decomposition of aqueous hydrogen peroxide, H_2O_2. This decomposition is believed to occur in two steps.

$$H_2O_2 + I^- \longrightarrow H_2O + IO^- \qquad \text{(elementary reaction)}$$
$$H_2O_2 + IO^- \longrightarrow H_2O + O_2 + I^- \qquad \text{(elementary reaction)}$$

What is the overall equation representing this decomposition? Note that IO^- is a reaction intermediate. The iodide ion is not an intermediate; it was added to the reaction mixture.

Wanted: net chemical equation.

Given: two elementary reactions.

Known: steps for obtaining the net equation.

Solution: Add both elementary reactions and cancel reaction intermediates or catalyst:

$$H_2O_2 + \cancel{I^-} \longrightarrow H_2O + \cancel{IO^-}$$
$$H_2O_2 + \cancel{IO^-} \longrightarrow H_2O + O_2 + \cancel{I^-}$$
$$\overline{}$$
$$2H_2O_2 \longrightarrow 2H_2O + O_2$$

> **Exercise 13.9**   The following is an elementary reaction that occurs in the decomposition of ozone in the stratosphere by nitric oxide.

$$O_3 + NO \longrightarrow O_2 + NO_2$$

What is the molecularity of this reaction? That is, is the reaction unimolecular, bimolecular, or termolecular?

Known: definitions of the three terms; there are two molecules on the reactant side.

Solution: The reaction is bimolecular.

> **Exercise 13.10**   Write the rate equation, showing the dependence of rate on concentrations, for the elementary reaction

$$NO_2 + NO_2 \longrightarrow N_2O_4$$

Known: We can write the rate law from the equation for an elementary reaction.

Solution: rate = $k\,[NO_2]^2$.

13.8   THE RATE LAW AND THE MECHANISM

Operational Skill

11.   Determining the rate law from a mechanism.   Given a mechanism with an initial slow step, obtain the rate law   (Example 13.11). Given a mechanism with an initial fast, equilibrium step, obtain the rate law   (Example 13.12).

Exercise 13.11   The iodide ion-catalyzed decomposition of hydrogen peroxide, H_2O_2, is believed to follow the mechanism

$$H_2O_2 + I^- \xrightarrow{\;k_1\;} H_2O + IO^- \qquad\qquad \text{(slow)}$$

$$H_2O_2 + IO^- \xrightarrow{\;k_2\;} H_2O + O_2 + I^- \qquad\qquad \text{(fast)}$$

What rate law is predicted by this mechanism? Explain.

Known: The slow step determines the rate; we write the rate law from the equation for the slow step.

Solution: rate = $k_1\,[H_2O_2]\,[I^-]$.

Exercise 13.12   Nitric oxide, NO, reacts with oxygen to produce nitrogen dioxide.

$$2NO(g) + O_2(g) \longrightarrow 2NO_2(g) \qquad\qquad \text{(overall equation)}$$

If the mechanism is

$$NO + O_2 \underset{k_{-1}}{\overset{k_1}{\rightleftharpoons}} NO_3 \qquad\qquad \text{(fast, equilibrium)}$$

$$NO_3 + NO \xrightarrow{\;k_2\;} NO_2 + NO_2 \qquad\qquad \text{(slow)}$$

what is the predicted rate law? Remember to express this in terms of substances in the chemical equation.

Known: All species in the rate law must appear in the net equation for the reaction.

Solution: Write the predicted rate law from the slow step:

$$\text{rate} = k_2 \, [NO_3] \, [NO]$$

However, $[NO_3]$ does not appear in the overall equation. Because the first step is fast, equilibrium will be reached when the forward and reverse rates are equal:

$$k_1 \, [NO] \, [O_2] = k_{-1} \, [NO_3]$$

Thus,

$$[NO_3] = \left(\frac{k_1}{k_{-1}} \right) [NO] \, [O_2]$$

Substitute into the predicted rate law to get

$$\text{rate} = k_2 \frac{k_1}{k_{-1}} \, [NO]^2 \, [O_2] = k[NO]^2 \, [O_2]$$

The observed rate constant, k, is the product $k_2 \dfrac{k_1}{k_{-1}}$.

13.9   CATALYSIS

An Element That Matters: IODINE (the Clock Reaction and the Oxidation States of Iodine)

Questions for Study

1.   Explain the iodine clock reaction.

2.   What is the origin of the name *iodine*?

3.   How is iodine recovered from Chilean nitrate?

4.   Write the equation for the reduction of triiodide ion by hydrogen sulfite ion in acidic solution.

5.   Give two uses of iodine compounds.

Answers to Questions for Study

1.   In the iodine clock reaction, the iodate ion is reduced by hydrogen sulfite in acidic solution. This relatively slow reaction occurs as long as hydrogen sulfite ion is available in solution. Once it is consumed, the iodide ion is quickly oxidized to

iodine by the excess iodate ion. The solution turns blue because of the blue-colored complex that is formed between the iodine and the starch.

2. The name iodine comes from the Greek word *iōdēs*, which means "violet-colored."

3. Chilean nitrate $(NaNO_3)$ contains calcium iodate $(Ca(IO_3)_2)$ as an impurity. The solution remaining after crystallization of sodium nitrate is treated with the stoichiometric amount of hydrogen sulfite ion to reduce the iodate ion to iodide ion. Then the necessary amount of solution containing iodate ion is added to react with the iodide ion to liberate elemental iodine.

4. The equation describing the reduction of I_3^- with HSO_3^- in acidic solution is

$$I_3^-(aq) + H_2O(l) + HSO_3^-(aq) \longrightarrow 3I^-(aq) + SO_4^{2-}(aq) + 3H^+(aq)$$

5. Potassium iodide is added to animal feeds and to salt (iodized salt) to supply iodide ion as a nutrient. Silver iodide is important in the manufacturing of photographic film.

ADDITIONAL PROBLEMS

1. Relate the rate of formation of the (underlined) product to the rate of disappearance of the (underlined) reactant in each of the following reactions.

(a) $N_2O_4(g) \longrightarrow 2\underline{NO_2}(g)$

(b) $2NO_2(g) + 7\underline{H_2}(g) \longrightarrow 2\underline{NH_3}(g) + 4H_2O(g)$

(c) $CO(g) + 3\underline{H_2}(g) \longrightarrow \underline{CH_4}(g) + H_2O(g)$

2. Hydrogen iodide is formed according to the following reaction:

$$H_2(g) + I_2(g) \longrightarrow 2HI(g)$$

If there are 4.35×10^{-2} mol HI in a 25.0-L vessel initially, and after 5.0 min there are 2.40×10^{-2} mol present, what is the rate of reaction during this time period? Express the answer in units of M/s.

3. The following data were taken for a reaction between substances X and Y:

	Initial Concentrations (M)		Initial Rate (M/s)
	X	Y	
Experiment 1	0.1000	1.0	1.4×10^{-6}
Experiment 2	0.1000	2.0	5.6×10^{-6}
Experiment 3	0.0500	1.0	7.0×10^{-7}

What is the rate law for this reaction?

4. A nonmetal oxide undergoes the following decomposition reaction upon heating:

$$M_xO_y(g) \longrightarrow M_xO_{y-1}(g) + \tfrac{1}{2}O_2(g)$$

Using the following data, calculate the rate constant for this first-order reaction at 45°C.

t (s)	$[M_xO_y]$
0	2.33
184	2.08
319	1.91
526	1.67
867	1.36
1198	1.11
1877	0.72
2315	0.55
3144	0.34

5. The hydrolysis of methyl acetate, CH_3COOCH_3, in hydrochloric acid is first order with a rate constant of 1.26×10^{-4} s^{-1}. If we start with a 0.664 M CH_3COOCH_3 solution, how much time will elapse before the hydrolysis is 80.0% complete?

6. The following data were collected for the decomposition of SO_2Cl_2

$$SO_2Cl_2(l) \longrightarrow SO_2(g) + Cl_2(g)$$

at 550 K:

t (s)	$[SO_2Cl_2]$ (M)
0	0.3130
280	0.3110
1155	0.3051
1690	0.3015
3040	0.2927
5060	0.2800

Using these data, determine (a) the rate law for the reaction, (b) the rate constant, and (c) how much SO_2Cl_2 remains after 3.00 h of reaction.

7. The activation energy for a particular decomposition reaction is determined to be 104.6 kJ/mol. What effect will a change in reaction temperature from 300°C to 335°C have on the rate constant for the reaction?

8. The overall chemical equation for the chlorination of trichloromethane is

$$CHCl_3(g) + Cl_2(g) \longrightarrow CCl_4(g) + HCl(g)$$

If the mechanism is

$$Cl_2(g) \underset{k_{-1}}{\overset{k_1}{\rightleftharpoons}} 2Cl(g) \qquad \text{(fast)}$$

$$Cl(g) + CHCl_3(g) \xrightarrow{k_2} HCl(g) + CCl_3(g) \qquad \text{(slow)}$$

$$Cl(g) + CCl_3(g) \xrightarrow{k_3} CCl_4(g) \qquad \text{(fast)}$$

what is the predicted rate law?

9. The rate constant for a reaction at exactly 800°C is 5.0×10^{-3} L/(mol • s). The rate constant at 875°C is 7.0×10^{-3} L/(mol • s). (a) Calculate the activation energy for the reaction.   (b) Determine the value of the rate constant at 925°C.

10. The reaction of hydrogen bromide and hydrogen peroxide in aqueous solution is

$$2H^+(aq) + 2Br^-(aq) + H_2O_2(aq) \longrightarrow 2H_2O(l) + Br_2(l)$$

The mechanism is believed to be

$$H^+ + H_2O_2 \underset{k_{-1}}{\overset{k_1}{\rightleftharpoons}} H_3O_2^+ \qquad \text{(fast)}$$

$$H_3O_2^+ + Br^- \xrightarrow{k_2} H_2O + HOBr \qquad \text{(slow)}$$

$$H^+ + HOBr \underset{k_{-3}}{\overset{k_3}{\rightleftharpoons}} H_2OBr^+ \qquad \text{(fast)}$$

$$Br^- + H_2OBr^+ \underset{k_{-4}}{\overset{k_4}{\rightleftharpoons}} Br_2 + H_2O \qquad \text{(fast)}$$

Verify that this mechanism agrees with the experimentally determined rate law:

$$\text{rate} = k\,[H_2O_2]\,[H^+]\,[Br^-]$$

ANSWERS TO ADDITIONAL PROBLEMS

If you missed an answer, study the text section and operational skill given in parentheses after the answer.

1. (a) $\Delta\,[NO_2]\,/\Delta t = -2\Delta\,[N_2O_4]\,/\Delta t$

    (b) $\Delta\,[NH_3]\,/\Delta t = -\frac{2}{7}\Delta\,[H_2]\,/\Delta t$

    (c) $\Delta\,[CH_4]\,/\Delta t = -\frac{1}{3}\Delta\,[H_2]\,/\Delta t$ (13.1, Op. Sk. 1)

2. $2.60 \times 10^{-6}\ M/s$ (13.1, Op. Sk. 2)

3. rate $= 1.4 \times 10^{-5}\ L^2/(mol^2 \cdot s)\ [X]\,[Y]^2$ (13.3, Op. Sk. 4)

4. $k = 6.17 \times 10^{-4}/s$ (13.2, 13.3, Op. Sk. 4)

5. 1.28×10^4 s, or 3.55 h   (13.4, Op. Sk. 5)

6. (a)   rate $= k [SO_2Cl_2]$

    (b) $k = 2.23 \times 10^{-5}/s$

    (c) $[SO_2Cl_2]$ after 3 h is 0.2460 M (13.3, 13.4, Op. Sk. 5)

7. The 35°C increase in temperature multiplies the rate constant, and thus the rate of reaction, by a factor of 3.5.   (13.6, Op. Sk. 7)

8. rate $= k [Cl_2]^{1/2} [CHCl_3]$ (13.8, Op. Sk. 11)

9. (a) $4.6 \times 10^4 J$ (b) $8.6 \times 10^{-3} L/(mol \cdot s)$ (13.6, Op. Sk. 7)

10. rate $= k_2 [H_3O_2^+] [Br^-]$

    $k_1 [H^+][H_2O_2] = k_{-1} [H_3O_2^+]$

    $[H_3O_2^+] = k_1/k_{-1} [H^+][H_2O_2]$

    Rate $= k_2 k_1/k_{-1} [H^+][H_2O_2][Br^-]$

    $k_2 k_1/k_{-1} = k$ (13.8, Op. Sk. 11)

CHAPTER POST-TEST

1. Complete the following sentences:
    (a)   The rate of a chemical reaction is expressed as the change in _____ of species per _____.
    (b)   Knowledge of the _____ is useful in explaining how a chemical reaction occurs at the molecular level.
    (c)   The reaction

$$2O_3(g) + 2NO(g) \longrightarrow N_2O_5(g) + O_2(g)$$

obeys the rate law: rate $= k [O_3]^{2/3} [NO]^{2/3}$. The reaction is _____ order with respect to O_3 and _____ order overall.

(d) In the mechanism for the chlorination of NO,

$$2NO + Br_2 \rightleftharpoons 2NOBr$$

$$2NOBr + Cl_2 \longrightarrow NOCl + Br_2$$

Br_2 acts as a _____.

(e) The half-life of a reaction-rate process is _____
_____ .

(f) In a reaction mechanism, the rate-determining step is _____
_____ .

(g) A reaction intermediate is _____ , whereas an activated complex is
_____ , even though both may appear in a reaction mechanism.

(h) An _____ reaction indicates what is occurring at the molecular level.

2. Using the following data from the rearrangement reaction of NH_4CNO, determine the average rate of reaction in that time interval:

t (min)	$[NH_4CNO]$ (M)
11<u>0</u>	0.289
239	0.205

3. Determine the order with respect to each reactant and the overall order of the reaction associated with each of the following empirical rate laws:

(a) rate $= k [NO]^2 [O_2]$

(b) rate $= k [Br_2]^{1/2} [H_2]$

(c) rate $= k [O_3]^2 [O_2]^{-1}$

4. During a 1-min interval, 141 nuclear disintegrations were counted in a sample of Pb-214. Three hours later, the number of disintegrations in a 20-min interval was 27. What are the decay constant and $t_{1/2}$ of Pb-214? (Nuclear-disintegration reactions are first order, and disintegrations are proportional to changes in concentration.)

5. Using the following kinetic data, calculate the rate constant and order of the following reaction at 670 K:

$$C_2H_4O \longrightarrow CH_4 + CO$$

t (min)	$[C_2H_4O]$ (M)
0	1.31×10^{-3}
14.9	1.09×10^{-3}
28.1	9.30×10^{-4}
54.3	6.70×10^{-4}
86.5	4.50×10^{-4}
148	2.10×10^{-4}
209	1.00×10^{-4}

6. On the following diagram, label the activated complex(es), activation energy(ies), reactant, reaction intermediate, and product.

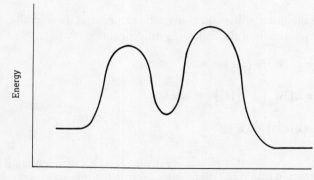

Reaction coordinate

7. A substance decomposes according to the second-order rate law. Assume that the rate constant is 6.8×10^{-4} L/(mol • s) and the half-life of the substance is 525 s.
    (a) What was the initial concentration of the substance?
    (b) How long would it take for the concentration of the substance to decrease to 0.140 M?
    (c) What is the concentration of the substance after 60.0 s?

8. The rate constant for the transformation of cyclopropane to propene at 500°C is 5.9×10^{-4}/s. Calculate the half-life of the reaction.

9. Which of the following does *not* result from increased temperature?
    (a) The number of reactant-molecule collisions increases.
    (b) The average molecular speed of the molecules increases proportional to $\sqrt{T}$.
    (c) The reaction rate is halved for every 10°C rise in temperature.
    (d) The fraction of collisions having an energy greater than the activation energy is $e^{-E_a/RT}$.
    (e) Both a and d.

10. The activation energy for the decomposition of acetonecarbonic acid is 9.77×10^4 J/mol. The rate of decomposition was measured to be 2.46×10^5 M/s at an unknown temperature. If the rate of decomposition at 313 K is 5.76×10^7 M/s, what is the unknown temperature?

11. The following are the proposed elementary reactions for a reaction mechanism:

    $$ClO^- + H_2O \longrightarrow HClO + OH^-$$

    $$HClO + OH^- \longrightarrow ClO^- + H_2O$$

    $$H_2O + IO^- \longrightarrow OH^- + HIO$$

    $$OH^- + HIO \longrightarrow H_2O + IO^-$$

    $$I^- + HClO \longrightarrow HIO + Cl^-$$

    What is the overall chemical equation for the reaction that occurs in aqueous solution?

12. Write the rate equation for the following elementary reactions and give the molecularity of each reaction:

    (a) $HClO + I^- \longrightarrow HIO + Cl^-$

    (b) $Co(CN)_5H_2O^{2-} \longrightarrow Co(CN)_5^{2-} + H_2O$

    (c) $2NO + Br_2 \longrightarrow 2BrNO$

13. Cl_2 and CO react to form phosgene, $COCl_2$, according to the postulated mechanism

$$Cl_2 \underset{k_{-1}}{\overset{k_1}{\rightleftharpoons}} 2Cl \qquad \text{(equilibrium)}$$

$$Cl + Cl_2 \underset{k_{-1}}{\overset{k_2}{\rightleftharpoons}} Cl_3 \qquad \text{(equilibrium)}$$

$$Cl_3 + CO \overset{k_3}{\longrightarrow} COCl_2 + Cl \qquad \text{(slow)}$$

Which of the following observed rate-law expressions conforms with that derived from the mechanism?

(a)   rate $= k_{obs} [Cl_3] [CO]$

(b)   rate $= k_{obs} [CO] [Cl] [Cl_2]$

(c)   rate $= k_{obs} [CO] [Cl_2]^{3/2}$

(d)   rate $= k_{obs} [CO] [Cl]^2$

(e)   none of the above

14. Which of the following mechanisms suggests that the overall reaction is influenced by catalytic action?

(a) $C_2D_6 \longrightarrow 2CD_3$

$$CD_3 + C_2D_6 \longrightarrow CD_4 + C_2D_5$$

$$C_2D_5 + CH_4 \longrightarrow C_2D_5H + CH_3$$

$$CH_3 + C_2D_6 \longrightarrow CH_3D + C_2D_5$$

(b) $N_2O_5 \longrightarrow NO_2 + NO_3$

$$NO_2 + NO_3 \longrightarrow N_2O_5$$

$$NO + NO_3 \longrightarrow 2NO_2$$

(c) $NO + O_2 \rightleftharpoons NO_3$

$NO + NO_3 \longrightarrow 2NO_2$

(d) $Ce^{4+} + Ag^+ \rightleftharpoons Ce^{3+} + Ag^{2+}$

$Ag^{2+} + Tl^+ \longrightarrow Tl^{2+} + Ag^+$

$Tl^{2+} + Ce^{4+} \longrightarrow Tl^{3+} + Ce^{3+}$

(e)   all of the above

15.   Given the following data, show graphically that the hypothetical reaction
$3A \longrightarrow B + C$ is first order. Calculate k from the plot.

t (s)	$[A]$ (M)
1.4	0.547
4.3	0.540
10.8	0.526
54.9	0.439
96.8	0.370
151.0	0.296

ANSWERS TO CHAPTER POST-TEST

If you missed an answer, study the text section and operational skill given in parentheses
after the answer.

1.   (a)   concentrations; unit time   (13.1)
(b)   rate law   (13.8)
(c)   2/3, 4/3   (13.3, Op. Sk. 3)
(d)   homogeneous catalyst   (13.9)
(e)   the time it takes for one-half the quantity of reactants to undergo the indicated
reaction   (13.4)
(f)   the slowest step   (13.8)
(g)   stable and capable of isolation; unstable and cannot be isolated   (13.5)
(h)   elementary   (13.7)

2.   Average rate $= 6.51 \times 10^{-4}\ M/s$ (13.1, Op. Sk. 2)

3.   (a)   second order in NO, first order in O_2; third order overall

     (b)   one-half order in Br_2, first order in H_2; three-halves order overall

     (c)   second order in O_3, -1 order in O_2; first order overall   (13.3, Op. Sk. 3)

4. $k = 2.58 \times 10^{-2}/min$; $t_{1/2} = 26.9\ min$ (13.4, Op. Sk. 5, 6)

5. $k = 1.23 \times 10^{-2}/min$; reaction is first order   (13.2, 13.3, 13.4, Op. Sk. 4, 5)

6.

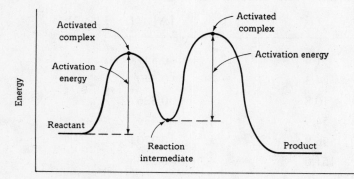

(13.5)

7.   (a)   2.8 mol/L

     (b) $1.0 \times 10^4\ s$

     (c)   2.5 mol/L   (13.4)

8. $1.2 \times 10^3\ s$ (13.4, Op. Sk. 6)

9.   c   (13.5)

10.   273 K   (13.6, Op. Sk. 7)

11. $I^- + HClO \longrightarrow HIO + Cl^-$ (13.7, Op. Sk. 8)

12. (a) rate $= k\,[\text{HClO}]\,[\text{I}^-]$, bimolecular

    (b) rate $= k\,[\text{Co(CN)}_5\text{H}_2\text{O}^{2-}]$, unimolecular

    (c) rate $= k\,[\text{NO}]^2\,[\text{Br}_2]$, termolecular   (13.7, Op. Sk. 9, 10)

13. c   (13.8, Op. Sk. 11)

14. d   (13.8, 13.9)

15. The plot is a straight line, showing that the reaction is first order in A;
    $k = 4.10 \times 10^{-3}/\text{s}$.   (13.4)

t (s)	log $[A]$
1.4	-0.262
4.3	-0.268
10.8	-0.279
54.9	-0.358
96.8	-0.432
151.0	-0.529

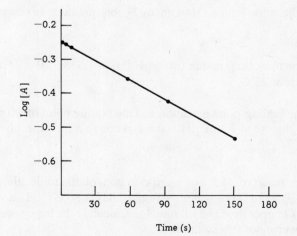

Time (s)

1.   Arrange the following in order of increasing melting point:

   (a)   HBr, HCl, HF, HI   (b) H_2O, H_2S, H_2Se, H_2Te

2.   Calculate the amount of heat required to melt 5.36 g of silver ($\Delta H_{fus} = 111$ J/g).

3.   A substance has a triple point of 0.33 atm and 15°C and a critical point of 200 atm and 253°C. The liquid state is less dense than the solid state. Sketch a phase diagram for this substance. Describe how the gas could be liquefied and give the conditions for melting of the solid.

4.   Nickel crystallizes in a face-centered cubic cell. The atomic weight of nickel is 58.7 amu and its density is 3.52 g/cm^3. Calculate the edge length of the unit cell.

5.   In NaF, the F$^-$ ions are in a face-centered cubic lattice, whereas in RbF, the F$^-$ ions form a simple cubic lattice. How many F$^-$ ions are there in one unit cell in NaF and in RbF?

6.   Calculate the osmotic pressure (in mmHg) of a 1.538×10^{-4} M aqueous solution of a polypeptide at 25°C.

7.   Calculate the boiling-point elevation and the boiling point of a 0.67 m aqueous solution of glucose at 1 atm. (Glucose is a nonvolatile molecular solid.) K_b for water is 0.521°C/m.

8.   A solution is made of 24.5 g of sucrose (a nonvolatile molecular substance, MW = 342.3) in 245 g of water (MW = 18.02) at 30.0°C. The vapor pressure of water at this temperature is 31.8 mmHg. Calculate the vapor pressure and vapor-pressure lowering of this solution.

9. Calculate the molecular weight of X if 1.10 g X in 27.4 g C_6H_6 (molar mass = 78.1 g/mol) formed a solution that froze at 2.61°C. The normal freezing point of C_6H_6 is 5.48°C. K_f for C_6H_6 is 5.065°C/m.

10. How would you prepare 25.3 g of an aqueous solution that is 14.7% sucrose by mass?

11. Calculate the molality and mole fractions of a solution consisting of 57.2 g Na_2CO_3 in 225 g of water. (Na_2CO_3: 106 g/mol; H_2O: 18.02 g/mol.)

12. Estimate the freezing point of a 0.024 m aqueous solution of K_2CO_3. Explain why the measured freezing point would not be this low.

13. For the equation

$$2H_2(g) + O_2(g) \longrightarrow 2H_2O(g)$$

how is the rate of formation of H_2O mathematically related to the rate of reaction of O_2?

14. In aqueous solutions, peroxydisulfate ions, $S_2O_8^{2-}$, oxidize iodide ions:

$$S_2O_8^{2-}(aq) + 2I^-(aq) \longrightarrow 2SO_4^{2-}(aq) + I_2(s)$$

The empirical rate law is

$$\text{Rate} = k[S_2O_8^{2-}][I^-]$$

(a) What is the order of reaction with respect to each reactant species?
(b) What is the overall order of reaction?

15. Rate data have been determined at a particular temperature for the overall reaction

$$2NO + 2H_2 \longrightarrow N_2 + 2H_2O$$

in which all reactants and products are gases. The data are

Trial	Initial [NO]	Initial [H$_2$]	Initial Rate (M/s)
1	0.20 M	0.20 M	0.060
2	0.20 M	0.30 M	0.090
3	0.40 M	0.20 M	0.24

   (a) Determine the rate law for this reaction.
   (b) What would be the initial rate of the reaction if the initial molar concentration of NO = 0.10 M and the initial molar concentration of H$_2$ = 0.30 M?

16. The mechanism for the reaction $3ClO^- \longrightarrow ClO_3^- + 2Cl^-$ is

$$ClO^- + ClO^- \longrightarrow ClO_2^- + Cl^- \qquad \text{(slow)}$$

$$ClO^- + ClO_2^- \longrightarrow ClO_3^- + Cl^- \qquad \text{(fast)}$$

   Derive the rate law for this reaction.

17. At a certain temperature the reaction $2B \longrightarrow C + D$ obeys the following rate law expression: rate = $(1.14 \times 10^{-3}/M \bullet s)\,[B]^2$. If 5.00 mol of B are initially present in a 1.00-L container at that temperature, how much B is left after 117 s?

18. Write a rate equation for each of the following elementary reactions.

   (a) $BrO_3^- + 2H^+ \longrightarrow H_2BrO_3^+$

   (b) $N_2O \longrightarrow N_2 + O$

   (c) $H_2 + 2NO \longrightarrow N_2O + H_2O$

ANSWERS TO UNIT EXAM 4

If you missed an answer, study the text section and operational skill given in parentheses after the answer.

1. (a)   HCl, HBr, HI, HF   (b) H_2S, H_2Se, H_2Te, H_2O (11.6, Op. Sk. 7)

2.   595 J   (11.2, Op. Sk. 1)

3.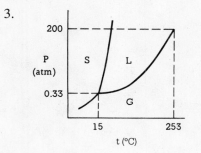

The gas could be liquefied by cooling it at pressures between 0.33 atm and 200 atm. The solid could be melted by heating it at a pressure above 0.33 atm.   (11.3, Op. Sk. 3)

4.   480 pm   (11.9, Op. Sk. 10)

5.   4 in NaF, 1 in RbF   (11.7, Op. Sk. 8)

6.   2.86 mmHg   (12.7, Op. Sk. 7)

7. $\Delta T_b = 0.35°C$; b.p. = 100.35°C   (12.6, Op. Sk. 5)

8.   V.P. = 31.6 mmHg, change in V.P. = -0.2 mmHg   (12.4, 12.5; Op. Sk. 2, 4)

9.   70.8 amu   (12.6, Op. Sk. 6)

10.   Dissolve 3.72 g of sucrose in 21.6 g H_2O.   (12.4, Op. Sk. 2)

11.   2.40 m Na_2CO_3; $X_{Na_2CO_3} = 0.0415$, $X_{H_2O} = 0.959$ (12.4, Op. Sk. 3)

12. $-0.13°C$. The measured freezing point would not be this low because the effective concentrations, or activities, of the ions are what determine the colligative properties. Due to electrical interactions of the ions in solution, the activities are less than the actual concentrations.   (12.8, Op. Sk. 5)

13. $$\frac{\Delta[H_2O]}{\Delta t} = \frac{-2\Delta[O_2]}{\Delta t}$$ (13.1, Op. Sk. 1)

14.   (a)   First order in $S_2O_8^{2-}$; first order in I^-

       (b)   Second order   (13.3, Op. Sk. 3)

15.   (a)   rate = 7.5 $m^{-2} \cdot s^{-1}[NO]^2[H_2]$ (b)   0.022 *M/s*   (13.3, Op. Sk. 4)

16.   rate = $k[ClO^-]^2$ (13.8, Op. Sk. 11)

17.   3.00 mol B   (13.4)

18.   (a)   rate = $k[BrO_3^-][H^+]^2$

       (b)   rate = $k[N_2O]$

       (c)   rate = $k[H_2][NO]^2$ (13.7, Op. Sk. 10)

CHAPTER 14   CHEMICAL EQUILIBRIUM

CHAPTER TERMS AND DEFINITIONS

Numbers in parentheses after definitions give the text sections in which the terms are explained. Starred terms are italicized in the text. Where a term does not fall directly under a text section heading, additional information is given for you to locate it.

reversible*   describes chemical reactions in which products formed can themselves react, giving back the original reactants   (chapter introduction)

catalytic methanation*   conversion of carbon monoxide and hydrogen to methane and water in the presence of a catalyst   (chapter introduction)

steam reforming*   preparing carbon monoxide and hydrogen by reacting hydrocarbons with steam   (chapter introduction)

dynamic equilibrium*   state in which the reactants and products of a reversible reaction or process are being formed at the same rate, such that there is no apparent change in the system   (14.1)

chemical equilibrium   state reached by a reaction mixture when the rates of forward and reverse reactions have become equal, so that net change no longer occurs   (14.1)

equilibrium constant*   quantity relating equilibrium compositions for a particular reaction at a given temperature   (14.2)

equilibrium-constant expression   arrangement of symbols showing multiplication of the concentrations of reaction products, and division by the concentrations of reactants, the concentration of each raised to a power equal to its coefficient in the chemical equation   (14.2)

equilibrium constant (K_c)   value obtained for the equilibrium-constant expression when equilibrium concentrations are substituted   (14.2)

law of mass action   relation stating that the values of the equilibrium-constant expression K_c are constant for a particular reaction at a given temperature whatever equilibrium concentrations are substituted   (14.2)

activities*   dimensionless quantities defining the equilibrium constant; for an ideal mixture, the activity of a substance is the ratio of its concentration (or partial pressure if a gas) to a standard concentration of 1 M (or partial pressure of 1 atm) so that units cancel   (14.2, marginal note)

361

K_p*   equilibrium constant for a gaseous reaction expressed in terms of partial pressures (14.2)

homogeneous equilibrium   equilibrium that involves reactants and products in a single phase   (14.3)

heterogeneous equilibrium   equilibrium that involves reactants and products in more than one phase   (14.3)

reaction quotient (Q_c)   expression identical to the equilibrium-constant expression, but with concentrations not necessarily those at equilibrium   (14.5)

quadratic formula*   solutions to a quadratic equation of the form $ax^2 + bx + c = 0$;

$$x = \left(-b \pm \sqrt{b^2 - 4ac}\right) / 2a \quad \text{(14.6, marginal note)}$$

Le Chatelier's principle   when a system in chemical equilibrium is disturbed by a change of temperature, pressure, or a concentration, the system shifts in equilibrium composition in a way that tends to counteract this change of variable (14.7)

contact process*   industrial method of preparing sulfuric acid by the catalytic oxidation of sulfur dioxide   (14.9)

catalyst*   substance that speeds up the attainment of equilibrium, is not consumed by the reaction, and has no effect on the equilibrium composition of the reaction mixture (14.9)

acid rain*   rain with increased acidity due to the presence of sulfuric and nitric acids (14.9, marginal note)

Ostwald process*   industrial method of preparing nitric acid by the catalytic oxidation of ammonia   (14.9)

water-gas reaction*   industrial process in which steam is passed over red-hot carbon (coal or coke) to give a gaseous mixture of carbon monoxide and hydrogen   (A Gas That Matters: Carbon Monoxide [Starting Material for Organic Compounds])

synthesis gas*   mixtures of carbon monoxide and hydrogen used in preparing various organic compounds   (A Gas That Matters: Carbon Monoxide [Starting Material for Organic Compounds])

CHAPTER DIAGNOSTIC TEST

1.   Write equilibrium-constant expressions for the following equilibria in terms of K_c:

   (a) $2HCl(g) + \frac{1}{2}O_2(g) \rightleftharpoons H_2O(g) + Cl_2(g)$

   (b) $2NO(g) + Br_2(g) \rightleftharpoons 2NOBr(g)$

   (c) $Ag^+(aq) + 2NH_3(aq) \rightleftharpoons Ag(NH_3)_2^+(aq)$

(d) $HCN(aq) + H_2O(l) \rightleftharpoons H_3O^+(aq) + CN^-(aq)$

(e) $4NH_3(g) + 3O_2(g) \rightleftharpoons 2N_2(g) + 6H_2O(g)$

(f) $I_3^-(aq) + H_2O(l) \rightleftharpoons HOI(aq) + 2I^-(aq) + H^+(aq)$

2.   The equilibrium concentrations for the decomposition of $PCl_5(g)$ at 433 K,

$$PCl_5(g) \rightleftharpoons PCl_3(g) + Cl_2(g)$$

are $[PCl_5] = 0.865$ mol/L, $[PCl_3] = [Cl_2] = 0.135$ mol/L. Calculate K_c.

3.   What effect would an increase in pressure have on the equilibrium of the system in Problem 2?

4.   A system containing nitrogen, hydrogen, and ammonia is allowed to come to equilibrium. The total equilibrium pressure is 5 atm. The partial pressures are $P_{N_2} = 1$ atm, $P_{H_2} = 2$ atm, and $P_{NH_3} = 2$ atm. Calculate K_p for the reaction

$$N_2(g) + 3H_2(g) \rightleftharpoons 2NH_3(g)$$

5.   Using the data from Problem 4, calculate K_p for

$$\tfrac{1}{2}N_2(g) + \tfrac{3}{2}H_2(g) \rightleftharpoons NH_3(g)$$

6.   If 1.00 mol CO_2 and 1.00 mol H_2 are placed in a 1.00-L flask at 825 K and react according to

$$CO_2(g) + H_2(g) \rightleftharpoons CO(g) + H_2O(g)$$

analysis of the equilibrium mixture shows 0.27 mol CO present. Determine K_c at this temperature.

7.   Consider the following equilibrium:

$$2H_2S(g) \rightleftharpoons 2H_2(g) + S_2(g)$$

For a 5.00-L vessel containing the following amounts of gases, determine whether the initial concentrations of these gases will remain fixed or change (and if they change, indicate which gases will show an increase in concentration):

0.0131 mol H_2, 0.00650 mol S_2, 0.0383 mol H_2S. K_c equals 2.3×10^{-4}.

8.   Consider the following reaction:

$$4HCl(g) + O_2(g) \rightleftharpoons 2H_2O(g) + 2Cl_2(g); \quad \Delta H = +28 \text{ kcal}$$

Describe what happens to the composition of the equilibrium mixture and to the equilibrium constant K with each of the following changes to the system at equilibrium:

(a)   addition of oxygen gas
(b)   an increase in temperature
(c)   reduction of the volume of the reaction container
(d)   addition of a catalyst
(e)   removal of $HCl(g)$ from the reaction vessel

9. I_2 vapor is a deep purple color. The dissociation of $HI(g)$ into $H_2(g)$ and $I_2(g)$ in a closed vessel can be qualitatively followed by observing changes in the relative intensity of the purple color of I_2 vapor. When H_2 gas is added to the reaction at equilibrium, the vapor slowly takes on a less intense purple color. Explain this observation in terms of Le Chatelier's principle.

10.   Consider the reaction

$$2SiO(g) \rightleftharpoons 2Si(l) + O_2(g); \quad K_c = 9.62 \times 10^{-1}$$

If 1.00 mol SiO is placed into a 1.00-L container, what are the equilibrium concentrations of SiO and O_2?

11.   Once the equilibrium in Problem 10 is reached, how would adding 10.0 g of Si affect that equilibrium?

12.   Consider the following reaction:

$$CO_2(g) + H_2(g) \rightleftharpoons CO(g) + H_2O(g); \quad K_c = 0.137$$

If 5.0 mol each of CO_2 and H_2 are placed in a 10.0-L flask, what are the equilibrium concentrations?

13.   Consider the following equilibrium:

$$2HI(g) \rightleftharpoons H_2(g) + I_2(g)$$

At equilibrium a 2.00-L vessel contains 1.25 mol I_2, 1.25 mol H_2, and an unknown amount of HI. K_c for this equilibrium is 0.0183. Calculate the equilibrium concentration of HI.

ANSWERS TO CHAPTER DIAGNOSTIC TEST

If you missed an answer, study the text section and operational skill given in parentheses after the answer.

1. (a) $\dfrac{[H_2O][Cl_2]}{[HCl]^2[O_2]^{1/2}}$ $\Big($Note that $[H_2O]$ is included, since the reaction is in the gas phase and $[H_2O]$ is not a constant.$\Big)$

(b) $\dfrac{[NOBr]^2}{[NO]^2[Br_2]}$ (d) $\dfrac{[H_3O^+][CN^-]}{[HCN]}$ (f) $\dfrac{[HOI][I^-]^2[H^+]}{[I_3^-]}$

(c) $\dfrac{[Ag(NH_3)_2^+]}{[Ag^+][NH_3]^2}$ (e) $\dfrac{[N_2]^2[H_2O]^6}{[NH_3]^4[O_2]^3}$ (14.2 , 14.3, Op. Sk. 2)

2. $K_c = 0.0211$ (14.2, Op. Sk. 3)

3. An increase in pressure would cause a shift to the side of the reaction that has the smaller number of moles of gaseous materials, which would reduce the increased pressure. Therefore, the rate of the reverse reaction would increase, and we say we would observe a shift to the left.   (14.8, Op. Sk. 7)

4. $K_p = 0.5$ (14.2, Op. Sk. 2, 3)   5. $K_p = 0.7$ (14.2, Op. Sk. 2, 3)

6. $K_c = 0.14$ (14.1, 14.2, Op. Sk. 1, 3)

7. $Q_c = 1.52 \times 10^{-4}$, which is $< K_c$. Therefore, initial concentrations will change, and the concentrations of H_2 and S_2 gases will increase as the equilibrium shifts to the right.   (14.4, 14.5, Op. Sk. 4)

8. (a) Shift right to consume the added oxygen; no change in K.   (14.7, Op. Sk. 7)
    (b) Favors endothermic reaction, so it shifts right; K increases.   (14.8, Op. Sk. 7)
    (c) This increases the pressure, so reaction shifts right, toward fewer moles of gas; no change in K.   (14.8, Op. Sk. 7)
    (d) No effect on the equilibrium composition; no change in K.   (14.9, Op. Sk. 7)
    (e) Shift left to replace the HCl lost; no change in K.   (14.7, 14.8, Op. Sk. 7)

9. The equilibrium may be expressed

$$2HI(g) \rightleftharpoons H_2(g) + I_2(g)$$

When H_2 is added to the reaction at equilibrium, the rate of the reverse reaction increases. We say the equilibrium shifts to decrease the concentration of H_2 and thus to form more HI. I_2 vapor is consumed in this equilibrium shift toward the left. A loss of I_2 vapor results in a less intense purple color.   (14.5, 14.7, Op. Sk. 7)

10. $[SiO] = 0.51$ mol/L, $[O_2] = 0.247$ mol/L   (14.6, Op. Sk. 6)

11.   It would not. The Si formed is a pure liquid with fixed density, and thus fixed concentration. Adding more of it will not change the Si concentration. This constant value is, in effect, included in the value of K_c for the reaction and is not part of the equilibrium-constant expression.   (14.3, 14.7, Op. Sk. 7)

12. $[CO_2] = [H_2] = 0.37$ mol/L

       $[CO] = [H_2O] = 0.14$ mol/L   (14.6, Op. Sk. 6)

13. $[HI] = 4.62$ mol/L   (14.6, Op. Sk. 5)

SUMMARY OF CHAPTER TOPICS

Chapter 14 is the first of four chapters devoted to the study of chemical equilibrium. Most of the problems in these chapters, although involving different species, are worked in essentially the same manner. The key to performing these calculations is to set up the table shown in Example 14.1 in the text and used in the solutions in this study guide. You may balk at using this table format, as it takes extra time to set it up. Take this extra time. It is extremely worthwhile.

        In preparing the table, there will usually be an unknown quantity, which we designate x. For each problem, be sure to define, and write down, what you are letting x be. This is another crucial step to ensure that you work the problem correctly.

        Pay close attention to the coefficients in the reaction when working with unknowns. If two molecules of a product substance are formed when one molecule reacts, then be sure your amount of substance that reacts is x and the amount of product formed is $2x$.

Make sure you then have the correct form of the equilibrium-constant expression. Many students forget exponents. Always go back and check that they are there if necessary. And remember that the form of the expression is always "products over reactants."

We will not follow our typical "Wanted, Given, etc.," format too closely in solving the exercises in these chapters, as the table gives an equally useful structure to problem solving. You should be quite familiar with the necessary steps of problem solving by now.

14.1 CHEMICAL EQUILIBRIUM — A DYNAMIC EQUILIBRIUM

Operational Skill

1. **Applying stoichiometry to an equilibrium mixture.** Given the starting amounts of reactants and the amount of one substance at equilibrium, find the equilibrium composition (Example 14.1).

Exercise 14.1 Synthesis gas (a mixture of CO and H_2) is increased in concentration of hydrogen by passing it with steam over a catalyst. This is the so-called water–gas shift reaction. Some of the CO is converted to CO_2, which can be removed:

$$CO(g) + H_2O(g) \rightleftharpoons CO_2(g) + H_2(g)$$

Suppose you start with a gaseous mixture containing 1.00 mol CO and 1.00 mol H_2O. When equilibrium is reached at 1000°C, the mixture contains 0.43 mol H_2. What is the molar composition of the equilibrium mixture?

Solution: Set up the table under the equation; let x = moles CO that react.

Amounts (mol)	$CO(g)$ +	$H_2O(g) \rightleftharpoons$	$CO_2(g)$ +	$H_2(g)$
Starting	1.00	1.00	0	0
Change	$-x$	$-x$	$+x$	$+x$
Equilibrium	$1.00-x$	$1.00-x$	x	0.43

For every mole of CO that reacts, 1 mole of H_2 is produced. Therefore, $x = 0.43$ mol.

Equilibrium amount of CO = $1.00 - 0.43 = 0.57$ mol CO
Equilibrium amount of H_2O = $1.00 - 0.43 = 0.57$ mol H_2O

Equilibrium amount of $CO_2 = x = 0.43$ mol

Equilibrium amount of $H_2 = 0.43$ mol (as given)

14.2   THE EQUILIBRIUM CONSTANT

Operational Skills

2.   Writing equilibrium-constant expressions.   Given the chemical equation, write the equilibrium-constant expression   (Example 14.2). (See also Section 14.3)

3.   Obtaining an equilibrium constant from reaction composition.   Given the equilibrium composition, find K_c (Example 14.3).

The equilibrium constant for a given reaction and equation is constant for that equation as long as the temperature remains unchanged. No matter what the starting mixture or how the equilibrium system is perturbed, the value of this constant will be the same as long as the temperature is not changed. This concept is called the law of mass action.

Exercise 14.2

(a)   Write the equilibrium-constant expression K_c for the equation

$$2NO_2(g) + 7H_2(g) \rightleftharpoons 2NH_3(g) + 4H_2O(g)$$

(b)   Write the equilibrium-constant expression K_c when this reaction is written

$$NO_2(g) + \frac{7}{2}H_2(g) \rightleftharpoons NH_3(g) + 2H_2O(g)$$

Known: The equilibrium-constant expression is equal to product concentrations over reactant concentrations, each to the coefficient power.

Solution: For (a):

$$K_c = \frac{[NH_3]^2[H_2O]^4}{[NO_2]^2[H_2]^7}$$

For (b):

$$K_c = \frac{[NH_3][H_2O]^2}{[NO_2][H_2]^{7/2}}$$

Exercise 14.3 When 1.00 mol each of carbon monoxide and water reach equilibrium at 1000°C in a 10.0-L vessel, the equilibrium mixture contains 0.57 mol CO, 0.57 mol H_2O, 0.43 mol CO_2, and 0.43 mol H_2. Write the chemical equation for the equilibrium. What is the value of K_c?

Solution: The equation is

$$CO(g) + H_2O(g) \rightleftharpoons CO_2(g) + H_2(g)$$

The equilibrium-constant expression is

$$K_c = \frac{[CO_2][H_2]}{[CO][H_2O]}$$

Each concentration must be calculated at equilibrium, then placed in this expression:

$$[CO] = [H_2O] = \frac{0.57 \text{ mol}}{10.0 \text{ L}} = 0.057 \ M$$

$$[CO_2] = [H_2] = \frac{0.43 \text{ mol}}{10.0 \text{ L}} = 0.043 \ M$$

$$K_c = \frac{(0.043)(0.043)}{(0.057)(0.057)} = 0.57$$

In this case, the units cancel. It is conventional, however, that even when they don't cancel, we do not write units for an equilibrium constant. The reason for this is discussed in text Section 18.6.

Exercise 14.4 Hydrogen sulfide, a colorless gas with a foul odor, dissociates on heating:

$$2H_2S(g) \rightleftharpoons 2H_2(g) + S_2(g)$$

When 0.100 mol H_2S was put into a 10.0-L vessel and heated to 1132°C, it gave an equilibrium mixture containing 0.0285 mol H_2. What is the value of K_c at this temperature?

Solution: First, change amounts to concentrations:

$$\text{Starting concentration of } H_2S = \frac{0.100 \text{ mol}}{10.0 \text{ L}} = 0.0100 \ M$$

$$\text{Equilibrium concentration of } H_2 = \frac{0.0285 \text{ mol}}{10.0 \text{ L}} = 0.00285 \ M$$

Second, set up a table under the equation, letting x = mol/L of H_2S that react.

Concentration (M)	$2H_2S(g)$ $\rightleftharpoons$	$2H_2(g)$	+	$S_2(g)$
Starting	0.0100	0		0
Change	$-2x$	$+2x$		$+x$
Equilibrium	$0.0100 - 2x$	$0.00285 = 2x$		x

Third, calculate equilibrium concentrations from the bottom line in the table:

$$[H_2S] = (0.0100 - 0.00285)\, M = 0.007\underline{1}5\, M$$

$$[H_2] = 0.00285\, M \text{ (as given)}$$

$$[S_2] = \frac{0.00285\, M}{2} = 0.00143\, M$$

Finally, write the equilibrium-constant expression from the equation and substitute in the calculated molarities:

$$K_c = \frac{[H_2]^2 [S_2]}{[H_2S]^2} = \frac{[0.00285]^2 [0.00143]}{[0.007\underline{1}5]^2} = \frac{[1.1\underline{5}3 \times 10^{-8}]}{[5.\underline{1}1 \times 10^{-5}]} = 2.3 \times 10^{-4}$$

Exercise 14.5   Phosphorus pentachloride dissociates on heating:

$$PCl_5(g) \rightleftharpoons PCl_3(g) + Cl_2(g)$$

If K_c equals 3.26×10^{-2} at 191°C, what is K_p at this temperature?

Known: $K_p = K_c(RT)^{\Delta n}$, where Δn is the sum of gaseous product coefficients minus gaseous reactant coefficients, and $R = 0.0821$ (L • atm)/(K • mol).

Solution:

$$\Delta n = 2 - 1 = 1$$

$$T = 191 + 273 = 464\, K$$

$$K_p = K_c(RT) = 3.26 \times 10^{-2} \times 0.0821 \times 464 = 1.24$$

Note that we do not use units, as K_c had no units. However, R must be expressed as (L • atm)/(K • mol), and T in kelvins.

14.3 HETEROGENEOUS EQUILIBRIA; SOLVENTS IN HOMOGENEOUS EQUILIBRIA

Operational Skill

2. Writing equilibrium-constant expressions. Given the chemical equation, write the equilibrium-constant expression   (Example 14.4).*

In working problems with gas reactions that may include heterogeneous equilibria, pay careful attention to the states of the substances in the equations. Only the gases are included in an equilibrium-constant expression.

Exercise 14.6   The Mond process for purifying nickel involves the formation of nickel tetracarbonyl, $Ni(CO)_4$, a volatile liquid, from nickel metal and carbon monoxide.

$$Ni(s) + 4CO(g) \rightleftharpoons Ni(CO)_4(g)$$

Write the expression for K_c for this reaction.

Solution:

$$K_c = \frac{[Ni(CO)_4]}{[CO]^4}$$

14.4 QUALITATIVELY INTERPRETING AN EQUILIBRIUM CONSTANT

Exercise 14.7   The equilibrium constant K_c for the reaction

$$2NO(g) + O_2(g) \rightleftharpoons 2NO_2(g)$$

equals 4.0×10^{13} at 25°C. Does the equilibrium mixture contain predominantly reactants or products? If $[NO] = [O_2] = 2.0 \times 10^{-6}$ M at equilibrium, what is the equilibrium concentration of NO_2?

---

*Note that Operational Skill 2 is used again in this section. Here, as well as in later chapters, operational skills are repeated as needed.

Solution: Since K_c is large, the equilibrium mixture contains mostly products. Determine $[NO_2]$ at equilibrium by substituting values in the equilibrium-constant expression and solving for $[NO_2]$:

$$K_c = \frac{[NO_2]^2}{[NO]^2[O_2]} = \frac{[NO_2]^2}{(2.0 \times 10^{-6})^2(2.0 \times 10^{-6})} = 4.0 \times 10^{13}$$

$$[NO_2] = 1.8 \times 10^{-2} \, M$$

14.5   PREDICTING THE DIRECTION OF REACTION

Operational Skill

4.   Using the reaction quotient.   Given the concentrations of substances in a reaction mixture, predict the direction of reaction   (Example 14.5).

Exercise 14.8   A 10.0-L vessel contains 0.0015 mol CO_2 and 0.10 mol CO. If a small amount of carbon is added to this vessel and the temperature raised to 1000°C, will more CO form? The reaction is

$$CO_2(g) + C(s) \rightleftharpoons 2CO(g)$$

The value of K_c for this reaction is 1.17 at 1000°C. Assume that the volume of gas in the vessel is 10.0 L.

Wanted: Will more CO form?

Given: 10.0-L vessel, 0.0015 mol CO_2, 0.10 mol CO; K_c at 1000°C = 1.17.

Known: Calculate the value of the reaction quotient Q_c and compare with K_c.

$$Q_c = \frac{[CO]^2_i}{[CO_2]_i}$$

Solution: Calculate concentrations:

$$[CO_2]_i = \frac{0.0015 \text{ mol}}{10.0 \text{ L}} = 0.00015 \, M \, CO_2$$

$$[CO]_i = \frac{0.10 \text{ mol}}{10.0 \text{ L}} = 0.010 \, M \, CO$$

$$Q_c = \frac{[0.010]^2}{[0.00015]} = 0.6\underline{6}7$$

Q_c is less than K_c. Thus, yes, the rate of the forward reaction will increase to produce more CO.

14.6 CALCULATING EQUILIBRIUM CONCENTRATIONS

Operational Skills

5. Obtaining one equilibrium concentration given the others. Given K_c and all concentrations of substances but one in an equilibrium mixture, calculate the concentration of this one substance (Example 14.6).

6. Solving equilibrium problems. Given the starting composition and K_c of a reaction mixture, calculate the equilibrium composition (Examples 14.7 and 14.8).

Exercise 14.9 Phosphorus pentachloride gives an equilibrium mixture of PCl_5, PCl_3, and Cl_2 when heated.

$$PCl_5(g) \rightleftharpoons PCl_3(g) + Cl_2(g)$$

A 1.00-L vessel contains an unknown amount of PCl_5 and 0.020 mol each of PCl_3 and Cl_2 at equilibrium at 250°C. How many moles of PCl_5 are in the vessel if K_c for this reaction is 0.0415 at 250°C?

Wanted: moles PCl_5.

Given: $V = 1.00$ L; 0.020 mol PCl_3, 0.020 mol Cl_2; $K_c = 0.0415$.

Known: $K_c = \dfrac{[PCl_3][Cl_2]}{[PCl_5]}$

Solution: Solve the above for $[PCl_5]$ and substitute in known values:

$$[PCl_5] = \frac{[PCl_3][Cl_2]}{K_c} = \frac{[0.020/1.00][0.020/1.00]}{0.0415}$$

$$= 0.0096$$

Moles $PCl_5 = 0.0096$

Note that in substituting values into the equilibrium-constant expression in Exercise 14.9, we did not write $[0.020/1.00]^2$. We wrote the number twice. The reason for doing this, when you are in a hurry, such as when taking a test, is you will not forget to square the value. If you write it twice, you will avoid making that error.

Exercise 14.10   What is the equilibrium composition of a reaction mixture if you start with 0.500 mol each of H_2 and I_2 in a 1.0-L vessel? The reaction is

$$H_2(g) + I_2(g) \rightleftharpoons 2HI(g); \quad K_c = 49.7 \text{ at } 458°C$$

Solution: First, set up a table of concentrations. Since it is a 1-L vessel, the concentrations are the mole amounts to two significant figures.

Concentrations (M)	$H_2(g)$	+	$I_2(g)$	$\rightleftharpoons$	$2HI(g)$
Starting	0.500		0.500		0
Change	$-x$		$-x$		$+2x$
Equilibrium	$0.500-x$		$0.500-x$		$2x$

Second, substitute values into the equilibrium-constant expression:

$$K_c = \frac{[HI]^2}{[H_2][I_2]} = \frac{(2x)(2x)}{(0.500-x)(0.500-x)} = \frac{(2x)^2}{(0.500-x)^2} = 49.7$$

Third, take the square root of both sides and solve for x:

$$\frac{2x}{(0.500-x)} = \pm 7.050 \quad \text{(only the positive value is possible)}$$

$$2x = 3.52 - 7.050x$$

$$9.050x = 3.52$$

$$x = \frac{3.52}{9.050} = 0.389 \, M$$

Thus, equilibrium concentrations are

$$[H_2] = [I_2] = 0.500 - 0.389 = 0.11 \, M$$

$$[HI] = 2(0.389) = 0.78 \, M$$

When calculating equilibrium concentrations given only starting amounts, be sure to look for ways to simplify your work so you do not have to deal with x^2. First, check to see whether the expression with the x is a perfect square. If it appears not to be, check to see that you've written the equilibrium-constant expression with the proper exponents. If you still do not have a perfect square, then see if you can eliminate the x that is subtracted from a beginning concentration. You can ignore this change when it is very small compared with the other number. This will often be the case when the equilibrium constant divided by the initial concentration is 10^{-3} or less. If you cannot simplify the work, then you will have to use the quadratic formula to solve for x. It is very useful but takes a long time to solve, and when you use it, you must take great care to avoid errors.

Exercise 14.11   Phosphorus pentachloride, PCl_5, decomposes when heated.

$$PCl_5(g) \rightleftharpoons PCl_3(g) + Cl_2(g)$$

If the initial concentration of PCl_5 is 1.00 mol/L, what is the equilibrium composition of the gaseous mixture at 160°C? The equilibrium constant K_c at 160°C is 0.0211.

Solution: Set up the table of concentrations.

Concentrations (M)	$PCl_5(g) \rightleftharpoons$	$PCl_3(g) +$	$Cl_2(g)$
Starting	1.00	0	0
Change	$-x$	$+x$	$+x$
Equilibrium	$1.00 - x$	x	x

Putting these values into the equilibrium-constant expression, we have

$$K_c = \frac{[PCl_3][Cl_2]}{[PCl_5]} = \frac{(x)(x)}{(1.00 - x)} = 0.0211$$

Since this is not a perfect square, and K_c divided by the initial concentration of PCl_5 is too large to ignore x, we must use the quadratic formula:

$$\frac{b \pm \sqrt{b^2 - 4ac}}{2a}$$

Put the K_c expression into the proper form:

$$x^2 = (0.0211)(1.00 - x) = 0.0211 - 0.0211x$$

We rearrange and assign values as follows:

$$x^2 + 0.0211x - 0.0211 = 0$$

$$\underbrace{}_{a=1} \quad \underbrace{}_{b} \quad \underbrace{}_{c}$$

Substitute values into the quadratic formula:

$$x = \frac{-0.0211 \pm \sqrt{(0.0211)^2 - 4(1)(-0.0211)}}{2} = \frac{-0.0211 \pm 0.2913}{2}$$

Since a negative value for x is impossible, the solution is

$$x = 0.13\underline{5}1$$

Referring to the bottom line of the table, we find that the equilibrium compositions are

$$[PCl_5] \;=\; 1.00 - 0.13\underline{5}1 = 0.86 \; M$$

$$[PCl_3] \;=\; [Cl_2] = 0.135 \; M$$

14.7   REMOVING PRODUCTS OR ADDING REACTANTS

Operational Skill

7.   Applying Le Chatelier's principle.   Given a reaction, use Le Chatelier's principle to decide the effect of adding or removing a substance   (Example 14.9).

It is important that you understand that Le Chatelier's principle is an observation, not an explanation. Recall from Chapter 13 that collision theory supposes that reaction occurs because of molecular collisions. Thus, it is easy to see why more product could form when the concentration of reactant is increased — because there is more chance for collision of reactants to form a product. If product is removed, there are fewer collisions to convert product back to reactant. Thus, Le Chatelier's principle supports collision theory.

Exercise 14.12   Consider each of the following equilibria that are disturbed in the manner indicated. Predict the direction of reaction.
(a)   The equilibrium

$$CaCO_3(s) \;\rightleftharpoons\; CaO(s) + CO_2(g)$$

is disturbed by increasing the pressure (that is, concentration) of carbon dioxide.

(b) The equilibrium

$$2Fe(s) + 3H_2O(g) \rightleftharpoons Fe_2O_3(s) + 3H_2(g)$$

is disturbed by increasing the concentration of hydrogen.

Solution: (a) Increasing the pressure of carbon dioxide in the reaction mixture is the same as increasing the concentration of CO_2. This would cause a shift left, or cause the rate of the reverse reaction to increase, which would reduce the added CO_2 concentration. (b) Increasing the concentration of hydrogen again increases the rate of the reverse reaction, which lowers the added H_2 concentration.

Remember that once disturbed and then left alone, a reaction system will again reach equilibrium; even though the concentrations will be different, the value of K_c will be the same as long as the temperature is not changed.

14.8   CHANGING THE PRESSURE AND TEMPERATURE

Operational Skill

7.   Applying Le Chatelier's principle.   Given a reaction, use Le Chatelier's principle to decide the effect of adding or removing a substance   (Example 14.9), changing the pressure   (Example 14.10), or changing the temperature   (Example 14.11).

Looking again at an equilibrium system from the viewpoint of molecular collisions, if we increase the pressure, we push molecules closer together. The side of the equation having more gas molecules will have more collisions. Thus, reaction will go toward the other side of the equation. On the other hand, the side with more gas molecules has to have more collisions for reaction to occur and is thus affected more by a decrease in pressure than is the side with fewer gas molecules. Thus, when the pressure is decreased, reaction is toward the side with more molecules. Again, Le Chatelier's principle describes what we expect.

Exercise 14.13   Can you increase the amount of product in each of the following reactions by increasing the pressure? Explain.

(a) $CO_2(g) + H_2(g) \rightleftharpoons CO(g) + H_2O(g)$

(b) $4CuO(s) \rightleftharpoons 2Cu_2O(s) + O_2(g)$

(c) $2SO_2(g) + O_2(g) \rightleftharpoons 2SO_3(g)$

Solution: (a)   No. According to Le Chatelier's principle, if the pressure is increased, reaction favors the direction to make fewer gas molecules. Both sides have the same number of gas molecules, so the rates of both forward and reverse reactions will be similarly affected, with no shift. (b)   No. Increasing the pressure will favor the reverse reaction, as there are no gas molecules on the left side. (c)   Yes. There are fewer gas molecules on the right — two, versus three on the left — so an increase in pressure will yield more product.

In the next exercise, to understand Le Chatelier's principle in relation to how temperature affects reactions, we must consider the reaction energetics. If the temperature is raised, the reaction that uses heat will be favored. If the temperature is lowered, the reaction that needs heat will be slowed down, and the opposite reaction will then go more quickly. In these instances, the value of the equilibrium constant will change. When equilibrium is again established after the heat perturbation, the value of K will not be the same as it was before the change. The value of K changes consistently with the rate change. If the rate of the forward reaction is increased, K will increase; if the rate of the reverse reaction is increased, K will decrease. This will be further discussed in text Chapter 17.

Exercise 14.14   Consider the possibility of converting carbon dioxide to carbon monoxide by the endothermic reaction

$$CO_2(g) + H_2(g) \rightleftharpoons CO(g) + H_2O(g)$$

Is a high or a low temperature more favorable to the production of carbon monoxide? Explain.

Solution: Endothermic reactions will be favored with an increase in temperature. So, a high temperature would be favorable.

Exercise 14.15   Consider the reaction

$$2CO_2(g) \rightleftharpoons 2CO(g) + O_2(g); \quad \Delta H° = 566 \text{ kJ}$$

Discuss the temperature and pressure conditions that would give the best yield of carbon monoxide.

Solution: High temperatures would favor the formation of products in this endothermic reaction. Low pressure would favor the products, as there are more gas molecules on the product side.

14.9   EFFECT OF A CATALYST

A Gas That Matters: CARBON MONOXIDE (Starting Material for Organic Compounds)

Questions for Study

1.   Carbon monoxide is a poisonous gas. Explain why this is so.

2.   How is carbon monoxide prepared commercially?

3.   List two organic compounds that can be produced from carbon monoxide. Give equations for the preparations.

4.   Describe the steps in producing synthetic gasoline by one method from coal and water.

Answers to Questions for Study

1.   CO acts as a poison by combining with the hemoglobin in red blood cells. It binds so strongly that it is not easily released. Thus, the hemoglobin is not available to combine with and transport oxygen to the cells.

2.   Carbon monoxide is prepared commercially by reacting natural gas $\left(\text{mostly } CH_4\right)$ with steam or by partially oxidizing natural gas with air.

3.   Methanol is prepared by the reaction of carbon monoxide with hydrogen using a zinc oxide–chromium(III) oxide catalyst.

$$CO(g) + 2H_2(g) \xrightarrow{\;\;ZnO\text{–}Cr_2O_3\;\;} CH_3OH(g)$$

Acetic acid is prepared by bubbling CO into liquid methanol at 215°C and 140 atm using NiI_2 as the catalyst.

$$CO(g) + CH_3OH(l) \xrightarrow{\;\;NiI_2\;\;} CH_3COOH(l)$$

4.   Synthetic gasoline can be produced from coal (C) and water by passing steam over a bed of red-hot coal to produce water-gas, a mixture of CO and H_2. This mixture can then be reacted in the presence of an Fe–Co catalyst to produce octane.

ADDITIONAL PROBLEMS

1. Calculate the concentrations of all substances present in the equilibrium mixture at a given temperature if 2.35 mol H_2 and 2.35 mol I_2 are placed in a 10.0-L flask and allowed to come to equilibrium, at which time 3.76 mol HI are present. What is the value of K_c? The reaction is

$$H_2(g) + I_2(g) \rightleftharpoons 2HI(g)$$

2. The following concentrations were found for the substances present in a reaction flask: $CS_2 = 0.48\ M$, $H_2 = 0.35\ M$, $CH_4 = 0.42\ M$, and $H_2S = 0.52\ M$. Is the reaction at equilibrium? If not, for which reaction will the rate be greater? K_c at this temperature is 0.28. The reaction is

$$CS_2(g) + 4H_2(g) \rightleftharpoons CH_4(g) + 2H_2S(g)$$

3. Write the equilibrium-constant expression for each of the following equations.

    (a) $SbBr_3(g) \rightleftharpoons Sb(s) + \frac{3}{2}Br_2(g)$

    (b) $SO_2(g) + \frac{1}{2}O_2(g) \rightleftharpoons SO_3(g)$

    (c) $2CaSO_4(s) \rightleftharpoons 2CaO(s) + 2SO_2(g) + O_2(g)$

    (d) $2Cl_2(g) + 2H_2O(g) \rightleftharpoons 4HCl(g) + O_2(g)$

    (e) $2Fe(s) + 4H_2O(g) \rightleftharpoons Fe_2O_4(s) + 4H_2(g)$

4. The value of K_p for the equilibrium

$$B_2H_6(g) + 4BF_3(g) \rightleftharpoons 6HBF_2(g)$$

    is 2.94 at 296 K. What is the value of K_c for this equilibrium at 296 K?

5. The equilibrium constant for the following reaction at a given temperature is 45.0. How many moles of each component are present in the reaction mixture at this temperature if 0.340 mol H_2 and 0.340 mol I_2 are placed in a 10.0-L vessel? The reaction is

$$H_2(g) + I_2(g) \rightleftharpoons 2HI(g)$$

6.   The equilibrium constant K_c for the following reaction at 150°C is 1.20×10^2. Calculate the concentrations of all components in the equilibrium mixture if 5.00 mol I_2 and 8.00 mol Br_2 are reacted in a 10.0-L vessel. The reaction is

$$I_2(g) + Br_2(g) \rightleftharpoons 2IBr(g)$$

7.   The decomposition of SO_3 proceeds according to the following equation:

$$2SO_3(g) \rightleftharpoons 2SO_2(g) + O_2(g)$$

At 298 K, K_p for this reaction is 8.30×10^{-23}. If 2.46 mol SO_3 are placed in a 1.00-L gas cylinder at 298 K, what will be the partial pressure of each species present at equilibrium?

8.   When a mixture of air and gasoline vapor explodes in the cylinder of a gasoline engine, some of the N_2 and O_2 from the air combine to form NO gas:

$$N_2(g) + O_2(g) \rightleftharpoons 2NO(g)$$

The NO is discharged in the exhaust gases and contributes to smog formation. K_c for the reaction is 1.21×10^{-4} at 1800 K. An experimental combustion chamber having a volume of 1.00 L is filled with air at 1.00 atm pressure at 0°C in the presence of a catalyst. It contains 0.0357 mol N_2 and 0.00892 mol O_2. The chamber is heated to 1800 K and held at this temperature until there is no observed change in pressure. How many moles of NO are present in the equilibrium mixture?

9.   Anhydrous $CaSO_4$ is commonly used as a drying agent. At 25°C, the heterogeneous equilibrium

$$CaSO_4(s) + 2H_2O(g) \rightleftharpoons CaSO_4 \cdot 2H_2O(s)$$

has a K_p value of 1.55×10^3. A 1.50×10^2-g sample of anhydrous $CaSO_4$ and 1.00×10^2 g $CaSO_4 \cdot 2H_2O$ are placed in the bottom of a small desiccator at 25°C and the desiccator is closed. What will be the equilibrium vapor pressure of the solid mixture?

10.   Predict the direction of increased reaction resulting from each of the following changes to the equilibrium system. Explain your answers. Also state any expected change in the value of K in each case.

$$2NO_2(g) + 7H_2(g) \rightleftharpoons 2NH_3(g) + 4H_2O(g) + 993 \text{ kJ}$$

(a)   an increase in the volume of the container
(b)   an increase in temperature
(c)   removal of NH_3
(d)   addition of a catalyst
(e)   addition of 0.15 mol of helium gas

ANSWERS TO ADDITIONAL PROBLEMS

If you missed an answer, study the text section and operational skill given in parentheses after the answer.

1. $[H_2] = [I_2] = 0.047 \ M$; $[HI] = 0.376 \ M$; $K_c = 64$ (14.1, Op. Sk. 1)

2.   The reaction is not at equilibrium because $Q_c = 16$, which is greater than K_c. The rate of the reverse reaction will be increased.   (14.5, Op. Sk. 4)

3.   (a) $\dfrac{[Br_2]^{3/2}}{[SbBr_3]}$;   (c) $[SO_2]^2 [O_2]$;   (e) $\dfrac{[H_2]^4}{[H_2O]^4}$ (14.2, Op. Sk. 2)

       (b) $\dfrac{[SO_3]}{[SO_2][O_2]^{1/2}}$;   (d) $\dfrac{[HCl]^4 [O_2]}{[Cl_2]^2 [H_2O]^2}$;

4. 1.21×10^{-1} (14.2)

5.   mol H_2 = mol I_2 = 0.078; mol HI = 0.524   (14.6, Op. Sk. 6)

6. $[I_2] = 0.023 \ M$; $[Br_2] = 0.32 \ M$; $[IBr] = 0.95 \ M$ (14.6, Op. Sk. 6)

7. $P_{SO_3} = 60.2$ atm; $P_{SO_2} = 8.44 \times 10^{-7}$ atm; $P_{O_2} = 4.22 \times 10^{-7}$ atm

                                                                      (14.2, 14.6; Op. Sk. 2, 6)

8. 1.95×10^{-4} mol   (14.6, Op. Sk. 6)

9. $P_{H_2O} = 2.54 \times 10^{-2}$ atm   (14.2, 14.6, Op. Sk. 6)

10. (a) An increase in volume would decrease the pressure. This would affect the reverse reaction (with 6 gas molecules) less than it would the forward reaction (with 9 gas molecules). Therefore, the rate of the forward reaction will be slowed, and increased reaction will occur to the left. The value of K will not change because the temperature remains constant.

    (b) Because the reaction is exothermic (with the heat of reaction on the product side), the reverse reaction will be favored with an increase in temperature, and the value of K will decrease.

    (c) Removal of product will decrease the rate of the reverse reaction; thus, reaction will increase to the right, or product, side; K will not change.

    (d) The addition of a catalyst will affect both forward and reverse reactions similarly; there will be no change.

    (e) Because helium is not part of the reacting system and will not react with other substances present, it will have no effect on the reaction; there will be no change.   (14.7, 14.8, Op. Sk. 7)

CHAPTER POST-TEST

1. Which of the following represents the correct equilibrium-constant expression for the reaction

$$3D(aq) + 5E(aq) \rightleftharpoons \tfrac{1}{2} F(aq) + \tfrac{2}{3} G(aq)$$

(a) $\dfrac{[F]^2 [G]^{3/2}}{[D]^{1/3} [E]^{1/5}}$ (c) $\dfrac{[F]^{1/2}}{[G]^{2/3}} \cdot \dfrac{[E]^5}{[D]^3}$ (e) $[F]^{1/2} [G]^{3/2} [D]^3 [E]^5$

(b) $\dfrac{[D]^3 [F]^{1/2}}{[E]^5 [G]^{2/3}}$ (d) $\dfrac{[F]^{1/2} [G]^{2/3}}{[E]^5 [D]^3}$

2. The equilibrium constant K_c for the dissociation

$$PCl_5(g) \rightleftharpoons PCl_3(g) + Cl_2(g)$$

equals 3.26×10^{-2} at 191°C. Does the equilibrium mixture contain predominantly reactant or products? When 0.285 mol PCl_5 is introduced into a 10.0-L vessel and heated to 191°C, the measured concentrations of PCl_3 and Cl_2 sometime later are both found to be [0.00853] or 0.00853 *M*. At this point has equilibrium been reached?

3. Determine whether the following statements are true or false. If a statement is false, change it so it is true.
   (a) For the equilibrium $2H_2(g) + O_2(g) \rightleftharpoons 2H_2O(l)$, a decrease in pressure will favor the increased formation of $H_2O(l)$. True/False: _____
   _____ .
   (b) Changes in either temperature or pressure of a system at equilibrium will alter the equilibrium constant for the system. True/False:_____
   _____ .

4. Write equilibrium expressions for the following reactions in terms of K_p:
   (a) $3O_2(g) \rightleftharpoons 2O_3(g)$
   (b) $4CuO(s) \rightleftharpoons 2Cu_2O(s) + O_2(g)$
   (c) $C(s) + S_2(g) \rightleftharpoons CS_2(g)$

5. Hydrogen sulfide gas, H_2S, when heated to 1132°C, dissociates to H_2 and S_2 gases. The equation is

$$2H_2S(g) \rightleftharpoons 2H_2(g) + S_2(g)$$

The equilibrium concentrations of H_2 and S_2 were measured to be 0.0384 *M* and 0.0192 *M*, respectively. If the K_c value is 2.3×10^{-4} at 1132°C, what is the equilibrium concentration of H_2S gas?

6. When 1.000 mol of gaseous HI is sealed in a 1.000-L flask at a specific temperature, it decomposes to form 0.182 mol each of hydrogen and iodine. Calculate K_c for the reaction

$$2HI(g) \rightleftharpoons H_2(g) + I_2(g)$$

7. The reaction

$$2NO(g) + O_2(g) \rightleftharpoons 2NO_2(g)$$

is endothermic. If we increase the temperature, will the K_c value increase or decrease?

8. If we suddenly decreased the volume of the container for the system in Problem 7, how would this affect the equilibrium?

9. If we inject more O_2 into the system in Problem 7, how would this affect the equilibrium? How would it affect the value of the equilibrium constant?

10. At 395 K the reaction

$$CO(g) + Cl_2(g) \rightleftharpoons COCl_2(g)$$

has equilibrium partial pressures for each species of 0.128 atm, 0.116 atm, and 0.334 atm, respectively. Determine K_p.

11. In a 1-L tank, 1 mol $MgCO_3$, 1 mol MgO, and 1 mol CO_2 are in equilibrium:

$$MgCO_3(s) \rightleftharpoons MgO(s) + CO_2(g)$$

If more CO_2 is added, will there be a change in the amounts of $MgCO_3$ and MgO present at equilibrium?

12. Consider the following reaction:

$$H_2(g) + I_2(g) \rightleftharpoons 2HI(g); \quad K_c = 54.7$$

If 2.00 mol H_2 and 2.00 mol I_2 are placed in a 1.00-L container, what will be the equilibrium concentrations?

13. Consider the following reaction:

$$2HI(g) \rightleftharpoons H_2(g) + I_2(g); \quad K_c = 0.0183$$

At equilibrium, $[H_2] = 0.32$ mol/L, $[I_2] = 0.32$ mol/L, and $[HI] = 2.36$ mol/L. Suppose that 0.70 mol/L HI is added and a shift in equilibrium occurs. Calculate the new equilibrium concentrations.

ANSWERS TO CHAPTER POST-TEST

If you missed an answer, study the text section and operational skill given in parentheses after the answer.

1. d (14.2, Op. Sk. 2)

2. Predominantly reactant; $Q_c = 3.64 \times 10^{-3}$, hence equilibrium has not been reached. (14.4, 14.5, Op. Sk. 4)

3. (a) False. The equilibrium will favor the increased formation of $H_2(g)$ and $O_2(g)$. (14.8, Op. Sk. 7)

   (b) False. Changes in temperature will alter the equilibrium constant. Changes in pressure will alter the equilibrium composition but not the equilibrium constant. (14.8, Op. Sk. 7)

4. (a) $K_p = \dfrac{P_{O_3}^2}{P_{O_2}^3}$ (b) $K_p = P_{O_2}$ (c) $K_p = \dfrac{P_{CS_2}}{P_{S_2}}$

   (14.2, 14.3, Op. Sk. 2)

5. $[H_2S] = 0.35$ mol/L (14.6, Op. Sk. 5)

6. $K_c = 0.0819$ (14.2, Op. Sk. 1, 3)

7. Increasing the temperature favors a shift to the right and increases K_c, which is temperature dependent. (14.8, Op. Sk. 7)

8. Decreasing the volume would increase the internal pressure. By Le Chatelier's principle, a shift should occur to decrease this pressure. The pressure will decrease if

there are fewer moles of gaseous particles, so the reaction would shift to the right. K_c does not change.   (14.8, Op. Sk. 7)

9.   With the addition of O_2 to this system, the rate of the forward reaction increases and additional O_2 is consumed. Eventually the product–reactant ratio once again is equal to the original value of the equilibrium constant. K_c does not change unless the temperature changes.   (14.7, Op. Sk. 7)

10. $K_p = 22.5$ (14.2, Op. Sk. 3)

11.   Yes. If CO_2 is added, the reaction shifts left, forming more $MgCO_3$ and consuming CO_2 and MgO.   (14.7, Op. Sk. 7)

12. $[H_2] = [I_2] = 0.43$ mol/L

       $[HI] = 3.15$ mol/L   (14.6, Op. Sk. 6)

13. $[HI] = 2.91$ mol/L

       $[H_2] = [I_2] = 0.39$ mol/L   (14.6, 14.7; Op. Sk. 6, 7)

CHAPTER 15   ACIDS AND BASES

CHAPTER TERMS AND DEFINITIONS

Numbers in parentheses after definitions give the text sections in which the terms are explained. Starred terms are italicized in the text. Where a term does not fall directly under a text section heading, additional information is given for you to locate it.

oxygen*   means "acid former" in Greek   (15.1, introductory section)

Arrhenius concept*   acid–base theory that defines an acid as a substance that, when dissolved in water, increases the concentration of hydrogen ion, $H^+(aq)$, and a base as a substance that, when dissolved in water, increases the concentration of hydroxide ion, $OH^-(aq)$ (15.1)

strong acid*   according to Arrhenius, a substance that completely ionizes in aqueous solution to give a hydronium ion $(H_3O^+(aq)]$ and an anion   (15.1)

strong base*   according to Arrhenius, a substance that completely ionizes in aqueous solution to give a hydroxide ion $[OH^-(aq)]$ and a cation   (15.1)

weak*   describes acids and bases that are not ionized completely in solution and exist in equilibrium with the corresponding ions   (15.1)

acid (Brønsted–Lowry)   species that donates the proton in a proton-transfer reaction   (15.2)

base (Brønsted–Lowry)   species that accepts the proton in a proton-transfer reaction   (15.2)

conjugate acid–base pair   in an acid–base equilibrium, two species that differ by the loss or gain of a proton   (15.2)

conjugate acid*   in a conjugate acid–base pair, the species that can donate a proton   (15.2)

conjugate base*   in a conjugate acid–base pair, the species that can accept a proton   (15.2)

amphiprotic species   species that can function as either a Brønsted–Lowry acid or base, i.e., can either lose or gain a proton   (15.2)

amphoteric*   describes a species, such as aluminum oxide, that can act as an acid or base   (15.2, marginal note)

Lewis acid   species that can form a covalent bond by accepting an electron pair from another species   (15.3)

Lewis base   species that can form a covalent bond by donating an electron pair to another species   (15.3)

complex ion (s)*   species produced when a metal ion forms a covalent bond by accepting an electron pair from a molecule or another ion   (15.3)

leveling effect*   phenomenon whereby, in water, the strong acids appear to be the same, although they show differing degrees of ionization in other solvents   (15.4)

self-ionization (autoionization)   reaction in which two like molecules react to give ions   (15.6)

ion-product constant for water (K_w)   equilibrium value of the ion product $[H_3O^+]$

$[OH^-]$, which equals 1.00×10^{-14} at 25°C   (15.6)

pH $-\log [H^+]$ (15.8)

CHAPTER DIAGNOSTIC TEST

1.   What is the Arrhenius concept of an acid?   *H containing substances yielding H^+.*

2.   Give the conjugate base for each of the following acids:

   (a)   HBr   *Br^-*   (c) HCO_3^- *CO_3^{2-}*   (e) H_2SO_4 *HSO_4^-*

   (b) NH_4^+ *NH_3*   (d) $HC_2H_3O_2$ *$C_2H_3O_2^-$*

3.   Give the conjugate acid for each of the following bases:

   (a) CN^- *HCN*   (c) HCO_3^- *H_2CO_3*   (e) $H_2PO_4^-$ *HPO_4*

   (b) H_2O *H_3O^+*   (d) SO_4^{2-} *HSO_4^-*

4.   Using text Table 15.2, predict the direction for each of the following acid–base reactions:

   (a) $H_3O^+ + CO_3^{2-} \rightleftharpoons HCO_3^- + H_2O$ *right*

   (b) $NO_3^- + HF \rightleftharpoons HNO_3 + F^-$ *left*

   (c) $HS^- + NH_4^+ \rightleftharpoons H_2S + NH_3$ *left*

   (d) $H_2S + CN^- \rightleftharpoons HS^- + HCN$ *right*

5. For each of the following reactions, what type of acid is present? Identify which species are the acids and which are the bases.

    (a) $Cu^{2+} + 4NH_3 \rightleftharpoons Cu(NH_3)_4^{2+}$

    (b) $COCl_2 + AlCl_3 \rightleftharpoons COCl^+ + AlCl_4^-$

6. A solution of NaOH is determined to be 7.6×10^{-3} *M* NaOH. What are the hydroxide and hydrogen ion concentrations of this solution?

7. The hydroxide ion concentration in skim milk was measured to be 4.2×10^{-8} *M*. What is the pH of this milk?

8. Will the following reaction occur as written?

    $$HClO(aq) + CO_3^{2-}(aq) \longrightarrow ClO^-(aq) + HCO_3^-(aq)$$

ANSWERS TO CHAPTER DIAGNOSTIC TEST

If you missed an answer, study the text section and operational skill given in parentheses after the answer.

1. According to the Arrhenius concept, an acid is any hydrogen-containing substance that yields a hydrogen ion in a water solution.   (15.1)

2. (a) Br^- (c) CO_3^{2-} (e) HSO_4^- (15.2, Op. Sk. 1)

    (b) NH_3 (d) $C_2H_3O_2^-$

3. (a) HCN   (c) H_2CO_3 (e) H_3PO_4 (15.2, Op. Sk. 1)

    (b) H_3O^+ (d) HSO_4^-

4. (a) to the right   (c) to the left
    (b) to the left   (d) to the right   (15.4, Op. Sk. 3)

5. Lewis acids

    (a) Cu^{2+}, acid; NH_3, base   (b) $COCl_2$, base; $AlCl_3$, acid   (15.3, Op. Sk. 2)

6. $[OH^-] = 7.6 \times 10^{-3} \, M$; $[H^+] = 1.3 \times 10^{-12} \, M$ (15.7, Op. Sk. 4)

7. pH = 6.62   (15.8, Op. Sk. 5)

8. yes   (15.4; Op. Sk. 3)

SUMMARY OF CHAPTER TOPICS

15.1   ARRHENIUS CONCEPT OF ACIDS AND BASES

15.2   BRØNSTED–LOWRY CONCEPT OF ACIDS AND BASES

Operational Skill

1.   Identifying acid and base species.   Given a proton-transfer reaction, label the acids and bases, and name the conjugate acid–base pairs   (Example 15.1).

Memorize the Brønsted–Lowry definition that an acid is a proton donor. This will help you remember that a base is a proton acceptor.

Exercise 15.1   For the reaction

$$H_2CO_3(aq) + CN^-(aq) \rightleftharpoons HCN(aq) + HCO_3^-(aq)$$

label each species as an acid or a base. For the base on the left, what is the conjugate acid?

Known: A Brønsted–Lowry acid is a proton donor; a conjugate acid is the base without its proton.

Solution: Write labels under the appropriate species.

$$H_2CO_3(aq) + CN^-(aq) \rightleftharpoons HCN(aq) + HCO_3^-(aq)$$

　　　acid　　　　　base　　　　　acid　　　　base

The base on the left is CN^-. Its conjugate acid is HCN.

15.3 LEWIS CONCEPT OF ACIDS AND BASES

Operational Skill

   2.   Identifying Lewis acid and base species.   Given a reaction involving the donation of an electron pair, identify the Lewis acid and the Lewis base   (Example 15.2).

       The Lewis acid is an electron-pair acceptor. Memorizing this definition will help you remember that a Lewis base is an electron-pair donor.

       Exercise 15.2   Identify the Lewis acid and the Lewis base in each of the following reactions. Write the chemical equations using electron-dot formulas.

$$\text{(a)}\quad BF_3 + CH_3OH \longrightarrow F_3B : \overset{\overset{\displaystyle H}{\displaystyle |}}{O}CH_3 \qquad \text{(b)}\quad O^{2-} + CO_2 \longrightarrow CO_3^{\,2-}$$

Solution:

   (a)   The equation, using labeled electron-dot formulas, is

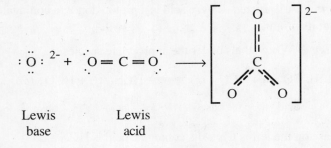

   (b)   The equation, using labeled electron-dot formulas, is

15.4   RELATIVE STRENGTHS OF ACIDS AND BASES

Operational Skill

3.   Deciding whether reactants or products are favored in an acid–base reaction.
Given an acid–base reaction and the relative strengths of acids (or bases), decide whether
reactants or products are favored   (Example 15.3).

> **Exercise 15.3**   Determine the direction of the following reaction from the
> relative strengths of acids and bases.
>
> $$H_2S(aq) + C_2H_3O_2^{-}(aq) \rightleftharpoons HC_2H_3O_2(aq) + HS^{-}(aq)$$
>
> *Known:*   Equilibrium favors the acid of lesser strength.
>
> *Solution:*   Text Table 15.2 shows that acetic acid is stronger than H_2S. Thus, the
> reaction goes from right to left.

15.5   MOLECULAR STRUCTURE AND ACID STRENGTH

> **Exercise 15.4**   Which member of each of the following pairs is the stronger
> acid?
>
> (a) NH_3, PH_3 (c) HSO_3^{-}, H_2SO_3 (e) HSO_4^{-}, $HSeO_4^{-}$
>
> (b) HI, H_2Te (d) H_3AsO_4, H_3AsO_3
>
> (a)   *Known:*   N and P are in the same periodic table group; acidity increases
> with size and thus down the group.
>
>   *Solution:* PH_3 is the stronger acid.
>
> (b)   *Known:*   I and Te are in the same period; acidity increases with electro-
> negativity and thus to the right.
>
>   *Solution:*   HI is the stronger acid.
>
> (c)   *Known:* HSO_3^{-} is an anion of H_2SO_3; the anion with its negative charge
> holds the hydrogen more tightly.
>
>   *Solution:* H_2SO_3 is the stronger acid.
>
> (d)   *Known:*   Both species are oxoacids of As; bond polarity and thus acid
> strength increase with the oxidation state of As.

Solution: H_3AsO_4, with the higher oxidation state of As, is the stronger acid.

(e)   *Known:*   Both species are anions of oxoacids of elements in the same periodic group with the same oxidation number; acidity increases with increasing electronegativity of the central atom.

Solution: HSO_4^- is the stronger acid, as S is more electronegative than Se.

15.6   SELF-IONIZATION OF WATER

Memorize the expression for and value of the ion-product constant for water, $K_w = [H^+][OH^-] = 1.0 \times 10^{-14}$ at 25°C. (As does the text author, we will use H^+ for H_3O^+.) You won't be able to solve acid–base problems without it. Although we do not report units for an equilibrium constant, in using K_w in calculations to find $[H^+]$ or $[OH^-]$, we attribute to K_w the units of M^2.

15.7   SOLUTIONS OF A STRONG ACID OR BASE

Operational Skill

4.   Calculating concentrations of H^+ and OH^- in solutions of a strong acid or base. Given the concentration of a strong acid or base, calculate the hydrogen-ion and hydroxide-ion concentrations   (Example 15.4).

Memorize the information relating hydrogen-ion concentration and acidity: in an acidic solution, $[H^+]$ is greater than (>) 1.00×10^{-7} M; in a neutral solution, $[H^+] = 1.00 \times 10^{-7}$ M; and in a basic solution, $[H^+]$ is less than (<) 1.00×10^{-7} M.

The negative exponents used here may give you trouble. Spend some time, if necessary, drilling on relative sizes of numbers with negative exponents so this will not slow you down or cause you to make mistakes. For example: 10^{-7} is greater than 10^{-9}, and 10^{-10} is less than 10^{-8}.

Exercise 15.5   A solution of barium hydroxide at 25°C is 0.125 M $Ba(OH)_2$. What are the concentrations of hydrogen ion and hydroxide ion in the solution?

Known: $Ba(OH)_2$ is a strong base; thus, it ionizes completely in water:

$$Ba(OH)_2(s) \longrightarrow Ba^{2+}(aq) + 2OH^-(aq)$$

so that for every $Ba(OH)_2$ that dissolves, two OH^- ions result. We also know that

$$K_w = [H^+][OH^-] = 1.00 \times 10^{-14} \text{ (attributed units are } M^2)$$

Solution:

$$[OH^-] = 2 \times 0.125\ M = 0.250\ M$$

Find $[H^+]$ using K_w:

$$[H^+] = \frac{K_w}{[OH^-]} = \frac{1.00 \times 10^{-14}}{0.250\ M} = 4.00 \times 10^{-14}\ M$$

Exercise 15.6 A solution has a hydroxide-ion concentration of $1.0 \times 10^{-5}\ M$ at 25°C. Is the solution acidic, neutral, or basic?

Known: In an acidic solution, $[H^+] > 1.00 \times 10^{-7}\ M$; in a neutral solution, $[H^+] = 1.00 \times 10^{-7}\ M$; in a basic solution, $[H^+] < 1.00 \times 10^{-7}\ M$; $K_w = [H^+][OH^-] = 1.00 \times 10^{-14}$.

Solution: Find $[H^+]$.

$$[H^+] = \frac{K_w}{[OH^-]} = \frac{1.00 \times 10^{-14}}{1.0 \times 10^{-5}\ M} = 1.0 \times 10^{-9}\ M$$

Since $[H^+]$ is less than ($<$) $1.00 \times 10^{-7}\ M$, the solution is basic.

15.8 THE pH OF A SOLUTION

Operational Skill

5. Calculating the pH from the hydrogen-ion concentration, or vice versa. Given the hydrogen-ion concentration, calculate the pH (Example 15.5); or given the pH, calculate the hydrogen-ion concentration (Example 15.6).

In the term *pH*, *p* means "–log of"; thus, pH = –log $[H^+]$, and pOH = –log $[OH^-]$. Memorize these definitions and that pH + pOH = 14.00. Also memorize the information relating pH to acidity: an acidic solution has a pH less than (<) 7, a neutral solution has a pH of 7, and a basic solution has a pH greater than (>) 7.

If you need a refresher on the use of logs and how to manipulate your calculator, refer to the manual for your calculator or consult a math book.

Exercise 15.7   What is the pH of a sample of gastric juice (digestive juice in the stomach) whose hydrogen-ion concentration is 0.045 M?

Known: $pH = -\log [H^+]$.

Solution: $pH = -\log 0.045 = -(-1.35) = 1.35$.

Exercise 15.8   A saturated solution of calcium hydroxide has a hydroxide-ion concentration of 0.025 M. What is the pH of the solution?

Known: $pH + pOH = 14.00$.

Solution: $pOH = -\log [OH^-] = -\log (0.025) = 1.60$

$pH = 14.00 - pOH = 14.00 - 1.60 = 12.40$

Exercise 15.9   A brand of carbonated beverage has a pH of 3.16. What is the hydrogen-ion concentration of the beverage?

Solution: Rearrange the definition of pH to give

$\log [H^+] = -pH = -3.16$

$[H^+] = $ antilog $-3.16 = 6.9 \times 10^{-4} M$

If you are using log tables, you must rewrite the negative log with a positive mantissa (decimal):

$\log [H^+] = -3.16 = 0.84 - 4.00$

$[H^+] = 10^{0.84} \times 10^{-4} = 6.9 \times 10^{-4} M$

Exercise 15.10   A 0.010 M solution of ammonia, NH_3, has a pH of 10.6 at 25°C. What is the concentration of hydroxide ion? (One way to solve this problem is to find the pOH first and then calculate the hydroxide-ion concentration.)

Known: $pH + pOH = 14.00$.

Solution: pOH = 14.00 − 10.6 = 3.4. Since pOH = −log [OH⁻], then

$$\log [\text{OH}^-] = -\text{pOH} = -3.4$$

$$[\text{OH}^-] = \text{antilog} -3.4 = 4 \times 10^{-4}\ M$$

or $$[\text{OH}^-] = 10^{0.6} \times 10^{-4} = 4 \times 10^{-4}\ M$$

A Base That Matters: SODIUM HYDROXIDE (a Strong Base)

Questions for Study

1. List some household uses for sodium hydroxide. Why is it effective?

2. Why must sodium hydroxide be kept in tightly stoppered containers?

3. Give equations for two methods of preparing sodium hydroxide.

4. Sodium hydroxide solutions should not be stored in aluminum cans. Why?

5. List some major commercial uses of sodium hydroxide.

Answers to Questions for Study

1. Sodium hydroxide is used as a drain opener and as an oven cleaner. It is effective in these uses because it will react chemically with grease and hair to form soluble compounds that can be washed away.

2. If not kept in tightly stoppered containers, sodium hydroxide becomes wet with moisture absorbed from the air and reacts with carbon dioxide in the air to form sodium hydrogen carbonate and sodium carbonate.

3. The preparation of sodium hydroxide by precipitation is shown in the following equation

$$\text{Ca(OH)}_2(aq) + \text{Na}_2\text{CO}_3(aq) \longrightarrow \text{CaCO}_3(s) + 2\text{NaOH}(aq)$$

The preparation of sodium hydroxide from electrolysis of sodium chloride in water is shown by

$$2\text{NaCl}(aq) + 2\text{H}_2\text{O}(l) \xrightarrow{\text{electrolysis}} \text{H}_2(g) + \text{Cl}_2(g) + 2\text{NaOH}(aq)$$

4. Sodium hydroxide should not be stored in aluminum cans because aluminum is dissolved in the solution formed from the hydroxide and water — either the moisture from the air that would get in from normal use, or water that was inadvertently added.

5.   Some major commercial uses of sodium hydroxide include the preparation of soaps and detergents, paper making, purification of aluminum ore, and petroleum refining.

ADDITIONAL PROBLEMS

1.   Give the conjugate acid of each of the following bases:

     (a) CN^- (b) NO_2^- (c) HSO_3^- (d) NH_3 (e) H_2O

2.   Using text Table 15.2, predict the direction of each of the following acid–base reactions:

     (a) $HS^- + ClO^- \rightleftharpoons S^{2-} + HClO$

     (b) $HC_2H_3O_2 + CN^- \rightleftharpoons HCN + C_2H_3O_2^-$

     (c) $HF + HCO_3^- \rightleftharpoons F^- + H_2CO_3$

     (d) $H_2O + NH_3 \rightleftharpoons NH_4^+ + OH^-$

3.   In the following reactions, name each of the reactants as a Lewis acid or base.

     (a) $Cu^{2+} + 4NH_3 \longrightarrow Cu(NH_3)_4^{2+}$

     (b) $Zn(OH)_2 + 2OH^- \longrightarrow Zn(OH)_4^{2-}$

4.   In the following reaction, name each of the reactants as a Brønsted–Lowry acid or base, and give the conjugate of each.

     $HCO_3^- + H_2O \rightleftharpoons H_2CO_3 + OH^-$

5.   Which acid in each of the following pairs is the stronger acid? Explain your answers.

     (a) H_2S and H_2Se (c) H_3PO_4 and $H_2PO_4^-$

     (b) $HClO_3$ and $HClO_4$ (d) H_2Se and HBr

6. $HClO_4$ is a strong acid. What concentration of $HClO_4$ is needed to give a solution of the acid with a hydrogen-ion concentration of 2.5×10^{-3} M?

7.   A formic acid solution of unknown concentration will be used in a reaction. The hydrogen-ion concentration is found to be 0.0052 M. What do you expect the pH measurement to be?

8.   Will the following reaction occur as written?

$$HS^-(aq) + HCO_3^-(aq) \longrightarrow H_2CO_3(aq) + S^{2-}(aq)$$

ANSWERS TO ADDITIONAL PROBLEMS

If you missed an answer, study the text section and operational skill given in parentheses after the answer.

1.   (a)   HCN   (c) H_2SO_3 (e) H_3O^+ (15.2, Op. Sk. 1)

      (b) HNO_2 (d) NH_4^+

2.   (a)   to the left   (c)   to the right
      (b)   to the right   (d)   to the left   (15.4, Op. Sk. 3)

3.   (a)   Lewis acid, Cu^{2+}; Lewis base, NH_3

      (b)   Lewis acid, $Zn(OH)_2$; Lewis base, OH^- (15.3, Op. Sk. 2)

4.   Brønsted–Lowry acid is H_2O, its conjugate base is OH^-; Brønsted–Lowry base is

      HCO_3^-, its conjugate acid is H_2CO_3.   (15.2, Op. Sk. 1)

5.   (a) H_2Se is the stronger acid, because acidity increases with size down a group in the periodic table.
      (b) $HClO_4$ is the stronger acid, because bond polarity and thus acid strength increase with increasing oxidation number of the central atom.
      (c) H_3PO_4 is stronger, because the hydrogens on the negatively charged anion will be held more tightly.
      (d)   HBr is stronger, because acidity increases with increasing electronegativity across a period.   (15.5)

6. $2.5 \times 10^{-3}\ M\ HClO_4$ (15.7, Op. Sk. 4)

7. pH = 2.28   (15.8, Op. Sk. 5)

8. No   (15.4, Op. Sk. 3)

CHAPTER POST-TEST

1. Identify the acid–base conjugate pairs in each of the following reactions:

    (a) $HSO_4^- + OH^- \longrightarrow H_2O + SO_4^{2-}$

    (b) $HCl + CN^- \longrightarrow HCN + Cl^-$

    (c) $H_3O^+ + HSO_4^- \longrightarrow H_2SO_4 + H_2O$

    (d) $Al(H_2O)_6^{3+} + C_6H_5O^- \longrightarrow Al(H_2O)_5OH^{2+} + C_6H_5OH$

2. Using text Table 15.2, decide the direction of reaction for each of the following equations:

    (a) $H_2O + H_2O_2 \rightleftharpoons H_3O^+ + HO_2^-$

    (b) $HCl + HCO_3^- \rightleftharpoons H_2CO_3 + Cl^-$

    (c) $NH_3 + HCl \rightleftharpoons NH_4Cl$

3. For each of the following reactions, identify which is the Lewis acid and the Lewis base:

    (a) $BF_3 + F^- \longrightarrow BF_4^-$

    (b) $S + SO_3^{2-} \longrightarrow S_2O_3^{2-}$

4. Thymol blue is used as an acid–base indicator. A solution containing this indicator has a hydrogen-ion concentration of 3.6×10^{-5} *M*. What is the hydrogen-ion concentration of this solution?

5. The pH of a lemonade drink is 2.35. What is the hydrogen-ion concentration of this drink?

6. Predict the direction of the following reaction

$$C_2H_3O_2^-(aq) + H_2SO_3(aq) \rightleftharpoons HSO_3^-(aq) + HC_2H_3O_2(aq)$$

ANSWERS TO CHAPTER POST-TEST

If you missed an answer, study the text section and operational skill given in parentheses after the answer.

1. | | | *Conjugate Acid* | *Conjugate Base* |
   |---|---|---|---|
   | (a) | | HSO_4^- | SO_4^{2-} |
   | | | H_2O | OH^- |
   | (b) | | HCl | Cl^- |
   | | | HCN | CN^- |
   | (c) | | H_3O^+ | H_2O |
   | | | H_2SO_4 | HSO_4^- |
   | (d) | | $Al(H_2O)_6^{3+}$ | $Al(H_2O)_5OH^{2+}$ |
   | | | C_6H_5OH | $C_6H_5O^-$ (15.2, Op. Sk. 1) |

2. (a) left   (b) right   (c) right   (15.4, Op. Sk. 3)

3. | | | *Lewis Acid* | *Lewis Base* |
   |---|---|---|---|
   | (a) | | BF_3 | F^- |
   | (b) | | S | SO_3^{2-} (15.3, Op. Sk. 2) |

4. $[OH^-] = 1.1 \times 10^{-3} M$ (15.7, Op. Sk. 4)

5. $[H^+] = 4.5 \times 10^{-3} M$ (15.8, Op. Sk. 5)

6. The reaction will proceed from left to right.   (15.4, Op. Sk. 3)

CHAPTER 16   ACID–BASE EQUILIBRIA

CHAPTER TERMS AND DEFINITIONS

Numbers in parentheses after definitions give the text sections in which the terms are explained. Starred terms are italicized in the text. Where a term does not fall directly under a text section heading, additional information is given for you to locate it.

acid ionization (dissociation)*   reaction of an Arrhenius acid with water to produce hydronium ion (hydrogen ion) and the conjugate base ion   (16.1)

acid-ionization constant (K_a)   equilibrium constant for the ionization of a weak acid   (16.1)

degree of ionization   fraction of molecules that react with water to give ions   (16.1)

percent ionization*   percentage of molecules that react with water to give ions   (16.1)

polyprotic acids*   acids with more than one mole of ionizable hydrogen atoms per mole of acid   (16.2)

acid rain*   rain with a pH lower than natural rain (5.6), resulting from the dissolving of sulfur oxides and nitrogen oxides   (A Chemist Looks At: Acid Rain)

base ionization*   reaction of an Arrhenius base with water to produce hydroxide ion and the conjugate acid ion   (16.3)

base-ionization constant (K_b)   equilibrium constant for the ionization of a weak base   (16.3)

hydrolysis   reaction of an ion with water to produce the conjugate acid and hydroxide ion or the conjugate base and hydrogen ion   (16.4)

common-ion effect   shift in an ionic equilibrium, caused by the addition of a solute that provides an ion that takes part in the equilibrium   (16.5)

buffer   solution characterized by the ability to withstand changes in pH when limited amounts of acid or base are added to it; usually contains a weak acid and its conjugate base, or a weak base and its conjugate acid   (16.6)

buffer capacity*   amount of acid or base with which a buffer can react before giving a significant pH change   (16.6)

Henderson–Hasselbalch equation $\text{pH} = pK_a + \log \dfrac{[\text{base}]}{[\text{acid}]}$ (16.6)

402

acid–base titration curve   plot of the pH of a solution of acid (or base) against the
    volume of added base (or acid)   (16.7)
equivalence point   point at which the stoichiometric amount of reactant has been added in
    a titration   (16.7)
thermite*   mixture of aluminum powder and iron(III) oxide, which on ignition produces
    molten iron for welding purposes   (A Metal That Matters: Aluminum [an
    Amphoteric Metal])
amphoteric*   describes a substance that reacts with both acids and bases   (A Metal That
    Matters: Aluminum [an Amphoteric Metal])

CHAPTER DIAGNOSTIC TEST

1. If an acid such as HNO_3 has an acid-dissociation constant of 24, which of the
    following statements is *not* true?

    (a)   The HNO_3 is essentially 100% dissociated.

    (b)   This acid would be classified as a very strong acid.

    (c)   The conjugate base, NO_3^-, is relatively strong in base strength.

    (d)   The value of K_b is 4.2×10^{-16}.

    (e)   A 1 M solution of HNO_3 would have $[H^+]$ equal to 1 M.

2. Two acids have K_a values of 5.6×10^{-2} and 6.8×10^{-4}. The stronger acid
    corresponds to which K_a value?

3. Consider this list of acids and their respective K_a values:

    H_2SO_3, 1.3×10^{-2} HF, 6.7×10^{-4}

    HClO, 1.1×10^{-8} $HClO_2$, 1.0×10^{-2}

    H_2CrO_4, 1.2 H_3BO_3, 5.9×10^{-10}

    Which of the following represents the correct ordering of acid strengths?

    (a) $H_2CrO_4 > H_2SO_3 > HClO$

    (b) $HF > HClO_2 > H_2CrO_4$

    (c) $H_3BO_3 > HClO > HF$

(d) $H_2SO_3 > H_3BO_3 > H_2CrO_4$

(e) $HClO > H_2SO_3 > HF$

4.   Ammonium nitrate is used in cold packs because its dissolution is an endothermic
      process. What are the $[H^+]$ and pH of a 0.35 M solution of NH_4NO_3? The K_b value
      for NH_3 is 1.8×10^{-5}.

5.   Phosphorous acid (H_2PHO_3) is a diprotic acid with $K_{a1} = 1.6 \times 10^{-2}$ and
      $K_{a2} = 7 \times 10^{-7}$. Calculate the concentration of H^+, $HPHO_3^-$, and PHO_3^{2-} for a
      0.48 M solution of phosphorous acid.

6.   If the pH of 0.82 mol of methylamine, CH_3NH_2, dissolved in 425 mL of water is
      12.46, what is the base-ionization constant for methylamine?

7.   In the reaction of boric acid with water,

$$H_3BO_3(aq) + H_2O(l) \rightleftharpoons B(OH)_4^-(aq) + H^+(aq)$$

      H_3BO_3 acts as a Lewis acid in accepting the OH^- from a water molecule. If K_a for
      H_3BO_3 is 5.9×10^{-10}, what is the pH of a 0.24 M solution of the acid?

8.   What is the concentration of salicylic acid in an aqueous solution of 3.4×10^{-2} M
      sodium salicylate? The K_a for salicylic acid, $HC_7H_5O_3$, is 1.1×10^{-3}.

9.   A 0.10 M solution of formic acid, HCOOH, is 4.0% dissociated. Calculate the K_a
      for formic acid.

10.   Calculate the $[H^+]$ of a solution prepared by dissolving 16.25 g KIO in 225 mL of a
      0.25 M HCl solution. The K_a for HIO is 2.95×10^{-6}.

11.   What is the pH of a buffer solution prepared from 0.375 mol $HC_2H_3O_2$ and 0.350
      mol $NaC_2H_3O_2$ in enough water to give 0.750 L of solution? K_a for $HC_2H_3O_2$ is
      1.7×10^{-5}.

12.   Calculate the change in pH produced by adding 10.0 mL of a 0.10 M HCl solution to
      0.200 L of a solution that is 0.20 M in cyanic acid and 0.080 M in sodium cyanate.
      K_a for cyanic acid is 3.5×10^{-4}.

13.   Calculate the pH of a solution after 5.00 mL of a 0.050 M NaOH solution have been added to 0.150 L of 0.010 M HCl.

14.   What is the pH at the equivalence point for a titration of 25 mL of 0.50 M benzoic acid, $HC_7H_5O_2$, with 1.0 M NaOH? K_a for benzoic acid is 6.3×10^{-5}.

ANSWERS TO CHAPTER DIAGNOSTIC TEST

If you missed an answer, study the text section and operational skill given in parentheses after the answer.

1.   c   (16.1)

2. 5.6×10^{-2} (16.1)

3.   a   (16.1)

4. $[H^+] = 1.4 \times 10^{-5} M$; pH = 4.85   (16.4, Op. Sk. 5, 6)

5. $[H^+] = [HPHO_3^-] = 0.080 M$
       $[PHO_3^{2-}] = 7 \times 10^{-7} M$ (16.2, Op. Sk. 2)

6. $K_b = 4.3 \times 10^{-4}$ (16.3, Op. Sk. 1)

7.   pH = 4.92   (16.1, Op. Sk. 2)

8. $[HC_7H_5O_3] = 5.6 \times 10^{-7} M$ (16.4, Op. Sk. 6)

9. $K_a = 1.7 \times 10^{-4}$ (16.1, Op. Sk. 1)

10. $[H^+] = 5.0 \times 10^{-6} M$ (16.5, Op. Sk. 7)

11.   pH = 4.74   (16.6, Op. Sk. 8)

12. $\Delta pH = -0.04$ (16.6, Op. Sk. 9)

13.   pH = 2.09   (16.7, Op. Sk. 10)

14.   pH = 8.86   (16.7, Op. Sk. 11)

SUMMARY OF CHAPTER TOPICS

The calculations in this chapter use the same equilibrium concepts that you learned previously. The reactions described here, however, are either the ionizations of weak acids or bases or the reaction of an ion of a salt with water.

The biggest problem many students have is in deciding how to write the reaction equation. Helpful tips are given with each section below. The key to success is to know the weak acids and bases so that you can immediately recognize one or recognize an ion as part of a weak acid or base. Since there are far fewer strong acids than weak, the easiest way to learn the acids is to memorize the strong ones (see text Table 15.1). Then you will be able to determine which are weak acids by process of elimination. The weak bases fall into recognizable categories (see text Table 16.2). They are all nitrogen-containing compounds — compounds in which nitrogen is covalently bonded to hydrogen, such as in ammonia, NH_3, and hydrazine, N_2H_4; to carbon, such as in methylamine, CH_3NH_2; or to oxygen, such as in hydroxylamine, NH_2OH.

We will again work with equilibrium constants, but in this chapter we symbolize them with special subscripts — *a* or *b* — because these convey more information than does K_c.

For each problem, write the table of initial concentrations, changes, and equilibrium concentrations, and fill it in as you did for the problems in Chapter 14. Simplify calculations whenever possible to avoid using the quadratic equation. Remember that when the value of K divided by the concentration is 10^{-3} or less, the degree of reaction, usually represented by x, will be very small; then x can be ignored when it is added to or subtracted from a starting concentration.

Another problem students frequently have is in determining the direction of change when something is added to an equilibrium system. Remember that Le Chatelier's principle tells us in what direction to expect a change if something has been added to, or subtracted from, the reaction. If we are given the starting amounts of all substances in an equilibrium system, we can tell the direction of change by evaluating the reaction quotient Q (see Section 14.5) and comparing this value to the value of K. However, a simpler way is to assume that the reaction goes to the right and to work the problem that way. You will get the correct value for the equilibrium concentrations regardless, as long as you are careful with your notations and calculations.

16.1   ACID-IONIZATION EQUILIBRIA

Operational Skills

1. Determining K_a (or K_b) from the solution pH. Given the molarity and pH of a solution of a weak acid, calculate K_a for the acid (Example 16.1).   The K_b for a base can be determined in a similar way (see Example 16.5).

2. Calculating concentrations of species in a weak acid solution using K_a. Given K_a, calculate the hydrogen-ion concentration and pH of a solution of a weak acid of known molarity   (Examples 16.2 and 16.3).   Given K_{a1}, K_{a2}, and the molarity of a diprotic acid solution, calculate the pH and the concentrations of H^+, HA^-, and A^{2-} (Example 16.4).

The reaction for this section is the ionization of a weak acid. The reactants are the acid and water, and the products are H_3O^+ and the anion of the acid. We simplify it even more by leaving out the water, so the reaction is merely the breaking apart of the acid into H^+ and the anion, written as

$$HA(aq) \rightleftharpoons H^+(aq) + A^-(aq), \text{ and } K_a = \frac{[H^+]\,[A^-]}{[HA]}$$

Exercise 16.1 Lactic acid, $HC_3H_5O_3$, is found in sour milk, where it is produced by the action of lactobacilli on lactose, or milk sugar. A 0.025 M solution of lactic acid has a pH of 2.75. What is the ionization constant K_a for this acid? What is the degree of ionization?

Wanted: K_a; degree of ionization.

Given: 0.025 M soln of $HC_3H_5O_3$, pH = 2.75.

Known: pH = $-\log[H^+]$, the $[H^+]$ is the equilibrium concentration;

$K_a = [H^+]\,[C_3H_5O_3^-]/[HC_3H_5O_3]$; degree of ionization is equal to:

$$\frac{\text{molarity of acid that reacts}}{\text{starting molarity of acid}}$$

Solution: Since the ionization reaction includes H^+, first solve for $[H^+]$:

$$pH = -\log [H^+] = 2.75$$

$$\log [H^+] = -2.75$$

$$[H^+] = 0.001\underline{7}8 \ M$$

Then write the ionization equation and set up the table of concentrations:

Concentration (M)	$HC_3H_5O_3(aq)$	$\rightleftharpoons$	$H^+(aq)$	$+$	$C_3H_5O_3^-(aq)$
Starting	0.025		0		0
Change	−0.001$\underline{7}$8		+0.001$\underline{7}$8		+0.001$\underline{7}$8
Equilibrium	0.02$\underline{3}$2		0.001$\underline{7}$8		0.001$\underline{7}$8

$$K_a = \frac{[H^+] \, [C_3H_5O_3^-]}{[HC_3H_5O_3]} = \frac{[0.001\underline{7}8] \, [0.001\underline{7}8]}{[0.02\underline{3}2]} = 1.4 \times 10^{-4}$$

$$\text{Degree of ionization} = \frac{\text{molarity of acid that reacts}}{\text{starting molarity of acid}} = \frac{0.001\underline{7}8}{0.025}$$

$$= 0.071$$

Exercise 16.2 What are the concentrations of hydrogen ion and acetate ion in a solution of 0.10 M acetic acid, $HC_2H_3O_2$? What is the pH of the solution? What is the degree of ionization? See Table 16.1 for the value of K_a.

Wanted: $[H^+]$, $[C_2H_3O_2^-]$; pH of soln; degree of ionization.

Given: 0.10 M soln of $HC_2H_3O_2$; text table 16.1.

Known: $K_a = 1.7 \times 10^{-5}$; requested concentrations are the equilibrium values from the table we set up; $pH = -\log [H^+]$; degree of ionization is equal to

$$\frac{\text{molarity of acid that reacts}}{\text{starting molarity of acid}}$$

Solution: If we let x equal the mol/L that ionize, the table is

Concentration (M)	$HC_2H_3O_2(aq) \rightleftharpoons$	$H^+(aq)$	$+$	$C_2H_3O_2^-(aq)$
Starting	0.10	0		0
Change	$-x$	$+x$		$+x$
Equilibrium	$0.10 - x$	x		x

We can solve for x by setting up the expression for K_a. Note that x is not only the molarity of acid that reacts but also the equilibrium concentrations of H^+ and $C_2H_3O_2^-$.

$$K_a = \frac{[H^+][C_2H_3O_2^-]}{[HC_2H_3O_2]} = \frac{(x)(x)}{(0.10 - x)} = \frac{x^2}{(0.10 - x)} = 1.7 \times 10^{-5}$$

We can simplify this expression, because K_a divided by 0.10 is less than 10^{-3}. Thus, x will be so small that $0.10 - x \approx 0.10$.

We then have

$$\frac{x^2}{0.10} = 1.7 \times 10^{-5}$$

$$x^2 = 1.7 \times 10^{-6}$$

$$x = 1.\underline{3}0 \times 10^{-3}$$

The requested concentrations are

$$[H^+] = 0.0013\ M$$

$$[C_2H_3O_2^-] = 0.0013\ M$$

Solving for pH, we get

$$pH = -\log [H^+] = -\log 0.001\underline{3}0 = 2.89$$

$$\text{Degree of ionization} = \frac{\text{molarity of acid that reacts}}{\text{starting molarity of acid}} = \frac{[0.001\underline{3}0]}{[0.10]} = 0.013$$

Exercise 16.3 What is the pH of an aqueous solution that is 0.0030 M pyruvic acid, $HC_3H_3O_3$?

Known: $K_a = 1.4 \times 10^{-4}$ (from text Table 16.1).

Solution: If we let x = mol/L of acid that ionize, the table is

Concentration (M)	$HC_3H_3O_3(aq) \rightleftharpoons$	$H^+(aq)$	+	$C_3H_3O_3^-(aq)$
Starting	0.0030	0		0
Change	$-x$	$+x$		$+x$
Equilibrium	$0.0030 - x$	x		x

$$K_a = \frac{[H^+]\,[C_3H_3O_3^-]}{[HC_3H_3O_3]} = \frac{(x)\,(x)}{0.0030 - x} = 1.4 \times 10^{-4}$$

Since K_a divided by 0.0030 is larger than 10^{-3}, we cannot ignore the change in concentration due to x. We will use the quadratic equation to solve for x:

$$x^2 = 1.4 \times 10^{-4}\,(0.0030 - x) = 4.2 \times 10^{-7} - 1.4 \times 10^{-4}x$$

$$x^2 + 1.4 \times 10^{-4}x - 4.2 \times 10^{-7} = 0$$

The quadratic equation is

$$x = \frac{-b \pm \sqrt{b^2 - 4ac}}{2a}$$

Substituting in the appropriate values gives

$$x = \frac{-1.4 \times 10^{-4} \pm \sqrt{(1.4 \times 10^{-4})^2 - 4(1)\,(-4.2 \times 10^{-7})}}{2(1)}$$

$$= \frac{-1.4 \times 10^{-4} \pm \sqrt{1.96 \times 10^{-8} + 1.68 \times 10^{-6}}}{2}$$

$$= \frac{-1.4 \times 10^{-4} \pm 1.304 \times 10^{-3}}{2}$$

As *x* must be positive, we use only the + value:

$$x = 5.82 \times 10^{-4} = [H^+]$$

$$pH = -\log [H^+] = -\log 5.82 \times 10^{-4} = 3.24$$

16.2 POLYPROTIC ACIDS

Operational Skill

2. Calculating concentrations of species in a weak acid solution using K_a. Given K_a, calculate the hydrogen-ion concentration and pH of a solution of a weak acid of known molarity. Given K_{a1}, K_{a2}, and the molarity of a diprotic acid solution, calculate the pH and the concentrations of H^+, HA^-, and A^{2-} (Example 16.4).

Exercise 16.4 Sulfurous acid, H_2SO_3, is a diprotic acid with $K_{a1} = 1.3 \times 10^{-2}$ and $K_{a2} = 6.3 \times 10^{-8}$. The acid forms when sulfur dioxide (a gas with a suffocating odor) dissolves in water. What is the pH of a 0.25 *M* solution of sulfurous acid? What is the concentration of sulfite ion, SO_3^{2-}, in the solution? Note that K_{a1} is relatively large.

Wanted: pH; $[SO_3^{2-}]$.

Given: 0.25 *M* soln of H_2SO_3; $K_{a1} = 1.3 \times 10^{-2}$; $K_{a2} = 6.3 \times 10^{-8}$.

Known: The pH can be calculated from the $[H^+]$ found from the first ionization because the second ionization is insignificant in comparison; we can use concentrations resulting from the first ionization to set up the table for the second ionization to show that $[SO_3^{2-}] \approx K_{a2}$.

Solution: The equations are

$$H_2SO_3(aq) \rightleftharpoons H^+(aq) + HSO_3^-(aq)$$

$$HSO_3^-(aq) \rightleftharpoons H^+(aq) + SO_3^{2-}(aq)$$

Setting up the table for the first calculation to get pH and $[HSO_3^-]$ (letting $x =$ mol/L of acid that ionize) gives

Concentration (M)	$H_2SO_3(aq)$	$\rightleftharpoons$	$H^+(aq)$	+	$HSO_3^-(aq)$
Starting	0.25		0		0
Change	$-x$		$+x$		$+x$
Equilibrium	$0.25 - x$		x		x

$$K_{a1} = \frac{[H^+][HSO_3^-]}{[H_2SO_3]} = \frac{(x)(x)}{0.25 - x} = 1.3 \times 10^{-2}$$

Because K_{a1} is large, we cannot ignore x in the denominator. We could, of course, use the quadratic equation to solve for x. Instead we will rearrange the above expression to

$$x^2 = (0.25 - x)(1.3 \times 10^{-2})$$

and use the *method of successive approximations*. This way of solving the problem involves choosing an approximate value of x, putting it into the right side of this equation, and solving for an improved value of x on the left. This value of x then becomes the new approximation to x. The method is repeated until the value used for x on the right equals the value obtained on the left (to the number of significant figures desired, usually two). To get the first approximation to x, we assume that x is insignificant and use $x = 0$ on the right side. The above equation then simplifies to

$$x^2 = 0.0032$$
$$x = 0.057$$

We now use this value of x for the next approximation.

$$x^2 = (0.25 - 0.057)(0.013) = 0.0025$$
$$x = 0.050$$

For the next approximation, we get

$$x^2 = (0.25 - 0.050)(0.013) = 0.0026$$
$$x = 0.051$$

Then, in the next approximation,

$$x^2 = (0.25 - 0.051)(0.013) = 0.0026$$
$$x = 0.051$$

Note that the approximation to x improves at each stage, with the changes in x becoming smaller. Because the value of x from the last approximation equals the value used in the right side, we can assume that $x = 0.051$, to two significant figures. Thus, the $[H^+] = 0.051 \; M$, and

$$pH = -\log [H^+] = -\log 0.051 = 1.29$$

The table for the second ionization (letting $x = $ mol/L of HSO_3^- that ionize) is as follows:

Concentration (*M*)	$HSO_3^-(aq)$ $\rightleftharpoons$	$H^+(aq)$ +	$SO_3^{-2}(aq)$
Starting	0.051	0.051	0
Change	$-x$	$+x$	$+x$
Equilibrium	$0.051 - x$	$0.051 + x$	x

Determine x from the second-step equilibrium expression:

$$K_{a2} = \frac{[H^+] [SO_3^{-2}]}{[HSO_3^-]} = \frac{(0.051 + x)(x)}{0.051 - x} = 6.3 \times 10^{-8}$$

We can ignore the change because K_{a2} is so small. Thus,

$$K_{a2} = \frac{(0.51)x}{(0.51)} = 6.3 \times 10^{-8}$$

and

$$x = [SO_3^{2-}] \approx 6.3 \times 10^{-8} \; M$$

16.3   BASE-IONIZATION EQUILIBRIA

Operational Skills

1. Determining K_a (or K_b) from the solution pH.   Given the molarity and pH of a solution of a weak acid, calculate the acid-ionization constant K_a (Example 16.1).   The K_b for a base can be determined in a similar way   (see Exercise 16.8).

3. Calculating concentrations of species in a weak base solution using K_b.   Given K_b, calculate the hydrogen-ion concentration and pH of a solution of a weak base of known molarity   (Example 16.5).

When we write the equation for the ionization of a weak base, we do not leave out the water. In these problems, we use the Brønsted–Lowry definition of a base: a base is a proton acceptor. In each case, the base takes the proton from water to form the conjugate acid ion, leaving the OH⁻ ion. The general reaction is

$$B(aq) + H_2O(l) \rightleftharpoons HB^+(aq) + OH^-(aq)$$

and

$$K_b = \frac{[HB^+][OH^-]}{[B]}$$

Recall that neither K_a nor K_b has the water in the ratio, because its concentration is included in each constant.

Exercise 16.5   Quinine is an alkaloid, or naturally occurring base, used to treat malaria. A 0.0015 M solution of quinine has a pH of 9.84. The basicity of alkaloids is due to a nitrogen atom that picks up protons from water in the same manner as ammonia does. What is K_b?

Wanted: K_b.

Given: 0.0015 M soln has pH = 9.84.

Known: pH = –log [H⁺]; pOH = 14 – pH = –log [OH⁻]; $K_b = \dfrac{[HQ^+][OH^-]}{[Q]}$.

Solution: If we let x = mol/L of Q that ionize, the table is

Concentration (M)	Q(aq) + H$_2$O(l) $\rightleftharpoons$ HQ$^+$(aq)	+	OH$^-$(aq)
Starting	0.0015	0	0
Change	$-x$	$+x$	$+x$
Equilibrium	$0.0015 - x$	x	x

But x can be determined from the pH:

$$\text{pOH} = 14.00 - \text{pH} = 14.00 - 9.84 = 4.16$$

$$[\text{OH}^-] = \text{antilog}\,(-4.16) = 6.\underline{9}2 \times 10^{-5} = x$$

Thus,

$$K_b = \frac{[\text{HQ}^+][\text{OH}^-]}{[\text{Q}]} = \frac{(x)(x)}{(0.0015 - x)} = \frac{(6.\underline{9}2 \times 10^{-5})(6.\underline{9}2 \times 10^{-5})}{(0.0015 - 6.\underline{9}2 \times 10^{-5})}$$

$$= 3.3 \times 10^{-6}$$

Exercise 16.6 What is the hydrogen-ion concentration of a 0.20 M solution of ammonia in water? See Table 16.2 for K_b.

Wanted: [H$^+$].

Given: 0.20 M soln of NH$_3$(aq); text Table 16.2 for K_b.

Known: $K_b = 1.8 \times 10^{-5}$ at 25°C; $K_w = [\text{H}^+]\,[\text{OH}^-] = 1.0 \times 10^{-14}$ at 25°C.

$$\text{Assume 25°C; } K_b = \frac{[\text{NH}_4^+][\text{OH}^-]}{[\text{NH}_3]}.$$

Solution: If we let x = mol/L of base that ionize, the table is

Concentration (M)	$NH_3(aq)$ + $H_2O(l)$ ⇌ $NH_4^+(aq)$ + $OH^-(aq)$		
Starting	0.20	0	0
Change	$-x$	$+x$	$+x$
Equilibrium	$0.20 - x$	x	x

Solve for x:

$$K_b = \frac{[NH_4^+][OH^-]}{[NH_3]} = \frac{(x)(x)}{(0.20 - x)} = 1.8 \times 10^{-5}$$

Because K_b divided by 0.20 is $< 10^{-3}$, x can be dropped from the denominator, giving

$$x^2 = 1.8 \times 10^{-5}(0.20) = 3.6 \times 10^{-6}$$

$$x = 1.\underline{9}0 \times 10^{-3}$$

Since $x = [OH^-]$, $[OH^-] = 1.\underline{9}0 \times 10^{-3}$. Thus,

$$[H^+] = \frac{K_w}{[OH^-]} = \frac{1.0 \times 10^{-14}}{1.\underline{9}0 \times 10^{-3}} = 5.3 \times 10^{-12} M$$

16.4   ACID–BASE PROPERTIES OF SALT SOLUTIONS

Operational Skills

  4. **Predicting whether a salt solution is acidic, basic, or neutral.** Decide whether an aqueous solution of a given salt is acidic, basic, or neutral   (Example 16.6).

  5. **Obtaining K_a from K_b or K_b from K_a.** Calculate K_a for a cation or K_b for an anion from the ionization constant of the conjugate base or acid   (Example 16.7).

  6. **Calculating concentrations of species in a salt solution**. Given the concentration of a solution of a salt in which one ion hydrolyzes, and given the ionization constant of the conjugate acid or base of this ion, calculate the H^+ concentration   (Example 16.8).

In order to do the problems in this section, you must know the strong and weak acids and bases and be able to recognize ions that hydrolyze. Refer to text Tables 15.1, 16.1, and 16.2.

Exercise 16.7 Consider solutions of the following salts: (a) NH_4NO_3; (b) KNO_3; (c) $Al(NO_3)_3$. Which solution is acidic? Which is basic? Which is neutral?

(a)   *Known:* NH_4^+ hydrolyzes to give H_3O^+ ions; NO_3^- is the anion of a strong acid, so it does not hydrolyze.

   Solution: This solution is acidic.

(b)   *Known:* K^+ is a Group I anion and does not hydrolyze; NO_3^- is the anion of a strong acid, so it does not hydrolyze.

   Solution: This solution is neutral.

(c)   *Known:* Al^{3+} hydrolyzes to give H_3O^+ ions; NO_3^- is the anion of a strong acid, so it does not hydrolyze.

   Solution: This solution is acidic.

The equations used in this section will show the reaction of an ion of a salt with water to produce a weak acid or base and either OH^- or H^+. To succeed in working these problems, you must first be able to recognize a salt by its formula. Then you must recognize one of the salt ions as being part of a weak acid or base. Again, you must know the weak acids and bases to be able to do this.

The word hydrolysis means "to cut with water." The equations we write for this section make it appear that the water is being cut, rather than doing the cutting, but remembering the meaning of the word should help you keep in mind what happens. Many of the K_a values we need for hydrolysis reactions can be found in the table of weak acid ionization constants (see text Table 16.1). To get the K_a values not listed in the table, and to get the K_b values for the anions that hydrolyze, use the fact that $K_a K_b = K_w = 1.0 \times 10^{-14}$ (at 25°C). We find the K_a or K_b of the conjugate of the ion from the table, then solve the relationship for the desired constant. This is exemplified in the solutions to the following exercises.

Exercise 16.8 Calculate the following, using Tables 16.1 and 16.2: (a) K_b for F^-; (b) K_a for $C_6H_5NH_3^+$ (conjugate acid of aniline, $C_6H_5NH_2$).

(a)   *Wanted:* K_b for F^-.

Given: text Tables 16.1 and 16.2.

Known: F^- is the anion of the weak acid HF with $K_a = 6.8 \times 10^{-4}$ (text Table 16.1); $K_a K_b = K_w = 1.0 \times 10^{-14}$ (25°C).

Solution: The equation, although not necessary, is written for clarification:

$$F^-(aq) + H_2O(l) \rightleftharpoons HF(aq) + OH^-(aq)$$

Solving $K_a K_b = K_w$ for K_b and substituting in known values gives

$$K_b = \frac{K_w}{K_a} = \frac{1.0 \times 10^{-14}}{6.8 \times 10^{-4}} = 1.5 \times 10^{-11}$$

(b)   *Wanted:* K_a for $C_6H_5NH_3^+$.

Known: This is the conjugate acid of aniline, for which $K_b = 4.2 \times 10^{-10}$ (text Table 16.2); other information given above.

Solution: The equation, for clarity, is

$$C_6H_5NH_3^+(aq) + H_2O(l) \rightleftharpoons H_3O^+(aq) + C_6H_5NH_2(aq)$$

Solving $K_a K_b = K_w$ for K_a and substituting in known values gives

$$K_a = \frac{K_w}{K_b} = \frac{1.0 \times 10^{-14}}{4.2 \times 10^{-10}} = 2.4 \times 10^{-5}$$

Exercise 16.9 Benzoic acid, $HC_7H_5O_2$, and its salts are used as food preservatives. What is the concentration of benzoic acid in an aqueous solution of 0.015 M sodium benzoate? What is the pH of the solution? K_a for benzoic acid is 6.3×10^{-5}.

Wanted: $[HC_7H_5O_2]$; pH of soln.

Given: 0.015 M soln of sodium benzoate; $K_a = 6.3 \times 10^{-5}$.

Known: $HC_7H_5O_2$ is a weak acid; the benzoate ion is formed by the loss of H^+ from $HC_7H_5O_2$; assume complete solubility of the salt so that ion concentration equals salt concentration; $K_b = K_w/K_a$; $[H^+][OH^-] = K_w = 1.0 \times 10^{-14}$ (25°C); pH $= -\log [H^+]$, will be > 7 as OH^- is formed.

Solution: If we let x = mol/L of benzoate ion that hydrolyze, the table is

Concentration (M)	$C_7H_5O_2^-(aq)$ + $H_2O(l)$ $\rightleftharpoons$	$HC_7H_5O_2(aq)$ +	$OH^-(aq)$
Starting	0.015	0	0
Change	$-x$	$+x$	$+x$
Equilibrium	$0.015 - x$	x	x

$$K_b = \frac{[HC_7H_5O_2][OH^-]}{[C_7H_5O_2^-]} = \frac{(x)(x)}{0.015 - x} = \frac{K_w}{K_a} = \frac{1.0 \times 10^{-14}}{6.3 \times 10^{-5}} = 1.\underline{5}9 \times 10^{-10}$$

Since K_b divided by 0.015 is $< 10^{-3}$, x can be ignored in the denominator, giving

$$\frac{x^2}{0.015} = 1.\underline{5}9 \times 10^{-10}$$

$$x^2 = (0.015)(1.\underline{5}9 \times 10^{-10}) = 2.\underline{3}8 \times 10^{-12}$$

$$x = 1.\underline{5}4 \times 10^{-6}$$

Thus,

$$[HC_7H_5O_2] = 1.5 \times 10^{-6}\,M$$

To solve for pH, we use the information that x is also equal to $[OH^-]$. Solving the ion product of water for $[H^+]$ and substituting the nonrounded value of x for $[OH^-]$ gives

$$[H^+] = \frac{K_w}{[OH^-]} = \frac{1.0 \times 10^{-14}}{1.\underline{5}4 \times 10^{-6}} = 6.\underline{4}9 \times 10^{-9}\,M$$

Thus,

$$pH = -\log[H^+] = -\log(6.\underline{4}9 \times 10^{-9}) = 8.19$$

16.5   COMMON-ION EFFECT

Operational Skill

7.   Calculating the common-ion effect on acid ionization.   Given K_a and the concentrations of weak acid and strong acid in a solution, calculate the degree of ionization of the weak acid   (Example 16.9).   Given K_a and the concentrations of weak acid and its salt in a solution, calculate the pH   (Example 16.10).

The key to solving these problems is recognizing which of the three types of reactions is going on. Careful analysis of the compounds that make up the solution and comparison of ionization constants will enable you to determine which equation to write. The table you set up will be slightly different in that now you will have starting molarities of one or more products.

Exercise 16.10   What is the concentration of formate ion, CHO_2^-, in a solution at 25°C that is 0.10 M $HCHO_2$ and 0.20 M HCl? What is the degree of ionization of formic acid, $HCHO_2$?

Wanted: $[CHO_2^-]$, degree of ionization.

Given: Solution consists of 0.10 M $HCHO_2$ and 0.20 M HCl.

Known: Since the given solution was made with the weak acid $HCHO_2$ and the strong acid HCl, the reaction for which to write the equation is the ionization of the weak acid; $K_a = 1.7 \times 10^{-4}$ (text Table 16.1); because HCl is a strong acid, the starting concentration of H^+ is the same as the molarity of the HCl; degree of ionization is equal to

$$\frac{\text{molarity of acid that reacts}}{\text{starting molarity of acid}}$$

Solution: If we let x = mol/L of acid that ionize, the table is

Concentration (M)	$HCHO_2(aq)$	$\rightleftharpoons$	$H^+(aq)$	+	$CHO_2^-(aq)$
Starting	0.10		0.20		0
Change	$-x$		$+x$		$+x$
Equilibrium	$0.10 - x$		$0.20 + x$		x

$$K_a = \frac{[\text{H}^+][\text{CHO}_2^-]}{[\text{HCHO}_2]} = \frac{(0.20 + x)(x)}{0.10 - x} = 1.7 \times 10^{-4}$$

K_a divided by 0.10 is 1.7×10^{-3}, but the presence of 0.20 M H^+ will suppress the ionization; thus, we will ignore the change of x, obtaining

$$\frac{0.20x}{0.10} = 1.7 \times 10^{-4}$$

$$x = [\text{CHO}_2^-] = 8.5 \times 10^{-5} \, M$$

$$\text{Degree of ionization} = \frac{\text{molarity of acid that reacts}}{\text{starting molarity of acid}} = \frac{8.5 \times 10^{-5}}{0.10}$$

$$= 8.5 \times 10^{-4}$$

Exercise 16.11 One liter of solution was prepared by dissolving 0.025 mol of formic acid, HCHO_2, and 0.018 mol of sodium formate, NaCHO_2, in water. What was the pH of the solution? K_a for formic acid is 1.7×10^{-4}.

Wanted: pH.

Given: 0.025 mol HCHO_2 and 0.018 mol NaCHO_2 in 1-L soln (*aq*); $K_a = 1.7 \times 10^{-4}$.

Known: We have present a weak acid and a salt of its conjugate base; assume complete solubility of the salt so that the concentration of the formate ion = salt concentration:

$$K_a = \frac{[\text{H}^+][\text{CHO}_2^-]}{[\text{HCHO}_2]}$$

Solution: If we let x = mol/L of acid that ionize, the table is

Concentration (M)	$\text{HCHO}_2(aq)$ $\rightleftharpoons$	$\text{H}^+(aq)$	+	$\text{CHO}_2^-(aq)$
Starting	0.025	0		0.018
Change	$-x$	$+x$		$+x$
Equilibrium	$0.025 - x$	x		$0.018 + x$

$$K_a = \frac{[H^+][CHO_2^-]}{[HCHO_2]} = \frac{(x)(0.018 + x)}{0.025 - x} = 1.7 \times 10^{-4}$$

Since K_a divided by the concentrations of acid, then of anion, are $\approx 10^{-3}$, we will ignore x as the change in concentration. The expression becomes

$$\frac{x(0.018)}{0.025} = 1.7 \times 10^{-4}, \text{ and } x = 2.\underline{3}6 \times 10^{-4} = [H^+]$$

Thus,

$$pH = -\log [H^+] = -\log (2.\underline{3}6 \times 10^{-4}) = 3.63$$

16.6 BUFFERS

Operational Skills

8. Calculating the pH of a buffer from given volumes of solution. Given concentrations and volumes of acid and conjugate base from which a buffer is prepared, calculate the buffer pH (Example 16.11).

9. Calculating the pH of a buffer when a strong acid or strong base is added. Calculate the pH of a given volume of buffer solution (given the concentrations of conjugate acid and base in the buffer) to which a specified amount of strong acid or strong base is added (Example 16.12).

Exercise 16.12 What is the pH of a buffer prepared by adding 30.0 mL of 0.15 M $HC_2H_3O_2$ (acetic acid) to 70.0 mL of 0.20 M $NaC_2H_3O_2$ (sodium acetate)?

Wanted: pH of the buffer solution.

Given: Solution contains 30.0 mL of 0.15 M $HC_2H_3O_2$ and 70.0 mL of 0.20 M $NaC_2H_3O_2$.

Known: The buffer contains an acid and its conjugate base in equilibrium. The equation can be written as

$$HC_2H_3O_2(aq) \rightleftharpoons H^+(aq) + C_2H_3O_2^-(aq)$$

and

$$K_a = \frac{[H^+][C_2H_3O_2^-]}{[HC_2H_3O_2]} = 1.7 \times 10^{-5}$$

Solution: We must first calculate the starting concentrations in the solution formed by mixing the $HC_2H_3O_2$ and $NaC_2H_3O_2$ solutions.

$$\text{Moles } HC_2H_3O_2 = \text{molarity of } HC_2H_3O_2 \times \text{liters } HC_2H_3O_2 \text{ solution}$$

$$= 0.15 \; \frac{mol}{\not{L}} \times 0.0300 \not{L} = 0.00450 \text{ mol}$$

Since the sodium acetate dissociated completely in water,

$$\text{Moles } C_2H_3O_2^- = \text{molarity of } NaC_2H_3O_2 \times \text{liters } NaC_2H_3O_2 \text{ solution}$$

$$= 0.20 \; \frac{mol}{\not{L}} \times 0.0700 \not{L} = 0.0140 \text{ mol}$$

Total volume of buffer equals the sum of the volumes of the two solutions.

$$\text{Buffer volume} = 30.0 \text{ mL} + 70.0 \text{ mL} = 100.0 \text{ mL} \; (0.1000 \text{ L})$$

Therefore, the concentrations of acid and conjugate base are

$$[HC_2H_3O_2] = \frac{0.00450 \text{ mol}}{0.1000 \text{ L}} = 0.0450 \; M$$

$$[C_2H_3O_2^-] = \frac{0.0140 \text{ mol}}{0.1000 \text{L}} = 0.140 \; M$$

Now we fill in the concentration table for the acid–base equilibrium. We let $x = $ mol/L of acetic acid that ionize.

Concentration (M)	$HC_2H_3O_2(aq) \rightleftharpoons$	$H^+(aq)$	$+$	$C_2H_3O_2^-(aq)$
Starting	0.0450	0		0.140
Change	$-x$	$+x$		$+x$
Equilibrium	$0.0450 - x$	x		$0.140 + x$

Substituting into the equilibrium-constant expression:

$$K_a = \frac{[H^+][C_2H_3O_2^-]}{[HC_2H_3O_2]} = \frac{(x)(0.140 + x)}{(0.0450 - x)} = 1.7 \times 10^{-5}$$

Since K_a divided by the initial concentrations of acid, and then anion, is $\leqq 10^{-3}$, we will ignore the change of x, and the expression reduces to

$$\frac{0.140\,x}{0.0450} = 1.7 \times 10^{-5}$$

$$x = 5.46 \times 10^{-6}$$

Since $x = [\text{H}^+]$, we can use it to solve for pH:

$$\text{pH} = -\log[\text{H}^+] = -\log\,(5.46 \times 10^{-6}) = 5.26$$

There are two reactions involved when an acid or base is added to a buffer solution. The first is the acid–base reaction. It is helpful to write out the equation for this reaction before you make the stoichiometric calculations. However, write only the net ionic equation (see text Section 3.2). The second is the equilibrium reaction, which is the ionization of the weak acid or base that is used to make the buffer. In making your table, use the concentrations of molecules and ions present in solution *after* the acid–base reaction occurs. Remember that not only is the weak electrolyte used up, but an ion of it is produced by the acid–base reaction.

Exercise 16.13   What is the pH of the solution described in Exercise 16.11 if 50.0 mL of 0.10 M sodium hydroxide is added to 1 L of solution?

Wanted: pH after addition of base.

Given: Original soln is 1 L of 0.025 M HCHO_2 and 0.018 M NaCHO_2; add 50.0 mL of 0.10 M NaOH; $K_a = 1.7 \times 10^{-4}$.

Known: Make stoichiometric calculation; then make equilibrium calculation with concentrations remaining in solution.

Solution: Stoichiometric Calculation: The net ionic equation for the acid–base reaction is

$$\text{OH}^-(aq) + \text{HCHO}_2(aq) \longrightarrow \text{CHO}_2^-(aq) + \text{H}_2\text{O}(l)$$

Find moles NaOH added:

$$0.0500 \ \text{L NaOH soln} \times \frac{0.10 \ \text{mol NaOH}}{\text{L NaOH soln}} = 0.00500 \ \text{mol NaOH}$$

The added NaOH will react with the HCHO_2, so we next find moles HCHO_2 left in solution after reaction with NaOH:

$$\text{Moles HCHO}_2 \ \text{left} \ = \ \text{moles}_{\text{initial}} \ - \ \text{moles}_{\text{reacted}}$$

Moles$_{initial}$ were given as 0.025. Moles$_{reacted}$ = moles NaOH added, as the ratio is 1:1. Thus,

$$\text{Moles } HCHO_2 \text{ left} = 0.025 - 0.005\underline{00} = 0.020$$

Now we find the moles of CHO_2^- ion in solution after the reaction. As the ion is formed in the reaction, the moles in solution = moles$_{initial}$ + moles$_{formed}$. Moles$_{initial}$ were given as 0.018. Moles$_{formed}$ = moles NaOH added, as the ratio is also 1:1. Thus,

$$\text{Moles } CHO_2^- \text{ in solution} = 0.018 + 0.005\underline{00} = 0.023$$

Equilibrium Calculation: First determine the concentrations. The total volume of solution is 1 L of original solution plus 50.0 mL NaOH(*aq*) = 1.0500 L. The concentrations are

$$[HCHO_2] = \frac{0.020 \text{ mol}}{1.0500 \text{ L}} = 0.01\underline{9}0 \text{ } M$$

$$[CHO_2^-] = \frac{0.023 \text{ mol}}{1.0500 \text{ L}} = 0.02\underline{1}9 \text{ } M$$

If we let x = mol/L of acid that ionize, the table is

Concentration (*M*)	$HCHO_2(aq)$ $\rightleftharpoons$	$H^+(aq)$	$+$	$CHO_2^-(aq)$
Starting	0.01$\underline{9}$0	0		0.02$\underline{1}$9
Change	$-x$	$+x$		$+x$
Equilibrium	0.01$\underline{9}$0 $- x$	x		0.02$\underline{1}$9 $+ x$

$$K_a = \frac{[H^+][CHO_2^-]}{[HCHO_2]} = \frac{x(0.02\underline{1}9 + x)}{(0.01\underline{9}0 - x)} = 1.7 \times 10^{-4}$$

We will ignore x as the change in concentration, which gives

$$\frac{x(0.02\underline{1}9)}{(0.01\underline{9}0)} = 1.7 \times 10^{-4}, \text{ and } x = 1.\underline{4}7 \times 10^{-4}$$

$$pH = -\log [H^+] = -\log (1.\underline{4}7 \times 10^{-4}) = 3.83$$

16.7   ACID–BASE TITRATION CURVES

Operational Skills

10.   Calculating the pH of a solution of a strong acid and a strong base. Calculate the pH during the titration of a strong acid by a strong base, given the volumes and concentrations of the acid and base   (Example 16.13).

11.   Calculating the pH at the equivalence point in the titration of a weak acid by a strong base. Calculate the pH at the equivalence point for the titration of a weak acid by a strong base   (Example 16.14).   Be able to do the same type of calculation for the titration of a weak base by a strong acid.

Exercise 16.14   What is the pH of a solution in which 15 mL of 0.10 M NaOH has been added to 25 mL of 0.10 M HCl?

Wanted: pH.

Given: 15 mL of 0.10 M NaOH added to 25 mL of 0.10 M HCl.

Known: pH = –log [H$^+$]; the solution was initially acid, then an acid–base reaction occurred, so we must determine whether acid or base is in excess.

Solution: The equation is

$$H^+(aq) + OH^-(aq) \longrightarrow H_2O(l)$$

First determine amounts of reactants:

$$\text{Mol H}^+ = 0.025 \text{ L soln} \times \frac{0.10 \text{ mol HCl}}{\text{L soln}} = 0.0025 \text{ mol}$$

$$\text{Mol OH}^- = 0.015 \text{ L soln} \times \frac{0.10 \text{ mol NaOH}}{\text{L soln}} = 0.0015 \text{ mol}$$

Because there is more H$^+$, it will be in excess when all the OH$^-$ reacts. Excess H$^+$ equals (0.0025 – 0.0015) mol = 0.0010 mol. Obtain [H$^+$] by dividing the moles by the total volume of the solution:

$$[\text{H}^+] = \frac{0.0010 \text{ mol}}{0.015 \text{ L} + 0.025 \text{ L}} = \frac{0.0010 \text{ mol}}{0.040 \text{ L}} = 0.025 \ M$$

Thus,

$$\text{pH} = -\log [\text{H}^+] = -\log (0.025) = 1.60$$

In titrations of a weak acid or base by a strong base or acid, the solution will not be neutral at the equivalence point due to the presence of an ion that hydrolyzes. Thus, Exercises 16.15 and 16.16 are hydrolysis, as well as stoichiometric, problems.

Exercise 16.15 What is the pH at the equivalence point when 25 mL of 0.10 *M* HF is titrated by 0.15 *M* NaOH?

Wanted: pH at equivalence point.

Given: 25 mL of 0.10 *M* HF plus 0.15 *M* NaOH.

Known: HF is a weak acid; F^-, which hydrolyzes, will be present in solution at the equivalence point; definition of equivalence point; K_a for HF = 6.8×10^{-4} (text Table 16.1); $K_a K_b = 1.00 \times 10^{-14} = [H^+][OH^-]$ (at 25°C); pH = $-\log [H^+]$.

Solution: The equation for the acid–base reaction is

$$HF(aq) + OH^-(aq) \longrightarrow H_2O(l) + F^-(aq)$$

First we determine moles HF:

$$\text{Mol HF} = 0.025 \, \text{L soln} \times \frac{0.10 \, \text{mol HF}}{\text{L soln}} = 0.0025$$

This will react with the same number of moles NaOH. To determine [F⁻], we need the volume of NaOH soln used:

$$\text{L soln} = 0.0025 \, \text{mol NaOH} \times \frac{\text{L soln}}{0.15 \, \text{mol NaOH}}$$

$$= 0.0167 \, \text{L}$$

The equation shows that moles F⁻$_{formed}$ = moles HF$_{reacted}$, so

$$[F^-] = \frac{0.0025 \, \text{mol}}{0.025 \, \text{L} + 0.0167 \, \text{L}} = \frac{0.0025 \, \text{mol}}{0.0417 \, \text{L}} = 0.0600 \, M$$

If we let x = mol/L of F^- that hydrolyze, the hydrolysis reaction and table are

Concentration (M)	$F^-(aq)$ + $H_2O(l)$ $\rightleftharpoons$	$HF(aq)$ +	$OH^-(aq)$
Starting	0.06$\underline{0}$0	0	0
Change	$-x$	$+x$	$+x$
Equilibrium	0.06$\underline{0}$0 $- x$	x	x

$$K_b = \frac{[HF][OH^-]}{[F^-]} = \frac{(x)(x)}{0.06\underline{0}0 - x} = \frac{1.0 \times 10^{-14}}{K_a} = \frac{1.0 \times 10^{-14}}{6.8 \times 10^{-4}} = 1.\underline{4}7 \times 10^{-11}$$

Because K_b is so small, x will be insignificant with respect to 0.06$\underline{0}$0 and the change can be ignored, giving

$$\frac{x^2}{0.06\underline{0}0} = 1.\underline{4}7 \times 10^{-11}$$

$$x^2 = (0.06\underline{0}0)(1.\underline{4}7 \times 10^{-11}) = 8.\underline{8}2 \times 10^{-13}$$

$$x = 9.\underline{3}9 \times 10^{-7}$$

Since $x = [OH^-]$, we solve for $[H^+]$ next by rearranging $[H^+][OH^-] = 1.0 \times 10^{-14}$ to get

$$[H^+] = \frac{1.0 \times 10^{-14}}{[OH^-]} = \frac{1.0 \times 10^{-14}}{9.\underline{3}9 \times 10^{-7}} = 1.\underline{0}6 \times 10^{-8} \ M$$

Thus,

$$pH = -\log [H^+] = -\log (1.\underline{0}6 \times 10^{-8}) = 7.97$$

Exercise 16.16 What is the pH at the equivalence point when 35 mL of 0.20 M ammonia is titrated by 0.12 M hydrochloric acid? K_b for ammonia is 1.8×10^{-5}.

Wanted: pH at equivalence point.

Given: 35 mL of 0.20 M NH_3 plus 0.12 M HCl; K_b for NH_3 = 1.8×10^{-5}.

Known: Ammonium ion, NH_4^+, which hydrolyzes, will be present at the equivalence point; other information as given with Exercise 16.15.

Solution: We can write the acid–base reaction as

$$NH_3(aq) + H^+(aq) \longrightarrow NH_4^+(aq)$$

Determine moles NH_3 initially:

$$\text{Moles } NH_3 = 0.035 \text{ L soln} \times \frac{0.20 \text{ mol } NH_3}{\text{L soln}} = 0.0070 \text{ mol}$$

This will react with 0.0070 mol H^+. To determine $[NH_4^+]$, we must find the volume of HCl soln used:

$$\text{L soln} = 0.0070 \text{ mol HCl} \times \frac{\text{L soln}}{0.12 \text{ mol HCl}} = 0.058\underline{3} \text{ L}$$

Thus, the total volume = $(0.035 + 0.058\underline{3})$ L = $0.09\underline{33}$ L. The equation shows that moles $NH_3 \text{ consumed}$ = moles $NH_4^+ \text{ formed}$, so

$$[NH_4^+] = \frac{0.0070 \text{ mol}}{0.09\underline{33} \text{ L}} = 0.07\underline{50} \, M$$

If we let x = mol/L of NH_4^+ that hydrolyze, the hydrolysis reaction and table are

Concentration (*M*)	$NH_4^+(aq)$ + $H_2O(l)$	$\rightleftharpoons$	$NH_3(aq)$ +	$H_3O^+(aq)$
Starting	0.07\underline{50}		0	0
Change	$-x$		$+x$	$+x$
Equilibrium	$0.07\underline{50} - x$		x	x

$$K_a = \frac{[NH_3][H_3O^+]}{[NH_4^+]} = \frac{(x)(x)}{0.07\underline{50} - x} = \frac{1.00 \times 10^{-14}}{K_b} = \frac{1.00 \times 10^{-14}}{1.8 \times 10^{-5}}$$

$$= 5.\underline{56} \times 10^{-10}$$

Since K_a is so small, the change in $[NH_4^+]$ can be ignored, giving

$$\frac{x^2}{0.07\underline{5}0} = 5.56 \times 10^{-10}$$

$$x^2 = (0.07\underline{5}0)(5.56 \times 10^{-10}) = 4.\underline{1}7 \times 10^{-11}$$

$$x = 6.\underline{4}6 \times 10^{-6}$$

Since $x = [H_3O^+]$, we use this to solve for pH:

$$pH = -\log [H^+] = -\log (6.\underline{4}6 \times 10^{-6}) = 5.19$$

A Metal That Matters: ALUMINUM (an Amphoteric Metal)

Questions for Study

1.   Give the chemical equation for the original commercial preparation of aluminum.

2.   How is aluminum prepared today? What is the ore used in this process?

3.   Aluminum is a reactive metal. Give chemical equations for the reaction of aluminum with (a) Fe_2O_3; (b) Cl_2; (c) Br_2; (d) O_2.

4.   Iron is also a reactive metal, and if unprotected, it oxidizes in moist air to form rust, eventually crumbling to powder. Although aluminum is even more reactive with oxygen than is iron, aluminum bars and sheets do not appear to corrode or oxidize so easily. Explain why.

5.   Aluminum is an amphoteric metal. Explain what that means. Then give equations for appropriate reactions that illustrate this property of aluminum.

Answers to Questions for Study

1.   The chemical equation for the original commercial preparation of aluminum is

$$AlCl_3(l) + 3Na(l) \longrightarrow Al(s) + 3NaCl(s)$$

2.   Aluminum is now obtained by electrolyzing aluminum oxide in a molten bath of cryolite, a mineral of aluminum, Na_3AlF_6. The ore used is bauxite, a hydrated aluminum oxide ore.

3. (a) $2Al(s) + Fe_2O_3(s) \longrightarrow 2Fe(l) + Al_2O_3(s)$

    (b) $2Al(s) + 3Cl_2(g) \longrightarrow 2AlCl_3(s)$

    (c) $2Al(s) + 3Br_2(g) \longrightarrow 2AlBr_3(s)$

    (d) $4Al(s) + 3O_2(g) \longrightarrow 2Al_2O_3(s)$

4. Aluminum readily reacts with oxygen in the air to form aluminum oxide, which adheres very tightly to the surface of the metal. This coating prevents further reaction of the aluminum.

5. As an amphoteric metal, aluminum exhibits certain chemical properties that are characteristic of metals and nonmetals. For example, aluminum reacts with both acids and bases. In both cases, hydrogen is formed. In the reaction with acids, aluminum displaces hydrogen:

$$2Al(s) + 6H^+(aq) \longrightarrow 2Al^{3+}(aq) + 3H_2(g)$$

With bases, aluminum forms the tetrahydroxoaluminum ion and hydrogen:

$$2Al(s) + 2OH^-(aq) + 6H_2O(l) \longrightarrow 2[Al(OH)_4^-(aq)] + 3H_2(g)$$

ADDITIONAL PROBLEMS

1. (a)   What are the equilibrium $H^+(aq)$ concentration and pH for a 1.12 M solution of hypochlorous acid?

    (b)   What is the percent dissociation of this acid in water? $K_a = 2.95 \times 10^{-8}$ for hypochlorous acid.

2. Calculate the pH and the concentrations of H^+, $HC_4H_2O_4^-$, and $C_4H_2O_4^{2-}$ in a 0.15 M solution of maleic acid, $H_2C_4H_2O_4$. The acid-ionization constants are $K_{a1} = 1.5 \times 10^{-2}$ and $K_{a2} = 2.6 \times 10^{-7}$.

3. Calculate the pH and degree of ionization of the following aqueous methylamine solutions ($K_b = 4.17 \times 10^{-4}$):
    (a)   1.85 M
    (b)   1.25 M

Discuss the reasons for the difference between the degrees of ionization for solutions (a) and (b).

4. Calculate the following at 25°C:

(a) K_b for the oxalate anion, $C_2O_4^{2-}$ (K_a for $HC_2O_4^-$ is 5.1×10^{-5})

(b) K_a for the anilinium ion, $C_6H_5NH_3^+$ (K_b for aniline is 4.2×10^{-10})

5. Calculate the OH^- concentration and pH of a 9.42×10^{-2} M solution of sodium formate, $NaCHO_2$. K_a for formic acid is 1.77×10^{-4} M.

6. Calculate the degree of ionization of cyanic acid, HCNO, and the concentration of the cyanate anion in a 0.12 M solution of the acid in which the concentration of HCl is 0.015 M. K_a for HCNO is 3.5×10^{-4} at 25°C.

7. Calculate the change in pH that occurs when a 0.0955 M HClO solution ($K_a = 2.95 \times 10^{-8}$) is made 0.0500 M in KClO.

8. How many grams of $NaCHO_2$ should be added to 1.50 L of a 0.600 M formic acid solution to buffer the solution at pH 3.20? K_a for formic acid is 1.77×10^{-4}.

9. What is the pH of the solution after the addition of 30.0 mL of 0.10 M NaOH to 50.0 mL of 0.10 M HBr?

10. Calculate the pH at the equivalence point for the titration of 75.0 mL of 0.100 M $NH_3(aq)$ with 0.110 M HCl. K_b for NH_3 is 1.8×10^{-5}.

ANSWERS TO ADDITIONAL PROBLEMS

If you missed an answer, study the text section and operational skill given in parentheses after the answer.

1. (a) $[H^+] = 1.82 \times 10^{-4}$ M

    pH = 3.740

    (b) 1.62×10^{-2}% (16.1, Op. Sk. 1)

2. $[H^+] = 0.041\ M$, pH = 1.39, $[HC_4H_2O_4^-] = 0.041\ M$,

   $[C_4H_2O_4^{2-}] = 2.6 \times 10^{-7}\ M$ (16.2, Op. Sk. 2)

3. (a) pH = 12.444; 0.0150

   (b) pH = 12.359; 0.0183

   In going to a more dilute solution of methylamine — from 1.85 M to 1.25 M — the concentrations of all species decrease. Because this dilution occurs through the addition of water, the equilibrium shifts to the right, producing more $CH_3NH_3^+$ and OH^- ions. This is reflected in the increase in the percent ionization of CH_3NH_2. (15.8, 16.1, 16.3)

4. (a) $K_b = 2.0 \times 10^{-10}$, (b) $K_a = 2.4 \times 10^{-5}$ (16.4, Op. Sk. 5)

5. $[OH^-] = 2.31 \times 10^{-6}\ M$

   pH = 8.363 (16.4, Op. Sk. 6)

6. 0.023, $[CNO^-] = 2.8 \times 10^{-3}\ M$ (16.5, Op. Sk. 7)

7. Initial pH = 4.275

   Final pH = 7.249 (16.5, Op. Sk. 7)

8. 17.2 g $NaCHO_2$ (16.6, Op. Sk. 9)

9. pH = 1.60 (16.7, Op. Sk. 10)

10. pH = 5.27 (16.7, Op. Sk. 11)

CHAPTER POST-TEST

1.   Consider this list of bases and their respective K_b values:

$(CH_3)_3N$	6.3×10^{-5}	$C_6H_5NH_2$	3.98×10^{-10}
AsO_2^-	1.66×10^{-5}	$C_2H_5NH_2$	4.27×10^{-4}
$(C_2H_4OH)_3N$	5.75×10^{-7}	$C_3H_5O_2^-$	7.41×10^{-10}

Which of the following represents a correct ordering of base strengths?

(a) $(CH_3)_3N < C_3H_5O_2^- < C_6H_5NH_2$

(b) $C_2H_5NH_2 < (C_2H_4OH)_3N < AsO_2^-$

(c) $AsO_2^- < (C_2H_4OH)_3N < C_3H_5O_2^-$

(d) $C_6H_5NH_2 < AsO_2^- < C_2H_5NH_2$

(e)   none of the above

2.   The pH of a 5.7×10^{-2} M propionic acid solution is 3.07. From these data, what are the calculated $[H^+]$ in this solution and the K_a for this acid?

3.   What is the pH of a 2.8×10^{-2} M solution of diprotic sulfurous acid (H_2SO_3)? $K_{a1} = 1.3 \times 10^{-2}$ and $K_{a2} = 6.3 \times 10^{-8}$ for sulfurous acid.

4.   A 0.348 M solution of the base aniline, $C_6H_5NH_2$, has a pH of 9.08. What is the K_b for aniline?

5.   What is the pH of a 0.50 M $NaNO_2$ solution? The K_a for HNO_2 is 4.5×10^{-4}.

6.   The reaction for the hydrolysis of I_3^- is

$$I_3^-(aq) + H_2O(l) \rightleftharpoons HIO(aq) + 2I^-(aq) + H^+(aq)$$

If the equilibrium concentrations are

$$[I_3^-] = 0.40 \ M$$

$$[HIO] = [H^+] = 7.95 \times 10^{-5} \ M$$

$$[I^-] = 1.59 \times 10^{-4} \ M$$

what is K_a for this reaction?

7. How many moles of hydrochloric acid, HCl, must be added to 0.100 L of 0.53 M chlorous acid, $HClO_2$, to produce 5.0% ionization? Ignore the change in volume due to the addition of HCl. K_a for $HClO_2$ is 1.1×10^{-2}.

8. What is the pH of a buffer solution composed of 0.125 mol of ammonia, NH_3, and 0.200 mol of ammonium chloride, NH_4Cl, in enough water to make a 0.750-L solution? K_b for $NH_3 = 1.8 \times 10^{-5}$.

9. If 25 mL of 0.10 M HNO_3 is added to a 0.100-L buffer solution that is 0.26 M HCNO and 0.13 M NaCNO, what is the pH of the final solution? K_a for HCNO = 3.5×10^{-4}.

10. What would be the pH at the equivalence point for the titration of 50.0 mL of a 0.10 M solution of weak acid HA with 0.10 M NaOH? K_a for HA = 1.0×10^{-7}.

11. For the titration of 50.0 mL of 0.020 M HI with 0.015 M NaOH, graph pH versus mL of base added from 0 to 100 mL. How many mL of NaOH are added at the equivalence point?

ANSWERS TO CHAPTER POST-TEST

If you missed an answer, study the text section and operational skill given in parentheses after the answer.

1. d   (16.3)

2. $[H^+] = 8.5 \times 10^{-4}$; $K_a = 1.3 \times 10^{-5}$ (15.8, 16.1, Op. Sk. 1)

3.   1.86   (16.2, Op. Sk. 2)

4. 4.2×10^{-10} (16.3, Op. Sk. 1)

5.   8.52   (16.4, Op. Sk. 4, 6)

6. 4.0×10^{-16} (16.1, 16.4, Op. Sk. 1)

7.   0.018 mol   (16.5, Op. Sk. 7)

8.   9.05   (16.6, Op. Sk. 8)

9.   3.02   (16.6, Op. Sk. 9)

10. 9.85   (16.7, Op. Sk. 11)

11. Graph follows; 67 mL NaOH added at equivalence point.   (16.7, Op. Sk. 10)

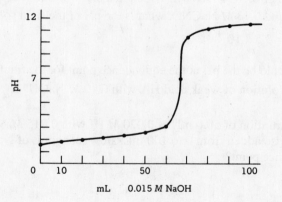

Values for graph in Problem 11

mL NaOH	pH	mL NaOH	pH
0	1.70	60	3.04
10	1.85	67	7.00
20	2.00	70	10.62
30	2.16	80	11.19
40	2.35	90	11.40
50	2.60	100	11.52

UNIT EXAM 5

1. Consider the following equation:

    $$2HBr(g) \rightleftharpoons H_2(g) + Br_2(g); K_c = 1.8 \times 10^{-9} \text{ at some temperature T}$$

    A 4.50-L container has 1.69 mol each of HBr and Br_2 and 5.45×10^{-5} mol H_2 at temperature T. Is the system at equilibrium?

2. Assume that 1.00 mol HI is placed in a 5.00-L container at 458°C. If the reaction is $2HI(g) \rightleftharpoons H_2(g) + I_2(g)$, what are the equilibrium concentrations of all species? ($K_c = 2.06 \times 10^{-2}$.)

3. Write the equilibrium-constant expression for each of the following:

    (a) $N_2(g) + 3H_2(g) \rightleftharpoons 2NH_3(g); K_p = ?$

    (b) $CaO(s) + CO_2(g) \rightleftharpoons CaCO_3(s); K_p = ?$

    (c) $HC_2H_3O_2(l) + CH_3OH(l) \rightleftharpoons C_2H_3O_2CH_3(l) + H_2O(l); K_c = ?$

4. A mixture of 1.47 mol NO and 0.270 mol Cl_2 is placed in a 1.0-L reaction vessel at a given temperature. After equilibrium conditions are established, the amount of NOCl present is estimated to be 0.380 mol. Calculate the equilibrium constant K_c for the reaction.

5. Give the conjugate base for each of the following acids:

    (a) HSO_3^- (c) HCO_2H (e) H_2CO_3

    (b) H_3O^+ (d) HS^-

437

6. What is the Lewis concept of a base?

7. Using the information in Table 15.2, determine the direction of the following reaction:

$$NO_3^-(aq) + NH_4^+(aq) \rightleftharpoons NH_3(aq) + HNO_3(aq)$$

8. What is meant by the term pH?

9. The pH of a 0.15 M solution of the potassium salt of a particular acid is measured to be 9.4. What is the hydroxide-ion concentration of this solution?

10. A thymol blue solution is yellow at pH 6.3. What are the hydrogen ion and hydroxide ion concentrations in the thymol blue solution at this pH?

11. Another name for $Na_2CO_3 \cdot 10H_2O$ is washing soda. If a student prepared a solution of washing soda having a hydrogen-ion concentration of 2.51×10^{-12} M, what is the hydroxide-ion concentration of this solution?

12. Which of the following statements is true concerning the hydrolysis of 0.10 mol of sodium acetate, $NaC_2H_3O_2$, in 1 L of water? (K_a of $HC_2H_3O_2$ is 1.7×10^{-5}.)
    (a) The pH of the aqueous solution is 4.8.

    (b) The conjugate acid, $HC_2H_3O_2$, which has a concentration of 7.7×10^{-6} M, is responsible for the acidic properties of the solution.

    (c) The OH⁻ ions produced during the hydrolysis immediately combine with Na⁺ ions to form NaOH.

    (d) The concentration of the conjugate base, $C_2H_3O_2^-$, is approximately 0.1 M at equilibrium.

13. Calculate the pH at the equivalence point for the titration of 50.00 mL of 0.10 M HA ($K_a = 1.0 \times 10^{-5}$) with 0.10 M NaOH.

14. The base-dissociation constant K_b for dimethylamine is 5.1×10^{-4} at 298 K. Calculate the pH and percent dissociation of a 0.225 M $(CH_3)_2NH$ solution.

15. The equilibrium concentrations for the ionization of hypochlorous acid at 298 K,

$$HClO(aq) \rightleftharpoons H^+(aq) + ClO^-(aq)$$

are [HClO] = 1.00 M; [H$^+$] = [ClO$^-$] = 1.79 × 10^{-4} M. Calculate K_a for hypochlorous acid.

16. Calculate the pH of a solution prepared from 60.0 mL of 0.150 M acetic acid, $HC_2H_3O_2$, and 40.0 mL of 0.125 M potassium acetate, $KC_2H_3O_2$. (K_a for $HC_2H_3O_2$ is 1.7 × 10^{-5}.)

ANSWERS TO UNIT EXAM 5

If you missed an answer, study the text section and operational skill given in parentheses after the answer. In some cases, we include text sections in which the important operational skills used were originally introduced.

1. No; $Q_c = 3.22 × 10^{-5} > K_c$.   (14.5, Op. Sk. 4)

2. [H$_2$] = [I$_2$] = 2.23 × 10^{-2} M; [HI] = 0.155 M (14.6, Op. Sk. 6)

3. (a) $K_p = \dfrac{P_{NH_3}^2}{P_{N_2} P_{H_2}^3}$ (b) $K_p = \dfrac{1}{P_{CO_2}}$

   (c) $K_c = \dfrac{[C_2H_3O_2CH_3][H_2O]}{[HC_2H_3O_2][CH_3OH]}$ (14.2, Op. Sk. 2)

4. 1.5 (14.1, 14.2, Op. Sk. 1, 2, 3)

5. (a) SO_3^{2-} (c) HCO_2^- (e) HCO_3^-

   (b) H_2O (d) S^{2-} (15.2, Op. Sk. 1)

6. A Lewis base is a species that can form a covalent bond by donating an electron pair to another species.   (15.3)

7. The reaction will proceed from right to left.   (15.4, Op. Sk. 3)

8.   The term pH is another way of referring to the numerical value for $-\log [H^+]$. (15.8)

9. $[OH^-] = 3 \times 10^{-5} M$ (15.8, Op. Sk. 5)

10. $[H^+] = 5 \times 10^{-7} M$; $[OH^-] = 2 \times 10^{-8} M$ (15.7, Op. Sk. 4; 15.8, Op. Sk. 5)

11. $3.98 \times 10^{-3} M$ (15.8, Op. Sk. 5)

12.   d (16.4, Op. Sk. 6)   13.   pH = 8.85   (16.7, Op. Sk. 11)

14.   pH = 12.02

      4.67% dissociation   (15.8, Op. Sk. 5; 16.3, Op. Sk. 2)

15. $K_a = 3.20 \times 10^{-8}$ (16.1, Op. Sk. 1)   16.   pH = 4.51   (16.6, Op. Sk. 8)

CHAPTER 17   SOLUBILITY AND COMPLEX-ION EQUILIBRIA

CHAPTER TERMS AND DEFINITIONS

Numbers in parentheses after definitions give the text sections in which the terms are explained. Starred terms are italicized in the text. Where a term does not fall directly under a text section heading, additional information is given for you to locate it.

solubility product constant (K_{sp})   equilibrium constant for the dissolution of a slightly soluble (or nearly insoluble) ionic compound   (17.1)

molar solubility*   moles of compound that dissolve to give a liter of saturated solution   (17.1)

common-ion effect*   shift in an ionic equilibrium caused by the addition of a substance that provides an ion that takes part in the equilibrium   (17.2)

reaction quotient (Q_c)*   expression identical to the equilibrium-constant expression with concentrations not necessarily equilibrium values   (17.3)

ion product   product of ion concentrations in a solution, each concentration raised to a power equal to the number of ions in the formula of the ionic compound   (17.3)

stoichiometric calculation*   calculation using moles, assuming that the reaction under study goes to completion   (17.3, Example 17.8)

equilibrium calculation*   calculation of concentrations by use of the equilibrium constant for the reaction under study   (17.3, Example 17.8)

fractional precipitation   technique of separating two or more ions from a solution by adding a reactant that first precipitates one ion, then another, and so forth   (17.3)

complex ion   ion formed from a metal ion with a Lewis base attached to it by a coordinate covalent bond   (17.5, introductory section)

complex*   compound containing complex ions   (17.5, introductory section)

ligand   Lewis base that bonds to a metal ion to form a complex ion   (17.5, introductory section)

formation (stability) constant (K_f)   equilibrium constant for the formation of a complex ion from the aqueous metal ion and the ligands   (17.5)

dissociation constant (of a complex ion) (K_d)   reciprocal, or inverse, of the formation constant (stability constant) for a complex ion   (17.5)

amphoteric hydroxide   metal hydroxide that reacts with both bases and acids   (17.5)

stepwise formation constants (K_{f_1}, K_{f_2})*   equilibrium constants for the individual steps of the formation of a complex ion from the aqueous metal ion and the ligands (17.5)

overall formation constant (K_f)*   product of the stepwise formation constants of a complex ion   (17.5)

qualitative analysis   determination of the identity of substances present in a mixture (17.7, introductory section)

plumbum*   Latin word for "lead," from which the word *plumbing* is derived   (A Metal That Matters: Lead [a Main-Group Metal])

CHAPTER DIAGNOSTIC TEST

1.   Write the solubility product expression for each of the following compounds: CdS; $Pb_3(AsO_4)_2$; Ag_2S.

2.   The solubility of CaF_2 is 1.6×10^{-2} g/L. Calculate K_{sp} for CaF_2.

3.   Calculate the mass of lead in grams that is dissolved in 125 mL of aqueous solution of $Pb_3(AsO_4)_2$. $K_{sp} = 4 \times 10^{-36}$.

4.   How many moles of CaF_2 will dissolve in 1 L of 0.10 M $Ca(OH)_2$? Use the value of K_{sp} given in text Table 17.1.

5.   To 0.740 L of a solution 0.25 M $AgNO_3$ and 0.36 M $Pb(NO_3)_2$ is added 0.130 L of a 5.35×10^{-11} M K_2CrO_4 solution. Will the yellow $PbCrO_4$ or red Ag_2CrO_4 precipitate from the solution? $K_{sp}(PbCrO_4) = 1.8 \times 10^{-14}$; $K_{sp}(Ag_2CrO_4) = 1.1 \times 10^{-12}$.

6.   Suppose 0.250 g KF is added to 0.250 L of 0.050 M $MgCl_2$. What is the concentration of the fluoride ion left in solution after the precipitation of MgF_2? Ignore the volume change. $K_{sp}(MgF_2) = 6.61 \times 10^{-9}$.

7. Arrange the following silver salts in order of increasing change in solubility as the pH of each solution is decreased: $AgC_2H_3O_2$, $K_a(HC_2H_3O_2) = 1.7 \times 10^{-5}$; $AgNO_2$, $K_a(HNO_2) = 4.5 \times 10^{-4}$; $AgCN$, $K_a(HCN) = 4.9 \times 10^{-10}$.

8. What pH range will separate Co^{2+} from Mn^{2+} in a solution that is 0.20 M Co^{2+}, 0.15 M Mn^{2+}, and saturated with H_2S gas? The concentration of H_2S in a saturated solution is 0.10 M. $K_{sp}(CoS) = 4.0 \times 10^{-21}$; $K_{sp}(MnS) = 2.5 \times 10^{-10}$; $K_a(H_2S) = 1.1 \times 10^{-20}$.

9. What is the concentration of Ag^+ in a solution that was prepared as 0.19 M $AgNO_3$ and 1.48 M CN^-? Ag^+ forms the complex ion $Ag(CN)_2^-$. Its formation constant is 5.6×10^{18}.

10. A solution is prepared that is 0.01 M $AgNO_3$, 0.10 M $NaCN$, and 0.010 M Na_2S. Will Ag_2S precipitate from the solution? $K_{sp}(Ag_2S) = 6 \times 10^{-50}$; K_f of $Ag(CN)_2^- = 5.6 \times 10^{18}$.

11. What is the molar solubility of $Ni(OH)_2$ in 10.0 M NH_3? K_{sp} of $Ni(OH)_2 = 2.0 \times 10^{-15}$; K_f of $Ni(NH_3)_6^{2+} = 5.6 \times 10^8$.

ANSWERS TO CHAPTER DIAGNOSTIC TEST

If you missed an answer, study the text section and operational skill given in parentheses after the answer.

1. $K_{sp} = [Cd^{2+}][S^{2-}]$; $[Pb^{2+}]^3[AsO_4^{3-}]^2$; $[Ag^+]^2[S^{2-}]$ (17.1, Op. Sk. 1)

2. 3.4×10^{-11} (17.1, Op. Sk. 2)

3. 3×10^{-6} g   (17.2, Op. Sk. 3)

4. 9.2×10^{-6} mol   (17.2, Op. Sk. 3)

5. Yellow $PbCrO_4$ will form.   (17.3, Op. Sk. 4)

6. $4.0 \times 10^{-4} M$ (17.3, Op. Sk. 5)

7. $AgNO_2 < AgC_2H_3O_2 < AgCN$ (17.4, Op. Sk. 6)

8.   CoS precipitates at a pH > 0.63; MnS precipitates at a pH > 6.09. Thus, a pH greater than 0.63 and less than 6.09 will separate the two ions by precipitation of the sulfide of cobalt.   (17.4, Op. Sk. 7)

9. $2.8 \times 10^{-20} M$ (17.5, Op. Sk. 8)

10.   Yes, Ag_2S will precipitate.   (17.6, Op. Sk. 9)

11. $0.65 M$ (17.6, Op. Sk. 10)

SUMMARY OF CHAPTER TOPICS

In this chapter we will again be dealing with equilibrium reactions and calculations. The first reaction we will discuss is the dissolution of a slightly soluble salt. It is a good idea to refer to text Section 3.3 and Table 3.1 to memorize the relatively few types of salts that are soluble, so that you can recognize a slightly soluble one when you see the formula. The equilibrium constant for the reaction is K_{sp}, *sp* meaning "solubility product." The constant is an ion product, since we write the expression as the product of the ion concentrations at equilibrium raised to the coefficient powers. (Recall the ion product of water, text Section 15.6.) The reactant does not appear in the expression. Many students make their first mistake here by writing the solid salt concentration in the expression! As in the previous chapter, a table of equilibrium-constant values (K_{sp}) is provided in the text for use in problem solving (Table 17.1).

The second type of reaction we will see is the formation of a complex ion from simpler ions. The equilibrium constant is denoted K_f. The equilibrium-constant expression appears as you first learned to write it in text Section 14.2, by including concentrations of all species that appear in the reaction. Table 17.2 in the text lists the values of K_f for some common complex ions. Some texts instead refer to the reverse reaction, which is the dissociation of the complex ion, and list a table of K_d values. As your author points out, $K_d = 1/K_f$, so you can use either table to solve any complex-ion problem.

Pay particular attention to your notations in working these problems. You will have ion charges as well as exponents to deal with, and it is easy to confuse them or to leave something out.

17.1 THE SOLUBILITY PRODUCT CONSTANT

Operational Skills

1. Writing solubility product expressions. Write the solubility product expression for a given ionic compound (Example 17.1).

2. Calculating K_{sp} from the solubility, or vice versa. Given the solubility of a slightly soluble ionic compound, calculate K_{sp} (Examples 17.2 and 17.3). Given K_{sp}, calculate the solubility of an ionic compound (Example 17.4).

> **Exercise 17.1** Give solubility product expressions for the following: (a) barium sulfate; (b) iron(III) hydroxide; (c) calcium phosphate.
>
> *Wanted*: K_{sp} expressions.
>
> *Given:* slightly soluble salts.
>
> *Known:* write the reaction first, do not include the solid salt in the K_{sp} expression; rules for writing formulas (text Section 2.6).
>
> *Solution:* (a) $BaSO_4(s) \rightleftharpoons Ba^{2+}(aq) + SO_4^{2-}(aq)$
>
> $$K_{sp} = [Ba^{2+}][SO_4^{2-}]$$
>
> (b) $Fe(OH)_3(s) \rightleftharpoons Fe^{3+}(aq) + 3OH^-(aq)$
>
> $$K_{sp} = [Fe^{3+}][OH^-]^3$$
>
> (c) $Ca_3(PO_4)_2(s) \rightleftharpoons 3Ca^{2+}(aq) + 2PO_4^{3-}(aq)$
>
> $$K_{sp} = [Ca^{2+}]^3[PO_4^{3-}]^2$$

> **Exercise 17.2** Silver ion may be recovered from used photographic fixing solution by precipitating it as silver chloride. The solubility of silver chloride is 1.9×10^{-3} g/L. Calculate K_{sp}.
>
> *Wanted:* K_{sp}.
>
> *Given:* Solubility of silver chloride = 1.9×10^{-3} g/L.
>
> *Known:* rules for writing formulas (text Section 2.6) and for writing the reaction and the K_{sp} expression; from the equation for the dissolution, we can see how many moles of ions are produced for each mole of salt that dissolves; AgCl = 143.3 g/mol.

Solution: First solve for the solubility in mol/L, as K_{sp} is in terms of concentration:

$$1.9 \times 10^{-3}\frac{\text{g}}{\text{L}} \times \frac{1 \text{ mol AgCl}}{143.3 \text{ g}} = 1.\underline{3}3 \times 10^{-5} \text{ mol / L}$$

This enables us to determine the equilibrium concentrations of the ions, using the equation coefficients. Now we solve the equilibrium problem, using the table

Concentration (*M*)	$AgCl(s) \rightleftharpoons$	$Ag^+(aq)$	+	$Cl^-(aq)$
Starting		0		0
Change		$1.\underline{3}3 \times 10^{-5}$		$1.\underline{3}3 \times 10^{-5}$
Equilibrium		$1.\underline{3}3 \times 10^{-5}$		$1.\underline{3}3 \times 10^{-5}$

$$K_{sp} = [Ag^+][Cl^-] = (1.\underline{3}3 \times 10^{-5})(1.\underline{3}3 \times 10^{-5})$$

$$= 1.8 \times 10^{-10}$$

Exercise 17.3 Lead(II) arsenate, $Pb_3(AsO_4)_2$, has been used as an insecticide. It is only slightly soluble in water. If the solubility is 3.0×10^{-5} g/L, what is the solubility product constant? Assume that the solubility equilibrium is the only important one.

Wanted: K_{sp}.

Given: Solubility of $Pb_3(AsO_4)_2 = 3.0 \times 10^{-5}$ g/L.

Known: Information listed in Exercise 17.2; $Pb_3(AsO_4)_2 = 899.4$ g/mol.

Solution: First solve for the solubility in mol/L:

$$3.0 \times 10^{-5} \frac{\text{g}}{\text{L}} \times \frac{1 \text{ mol } Pb_3(AsO_4)_2}{899.4 \text{ g}} = 3.\underline{3}4 \times 10^{-8} \text{ mol / L}$$

In writing the reaction and setting up the table, be sure to get the right coefficients for the ions and use them in calculating ion concentrations at equilibrium from solubility.

Concentration (M) $Pb_3(AsO_4)_2(s) \rightleftharpoons$	$3Pb^{2+}(aq)$ +	$2AsO_4^{3-}(aq)$
Starting	0	0
Change	$3 \times 3.\underline{34} \times 10^{-8}$	$2 \times 3.\underline{34} \times 10^{-8}$
Equilibrium	$1.\underline{00} \times 10^{-7}$	$6.\underline{68} \times 10^{-8}$

$$K_{sp} = [Pb^{2+}]^3[AsO_4^{3-}]^2 = (1.\underline{00} \times 10^{-7})^3(6.\underline{68} \times 10^{-8})^2$$
$$= 4.5 \times 10^{-36}$$

In doing problems like Example 17.4 and Exercise 17.4 below, where you are asked to determine the solubility, it is very important that you use the coefficient of the ion twice. Take, for instance, the fluoride ion in Example 17.4. The balanced equation is

$$CaF_2(s) \rightleftharpoons Ca^{2+}(aq) + 2F^-(aq)$$

The coefficient of the F^- ion is 2. When we set up the table and let x be the mol/L of CaF_2 that dissolve, the concentration of F^- at equilibrium will be $2x$. Then, when we set up the expression for K_{sp}, we must also square that concentration:

$$K_{sp} = [Ca^{2+}][F^-]^2 = (x)(2x)^2 = 4x^3$$

Forgetting to use the coefficient twice is the most common source of error in these problems.

Exercise 17.4 Anhydrite is a calcium sulfate mineral deposited when seawater evaporates. What is the solubility of calcium sulfate, in grams per liter? Table 17.1 gives the solubility product for calcium sulfate.

Wanted: solubility of calcium sulfate in g/L.

Given: $K_{sp} = 2.4 \times 10^{-5}$ (text Table 17.1).

Known: how to set up an equilibrium problem; how to write formulas; molar mass $CaSO_4$ = 136.2 g/mol.

Solution: Letting x = mol/L that dissolve, the table is

Concentration (M)	$CaSO_4(s)$ $\rightleftharpoons$ $Ca^{2+}(aq)$	+	$SO_4^{2-}(aq)$
Starting	0		0
Change	+x		+x
Equilibrium	x		x

$$K_{sp} = [Ca^{2+}][SO_4^{2-}] = (x)(x) = 2.4 \times 10^{-5}$$

$$x^2 = 2.4 \times 10^{-5}$$

$$\text{so} \quad x = 4.\underline{9}0 \times 10^{-3} \, M$$

Now we convert the solubility in mol/L to g/L:

$$4.90 \times 10^{-3} \, \frac{\cancel{\text{mol}}}{\text{L}} \times \frac{136.2 \text{ g}}{\cancel{\text{mol}}} = 0.67 \text{ g} / \text{L}$$

17.2 SOLUBILITY AND THE COMMON-ION EFFECT

Operational Skill

3. Calculating the solubility of a slightly soluble salt in a solution of a common ion.
Given the solubility product constant, calculate the molar solubility of a slightly soluble
ionic compound in a solution that contains a common ion   (Example 17.5).

Exercise 17.5   (a) Calculate the molar solubility of barium fluoride, BaF_2, in water
at 25°C. The solubility product constant for BaF_2 at this temperature is
1.0×10^{-6}.
(b)   What is the molar solubility of barium fluoride in 0.15 M NaF at 25°C?
Compare the solubility in this case with that of BaF_2 in pure water.

Wanted: solubility of BaF_2 (mol/L) (a) in water, (b) in 0.15 M NaF; comparison.

Given: K_{sp} for $BaF_2 = 1.0 \times 10^{-6}$.

Known: When the table is set up for (b), there will be an initial concentration of F^- ion $= 0.15\ M$ (as NaF is assumed to be completely soluble).

Solution: Letting $x = \text{mol/L}$ that dissolve, the tables and K_{sp} calculations are

(a)

Concentration (M)	$BaF_2(s) \rightleftharpoons$	$Ba^{2+}(aq)$	$+$	$2F^-(aq)$
Starting		0		0
Change		$+x$		$+2x$
Equilibrium		x		$2x$

$$K_{sp} = [Ba^{2+}][F^-]^2 = (x)(2x)^2\ = 1.0 \times 10^{-6}$$

$$4x^3 = 1.0 \times 10^{-6}$$

$$\text{so} \quad x = 6.3 \times 10^{-3}\ M$$

(b)

Concentration (M)	$BaF_2(s) \rightleftharpoons$	$Ba^{2+}(aq)$	$+$	$2F^-(aq)$
Starting		0		0.15
Change		$+x$		$+2x$
Equilibrium		x		$0.15 + 2x$

$$K_{sp} = [Ba^{2+}][F^-]^2 = (x)(0.15 + 2x)^2 = 1.0 \times 10^{-6}$$

Since K_{sp} is less than 10^{-5} and the $0.15\ M\ F^-$ ion will suppress the solubility, we ignore the change of $2x$. Thus,

$$x(0.15)^2 = 1.0 \times 10^{-6}$$

$$x = 4.4 \times 10^{-5}\ M$$

The solubility in pure water is $\dfrac{6.3 \times 10^{-3}}{4.4 \times 10^{-5}} \approx 140$ times the solubility with $0.15\ M\ F^-$ present.

17.3   PRECIPITATION CALCULATIONS

Operational Skills

   4. Predicting whether precipitation will occur. Given the concentrations of ions
originally in solution, determine whether a precipitate is expected to form   (Example 17.6).
Determine whether a precipitate is expected to form when two solutions of known volume
and molarity are mixed   (Example 17.7).   For both problems, you will need the solubility
product constant.

   5. Determining the completeness of precipitation. Calculate the concentration and
percentage of an ion remaining after the corresponding ionic compound precipitates from a
solution of known concentrations of ions   (Example 17.8). K_{sp} is required.

      A concept that many students find difficult to grasp is that when a precipitate
forms, a certain concentration of each ion is left in solution. K_{sp} is the product of the
concentrations of those ions left in solution when the precipitate forms. It is not the
concentrations that precipitate. This is why we use the term "slightly soluble" rather than
"insoluble" to describe a salt that we do not consider soluble. Recall the equation for the
dissolution of a slightly soluble salt. The equation is an equilibrium between the solid and
these ions.

      Just as the reaction quotient, Q, can be calculated at any point in a reaction (text
Section 14.5), so may the ion product Q. Comparing the value for Q with K_{sp} allows us to
see whether a precipitate will form.

      Exercise 17.6   Anhydrite is a mineral composed of $CaSO_4$ (calcium sulfate). An
inland lake has Ca^{2+} and SO_4^{2-} concentrations of 0.0052 M and 0.0041 M,
respectively. If these concentrations were doubled by evaporation, would you
expect calcium sulfate to precipitate?

      Wanted: Will $CaSO_4$ precipitate?

      Given: $[Ca^{2+}]_{final} = 2 \times 0.0052\ M = 0.0104\ M$

               $[SO_4^{2-}]_{final} = 2 \times 0.0041\ M = 0.0082\ M$

      Known: If the ion product, Q, is greater than K_{sp}, a precipitate will form;
$K_{sp} = 2.4 \times 10^{-5}$ (text Table 17.1).

Solution: The equation and calculation are

$$CaSO_4(s) \rightleftharpoons Ca^{2+}(aq) + SO_4^{2-}(aq)$$

$$Q = [Ca^{2+}][SO_4^{2-}] = (0.01\underline{0}4)(0.0082)$$

$$= 8.\underline{5}3 \times 10^{-5}$$

Since $K_{sp} = 2.4 \times 10^{-5}$, the ion product is larger than K_{sp}; thus, precipitation should occur.

Exercise 17.7 A solution of 0.00016 *M* lead(II) nitrate, $Pb(NO_3)_2$, was poured into 456 mL of 0.00023 *M* sodium sulfate, Na_2SO_4. Would a precipitate of lead(II) sulfate, $PbSO_4$, be expected to form if 255 mL of the lead nitrate solution were added?

Wanted: Will $PbSO_4$ precipitate?

Given: 456 mL of 0.00023 *M* Na_2SO_4 plus 255 mL of 0.00016 *M* $Pb(NO_3)_2$.

Known: $[Pb^{2+}]$ and $[SO_4^{2-}]$ must be calculated; a precipitate will form if $Q > K_{sp}$; $K_{sp} = 1.7 \times 10^{-8}$.

Solution: First we must find the moles of each ion present:

$$0.255 \, \cancel{L} \times \frac{0.00016 \text{ mol } Pb(NO_3)_2}{\cancel{L}} = 4.\underline{0}8 \times 10^{-5} \text{ mol } Pb(NO_3)_2$$

$$= \text{mol } Pb^{2+}$$

$$0.456 \, \cancel{L} \times \frac{0.00023 \text{ mol } Na_2SO_4}{\cancel{L}} = 1.\underline{0}5 \times 10^{-4} \text{ mol } Na_2SO_4 = \text{mol } SO_4^{2-}$$

Now, using a total volume of $0.255 + 0.456 = 0.711$ L, we can calculate the starting ion concentrations.

$$\frac{4.\underline{0}8 \times 10^{-5} \text{ mol } Pb^{2+}}{0.711 \text{ L}} = 5.\underline{7}4 \times 10^{-5} \, M$$

$$\frac{1.\underline{0}5 \times 10^{-4} \text{ mol } SO_4^{2-}}{0.711 \text{ L}} = 1.\underline{4}8 \times 10^{-4} \, M$$

Now we compute the ion product:

$$Q = [Pb^{2+}][SO_4^{2-}] = (5.\underline{7}4 \times 10^{-5})(1.\underline{4}8 \times 10^{-4})$$

$$= 8.\underline{5}0 \times 10^{-9}$$

K_{sp} for $PbSO_4$ is 1.7×10^{-8}. Since the ion product is less than K_{sp}, no precipitation should occur.

Exercise 17.8   Lead chromate, $PbCrO_4$, is a yellow pigment used in paints. Suppose 0.50 L of 1.0×10^{-5} M $Pb(C_2H_3O_2)_2$ and 0.50 L of 1.0×10^{-3} M K_2CrO_4 are mixed. Calculate the equilibrium concentration of Pb^{2+} ion remaining in solution after $PbCrO_4$ precipitates. What is the percentage of Pb^{2+} remaining in solution after precipitation of $PbCrO_4$?

Wanted: $[Pb^{2+}]$ at equilibrium; % of Pb^{2+} remaining.

Given: 0.50 L of 1.0×10^{-5} M $Pb(C_2H_3O_2)_2$ plus 0.50 L of 1.0×10^{-3} M K_2CrO_4; $PbCrO_4$ precipitate.

Known: $K_{sp}(PbCrO_4) = 1.8 \times 10^{-14}$ (text Table 17.1);   assume complete solubility of given salts. Assume complete precipitation of Pb^{2+}, the limiting ion (as it has a lower concentration), to find the amount of CrO_4^{2-} still in solution, then solve the K_{sp} expression to find $[Pb^{2+}]_{equilibrium}$.

$$\% \ Pb^{2+} \ remaining = \frac{[Pb^{2+}]_{equilibrium}}{[Pb^{2+}]_{initial}} \times 100$$

Solution: Find initial moles of each ion that precipitates:

$$0.50\ \cancel{L} \times 1.0 \times 10^{-5}\ \frac{mol\ Pb^{2+}}{\cancel{L}} = 5.0 \times 10^{-6}\ mol$$

$$0.50\ \cancel{L} \times 1.0 \times 10^{-3}\ \frac{mol\ CrO_4^{2-}}{\cancel{L}} = 5.0 \times 10^{-4}\ mol$$

The precipitation reaction and calculation of moles CrO_4^{2-} remaining are

Moles	$Pb^{2+}(aq)$	+	$CrO_4^{2-}(aq)$	$\longrightarrow$	$PbCrO_4(s)$
Initial	5.0×10^{-6}		5.0×10^{-4}		0
Change	-5.0×10^{-6}		-5.0×10^{-6}		$+5.0 \times 10^{-6}$
Final	0		$5.0 \times 10^{-4} - 5.0 \times 10^{-6}$ $= 4.\underline{9}5 \times 10^{-4}$		5.0×10^{-6}

If we let x = mol/L that dissolve, the table is

Concentration (M)	$PbCrO_4(s)$ $\rightleftharpoons$	$Pb^{2+}(aq)$	+	$CrO_4^{2-}(aq)$
Starting		0		$\dfrac{4.\underline{9}5 \times 10^{-4} \text{ mol}}{(0.50 + 0.50) \text{ L}}$
Change		$+x$		$+x$
Equilibrium		x		$0.0004\underline{9}5 + x$

$$K_{sp} = [Pb^{2+}][CrO_4^{2-}] = x(0.0004\underline{9}5 + x) = 1.8 \times 10^{-14}$$

Because K_{sp} is so very small, we will ignore the change in $[CrO_4^{2-}]$, giving

$$x(0.0004\underline{9}5) = 1.8 \times 10^{-14}$$

$$x = 3.6 \times 10^{-11} \ M$$

Note that x is very small and will not show up as a change in $[CrO_4^{-2}]$. Thus,

$$[Pb^{2+}] \text{ at equilibrium } = x = 3.6 \times 10^{-11} \ M$$

$$[Pb^{2+}]_{initial} = \frac{5.0 \times 10^{-6} \text{ mol}}{0.50 + 0.50 \text{ L}} = 5.0 \times 10^{-6} \ M$$

The % of Pb^{2+} remaining equals

$$\frac{[Pb^{2+}]_{equilibrium}}{[Pb^{2+}]_{initial}} \times 100 = \frac{3.6 \times 10^{-11} \, \cancel{M}}{5.0 \times 10^{-6} \, \cancel{M}} \times 100 = 7.2 \times 10^{-4}\%$$

17.4 EFFECT OF pH ON SOLUBILITY

Operational Skills

6. Determining the qualitative effect of pH on solubility. Decide whether the solubility of a salt will be greatly increased by decreasing the pH   (Example 17.9).

7. Separating metal ions by sulfide precipitation. Given the metal-ion concentrations in solution, calculate the range of pH required to separate a mixture of two metal ions by precipitating one as the metal sulfide   (Example 17.10). K_{sp} values are required.

Exercise 17.9   Which salt would have its solubility more affected by changes in pH: silver chloride or silver cyanide?

Known: The weaker the conjugate acid, the greater effect pH has.

Solution: The solubility equilibrium of AgCl is

$$AgCl(s) \rightleftharpoons Ag^+(aq) + Cl^-(aq)$$

Since the conjugate acid of Cl^- is HCl, a strong acid, no ion would be taken from solution on the addition of H^+. The solubility equilibrium of AgCN is

$$AgCN(s) \rightleftharpoons Ag^+(aq) + CN^-(aq)$$

The conjugate acid of the cyanide ion is hydrocyanic acid. The addition of hydrogen ion would remove cyanide ion from solution, which would shift the equilibrium to the right for increased solubility. Thus, the solubility of silver cyanide would be more affected by changes in pH.

Exercise 17.10   Find the range of pH that will allow only one of the metal ions in a solution that is 0.050 M Cu^{2+}, 0.050 M Fe^{2+}, and saturated with H_2S to precipitate as a sulfide.

Wanted: pH range.

Given: solution 0.050 M in Cu^{2+} and 0.050 M in Fe^{2+}; 0.10 M H_2S (Example 17.10).

Known: By adjusting the concentration of hydrogen ion, H^+, we can control the concentration of S^{2-} in solution so only one sulfide will precipitate: $K_{sp}(CuS) = 6 \times 10^{-36}$; $K_{sp}(FeS) = 6 \times 10^{-18}$; the equilibrium expression for the double ionization of H_2S is $[H^+]^2[S^{2-}]/[H_2S]$ and the K value is 1.1×10^{-20}; pH = $-\log [H^+]$.

Solution: Determine the minimum S^{2-} ion concentration that must be present before each metal sulfide can precipitate. For Cu^{2+} the calculation is

$$[Cu^{2+}][S^{2-}] = K_{sp}$$

$$[S^{2-}] = \frac{K_{sp}}{[Cu^{2+}]} = \frac{6 \times 10^{-36}}{0.050} = 1.2 \times 10^{-34}\ M$$

For Fe^{2+} the calculation is

$$[Fe^{2+}][S^{2-}] = K_{sp}$$

$$[S^{2-}] = \frac{K_{sp}}{[Fe^{2+}]} = \frac{6 \times 10^{-18}}{0.050} = 1.2 \times 10^{-16}\ M$$

Now we substitute these concentrations into the equilibrium expression for the double ionization of H_2S to obtain the corresponding H^+ ion concentration. For Cu^{2+} the calculation is

$$\frac{[H^+]^2[S^{2-}]}{[H_2S]} = K$$

$$[H^+]^2 = \frac{K[H_2S]}{[S^{2-}]} = \frac{(1.1 \times 10^{-20})(0.10)}{1.2 \times 10^{-34}}$$

$$[H^+] = 3.0 \times 10^6\ M$$

At a H^+ ion concentration less than this, the S^{2-} ion concentration will be greater than 1.2×10^{-34} and CuS will precipitate. As it is impossible to obtain $[H^+]$ of this large value, CuS will precipitate in any solution.

For Fe^{2+} the calculation is

$$[H^+]^2 = \frac{K[H_2S]}{[S^{2-}]} = \frac{(1.1 \times 10^{-20})(0.10)}{1.2 \times 10^{-16}}$$

$$[H^+] = 3.0 \times 10^{-3}\ M$$

$pH = -\log [H^+] = -\log [3.0 \times 10^{-3}] = 2.5$. So, at any pH greater than 2.5, the FeS will precipitate. As long as the pH is 2.5 or less, only the CuS will precipitate.

17.5   COMPLEX-ION FORMATION

Operational Skill

8. Calculating the concentration of a metal ion in equilibrium with a complex ion. Calculate the concentration of aqueous metal ion in equilibrium with the complex ion, given the original metal-ion and ligand concentrations   (Example 17.11).   The formation constant K_f of the complex ion is required.

The problems in this section are stoichiometric, then equilibrium calculations. We use the table of starting, change, and final concentrations written under the reaction, which first is the formation of a complex ion, and then its dissociation.

Exercise 17.11   What is the concentration of $Cu^{2+}(aq)$ in a solution that was originally 0.015 M $Cu(NO_3)_2$ and 0.100 M NH_3? The Cu^{2+} ion forms the complex ion $Cu(NH_3)_4^{2+}$. Its formation constant is given in Table 17.2.

Wanted: $[Cu^{2+}]$.

Given: solution that is 0.015 M in $Cu(NO_3)_2$ and 0.100 M in NH_3; the complex ion is $Cu(NH_3)_4^{2+}$; text Table 17.2.

Known: $K_f = 4.8 \times 10^{12}$ from text Table 17.2; assume complete formation, then slight dissociation.

Solution: First calculate solution concentrations after the formation. Because NH_3 is in excess, the Cu^{2+} is the limiting reagent, and all of it will be used.

The formation reaction and concentrations are

Concentration (M)	$Cu^{2+}(aq)$	$+$	$4NH_3(aq)$	$\rightleftharpoons$	$Cu(NH_3)_4^{2+}(aq)$
Starting	0.015		0.100		0
Change	−0.015		−0.060		+0.015
Final	0.00		0.040		0.015

Letting x = mol/L of complex ion that dissociate, the table is

Concentration (M)	$Cu(NH_3)_4^{2+}(aq)$	$\rightleftharpoons$	$Cu^{2+}(aq)$	$+$	$4NH_3(aq)$
Starting	0.015		0		0.040
Change	$-x$		$+x$		$+4x$
Equilibrium	$0.015 - x$		$+x$		$0.040 + 4x$

$$K_d = \frac{1}{K_f} = \frac{[Cu^{2+}][NH_3]^4}{[Cu(NH_3)_4^{2+}]} = \frac{x(0.040 + 4x)^4}{(0.015 - x)} = \frac{1}{4.8 \times 10^{12}} = 2.\underline{0}8 \times 10^{-13}$$

Since K_d is so small, x will be very small, and the changes in starting concentrations of $Cu(NH_3)_4^{2+}$ and NH_3 can be ignored, giving

$$\frac{x(0.040)^4}{(0.015)} = 2.\underline{0}8 \times 10^{-13}$$

Thus,

$$x = [Cu^{2+}] = 1.2 \times 10^{-9} \, M$$

17.6   COMPLEX IONS AND SOLUBILITY

Operational Skills

9.   Predicting whether a precipitate will form in the presence of the complex ion. Predict whether an ionic compound will precipitate from a solution of known concentrations of cation, anion, and ligand that complexes with the cation   (Example 17.12). K_f and K_{sp} are required.

10.   Calculating the solubility of a slightly soluble ionic compound in a solution of the complex ion.   Calculate the molar solubility of a slightly soluble ionic compound in a solution of known concentration of a ligand that complexes with the cation   (Example 17.13). K_{sp} and K_f are required.

Exercise 17.12   Will silver iodide precipitate from a solution that is 0.0045 M $AgNO_3$, 0.15 M NaI, and 0.20 M KCN?

Wanted: Will AgI precipitate?

Known: Precipitate will form if the ion product, Q, is greater than K_{sp}. K_{sp} (AgI) = 8.3×10^{-17} (text Table 17.1); Ag^+ forms a complex ion with CN^-; K_f of $Ag(CN)_2^-$ is 5.6×10^{18} (text Table 17.2).

Solution: First determine the solution concentration after complex formation. Since we need twice as much CN^- as Ag^+, the Ag^+ is the limiting reagent.

The complex formation reaction and concentrations are

Concentration (M)	$Ag^+(aq)$	+	$2CN^-(aq) \longrightarrow$	$Ag(CN)_2^-(aq)$
Starting	0.0045		0.20	0
Change	−0.0045		−0.0090	+ 0.0045
Final	0		0.191	0.0045

Now we determine $[Ag^+]$ in solution after slight dissociation of the complex. Letting x = mol/L of complex that dissociate, the table is

Concentration (M)	$Ag(CN)_2^-(aq)$ $\rightleftharpoons$	$Ag^+(aq)$ +	$2CN^-(aq)$
Starting	0.0045	0	0.191
Change	$-x$	$+x$	$+2x$
Equilibrium	$0.0045 - x$	x	$0.191 + 2x$

$$K_d = \frac{1}{K_f} = \frac{[Ag^+][CN^-]^2}{[Ag(CN)_2^-]} = \frac{x(0.191 + 2x)^2}{(0.0045 - x)} = \frac{1}{5.6 \times 10^{18}} = 1.79 \times 10^{-19}$$

Because K_d is so small, we can ignore the changes in $[Ag(CN)_2^-]$ and $[CN^-]$, giving

$$\frac{x(0.191)^2}{(0.0045)} = 1.79 \times 10^{-19}$$

$$[Ag^+] = x = 2.21 \times 10^{-20} \, M$$

Now we solve for the ion product of AgI:

$$\text{Ion product} = [Ag^+][I^-]$$
$$= (2.21 \times 10^{-20})(0.15)$$
$$= 3.32 \times 10^{-21}$$

Since this value is less than K_{sp}, a precipitate of AgI will not form.

Exercise 17.13   What is the molar solubility of AgBr in 1.0 M $Na_2S_2O_3$ (sodium thiosulfate)? Silver ion forms the complex ion $Ag(S_2O_3)_2^{3-}$. See text Tables 17.1 and 17.2 for data.

Known: We solve this problem by combining the equations for the solution of AgBr and for complex formation. K_c for this combined equation is $K_{sp}K_f$; $K_{sp}(\text{AgBr}) = 5.0 \times 10^{-13}$; K_f for $Ag(S_2O_3)_2^{3-} = 2.9 \times 10^{13}$.

Solution: Letting x = mol/L AgBr that dissolve, the overall equation and table are written as follows:

$$AgBr(s) \rightleftharpoons Ag^+(aq) + Br^-(aq)$$

$$Ag^+(aq) + 2S_2O_3^{2-}(aq) \rightleftharpoons Ag(S_2O_3)_2^{3-}(aq)$$

Concentration (M)	AgBr(s) +	$2S_2O_3^{2-}(aq)$	$\rightleftharpoons$ $Ag(S_2O_3)_2^{3-}(aq)$ +	$Br^-(aq)$
Starting		1.0	0	0
Change		$-2x$	$+x$	$+x$
Equilibrium		$1.0 - 2x$	$+x$	$+x$

$$K_c = \frac{[Ag(S_2O_3)_2^{3-}][Br^-]}{[S_2O_3^{2-}]^2} = K_{sp}K_f$$

Putting in values gives

$$\frac{(x)(x)}{(1.0 - 2x)^2} = (5.0 \times 10^{-13})(2.9 \times 10^{13}) = 14.5$$

Taking the square root of both sides gives

$$\frac{x}{1.0 - 2x} = 3.81$$

$$x = (1.0 - 2x)3.81 = 3.81 - 7.62x$$

$$8.62\,x = 3.81$$

$$x = 0.44 \text{ mol/L of AgBr that dissolves}$$

17.7   QUALITATIVE ANALYSIS OF METAL IONS

A Metal That Matters: LEAD (a Main-Group Metal)

Questions for Study

1.   What metals other than lead were known to the ancients? Why was lead discovered so early?

2.   What are some of the principal uses of lead?

3. Give the chemical equations involved in the preparation of lead from its ore galena.

4. List the formulas and colors of three oxides of lead.

5. Which oxide is used commercially to prepare other lead compounds? How could you prepare lead nitrate from it?

6. How could you prepare lead(II) chromate?

Answers to Questions for Study

1. Six other metals were known to the ancients: gold, silver, mercury, copper, tin, and iron. Lead was probably discovered very early because its ore (galena, PbS) is very easily reduced to lead by moderate heating of galena in open charcoal fires.

2. Lead is used to make storage-battery plates, paint pigments, and ammunition. Although a major use was in the production of tetraethyllead for gasoline antiknock compounds, this is declining due to the metal's toxicity.

3. The following equations describe the reactions involved in the preparation of lead from PbS (galena):

$$2PbS(s) + 3O_2(g) \longrightarrow 2PbO(s) + 2SO_2(g)$$

$$PbO(s) + C(s) \longrightarrow Pb(l) + CO(g)$$

$$PbO(s) + CO(g) \longrightarrow Pb(l) + CO_2(g)$$

4. PbO is yellow; Pb_3O_4 is red-orange; PbO_2 is dark brown.

5. Most lead compounds are prepared from PbO. Lead nitrate could be prepared from PbO by reacting it with nitric acid.

6. Lead(II) chromate could be prepared from lead nitrate by the metathesis reaction:

$$Pb(NO_3)_2(aq) + Na_2CrO_4(aq) \longrightarrow 2NaNO_3(aq) + PbCrO_4(s)$$

The lead chromate is very insoluble and will precipitate from solution.

ADDITIONAL PROBLEMS

1. Write an ionic equation and solubility product expression for each of these slightly soluble salts: (a) AgCl, (b) BaF_2, (c) Ag_2CrO_4.

2.   A saturated solution of barium fluoride contains 6.3×10^{-3} mol/L of the salt at 25°C. What is its solubility product?

3.   Calculate the solubility of PbI_2 in (a) water, and (b) a solution that is $0.013\ M\ MgI_2$. (c) By what factor are the solubilities different? K_{sp} for PbI_2 is 6.5×10^{-9}.

4.   What concentration of Ag^+ is needed to initiate precipitation of Ag_2CrO_4 in 255 mL of a solution that is $2.2 \times 10^{-6}\ M\ CrO_4^{2-}$? K_{sp} for Ag_2CrO_4 is 1.1×10^{-12}.

5.   (a) Will a precipitate form if 3.8 mg $AgNO_3$ and 2.5 mg NaCl are added to 275 mL H_2O? Explain your answer. If so, what is (b) the limiting ion, and (c) the percentage of that ion remaining in solution after precipitation? (d) How could you lower the percent remaining?

6.   Which of the following salts would you expect to dissolve readily in acid solution: lead iodide, PbI_2, or lead sulfide, PbS? Explain.

7.   A solution is $0.050\ M\ Zn^{2+}$ and $0.050\ M\ Ni^{2+}$. Calculate the range of pH in which only one of the metal sulfides will precipitate when H_2S is bubbled through the solution.

8.   (a)   Give the concentration of $Ag^+(aq)$ in a solution that was originally $0.025\ M$ $AgNO_3$ and $0.095\ M\ Na_2S_2O_3$.

      (b)   By what factor is its concentration changed over that in the same solution without the thiosulfate ion? The formation constant of the $Ag(S_2O_3)_2^{3-}$ complex ion is 2.9×10^{13}.

9.   A 225-mL portion of an aqueous solution contains 1.00×10^{-3} mol Cl^- and 2.50×10^{-4} mol CrO_4^{2-}. When this solution is titrated with $AgNO_3$, will AgCl or Ag_2CrO_4 precipitate first? K_{sp} for AgCl is 1.8×10^{-10}; K_{sp} for Ag_2CrO_4 is 1.1×10^{-12}.

10. Lead nitrate, $Pb(NO_3)_2$, is slowly added to a solution that is $9.4 \times 10^{-3}\ M\ Cl^-$ and $7.2 \times 10^{-2}\ M\ SO_4^{2-}$. Will $PbCl_2$ or $PbSO_4$ precipitate first? K_{sp} for $PbCl_2$ is 1.6×10^{-5}; K_{sp} for $PbSO_4$ is 1.7×10^{-8}.

11. (a) Calculate the molar solubility of $Zn(OH)_2$ in a solution that is 0.12 *M* NaOH. K_{sp} for $Zn(OH)_2$ is 2.1×10^{-16}; K_f for $Zn(OH)_4{}^{2-}$ is 2.8×10^{15}.

     (b) By what factor is its solubility changed in this solution of hydroxide ion from its solubility in water?

ANSWERS TO ADDITIONAL PROBLEMS

If you missed an answer, study the text section and operational skill given in parentheses after the answer.

1. (a) $AgCl(s) \rightleftharpoons Ag^+(aq) + Cl^-(aq)$; $K_{sp} = [Ag^+][Cl^-]$

    (b) $BaF_2(s) \rightleftharpoons Ba^{2+}(aq) + 2F^-(aq)$; $K_{sp} = [Ba^{2+}][F^-]^2$

    (c) $Ag_2CrO_4(s) \rightleftharpoons 2Ag^+(aq) + CrO_4{}^{2-}(aq)$; $K_{sp} = [Ag^+]^2[CrO_4{}^{2-}]$
    (17.1, Op. Sk. 1)

2. $K_{sp} = 1.0 \times 10^{-6}$ (17.1, Op. Sk. 2)

3. The solubilities are (a) 1.2×10^{-3} *M*, and (b) 9.6×10^{-6} *M*. (c) PbI_2 is ≈ 120 times more soluble in water than in the MgI_2 solution. (17.1, 17.2, Op. Sk. 2, 3)

4. $[Ag^+] = 7.1 \times 10^{-4}$ *M* (17.3, Op. Sk. 4)

5. (a) Yes; the ion product is 1.3×10^{-8}, $> K_{sp}$.

    (b) Ag^+

    (c) 3.0%

    (d) Increase the concentration of Cl^- ion. (17.3, Op. Sk. 4, 5)

6. Lead sulfide; H^+ ions in acid solution will react with the sulfide ion to form the weak acid hydrosulfuric acid, shifting the solubility equilibrium to the right. Iodide is the anion of a strong acid and will not be removed from solution in the presence of H^+ ion. (17.4, Op. Sk. 6)

7.   At pH 0.65 or higher, ZnS precipitates; at pH 1.9 or higher, NiS precipitates. At pH values greater than 0.65 but less than 1.9, ZnS precipitates but NiS does not.

(17.4, Op. Sk. 7)

8.   (a) $[Ag^+] \simeq 4.3 \times 10^{-13}\ M$

      (b) $[Ag^+]$ is 5.8×10^{10} (58 billion) times smaller.   (17.5, Op. Sk. 8)

9.   The $[Ag^+]$ required for precipitation of AgCl is $4.1 \times 10^{-8}\ M$; that required for precipitation of Ag_2CrO_4 is $3.1 \times 10^{-5}\ M$. Thus, AgCl will precipitate initially.

(17.6, Op. Sk. 9)

10.   Since $PbSO_4$ requires $[Pb^{2+}] = 2.4 \times 10^{-7}\ M$, whereas $PbCl_2$ requires $[Pb^{2+}] \simeq 1.8 \times 10^{-1}\ M$, $PbSO_4$ will precipitate initially.   (17.6, Op. Sk. 9).

11.   (a) $6.7 \times 10^{-3}\ M$

      (b)   Solubility is increased 1,800 times.   (17.6, 17.1, Op. Sk. 10, 2)

CHAPTER POST-TEST

1.   Write the solubility product expression for each of the following compounds:

      (a) $Cu(OH)_2$ (b) BaF_2 (c) $Mg_3(AsO_4)_2$

2. $Ca_3(PO_4)_2$ has a K_{sp} value of 1×10^{-26}. Calculate the solubility in grams per liter.

3.   Calculate the molar solubility of $Ni(OH)_2$ in a solution with a pH of 11.00. Compare this with the molar solubility in pure water with a pH of 7.00.   (Recall that pH = $-\log [H^+]$ and that pH + pOH = 14.00.) K_{sp} for $Ni(OH)_2 = 2.0 \times 10^{-15}$.

4.   Suppose 0.150 L of a $6.5 \times 10^{-10}\ M$ AgNO$_3$ solution is added to 1.6 L of a solution that is $2.4 \times 10^{-4}\ M$ KI and $2.9 \times 10^{-4}\ M$ K$_2$CrO$_4$. K_{sp} (AgI) = 8.3×10^{-17}; K_{sp} (Ag$_2$CrO$_4$) = 1.1×10^{-12}. Which of the following is a correct statement of what will occur?

      (a)   Not enough AgNO$_3$ has been added to cause precipitation of AgI and/or Ag$_2$CrO$_4$.

(b)   AgI will precipitate.

(c)   Both AgI and Ag_2CrO_4 will precipitate.

(d) Ag_2CrO_4 will precipitate.

(e)   More $AgNO_3$ must be added to precipitate AgI.

5.   A solution is 0.010 M NaCl and 1.0×10^{-3} M NaBr. $AgNO_3$ is then added. Does AgCl or AgBr precipitate first? What percentage of the precipitated anion remains in solution when the second anion begins to precipitate? Note that K_{sp} (AgCl) = 1.8×10^{-10}; K_{sp} (AgBr) = 5.0×10^{-13}.

6.   Arrange the following salts in order of increasing change in solubility as the pH of each solution is decreased:

silver benzoate, $AgC_7H_5O_2$; $K_a(HC_7H_5O_2)$ = 6.3×10^{-5}

silver cyanide, AgCN; $K_a(HCN)$ = 4.9×10^{-10}

silver propionate, $AgC_3H_5O_2$; $K_a(HC_3H_5O_2)$ = 1.3×10^{-5}

7.   A solution containing 0.75 M Cu^{2+} and 0.50 M Cd^{2+} is saturated with H_2S. Can these two cations be separated if the pH of the solution is adjusted to 4.5? Note that $[H_2S]$ = 0.10 M in a saturated solution; K_{sp}(CuS) = 6×10^{-36}; K_{sp}(CdS) = 8×10^{-27}; $K_a(H_2S)$ = 1.1×10^{-20}.

8.   A solution is made with 0.050 L of 1.1 M KCN and 0.050 L of 0.15 M $Fe(NO_3)_3$. What is the concentration of Fe^{3+} at equilibrium? K_f for $Fe(CN)_6^{3-}$ = 9.1×10^{41}. Comment on the answer.

9.   Suppose 0.050 L of 2.2×10^{-3} M NaI is added to 0.100 L of a solution that contains 4.7×10^{-3} M $AgNO_3$ and 0.015 M $Na_2S_2O_3$. Will AgI precipitate? K_{sp}(AgI) = 8.3×10^{-17}; K_f for $Ag(S_2O_3)_2^{3-}$ = 2.9×10^{13}. Assume the volumes are additive.

10.   What is the molar solubility of AgI in 1.0 M $Na_2S_2O_3$? K_{sp}(AgI) = 8.3×10^{-17}; K_f for $Ag(S_2O_3)_2^{3-}$ = 2.9×10^{13}.

ANSWERS TO CHAPTER POST-TEST

If you missed an answer, study the text section and operational skill given in parentheses after the answer.

1. (a) $K_{sp} = [Cu^{2+}][OH^-]^2$

    (b) $K_{sp} = [Ba^{2+}][F^-]^2$

    (c) $K_{sp} = [Mg^{2+}]^3[AsO_4^{3-}]^2$ (17.1, Op. Sk. 1)

2. 8×10^{-4} g/L   (17.1, Op. Sk. 2)

3. 2.0×10^{-9} M; in pure water, solubility is 7.9×10^{-6} M, which is about 4000 times greater   (17.2, Op. Sk. 3)

4. b   (17.3, Op. Sk. 4)

5. AgBr precipitates first; 2.8% of the Br^- remains in solution when Cl^- begins to precipitate.   (17.3, Op. Sk. 5)

6. $AgC_7H_5O_2 < AgC_3H_5O_2 < AgCN$ (17.4, Op. Sk. 6)

7. no   (17.4, Op. Sk. 7)

8. 8.2×10^{-38} M. This meaningless value shows the great stability of the $Fe(CN)_6^{3-}$ ion.   (17.5, Op. Sk. 8)

9. yes   (17.6, Op. Sk. 9)

10. 4.5×10^{-2} M (17.6, Op. Sk. 10)

CHAPTER 18   THERMODYNAMICS AND EQUILIBRIUM

CHAPTER TERMS AND DEFINITIONS

Numbers in parentheses after definitions give the text section in which the terms are
explained. Starred terms are italicized in the text. Where a term does not fall directly under
a text section heading, additional information is given for you to locate it.

thermodynamics*   study of the relationship between heat and other forms of energy
      involved in a chemical or physical process   (chapter introduction)

internal energy (U)   sum of the kinetic and potential energies of the particles making up a
      system   (18.1)

state function   property of a system, such as temperature and pressure, that depends only
      on its present condition and that is a defining factor of that condition   (18.1)

work (w)   energy exchange that results when a force F moves an object through a distance
      d; $w = F \times d$ (18.1)

first law of thermodynamics   the change in internal energy of a system, ΔU, equals the
      heat absorbed or lost by the system, $\pm q$, plus the work done on or by the system,
      $\pm w$; $\Delta U = q + w$ (18.1)

enthalpy (H)   thermodynamic quantity defined by the equation $H = U + PV$ (18.1)

spontaneous process   physical or chemical change that occurs by itself   (18.2,
      introductory section)

nonspontaneous*   describes a physical or chemical change that occurs only with the aid
      of an outside force or agent   (18.2, introductory section)

entropy (S)   thermodynamic quantity that is a measure of the randomness or disorder in a
      system   (18.2)

second law of thermodynamics   the total entropy of a system and its surroundings
      always increases for a spontaneous process; $\Delta S > q/T$ (18.2)

reversible*   describes a process that can go in one direction or the other   (18.2, marginal
      note)

third law of thermodynamics   a substance that is perfectly crystalline at 0 K has an
      entropy of zero   (18.3)

standard (absolute) entropy ($S°$)   entropy value for the standard state of a species
      (18.3)

free energy (G)   thermodynamic quantity defined by the equation $G = H - TS$ (18.4)

standard states*   conditions under which to report the properties of matter: 1 atm pressure for pure substances, 1 M concentration for ions in solution   (18.4)

standard free energy of formation (ΔG_f°)   free-energy change when 1 mol of substance is formed from its elements in their stablest states at 1 atm pressure and at a specified temperature (usually 25°C)   (18.4)

thermodynamic equilibrium constant (K)   equilibrium constant in which the concentrations of gases are expressed in partial pressures in atmospheres, whereas the concentrations of solutes in liquid solutions are expressed in molarities   (18.6)

slag*   glassy material formed in the blast furnace from CaO from limestone and iron-ore impurities, and which floats on top of the molten iron   (An Element That Matters: Iron [a Transition Metal])

basic oxygen process*   method of making steel that involves blowing oxygen into the molten iron to oxidize the impurities and decrease the amount of carbon present (An Element That Matters: Iron [a Transition Metal])

CHAPTER DIAGNOSTIC TEST

1.   Determine whether each of the following statements is true or false. If the statement is false, change it so it is true.

(a)   A spontaneous process is a physical or chemical change that occurs without warning. True/False: _____

_____ .

(b)   Entropy is a measure of disorder in a physical or chemical system. True/False: _____

_____ .

(c)   Entropy increases going from solid to liquid to gas.   True/False: _____

_____ .

(d)   A reaction is spontaneous as written if $\Delta G°$ is a large positive number. True/False: _____ .

2.   Calculate the change in entropy, $\Delta S°$, for the following precipitation reaction:

$$Ag^+(aq) + Cl^-(aq) \longrightarrow AgCl(s)$$

Use text Table 18.1 for standard entropy values.

3. Predict by inspection the sign of $\Delta S°$ for each of the following reactions, and explain your answers.

   (a) $H_2(g) + C_2H_6(g) \rightleftharpoons 2CH_4(g)$

   (b) $2SO_2(g) + O_2(g) \rightleftharpoons 2SO_3(g)$

   (c) $2NO_2(g) \rightleftharpoons 2NO(g) + O_2(g)$

   (d) $Pb(NO_3)_2(aq) + Na_2SO_4(aq) \longrightarrow PbSO_4(s) + 2NaNO_3(aq)$

4. The heat of vaporization of liquid bromine, $Br_2(l)$, at 25°C is 30.7 kJ/mol:

$$Br_2(l) \rightleftharpoons Br_2(g)$$

   If liquid bromine at 25°C has an entropy of 152.2 J/(K • mol), what is the entropy of one mole of the vapor in equilibrium with the liquid at that temperature?

5. Calculate $\Delta G°$ for the following precipitation reaction at 25°C:

$$2Ag^+(aq) + S^{2-}(aq) \longrightarrow Ag_2S(s)$$

   Use values of $\Delta H_f°$ and $S°$ from text Appendix C.

6. Calculate the free-energy change, $\Delta G°$, for the following reaction, using values of $\Delta G_f°$ from text Appendix C:

$$2Li(s) + 2H_2O(l) \longrightarrow 2Li^+(aq) + 2OH^-(aq) + H_2(g)$$

7. Give the expression for K for the following reactions:

   (a) $2NaHCO_3(s) \longrightarrow Na_2CO_3(s) + H_2O(g) + CO_2(g)$

   (b) $NH_4OH(aq) \rightleftharpoons NH_3(g) + H_2O(l)$

   (c) $2H_2O_2(l) \rightleftharpoons 2H_2O(l) + O_2(g)$

8. Determine whether each of the reactions in Problem 3 is spontaneous or nonspontaneous as written. Use values in text Appendix C and give reasons for your answers.

9. Calculate K_a for the ionization of formic acid, HCOOH, in aqueous solution. Use values of $\Delta G_f°$ in text Appendix C.

10.   Calculate the value of the thermodynamic equilibrium constant at 648 K for the following reaction:

$$N_2(g) + 3H_2(g) \rightleftharpoons 2NH_3(g)$$

$\Delta G°$ for this reaction at the given temperature is 43.2 kJ.

ANSWERS TO CHAPTER DIAGNOSTIC TEST

If you missed an answer, study the text section and operational skill given in parentheses after the answer.

1.   (a)   False. A spontaneous process is a physical or chemical change that can occur by itself, requiring no continuing outside agency to make it happen. (18.2, introductory section)

      (b)   True. (18.2)   (c)   True. (18.2)

      (d)   False. A reaction is spontaneous as written if $\Delta G°$ is a large negative number.
                                                                                    (18.4)

2. $\Delta S° = -32.9$ J/K   (18.3, Op. Sk. 3)

3.   (a) $\Delta S° \approx 0$; there is no change in the number of gas molecules.
      (b) $\Delta S° < 0$; there are three molecules of gas on the left but only two on the right.
      (c) $\Delta S° > 0$; there are two molecules of gas on the left and three on the right.
      (d) $\Delta S° < 0$; a solid is formed on the right.   (18.3, Op. Sk. 2)

4. $\Delta S° = 255$ J/(K • mol)   (18.2, Op. Sk. 1)

5. $\Delta G° = -278.3$ kJ   (18.4, Op. Sk. 4)

6. $\Delta G° = -427.8$ kJ   (18.4, Op. Sk. 5)

7.   (a) $K = P_{CO_2} P_{H_2O}$ (b) $K = \dfrac{P_{NH_3}}{[NH_4OH]}$

      (c)   Water is not a solvent and must be included in the expression:

$$K = \frac{[H_2O]^2 P_{O_2}}{[H_2O_2]^2}$$ (18.6, Op. Sk. 7)

8. (a) spontaneous; $\Delta G° < 0$ (c) nonspontaneous; $\Delta G° > 0$
   (b) spontaneous; $\Delta G° < 0$ (d) spontaneous; $\Delta G° < 0$ (18.4, Op. Sk. 6)

9. $K_a = K = 2.1 \times 10^{-4}$ (18.6, Op. Sk. 8) 10. $K = 3.3 \times 10^{-4}$ (18.6, Op. Sk. 8)

SUMMARY OF CHAPTER TOPICS

18.1 FIRST LAW OF THERMODYNAMICS; ENTHALPY

Exercise 18.1 A gas is enclosed in a system similar to that shown in Figure 18.2. More weights are added to the piston, giving a total mass of 2.20 kg. As a result, the gas is compressed and the weights are lowered 0.250 m. At the same time, 1.50 J of heat evolves from (leaves) the system. What is the change in internal energy of the system, ΔU? The force of gravity on a mass m is mg, where g is the constant acceleration of gravity ($g = 9.80$ m/s^2).

Wanted: ΔU, in joules.

Given: Mass of weights = 2.20 kg, piston dropped 0.250 m, heat evolved $(q) = -1.50$ J; $F = mg$, $g = 9.80$ m/s^2.

Known: $q = -1.50$ J, as heat is evolved; $\Delta U = q + w$; $w = F \times d$, the sign of w is + because the work is done on the system; 1 J = 1 kg • m^2/s^2.

Solution: Substitute algebraically, then with numbers, into

$$\Delta U = q + w = q + (F \times d) = q + (mg \times d).$$

$$= -1.50 \text{ J} + 2.20 \,\cancel{\text{kg}} \times 9.80 \,\cancel{\text{m/s}^2} \times 0.250 \,\cancel{\text{m}} \times \frac{\text{J}}{\cancel{\text{kg} \bullet \text{m}^2/\text{s}^2}}$$

$$\Delta U = 3.89 \text{ J}$$

Exercise 18.2 Consider the combustion (burning) of methane, CH_4, in oxygen.

$$CH_4(g) + 2O_2(g) \longrightarrow CO_2(g) + 2H_2O(l)$$

The heat of reaction at 25°C and 1.00 atm is –890.2 kJ. What is the change in volume when 1.00 mol CH_4 reacts with 2.00 mol O_2? (You can ignore the volume of liquid water, which is insignificant compared with volumes of gases.) What is w for this change? Calculate ΔU for the change indicated by the chemical equation.

Wanted: ΔV, in L; w, in kJ; ΔU, in kJ.

Given: equation; $t = 25°C$; $P = 1$ atm; $q = -890.2$ kJ; 1 mol CH_4; 2.00 mol O_2.

Known: $P\Delta V = \Delta nRT$; $w = P\Delta V$ and is + because work is done on the system (the volume of the system decreases); $\Delta U = q + w$, 1 L = 10^{-3} m^3; 1 Pa • m^3 = 1 J; 1 atm = 1.01×10^5 Pa.

Solution: Since there are 3 mol of gas on the left of the equation and 1 mol on the right, $\Delta n = 2$ mol, and

$$\Delta V = \frac{\Delta nRT}{P} = \frac{2 \text{ mol} \times 0.0821 \text{ L} \cdot \text{atm} \times 298 \text{ K}}{1.00 \text{ atm} (\text{K} \cdot \text{mol})} = 48.9 \text{ L}$$

$$w = +P\Delta V = 1.01 \times 10^5 \text{ Pa} \times 48.9 \text{ L} \times \frac{10^{-3} \text{ m}^3}{\text{L}} = \frac{1 \text{ J}}{1 \text{ Pa} \cdot \text{m}^3}$$

$$= 4.94 \times 10^3 \text{ J} = 4.94 \text{ kJ}$$

$$\Delta U = q + w = -890.2 \text{ kJ} + 4.94 \text{ kJ} = -885.3 \text{ kJ}$$

18.2 ENTROPY AND THE SECOND LAW OF THERMODYNAMICS

Operational Skill

1. Calculating the entropy change for a phase transition. Given the heat of phase transition and the temperature of the transition, calculate the entropy change, ΔS, of the system (Example 18.1).

Stop a minute and think of your room at home or in the dorm. If you are like most of us, you envision clothes hanging over a chair, books scattered, maybe a sock or two half-hidden under the bed. Such a room gives a vivid example of the second law of thermodynamics — that entropy increases in a spontaneous process. It takes work to hang up the clothes, to arrange the books neatly, and to put the socks in the hamper.

It is a good idea to memorize the expression for entropy, $S = q/T$, and to remember that the entropy change in a system during a spontaneous change is $\Delta S > q/T$ (i.e., non-reversible).

Exercise 18.3 Liquid ethanol, $C_2H_5OH(l)$, at 25°C has an entropy of 161 J/(K · mol). If the heat of vaporization, ΔH_{vap}, at 25°C is 42.3 kJ/mol, what is the entropy of the vapor in equilibrium with the liquid at 25°C?

Wanted: entropy of the vapor, in J/(K • mol).

Given: $S°$ for liquid = 161 J/(K • mol); ΔH_{vap} = 42.3 kJ/mol; t = 25°C.

Known: The entropy of the vapor = entropy of the liquid plus ΔS of vaporization; $q_p = \Delta H$; $\Delta S_{vap} = \Delta H_{vap}/T$; 25°C = 298 K.

Solution: The calculation is

$$\Delta S_{vap} = \frac{42.3 \times 10^3 \text{ J}}{298 \text{ K} \cdot \text{mol}} = 1.4\underline{2}0 \times 10^2 \text{ J/(K} \cdot \text{mol)}$$

$$S_{vapor} = (161 + 142) \text{ J/(K} \cdot \text{mol)} = 303 \text{ J/(K} \cdot \text{mol)}$$

18.3   STANDARD ENTROPIES AND THE THIRD LAW OF THERMODYNAMICS

Operational Skills

2. Predicting the sign of the entropy change of a reaction. Predict the sign of $\Delta S°$ for a reaction to which the rules listed in the text can be clearly applied   (Example 18.2).

3. Calculating $\Delta S°$ for a reaction. Given the standard entropies of reactants and products, calculate the entropy of reaction, $\Delta S°$ (Example 18.3).

It is important to note that the zero point of entropy is at 0 K and that we talk of absolute entropies, $S°$, of substances rather than of entropies of formation. Recall that substances have an enthalpy of formation, $\Delta H_f°$, and that the zero point for these values is the element in its most stable state at standard conditions.

Exercise 18.4   Predict the sign of $\Delta S°$ for each of the following reactions.

(a) $CaCO_3(s) \longrightarrow CaO(s) + CO_2(g)$

(b) $CS_2(l) \longrightarrow CS_2(g)$

(c) $2Hg(l) + O_2(g) \longrightarrow 2HgO(s)$

(d) $2Na_2O_2(s) + 2H_2O(l) \longrightarrow 4NaOH(aq) + O_2(g)$

Known: Entropy is a measure of disorder; solids are more ordered than liquids, which are more ordered than gases.

Solution: (a) $\Delta S° > 0$; a molecule of gas is formed.
   (b) $\Delta S° > 0$; a molecule of gas is formed.
   (c) $\Delta S° < 0$; the solid product is more ordered than the reactants.
   (d) $\Delta S° > 0$; a molecule of gas is formed.

474

Exercise 18.5   Calculate the change of entropy, $\Delta S°$, for the reaction given in Example 18.2(a). The standard entropy of glucose, $C_6H_{12}O_6(s)$, is 212 J/(K • mol). See Table 18.1 for other values.

Wanted: $\Delta S°$ in J/K

Given: $S°$ for $C_6H_{12}O_6$ = 212 J/(K • mol).

Known: reaction from Example 18.2(a); text Table 18.1 for $S°$ values.

Solution: The calculation is

$$C_6H_{12}O_6(s) \longrightarrow 2C_2H_5OH(l) + 2CO_2(g)$$

$S°$ 212   2(161)   2(213.7) J/(K • mol)

$\Delta S° = [2(161) + 2(213.7) - 212]$ J/K

$\Delta S° = 537$ J/K

It is important to recall that in Section 14.2, you learned that the equilibrium-constant expression and its value depend on the coefficients used to write the reaction. It is customary to write the coefficients using the smallest ratio of whole numbers. But the reaction is still correct as long as we use any numbers in that ratio.

Thus, the values for the entropy of reaction calculated in Example 18.3 and in Exercise 18.3 both depend on the coefficients used to write the individual reactions. The entropies would be different if different coefficients were used.

This is also true of the enthalpies of reaction, $\Delta H°$, calculated in Chapter 6. Sometimes the term *mole of reaction* is used to describe this idea. This use of the word *mole* does not designate 6.02×10^{23} particles. It merely means "the reaction as written" or "the quantities specified in the reaction as written." In Example 18.3, where the reaction is written as

$$2NH_3(g) + CO_2(g) \rightleftharpoons NH_2CONH_2(aq) + H_2O(l)$$

"one mole of reaction" means "$2 \times 6.02 \times 10^{23}$ molecules of NH_3 and 6.02×10^{23} molecules of CO_2 react to form 6.02×10^{23} molecules of urea and 6.02×20^{23} molecules of water." In Exercise 18.5, "one mole of reaction" means "6.02×10^{23} molecules of glucose yield $2 \times 6.02 \times 10^{23}$ molecules of ethyl alcohol and $2 \times 6.02 \times 10^{23}$ molecules of CO_2." So the units for the entropies of reaction could be written J/(K • mol), mol meaning "mole of reaction." We will refer to this concept again in the next section and in the following chapter.

18.4 FREE ENERGY AND SPONTANEITY

Operational Skills

4. Calculating $\Delta G°$ from $\Delta H°$ and $\Delta S°$. Given enthalpies of formation and standard entropies of reactants and products, calculate the standard free-energy change $\Delta G°$ for a reaction (Example 18.4).

5. Calculate $\Delta G°$ from standard free energies of formation. Given the free energies of formation of reactants and products, calculate the standard free-energy change $\Delta G°$ for a reaction (Example 18.5).

6. Interpreting the sign of $\Delta G°$. Use the standard free-energy change to determine the spontaneity of a reaction (Example 18.6).

Note that the zero point of free energy of formation, $\Delta G_f°$, is the element in its most stable state at 1 atm pressure and usually 25°C, as with the enthalpy of formation. It is a good idea to memorize the expression for the free-energy change of a reaction: $\Delta G = \Delta H - T\Delta S$. You must know that ΔG for a spontaneous reaction is negative.

Exercise 18.6 Calculate $\Delta G°$ for the following reaction at 25°C. Use data given in Tables 6.2 and 18.1.

$$CH_4(g) + 2O_2(g) \longrightarrow CO_2(g) + 2H_2O(g)$$

Known: $\Delta G° = \Delta H° - T\Delta S°$; $\Delta H_f°$ values are in text Table 6.2; $S°$ values are in text Table 18.1.

Solution: Calculations follow.

$$CH_4(g) \; + \; 2O_2(g) \; \longrightarrow \; CO_2(g) \; + \; 2H_2O(g)$$

$\Delta H_f°$ −74.9   2(0)   −393.5   2(−241.8) kJ/mol

$S°$ 186.1   2(205.0)   213.7   2(188.7) J/(K • mol)

$\Delta H° = [-393.5 + 2(-241.8) - (-74.9)]$ kJ

$\qquad = -802.2$ kJ

$\Delta S° = \{[213.7 + 2(188.7)] - [186.1 + 2(205.0)]\}$ J/K

$\qquad = -5.0$ J/K

$\Delta G° = \Delta H° - T\Delta S°$

$$\Delta G^\circ = -802.2 \text{ kJ} - \left(298 \cancel{K} \times \frac{-5.0 \cancel{J}}{\cancel{K}} \times \frac{10^{-3} \text{ kJ}}{\cancel{J}} \right)$$

$$\Delta G^\circ = -800.7 \text{ kJ}$$

Let us again refer to the term *mole of reaction*. The free-energy change could also be thought of as per "mole of reaction" for the reaction with coefficients specified. This means that the units for the ΔH° of the above reaction could be calculated as

$$\Delta H^\circ = \frac{1 \cancel{\text{mol } CO_2}}{\text{mol (of } rxn)} \times \frac{-393.5 \text{ kJ}}{\cancel{\text{mol } CO_2}} + \frac{2 \cancel{\text{mol } H_2O(g)}}{\text{mol (of } rxn)} \times \frac{-241.8 \text{ kJ}}{\cancel{\text{mol } H_2O(g)}}$$

$$- \frac{1 \cancel{\text{mol } CH_4}}{\text{mol (of } rxn)} \times \frac{-74.9 \text{ kJ}}{\cancel{\text{mol } CH_4}}$$

$$= -802.2 \text{ kJ / mol (of } rxn)$$

We could do the same for the calculation of ΔS° for the reaction.

In another text you might see ΔH° or ΔG° written as kJ/mol, where *mol* means "mole of reaction."

Exercise 18.7 Calculate ΔG° at 25°C for the following reaction, using values of ΔG_f°.

$$CaCO_3(s) \longrightarrow CaO(s) + CO_2(g)$$

Known: ΔG_f° values given in text Table 18.2.

Solution: The calculation is

$$CaCO_3(s) \longrightarrow CaO(s) + CO_2(g)$$

ΔG_f° −1128.8   −603.5   −394.4 kJ/mol

$$\Delta G^\circ = [-603.5 + (-394.4) - (-1128.8)] \text{ kJ}$$

$$\Delta G^\circ = 130.9 \text{ kJ}$$

Exercise 18.8 Which of the following reactions are spontaneous in the direction written? See Table 18.2 for data.

(a) $C(\text{graphite}) + 2H_2(g) \longrightarrow CH_4(g)$

(b) $2H_2(g) + O_2(g) \longrightarrow 2H_2O(l)$

(c) $4HCN(g) + 5O_2(g) \longrightarrow 2H_2O(l) + 4CO_2(g) + 2N_2(g)$

(d) $Ag^+(aq) + I^-(aq) \longrightarrow AgI(s)$

Wanted: spontaneity of reaction.

Given: ΔG_f° values in text Table 18.2.

Known: If $\Delta G^\circ > 0$, reaction is nonspontaneous; if $\Delta G^\circ < 0$, reaction is spontaneous.

Solution: All the reactions are spontaneous, as the following calculations show:

(a) $\qquad\qquad C(graphite) + 2H_2(g) \longrightarrow CH_4(g)$

$\Delta G_f^\circ \qquad\qquad 0 \qquad\quad 2(0) \qquad -50.8 \text{ kJ/mol}$

$\Delta G^\circ = -50.8 \text{ kJ}$

(b) $\qquad\qquad 2H_2(g) + O_2(g) \longrightarrow 2H_2O(l)$

$\Delta G_f^\circ \qquad\quad 2(0) \qquad 2(0) \qquad 2(-237.2) \text{ kJ/mol}$

$\Delta G^\circ = [2(-237.2)] \text{ kJ}$

$\Delta G^\circ = -474.4 \text{ kJ}$

(c) $\qquad\quad 4HCN(g) + 5O_2(g) \longrightarrow 2H_2O(l) + 4CO_2(g) + 2N_2(g)$

$\Delta G_f^\circ \quad 4(125) \qquad 5(0) \qquad\quad 2(-237.2) \quad 4(-394.4) \quad 2(0) \text{ kJ/mol}$

$\Delta G^\circ = [2(-237.2) + 4(-394.4) - 4(125)] \text{ kJ}$

$\Delta G^\circ = -2552 \text{ kJ}$

(d) $\qquad\qquad Ag^+(aq) + I^-(aq) \longrightarrow AgI(s)$

$\Delta G_f^\circ \qquad\quad 77.1 \qquad -51.7 \qquad -66.3 \text{ kJ/mol}$

$\Delta G^\circ = [-66.3 - (-51.7 + 77.1)] \text{ kJ}$

$\Delta G^\circ = -91.7 \text{ kJ}$

18.5   INTERPRETATION OF FREE ENERGY

18.6   RELATING $\Delta G°$ TO THE EQUILIBRIUM CONSTANT

Operational Skills

   7.   Writing the expression for a thermodynamic equilibrium constant.   For any
balanced chemical equation, write the expression for the thermodynamic equilibrium
constant (Example 18.7).

   8.   Calculating K from the standard free-energy change.   Given the standard free-
energy change for a reaction, calculate the thermodynamic equilibrium constant
(Examples 18.8 and 18.9).

        Memorize the relationship between $\Delta G°$ and the thermodynamic equilibrium
constant: $\Delta G° = -2.303\ RT \log K$. Remember also that $K = K_c$ for reactions of solutes in
liquid solution, $K = K_p$ when the reaction involves only gases, and K includes both
pressures and concentrations in heterogeneous systems.

        You may well ask why R, the gas constant, shows up here, when many reactions
do not involve gases. The presence of R in the expression for $\Delta G°$ is due to the intimate
relationship between R and how the mole is defined. Recall from your study of the ideal
gas law (text Section 5.3) that the amount of gas in a given volume (22.41 L) at a specified
temperature (273.15 K) and pressure (1 atm) was called a mole. Also recall that R is the
constant ratio of these standard conditions, PV/T. Thus, R defines the mole. In this
expression for $\Delta G°$, the word *mole* in the units of R, J/(K • mol), means "mole of
reaction," explained previously in this chapter.

        Exercise 18.9   Give the expression for K for each of the following reactions.

        (a) $CaCO_3(s) \rightleftharpoons CaO(s) + CO_2(g)$

        (b) $PbI_2(s) \rightleftharpoons Pb^{2+}(aq) + 2I^-(aq)$

        (c) $H^+(aq) + HCO_3^-(aq) \rightleftharpoons H_2O(l) + CO_2(g)$

        Solution: The calculations are

        (a) $K = P_{CO_2} = K_p$

        (b) $K = [Pb^{2+}][I^-]^2 = K_{sp}$

        (c) $K = \dfrac{P_{CO_2}}{[H^+][HCO_3^-]}$

Exercise 18.10 Use data from Table 18.2 to obtain the equilibrium constant K_p at 25°C for the reaction

$$CaCO_3(s) \rightleftharpoons CaO(s) + CO_2(g)$$

Note that values of ΔG_f° are needed for $CaCO_3$ and CaO, even though the substances do not appear in $K_p = P_{CO_2}$.

Wanted: K_p (from ΔG°).

Given: reaction; text Table 18.2 (ΔG_f°).

Known: $\Delta G^\circ = -2.303\ RT \log K$; K for this reaction $= K_p$; $R = 8.31$ J/(K • mol), so in order to have the units work out properly, we must report ΔG° as kJ/mol; $T = 298$ K.

Solution: Calculations follow:

$$CaCO_3(s) \longrightarrow CaO(s) + CO_2(g)$$

ΔG_f° −1128.8   −603.5   −394.4 kJ/mol

$\Delta G^\circ = [-603.5 + (-394.4) - (-1128.8)]$ kJ

$\Delta G^\circ = 130.9$ kJ/mol

Rearrange the expression for ΔG° to solve for K_p:

$$\log K_p = \frac{-\Delta G^\circ}{2.303\ RT} = \frac{-130.0 \times 10^3\ \cancel{J} \times \cancel{K} \bullet \cancel{mol}}{2.303 \times \cancel{mol} \times 8.31\ \cancel{J} \times 298\ \cancel{K}} = -22.\underline{78}$$

$$K_p = 2 \times 10^{-23}$$

Exercise 18.11 Calculate the solubility product constant for $Mg(OH)_2$ at 25°C. The ΔG_f° values (in kJ/mol) are as follows: $Mg^{2+}(aq)$, −456.0; $OH^-(aq)$, −157.3; $Mg(OH)_2(s)$, −933.9.

Known: $\Delta G^\circ = -2.303\ RT \log K_{sp}$; R (see Exercise 18.10) and T given above.

Solution: Calculations follow:

$$Mg(OH)_2(s) \rightleftharpoons Mg^{2+}(aq) + 2OH^-(aq)$$

$$\Delta G_f^\circ \qquad -933.9 \qquad\qquad -456.0 \quad 2(-157.3) \text{ kJ/mol}$$

$$\Delta G^\circ = [-456.0 + 2(-157.3) - (-933.9)] \text{ kJ}$$

$$\Delta G^\circ = 163.3 \text{ kJ/mol}$$

$$\log K_{sp} = \frac{-\Delta G^\circ}{2.303\, RT} = \frac{-163.3 \times 10^3\, \cancel{J} \times \cancel{K} \cdot \cancel{mol}}{2.303 \times \cancel{mol} \times 8.31\, \cancel{J} \times 298\, \cancel{K}} = -28.\underline{63}$$

$$K = 2 \times 10^{-29}$$

18.7   CHANGE OF FREE ENERGY WITH TEMPERATURE

Operational Skill

9. Calculating ΔG° and K at various temperatures. Given ΔH° and ΔS° at 25°C, calculate ΔG° and K for a reaction at a temperature other than 25°C   (Example 18.10).

> **Exercise 18.12** The thermodynamic equilibrium constant for the vaporization of water,
>
> $$H_2O(l) \rightleftharpoons H_2O(g)$$
>
> is $K_p = P_{H_2O}$. Use thermodynamic data to calculate the vapor pressure of water at 45°C. Compare your answer with the value given in Table 5.6.
>
> *Wanted:* P_{H_2O} at 45°C.
>
> *Known:* $P_{H_2O} = K_p$; $\Delta G^\circ = -2.303\, RT \log K_p$; need values from text Tables 6.2 and 18.1; $R = 8.31$ J/(K • mol); $T = 318$ K.

Solution: Calculations follow:

$$H_2O(l) \longrightarrow H_2O(g)$$

ΔH_f°	-285.8	-241.8 kJ/mol
S°	69.9	188.7 J/(K • mol)

$\Delta H^\circ = [-241.8 - (-285.8)]$ kJ $= 44.0$ kJ

$\Delta S^\circ = (188.7 - 69.9)$ J/K $= 118.8$ J/K

$\Delta G^\circ = \Delta H^\circ - T\Delta S^\circ$

$$\Delta G^\circ = 44.0 \cancel{kJ} \times \frac{10^3 \text{ J}}{\cancel{kJ}} - 318 \cancel{K} \times \frac{118.8 \text{ J}}{\cancel{K}} = 6.2\underline{2}2 \times 10^3 \text{ J}$$

which we will express as $6.2\underline{2}2 \times 10^3$ J/mol.

$$\log K_p = \frac{-\Delta G^\circ}{2.303 \, RT} = \frac{-6.2\underline{2}2 \times 10^3 \cancel{J} \times \cancel{K} \cdot \cancel{mol}}{2.303 \times \cancel{mol} \times 8.31\cancel{4} \times 318\cancel{K}} = -1.0\underline{2}2$$

$$K_p = 9.5 \times 10^{-2} \text{ atm}$$

To compare with the value in text Table 5.6, convert to mmHg:

$$9.5 \times 10^{-2} \cancel{\text{atm}} \times \frac{760 \text{ mmHg}}{1 \cancel{\text{atm}}} = 72 \text{ mmHg}$$

The text table value is 71.9 mmHg. The calculated value is the text table value to two significant figures.

Exercise 18.13 To what temperature must magnesium carbonate be heated to decompose it to MgO and CO_2 at 1 atm? Is this higher or lower than the temperature required to decompose $CaCO_3$? Values of ΔH_f° (in kJ/mol) are as follows: MgO(s), -601.2; $MgCO_3(s)$, -1112. Values of S° (in J/K) are as follows: MgO(s), 26.9; $MgCO_3(s)$, 65.9. Data for CO_2 are given in Tables 6.2 and 18.1.

Wanted: T at which $MgCO_3$ will decompose; compare with T for $CaCO_3$.

Given: ΔH_f° values; S° values; see text Tables 6.2 and 18.1.

Known: $\Delta G^\circ = 0$ at that T; $\Delta G^\circ = \Delta H^\circ - T\Delta S^\circ$; assume ΔH° and ΔS° are constant, so we can use their values at 25°C for the calculation.

Solution: If $\Delta G° = 0$, then $\Delta H° = T\Delta S°$, and $T = \Delta H°/\Delta S°$. Calculations of $\Delta H°$ and $\Delta S°$ follow.

$$MgCO_3(s) \longrightarrow MgO(s) + CO_2(g)$$

$\Delta H_f°$	-1112	-601.2	-393.5 kJ/mol
$\Delta S°$	65.9	26.9	213.7 J/(K • mol)

$\Delta H° = [-601.2 + (-393.5) - (-1112)]$ kJ

$\Delta H° = 11\underline{7}.3$ kJ

$\Delta S° = [(26.9 + 213.7) - 65.9]$ J/K

$\Delta S° = 174.7$ J/K

To solve for temperature:

$$T = \frac{11\underline{7}.3 \times 10^3 \text{J}}{174.7 \text{J} / \text{K}}$$

$T = 6.71 \times 10^2$ K; $t = 398°$C

This temperature is considerably lower than the 848°C decomposition temperature of $CaCO_3$, given in the text immediately preceding this exercise.

An Element That Matters: IRON (a Transition Metal)

Questions for Study

1. Describe how iron ore is reduced in a blast furnace. What is the chemical equation for the overall reduction of Fe_2O_3 by CO?

2. What is the purpose of the basic oxygen process?

3. What are the products of the reaction of iron and hydrochloric acid?

4. Iron(II) ion is a mild reducing agent. Write the balanced equation for (a) the reduction of Sn^{4+} to Sn^{2+} by Fe^{2+}, and (b) the oxidation of Fe^{2+} by $Cr_2O_7^{2-}$ to give Cr^{2+} in acidic solution.

5. Calculate $\Delta G°$ for the following reaction:

$$Fe_2O_3(s) + 3H_2(g) \longrightarrow 2Fe(s) + 3H_2O(g)$$

What is the value of K at 25°C?

6. How is Prussian blue prepared? What is it used for?

Answers to Questions for Study

1. In the blast furnace, a mixture of iron ore, coke (carbon produced by heating coal), and limestone is added at the top of the furnace. A blast of heated air enters at the bottom. Near the bottom of the furnace, the coke burns to carbon dioxide. As the CO_2 rises through the heated coke, it is reduced to carbon monoxide. The carbon monoxide then reduces the iron oxides in the ore to metallic iron. The molten iron then flows to the bottom of the blast furnace, where it is drawn off. The chemical equation for the overall reduction process is

$$Fe_2O_3(s) + 3CO(g) \longrightarrow 2Fe(l) + 3CO_2(g)$$

2. In the basic oxygen process, steel is made by blowing oxygen into molten iron to oxidize impurities and decrease the amount of carbon present.

3. Iron reacts with hydrochloric acid to form hydrogen gas and a solution of $FeCl_2$.

4. (a) $2Fe^{2+}(aq) + Sn^{4+}(aq) \longrightarrow 2Fe^{3+}(aq) + Sn^{2+}(aq)$

    (b) $6Fe^{2+}(aq) + Cr_2O_7^{2-}(aq) + 14H^+(aq) \longrightarrow 6Fe^{3+}(aq) + 2Cr^{3+}(aq) + 7H_2O(l)$

5. $\Delta G° = \Sigma n \Delta G_f°(\text{products}) - \Sigma m \Delta G_f°(\text{reactants})$

   $= [2 \Delta G_f°(\text{Fe}) + 3 \Delta G_f°(\text{H}_2\text{O}, g)] - [\Delta G_f°(\text{Fe}_2\text{O}_3) + 3 \Delta G_f°(\text{H}_2)]$

   $= [0 + 3 \text{ mol}(-228.6 \text{ kJ/mol})] - [1 \text{ mol}(-743.6 \text{ kJ/mol}) + 0]$

   $= 57.8 \text{ kJ} = 5.78 \times 10^4 \text{ J}$

   Since $\Delta G° = -2.303 \, RT \log K$

   $$\log K = \frac{\Delta G°}{-2.303 \, RT} = \frac{5.78 \times 10^4}{-2.303 \times 8.31 \times 298} = -10.\underline{1}4$$

   $K = 7 \times 10^{-11}$

6. The addition of potassium ferrocyanide, $K_4[Fe(CN)_6]$, to an aqueous solution of Fe^{3+} yields a dark-blue precipitate, known as Prussian blue. It is used as a pigment in paints and inks.

ADDITIONAL PROBLEMS

1.   As a sample of gas cools at 1 atm pressure, it loses 54 J of heat, and the volume changes by 0.200 L. What are the values of q, w, and ΔU for this process?

2.   Gray tin, which exhibits the diamond structure, is transformed above 13.2°C to white tin, which has a metallic nature. The $S°$ values at 25°C are 44.8 and 51.5 J/(K · mol), respectively. Predict $\Delta H°$ for the phase transition.

3.   The $\Delta H°$ value for the transition of $PCl_3(l)$ to $PCl_3(g)$ is 32.7 kJ. The observed boiling point is 75°C. Predict the entropy change on boiling 2 mol of the liquid.

4.   Will the entropy change for each of the following processes probably be positive or negative? Explain why.

   (a) $C(diamond) \longrightarrow C(graphite)$

   (b) $NH_3(g) + HCl(g) \longrightarrow NH_4Cl(s)$

   (c) $NaCl(s) \xrightarrow{\text{water}} Na^+(aq) + Cl^-(aq)$

   (d) $CaO(s) + CO_2(g) \longrightarrow CaCO_3(s)$

5.   The enthalpy of formation of $ICl(g)$ is 17.8 kJ/mol at 25°C. Calculate the absolute entropy change for the reaction

$$\frac{1}{2} I_2(s) + \frac{1}{2} Cl_2(g) \longrightarrow ICl(g)$$

$S°$ for $I_2(s) = 116.1$ J/(K • mol); $S°$ for $Cl_2(g) = 223.0$ J/(K • mol); $S°$ for $ICl(g) = 247.4$ J/(K • mol).

6.   (a)   Using the following values for $\Delta H_f°$ and $S°$:

   $H_2O(g)$: $\Delta H_f° = -241.8$ kJ/mol; $S° = 188.7$ J/(K • mol)

   $H_2O(l)$: $\Delta H_f° = -285.8$ kJ/mol; $S° = 69.9$ J/(K • mol)

   calculate $\Delta H°$ and $\Delta S°$ for the process

   $$H_2O(l) \longrightarrow H_2O(g)$$

   at 1 atm and 25°C.

   (b)   Use your calculated values to determine $\Delta G°$ for this process.

    (c)   Is this process spontaneous at 25°C?

    (d)   If not, how could you make it spontaneous?

7.   Calculate $\Delta G°$ for the complete combustion of 1 mol of methanol, $CH_3OH(l)$, to $CO_2(g)$ and $H_2O(g)$. The $\Delta G_f°$ values are -166.2, -394.4, and -228.6 kJ/mol, respectively.

8.   Use the following information to calculate $\Delta G°$ at 325.37 K for the reaction given. What is important about this temperature?

$$2NO_2(g) \rightleftharpoons N_2O_4(g)$$

$\Delta H_f°$	33.18	9.16	kJ/mol
$S°$	240.0	304.2	J/(K • mol)

9.   The reaction

$$CaCO_3(s) \longrightarrow CaO(s) + CO_2(g)$$

is nonspontaneous at 25°C and 1 atm. This does not surprise us when we know that marble, which is essentially $CaCO_3$ with small amounts of impurities present, is an excellent building stone. The white marble Parthenon on the Acropolis in Athens has stood for 2500 years and has not decomposed to CaO and CO_2. Limestone ($CaCO_3$ with more impurities than occur in white marble) is used extensively for making quicklime, CaO, an important ingredient in building plaster. The limestone is mixed with fuel that burns and heats the stone, decomposing it according to the foregoing equation. How hot must we heat the limestone to make quicklime? Use the following values for $\Delta H_f°$:

    $CaCO_3(s)$, -1206 kJ/mol; $CaO(s)$, -635.5 kJ/mol; $CO_2(g)$, -393.5 kJ/mol

and for $S°$:

    $CaCO_3(s)$, 92.9 J/(K • mol); $CaO(s)$, 39.7 J/(K • mol);

    $CO_2(g)$, 214 J/(K • mol).

10.   For each of the following systems, calculate the thermodynamic equilibrium constant at 25°C and 1 atm. From this value of the constant, predict whether the reaction will produce only negligible amounts of the products, will yield moderate amounts of products, or will go essentially to completion toward the right.

    (a) $2HI(g) + Cl_2(g) \longrightarrow 2HCl(g) + I_2(g)$; $\Delta G° = -174.6$ kJ

(b) $2NO_2(g) \longrightarrow N_2O_4(g); \quad \Delta G° = -4.78 \text{ kJ}$

(c) $N_2(g) + 2O_2(g) \longrightarrow 2NO_2(g); \quad \Delta G° = 103 \text{ kJ}$

ANSWERS TO ADDITIONAL PROBLEMS

If you missed an answer, study the text section and operational skill given in parentheses after the answer.

1. $q = -54 \text{ J}, w = +20.2 \text{ J}, \Delta U = -34 \text{ J}$ (18.1)

2. $\Delta H° \simeq 1.9 \text{ kJ/mol}$ (18.2, Op. Sk. 1)

3. $\Delta S° \simeq 188 \text{ J/K}$ (18.1, 18.2, Op. Sk. 1)

4.   (a) ΔS should be positive, because graphite is softer and the disorder greater than in diamond.

     (b) ΔS will be negative; the disorder in $NH_4Cl(s)$ crystals is much less than the disorder in $NH_3(g)$ and $HCl(g)$. Furthermore, the reaction proceeds from 2 mol of a gas to 1 mol of a solid — with a decrease in the number of individual molecular units.

     (c) ΔS will be positive, because NaCl in solution is dissociated into 2 mol of ions per mole of NaCl dissolved, and these ions, with their freedom of motion, are more disordered than the highly ordered Na^+ and Cl^- ions in crystalline NaCl.

     (d) ΔS will be negative, because a gas with large entropy becomes incorporated into a solid with much less entropy, and the reaction proceeds with 2 mol of reactants forming 1 mol of product.   (18.3, Op. Sk. 2)

5. $\Delta S° = 77.8 \text{ J/K}$ (18.3, Op. Sk. 3)

6.   (a) $\Delta H° = 44.0 \text{ kJ}$ (6.8, 18.1)

           $\Delta S° = 118.8 \text{ J/K}$ (18.3, Op. Sk. 3)

     (b) $\Delta G° = 8.60 \text{ kJ}$ (18.4, Op. Sk. 4)

     (c)   Because $\Delta G°$ is positive, the process is not spontaneous at 25°C.

                                                                        (18.4, Op. Sk. 6)

(d)   The process can be made spontaneous by increasing the temperature until $-T\Delta S$ exceeds ΔH, giving a negative value of ΔG. We know that water boils at 100°C. At this temperature, $H_2O(l)$ and $H_2O(g)$ are in equilibrium, and ΔG for the vaporization process must be zero. For the process to become spontaneous, the temperature must be raised an infinitesimal amount above 100°C.   (18.7, Op. Sk. 9)

7. $\Delta G° = -685.4$ kJ   (18.4, Op. Sk. 5)

8. $\Delta G° = \simeq 0$. This is the temperature at which the stoichiometry of the balanced equation represents the relative amounts of the substances at equilibrium.
                                                                    (18.7, Op. Sk. 9)

9.   The temperature must be kept above 1100 K.
                                           (18.2, 18.3, 18.4, 18.7, Op. Sk. 3, 4, 6, 9)

10.   (a) $K = K_p = 4 \times 10^{30}$; the reaction goes essentially to completion.

(b) $K = K_p = 6.89$; only moderate amounts of products form.

(c) $K = K_p = 9 \times 10^{-19}$; negligible amounts of products form.
                                                                    (18.6, Op. Sk. 8)

CHAPTER POST-TEST

1.   The enthalpy change when liquid dimethyl sulfide, $(CH_3)_2S(l)$, vaporizes at 25°C is 28.0 kJ/mol. What is the entropy change when one mole of vapor in equilibrium with the liquid condenses to liquid at 25°C if the entropy of this vapor at 25°C is 285.7 J/(K • mol)?

2.   Predict by inspection the sign of $\Delta S°$ for each of the following reactions, and explain your answers.

(a) $2NOCl(g) \rightleftharpoons 2NO(g) + Cl_2(g)$

(b) $Zn(s) + 2HCl(aq) \longrightarrow ZnCl_2(aq) + H_2(g)$

(c) $N_2(g) + 3H_2(g) \rightleftharpoons 2NH_3(g)$

(d) $HCOOH(aq) + OH^-(aq) \longrightarrow HCOO^-(aq) + H_2O(l)$

3.   Calculate the change in entropy, $\Delta S°$, for the following neutralization reaction:

$$Ca^{2+}(aq) + 2OH^-(aq) + H_2SO_4(aq) \longrightarrow CaSO_4(s) + 2H_2O(l)$$

Use text Appendix C for standard entropy values.

4.   Calculate the $\Delta G°$ for the following reaction at 25°C:

$$2NO_2(g) \rightleftharpoons 2NO(g) + O_2(g)$$

Is the reaction spontaneous as written? Use values of $\Delta H_f°$ and $S°$ from text Appendix C.

5.   Determine whether each of the following statements is true or false. If the statement is false, change it so it is true.

(a)   A flow of entropy, a transfer of disorder, always follows a temperature change. True/False: _____ _____ .

(b)   The second law of thermodynamics states that the total entropy of a system and surroundings always increases for a spontaneous process.   True/False: _____ _____ .

(c)   The free-energy change of a reacting system is the amount of energy available to do work.   True/False: _____ _____

(d)   The zero point for free energy is the same as that for entropy.   True/False: _____ _____ .

6.   Calculate the free-energy change, $\Delta G°$, for the following reaction, using values of $\Delta G_f°$ from text Appendix C:

$$Mg(s) + 2HCl(aq) \longrightarrow Mg^{2+}(aq) + 2Cl^-(aq) + H_2(g)$$

7.   Determine whether each of the reactions in Problem 2 is spontaneous or nonspontaneous as written. Give reasons for your answers.   (Use text Appendix C. $\Delta G_f°(NOCl) = 66.36$ kJ/mol.)

8. The work that a reacting system open to the atmosphere can do is $P\Delta V$ work, where the atmosphere is pushed away. Using the ideal gas law (text Section 5.3), express this work as a more useful term, and calculate the maximum amount of work that could be done in the following reaction at 25°C:

$$2C_8H_{18}(l) + 25O_2(g) \longrightarrow 16CO_2(g) + 18H_2O(g)$$

9. Use data from text Table 18.2 to calculate the thermodynamic equilibrium constant for the reaction

$$2H_2O_2(l) \longrightarrow 2H_2O(l) + O_2(g)$$

at 25°C. ΔG_f° for $H_2O_2(l) = -120.4$ kJ/mol.

10. To what temperature must potassium chlorate be heated to decompose it to potassium chloride and oxygen gas? Use values of ΔH_f° and S° from text Appendix C. Values for $KClO_3$ are $\Delta H_f^\circ = -391.20$ kJ/mol, $S^\circ = 142.97$ J/K • mol.

ANSWERS TO CHAPTER POST-TEST

If you missed an answer, study the text section and operational skill given in parentheses after the answer.

1. $\Delta S^\circ = -94.0$ J/K (18.2, Op. Sk. 1)

2. (a) $\Delta S > 0$; there are two molecules of gas on the left and three on the right.
    (b) $\Delta S > 0$; there is one molecule of gas on the right.
    (c) $\Delta S < 0$; there are four molecules of gas on the left and only two on the right.
    (d) $\Delta S \approx 0$. One would expect the molecular HCOOH to be associated with water molecules and the small OH^- to be tightly surrounded by water molecules. In the products, the $COOH^-$ ion would not have quite the association with water the OH^- does, but water associates with itself. (The calculated value is 8.08 J/K.) (18.3, Op. Sk. 2)

3. $\Delta S^\circ = 305$ J/K (18.3, Op. Sk. 3)

4. $\Delta G^\circ = 70.5$ kJ; reaction is nonspontaneous. (18.4, Op. Sk. 4)

5. (a) True. (18.2)   (b) True. (18.2)   (c) True. (18.5)

(d)   False. The zero point for free energy is the element in its standard state at 1 atm pressure and a specified temperature (usually 25°C). The zero point for entropy is a perfect crystalline substance at 0 K.   (18.3, 18.4)

6. $\Delta G° = -456.01$ kJ   (18.4, Op. Sk. 5)

7.   (a)   nonspontaneous; $\Delta G° > 0$ (c)   spontaneous; $\Delta G° < 0$
      (b)   spontaneous; $\Delta G° < 0$ (d)   spontaneous; $\Delta G° < 0$ (18.4, Op. Sk. 5, 6)

8. $w_{max} = \Delta nRT = 22.3$ kJ   (18.1, 18.5)

9. $K = K_p = 9 \times 10^{40}$ (18.6, Op. Sk. 8)

10.   The reaction is spontaneous at any temperature.   (18.7, Op. Sk. 9)

CHAPTER 19   ELECTROCHEMISTRY

CHAPTER TERMS AND DEFINITIONS

Numbers in parentheses after definitions give the text sections in which the terms are explained. Starred terms are italicized in the text. Where a term does not fall directly under a text section heading, additional information is given for you to locate it.

oxidation half-reaction*   half-reaction in which there is a loss of electrons by a species (19.1)

reduction half-reaction*   half-reaction in which there is a gain of electrons by a species (19.1)

oxidized*   describes species that loses electrons   (19.1)

oxidizing agent*   species in the reaction that is reduced   (19.1)

reduced*   describes species that gains electrons   (19.1)

reducing agent*   species in the reaction that is oxidized   (19.1)

electrochemical cell   system consisting of electrodes that dip into an electrolyte and in which a chemical reaction either uses or generates an electric current   (19.2, introductory section)

voltaic (galvanic) cell   electrochemical cell in which a spontaneous reaction generates an electric current   (19.2, introductory section)

electrolytic cell   electrochemical cell in which an electric current drives an otherwise nonspontaneous reaction   (19.2, introductory section)

half-cell   portion of an electrochemical cell in which a half-reaction (oxidation or reduction) takes place   (19.2)

salt bridge   tube of gelled electrolyte that connects two half-cells of a voltaic cell, allowing the flow of ions but preventing mixing of the different solutions that would allow direct reaction of the cell reactants   (19.2)

anode   electrode at which oxidation occurs   (19.2)

cathode   electrode at which reduction occurs   (19.2)

cell reaction   net reaction that occurs in a voltaic cell   (19.2)

potential difference   difference in electric potential (electrical pressure) between two points; measured in volts   (19.4)

volt (*V*)   SI unit of potential difference   (19.4)

Faraday constant (*F*)   magnitude of charge on one mole of electrons; 9.65×10^4 C/mol (19.4)

faraday*   unit of charge equal to 9.65×10^4 C   (19.4)

electromotive force (emf) (*E_{cell}*)   maximum potential difference between the electrodes of a voltaic cell   (19.4)

oxidation potential*   measure of the tendency for a reduced species to lose electrons in the oxidation (anode) half-reaction   (19.5)

reduction potential (electrode potential)* (*E*)   measure of the tendency for an oxidized species to gain electrons in the reduction (cathode) half-reaction   (19.5)

standard emf (*E°_{cell}*)   emf of a voltaic cell operating under standard-state conditions (solute concentrations are each 1 *M*, gas pressures are each 1 atm, and the temperature has a specified value — usually 25°C)   (19.5)

standard hydrogen electrode*   reference electrode assigned a potential of zero, in which the H^+ concentration is 1 *M*, the H_2 pressure is 1 atm, and the temperature is 25°C (19.5)

standard electrode potential (*E°*)   electrode potential of a half-cell reaction when the concentrations of solutes are 1 *M*, the gas pressures are 1 atm, and the temperature has a specified value (usually 25°C)   (19.5)

strongest oxidizing agents*   oxidized species corresponding to half-reactions with the largest (most positive) E° values   (19.5)

strongest reducing agents*   reduced species corresponding to half-reactions with the smallest (most negative) E° values   (19.5)

Nernst equation   equation relating cell emf to standard emf and to the reaction quotient;

$$E_{cell} = E^\circ_{cell} - \frac{RT}{nF} \ln Q \quad \text{or} \quad E_{cell} = E^\circ_{cell} - \frac{2.303\,RT}{nF} \log Q \quad (19.7)$$

glass electrode*   compact, stable electrode used to determine pH by emf measurements (19.7)

ion-selective electrode*   specially constructed electrode that is responsive to a particular ion in solution   (19.7)

zinc–carbon (Leclanché) dry cell   common voltaic cell (battery) in which zinc is the anode; a graphite rod in the center, surrounded by a paste of manganese dioxide, ammonium and zinc chlorides, and carbon black, is the cathode   (19.8)

alkaline dry cell   similar to the Leclanché dry cell, but has potassium hydroxide in place of ammonium chloride   (19.8)

lithium–iodine battery   voltaic cell in which the anode is lithium metal and the cathode is an I_2 complex   (19.8)

lead storage cell   common voltaic cell (car battery) in which a lead-alloy grid packed with a spongy lead is the anode and a lead-alloy grid packed with lead dioxide is the cathode   (19.8)

maintenance-free (battery)* lead storage battery using lead electrodes containing some calcium metal, which resists the decomposition of water, and is thus sealed (19.8)

nickel–cadmium cell (nicad cell) voltaic cell consisting of an anode of cadmium, a cathode of hydrated nickel oxide on nickel, and KOH as the electrolyte (19.8)

fuel cell voltaic cell operating with a continuous supply of energetic reactants or fuel (19.8)

cathodic protection* use of an active metal (e.g., magnesium), which acts as an anode, to protect another metal (e.g., iron), which acts as the cathode, from corrosion (19.8)

electrolysis process of producing a chemical change in an electrolytic cell (19.9, introductory section)

Downs cell commercial electrochemical cell used to obtain sodium metal by the electrolysis of molten NaCl (19.9)

overvoltage* additional voltage needed in electrolysis when the half-reactions are far from equilibrium (19.10)

chlor-alkali membrane cell cell for the electrolysis of aqueous sodium chloride in which the anode and cathode compartments are separated by a special plastic membrane that allows only cations to pass through it (19.10)

chlor-alkali mercury cell cell for the electrolysis of aqueous sodium chloride in which mercury metal is used as the cathode (19.10)

amalgam* an alloy of mercury with any of various other metals (19.10)

electrogalvanizing* deposition of a thin zinc coating on a steel object that has been used as the cathode in an electrolytic cell that contains dissolved zinc salts (19.10)

ampere (A) SI base unit of current (19.11)

CHAPTER DIAGNOSTIC TEST

1. Sketch a cell in which a zinc electrode is in a solution of zinc sulfate and a copper electrode is in a solution of copper sulfate. The cells are connected by a salt bridge. Copper is reduced when the cell is operating. Label the anode, cathode, and direction of ion and electron movement, and write the electrode half-reactions.

2.   Complete the following table:

Voltaic Cell Notation	Anode Reaction	Cathode Reaction
(a)	$2Br^-(aq) \rightleftharpoons Br_2(l) + 2e^-$	$MnO_4^-(aq) + 8H^+(aq) + 5e^-$ $\rightleftharpoons Mn^{2+}(aq) + 4H_2O(l)$
$Pb(s) \mid PbSO_4(s) \mid\mid$ $NO_3^-(aq) \mid NO(g) \mid Pt$	(b)	(c)
$Pt \mid H_2O(l) \mid O_2(g) \mid\mid$ $Cr_2O_7^{2-}(aq), Cr^{3+}(aq) \mid Pt$	$2H_2O(l) \rightleftharpoons$ $O_2(g) + 4H^+(aq) + 4e^-$	(d)
$In(s) \mid In^{3+}(aq) \mid\mid$ (e)	(f)	$UO_2^{2+}(aq) + e^- \rightleftharpoons$ $UO_2^+(aq)$
(g)	$Sn^{2+}(aq) \rightleftharpoons$ $Sn^{4+}(aq) + 2e^-$	$2IO_3^-(aq) + 12H^+ + 10e^-$ $\rightleftharpoons I_2(aq) + 6H_2O(l)$

3.   The emf of the voltaic cell with the cell reaction

$$H_3AsO_4(aq) + 2H^+(aq) + 2I^-(aq) \rightleftharpoons HAsO_2(aq) + I_2(aq) + 2H_2O(l)$$

is 0.020 V. Calculate the maximum electrical work that can be obtained when 0.750 g I_2 is produced.

4.   Arrange the following in order of increasing oxidizing strength: $Ag^+(aq)$, $Sn^{4+}(aq)$, $MnO_4^-(aq)$, $O_2(g)$.

5.   Calculate the emf for the cell

$$Pt(s) \mid Sn^{2+}(aq), Sn^{4+}(aq) \mid\mid ClO_3^-(aq), ClO_2^-(aq), OH^-(aq) \mid Pt(s)$$

For $Sn^{4+}(aq)$, $Sn^{2+}(aq) \mid Pt(s)$, $E° = 0.15$ V; for $ClO_3^-(aq)$, $ClO_2^-(aq)$, $OH^-(aq) \mid Pt(s)$, $E° = 0.35$ V.

6. Calculate K_c at 298 K for the reaction

$$3Sn^{2+}(aq) + 2NO_3^-(aq) + 8H^+(aq) \longrightarrow 3Sn^{4+}(aq) + 2NO(g) + 4H_2O(l)$$

The standard cell emf is +0.71 V.

7. Calculate the emf of the cell

$$Pt\,|\,Cr^{2+}(1.0 \times 10^{-2}\,M),\, Cr^{3+}(1.0 \times 10^{-2}\,M)\,||\,Pb^{2+}(4.0 \times 10^{-4}\,M)\,|\,Pb(s)$$

For $Cr^{3+}(aq),\, Cr^{2+}(aq)\,|\,Pt$, $E° = -0.41$ V; for $Pb^{2+}(aq)\,|\,Pb(s)$, $E° = -0.13$ V.

8. When an aqueous solution of H^+(1 *M*), Fe^{2+}(1 *M*), I^-(1 *M*), and Br^-(1 *M*) is electrolyzed, what are the products at the cathode and anode? (Use text Table 19.1.)

9. Of what strength must the electric current be in order to plate out 50.0 mg Ag in 3.0 h? The electrode reaction is $Ag^+ + e^- \longrightarrow Ag$.

10. Determine whether each of the following statements is true or false. If the statement is false, change it so it is true.
    (a) For a voltaic cell, the electrical work equals $-nFE$. True/False: _____
    _____ .

    (b) The reaction $NO_3^-(aq) + 4H^+(aq) + 3Ce^{3+}(aq) \longrightarrow 3Ce^{4+}(aq) + NO(g) + 2H_2O(l)$ goes in the direction indicated. The standard electrode potentials are: $NO_3^-(aq),\, H^+(aq)\,|\,NO(g)\,|\,Pt$, $E° = 0.96$ V; $Ce^{4+}(aq),\, Ce^{3+}(aq)\,|\,Pt$, $E° = 1.44$ V. True/False: _____
    _____ .

    (c) In a voltaic cell, reduction occurs at the anode; in an electrolytic cell, oxidation occurs at the anode. True/False: _____
    _____ .

    (d) The voltaic cell electrode with the higher reduction potential is said to have positive polarity. True/False: _____
    _____ .

    (e) An ion-selective electrode must have a potential that is sensitive to the concentration of a particular ion. True/False: _____
    _____ .

ANSWERS TO CHAPTER DIAGNOSTIC TEST

If you missed an answer, study the text section and operational skill given in parentheses after the answer.

1.

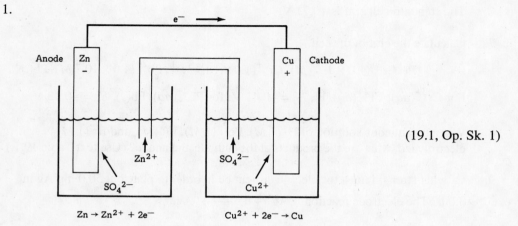

(19.1, Op. Sk. 1)

$$Zn \rightarrow Zn^{2+} + 2e^- \qquad\qquad Cu^{2+} + 2e^- \rightarrow Cu$$

2. (a) $Pt|Br^-(aq)|Br_2(l)||MnO_4^-(aq), Mn^{2+}(aq)|Pt$

    (b) $Pb(s) + SO_4^{2-}(aq) \rightleftharpoons PbSO_4(s) + 2e^-$

    (c) $NO_3^-(aq) + 4H^+(aq) + 3e^- \rightleftharpoons NO(g) + 2H_2O(l)$

    (d) $Cr_2O_7^{2-}(aq) + 14H^+(aq) + 6e^- \rightleftharpoons 2Cr^{3+}(aq) + 7H_2O(l)$

    (e) $||UO_2^{2+}(aq), UO_2^+(aq)|Pt$

    (f) $In(s) \rightleftharpoons In^{3+}(aq) + 3e^-$

    (g) $Pt|Sn^{2+}(aq), Sn^{4+}(aq)||IO_3^-(aq)|I_2(s)|Pt$ (19.2, Op. Sk. 2)

3. 11 J (19.3, Op. Sk. 3)

4. $Sn^{4+}(aq), Ag^+(aq), O_2(g), MnO_4^-(aq)$ (19.4, Op. Sk. 4)

5. 0.20 V (19.2, 19.3, 19.4, Op. Sk. 6)

6. 10^{72} (19.5, Op. Sk. 9)

7. 0.18 V (19.6, Op. Sk. 10)

8. Cathode, H_2; anode, I_2 (19.9, Op. Sk. 11)

9. 4.1×10^{-3} A (19.10, Op. Sk. 12)

10. (a) True (19.3)

    (b) False. The reaction $NO_3^-(aq) + 4H^+(aq) + 3Ce^{3+}(aq) \longrightarrow$ $3Ce^{4+}(aq) + NO(g) + 2H_2O(l)$ goes in the reverse direction. (E°_{cell} as written is 0.96 V – 1.44 V = –0.48 V. This gives a positive ΔG° value, which indicates the reaction is nonspontaneous.) (19.4, Op. Sk. 5)

    (c) False. In both kinds of electrochemical cells, reduction occurs at the cathode and oxidation at the anode. (19.1)

    (d) True (19.14)   (e) True (19.6)

SUMMARY OF CHAPTER TOPICS

19.1 OXIDATION–REDUCTION REACTIONS: A REVIEW

This section reviews some of the basic concepts and terminology used for oxidation–reduction reactions. You should go back and review the material in Sections 3.5 and 3.6 to refresh your skills in working with these reactions.

19.2 CONSTRUCTION OF VOLTAIC CELLS

Operational Skill

1. Sketching and labeling a voltaic cell. Given a verbal description of a voltaic cell, sketch the cell, labeling the anode and cathode, and give the directions of electron flow and ion migration (Example 19.1).

You must memorize that oxidation always occurs at the anode and reduction at the cathode. One way to remember this is to recall that *anode* and *oxidation* begin with vowels, whereas *cathode* and *reduction* begin with consonants. You may be able to come up with an even better memory aid. Also, anions always move toward the anode, and cations always move toward the cathode, just as the names imply. In the external circuit of the voltaic cell, the electrons always move toward the less electropositive (more electronegative) metal or electrode. Be sure to realize that the direction of the electron flow determines which electrode is labeled cathode or anode.

> **Exercise 19.1** A voltaic cell consists of a silver–silver ion half-cell and a nickel–nickel(II) ion half-cell. Silver ion is reduced during operation of the cell. Sketch the cell, labeling the anode and cathode and indicating the corresponding electrode reactions. Show the direction of electron flow in the external circuit and the direction of cation movement in the half-cells.
>
> *Wanted:* a labeled cell drawing.
>
> *Given:*   two half-cells for the drawing; cell is voltaic; silver ion is reduced.
>
> *Known:*   Ni is more electropositive than Ag, so electrons will flow from Ni to Ag.
>
> *Solution:*

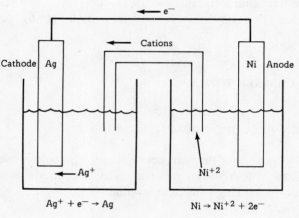

$$Ag^+ + e^- \rightarrow Ag \qquad\qquad Ni \rightarrow Ni^{+2} + 2e^-$$

19.3 NOTATION FOR VOLTAIC CELLS

Operational Skill

  2. Writing the cell reaction from the cell notation. Given the notation for a voltaic cell, write the overall cell reaction (Example 19.2). Alternatively, given the cell reaction, write the cell notation.

     In writing cell notation, a vertical double bar is used to separate the anode on the left from the cathode on the right. A single vertical bar separates two different phases, and a comma separates two ionic species that are both in the aqueous phase.

Exercise 19.2   Write the notation for a cell in which the electrode reactions are

$$2H^+(aq) + 2e^- \longrightarrow H_2(g)$$

$$Zn(s) \longrightarrow Zn^{2+}(aq) + 2e^-$$

Known: Oxidation occurs at the anode; notation for hydrogen electrode is given in text; conventions for cell notations were given previously.

Solution: The notation is

$$Zn(s)\,|\,Zn^{2+}(aq)\,|\,|\,H^+(aq)\,|\,H_2(g)\,|\,Pt$$

Exercise 19.3   Give the overall cell reaction for the voltaic cell

$$Cd(s)\,|\,Cd^{2+}(aq)\,|\,|\,H^+(aq)\,|\,H_2(g)\,|\,Pt$$

Known: The anode reaction (oxidation) is on the left.

Solution: The overall reaction is

$$Cd(s) \longrightarrow Cd^{2+}(aq) + 2e^-$$

$$\underline{2H^+(aq) + 2e^- \longrightarrow H_2(g)}$$

$$2H^+(aq) + Cd(s) \longrightarrow Cd^{2+}(aq) + H_2(g)$$

19.4   ELECTROMOTIVE FORCE

Operational Skill

 3. Calculating the quantity of work from a given amount of cell reactant.   Given the emf and overall reaction for a voltaic cell, calculate the maximum work that can be obtained from a given amount of reactant   (Example 19.3).

Exercise 19.4   What is the maximum electrical work that can be obtained from 6.54 g of zinc metal that reacts in a Daniell cell, described in the chapter opening, whose emf is 1.10 V? The overall cell reaction is

$$Zn(s) + Cu^{2+}(aq) \longrightarrow Zn^{2+}(aq) + Cu(s)$$

Wanted: maximum electrical work, w_{max} (joules or kilojoules).

Given: 6.54 g Zn in Daniell cell, emf = 1.10 V = E_{cell}, cell reaction.

Known: $w_{max} = -nFE$, atomic weight of Zn = 65.4 amu. Also,
$F = 9.65 \times 10^4$ C/mol e$^-$ = 9.65×10^4 J/(V • mol e$^-$), since 1 J = 1 V • C.

Solution: Writing the cell half-reactions:

$$Zn \longrightarrow Zn^{2+} + 2e^- \quad \text{and} \quad Cu^{2+} + 2e^- \longrightarrow Cu$$

gives $n = 2$ (which is 2 mol e$^-$ per mole of Zn reacted). Solve for w_{max} per mole of Zn:

$$\frac{w_{max}}{\text{mol Zn}} = -nFE_{cell} = -\frac{2 \text{ mol e}^-}{\text{mol Zn}} \times \frac{9.65 \times 10^4 \text{ J}}{\text{V} \cdot \text{mol e}^-} \times 1.10 \text{ V}$$

$$= -2.1\underline{2}3 \times 10^5 \frac{\text{J}}{\text{mol Zn}}$$

Find moles of Zn that react:

$$6.54 \text{ g Zn} \times \frac{\text{mol Zn}}{65.4 \text{ g Zn}} = 0.10\underline{0}0 \text{ mol Zn}$$

And finally,

$$w_{max} = 0.10\underline{0}0 \text{ mol Zn} \times -2.1\underline{2}3 \times 10^5 \frac{\text{J}}{\text{mol Zn}}$$

$$= -2.12 \times 10^4 \text{ J} \times \frac{\text{kJ}}{10^3 \text{ J}} = -21.2 \text{ kJ}$$

Once you understand the steps involved in the solution, you can do the entire calculation in one line using dimensional analysis, putting the negative sign at the beginning, as work is done by the system:

$$-6.54 \text{ g Zn} \times \frac{1 \text{ mol Zn}}{65.4 \text{ g Zn}} \times \frac{2 \text{ mol e}^-}{\text{mol Zn}} \times \frac{9.65 \times 10^4 \text{ J}}{\text{V} \cdot \text{mol e}^-} \times 1.10 \text{ V} \times \frac{\text{kJ}}{10^3 \text{ J}}$$

$$= -21.2 \text{ kJ}$$

19.5   STANDARD CELL emf's AND STANDARD ELECTRODE POTENTIALS

Operational Skills

4.   Determining the relative strengths of oxidizing and reducing agents.   Given a table of standard electrode potentials, list oxidizing or reducing agents by increasing strength   (Example 19.4).

5.   Determining the direction of spontaneity from electrode potentials.   Given standard electrode potentials, decide the direction of spontaneity for an oxidation–reduction reaction under standard conditions   (Example 19.5).

6.   Calculating the emf from standard potentials.   Given standard electrode potentials, calculate the standard emf of a voltaic cell   (Example 19.6).

An acronym to help you remember the negative polarity of the anode in a voltaic cell is VAN for Voltaic, Anode, Negative.

It is often confusing to students that in a voltaic cell the anode, to which the anions flow, has negative polarity. (Recall that anions are negatively charged.) To understand this, it is helpful to think of what occurs in solution. Consider, for example, the solution in Exercise 19.1. The Ni gives up electrons more easily than the Ag, so electrons pile up on the Ni electrode as Ni^{2+} ions form about the electrode·in solution. Thus, the Ni electrode becomes negative in relation to the Ag electrode. In solution, however, the formation of Ni^{2+} ions gives positive character to the Ni electrode, to which the anions are drawn. The Ag metal has more attraction for electrons than Ni does, so the electrons are pulled toward it through the wire. In solution, Ag^{+} ions move to the Ag electrode to take the incoming electrons, so the Ag electrode surface becomes depleted in positive ions, causing net cation movement toward it.

It is well worth your time to look carefully at text Table 19.1. Note that the element that is most easily oxidized, Li, is at the top and has the most negative reduction potential. Thus, it is the best reducing agent. At the bottom of the table is the element that is most easily reduced, F_2, which has the highest reduction potential and is therefore the best oxidizing agent. It is conventional today to list electrode potentials as reductions. However, it was not always done this way, and you may come across an older text that lists the potentials as oxidation potentials. In that case, all the signs will be the opposite of those in the reduction tables. Be sure to note what reaction is given in the table you are using.

Remember that when you calculate a cell potential, $E°$, you do *not* multiply the table values by the number of electrons that are transferred in the reaction. This is a major source of student error. Electrode potential — the "pull" that electrons experience — does not depend on the number of electrons that experience the pull.

Exercise 19.5 Which is the stronger oxidizing agent, $NO_3^-(aq)$ in acidic solution (to NO) or $Ag^+(aq)$?

Known: table of electrode potentials; species are ordered from top to bottom in order of increasing oxidizing power.

Solution: The half-reactions and corresponding electrode potentials are as follows:

$$Ag^+(aq) + e^- \longrightarrow Ag(s) \qquad\qquad 0.80 \text{ V}$$

$$NO_3^-(aq) + 4H^+(aq) + 3e^- \longrightarrow NO(g) + 2H_2O(l) \quad 0.96 \text{ V}$$

Thus, the stronger oxidizing agent is $NO_3^-(aq)$.

Exercise 19.6 Does the following reaction occur spontaneously in the direction indicated, under standard conditions?

$$Cu^{2+}(aq) + 2I^-(aq) \longrightarrow Cu(s) + I_2(s)$$

Known: Reaction is spontaneous if E°_{cell} is positive; E° values are in text Table 19.1.

Solution: Breaking the cell into two half-reactions gives

$$Cu^{2+}(aq) + 2e^- \longrightarrow Cu(s) \qquad\qquad E^\circ = 0.34 \text{ V}$$

and

$$2I^-(aq) \longrightarrow 2e^- + I_2(s) \qquad\qquad E^\circ = -0.54 \text{ V}$$

$E^\circ_{cell} = 0.34 \text{ V} + (-0.54 \text{ V}) = -0.20 \text{ V}$. As E°_{cell} is negative, the reaction is nonspontaneous as written.

Exercise 19.7 Using standard electrode potentials, calculate E°_{cell} at 25°C for the following cell.

$$Zn(s)\,|\,Zn^{2+}(aq)\,||\,Cu^{2+}(aq)\,|\,Cu(s)$$

Wanted: E°_{cell}.

Known: Reduction potentials are in text Table 19.1; anode (oxidation) is on the left in the given notation.

Solution: The calculation is

Anode reaction:	$Zn(s) \longrightarrow Zn^{2+}(aq) + 2e^-$	$E° = 0.76$ V
Cathode reaction:	$Cu^{2+}(aq) + 2e^- \longrightarrow Cu(s)$	$E° = 0.34$ V
Cell reaction:	$Cu^{2+}(aq) + Zn(s) \longrightarrow Zn^{2+}(aq) + Cu(s)$	$E° = 1.10$ V

19.6 EQUILIBRIUM CONSTANTS FROM emf's

Operational Skills

7. Calculating the free-energy change from electrode potentials. Given standard electrode potentials, calculate the standard free-energy change for an oxidation–reduction reaction   (Example 19.7).

8. Calculating the cell emf from free-energy change. Given a table of standard free energies of formation, calculate the standard emf of a voltaic cell   (Example 19.8).

9. Calculating the equilibrium constant from cell emf. Given standard potentials (or standard emf), calculate the equilibrium constant of an oxidation–reduction reaction (Example 19.9).

Exercise 19.8 What is $\Delta G°$ at 25°C for the reaction

$$Sn^{2+}(aq) + 2Hg^{2+}(aq) \longrightarrow Sn^{4+}(aq) + Hg_2{}^{2+}(aq)$$

For data, see Table 19.1.

Wanted: $\Delta G°$.

Given: reaction; refer to text Table 19.1.

Known: $\Delta G° = -nFE°_{cell}$; $F = 9.65 \times 10^4$ C/mol e$^-$ = 9.65×10^4 J/(V • mol e$^-$), since 1 J = 1 V • C.

Solution: The calculation is

Anode reaction:	$Sn^{2+}(aq) \longrightarrow Sn^{4+}(aq) + 2e^-$	$-E° = -0.15$ V
Cathode reaction:	$2Hg^{2+}(aq) + 2e^- \longrightarrow Hg_2{}^{2+}(aq)$	$E° = 0.90$ V
	$Sn^{2+}(aq) + 2Hg^{2+}(aq) \longrightarrow Sn^{4+}(aq) + Hg_2{}^{2+}(aq)$	$E° = 0.75$ V

Since $n = 2$,

$$\Delta G^\circ = -nFE^\circ = -2 \, \cancel{mol \, e^-} \left(9.65 \times 10^4 \, \frac{J}{\cancel{V} \cdot \cancel{mol \, e^-}} \right)(0.75 \, \cancel{V})$$

$$= -1.4 \times 10^5 \, J$$

We must again speak of the "mole of reaction" that was discussed in Chapter 18. In Exercise 19.8, the electrons transferred were for the reaction as written. If the coefficients of the reaction had been doubled, there would have been 4 moles of electrons transferred. Thus, n depends on what coefficients are used. We could think of units of n as moles of "electrons per mole of reaction." Thus, the units of ΔG° would come out J/mol, where "mol" means "mole of reaction." We must use this idea in order for our units to cancel properly in later problems in this section.

Exercise 19.9 Use standard free energies of formation (Appendix C) to obtain the standard emf of a cell at 25°C with the reaction

$$Mg(s) + Cu^{2+}(aq) \longrightarrow Mg^{2+}(aq) + Cu(s)$$

Wanted: E°_{cell}.

Given: reaction; Appendix C.

Known: $\Delta G^\circ_{rxn} = \Sigma n \Delta G^\circ_f$ (products) $- \Sigma m \Delta G^\circ_f$ (reactants); $\Delta G^\circ = -nFE^\circ_{cell}$;

$F = 9.65 \times 10^4 \, \dfrac{J}{V \cdot mol \, e^-}$

Solution: First write the equation with ΔG°_f values:

$$Mg(s) + Cu^{2+}(aq) \longrightarrow Mg^{2+}(aq) + Cu(s)$$

$\Delta G^\circ_f:$ 0   65.0   −456.0   0   kJ/mol

Then,

$$\Delta G^\circ_{rxn} = 1 \text{ mol } Mg^{2+}(-456.0 \text{ kJ/mol}) - 1 \text{ mol } Cu^{2+}(65.0 \text{ kJ/mol})$$

$$= -521.0 \text{ kJ}$$

It can be seen from the reaction that $n = 2$.

Solving $\Delta G^\circ = -nFE^\circ$ for E° gives

$$E^\circ = \frac{-\Delta G^\circ}{nF} = \frac{-\left(-521.0 \times 10^3 \, \cancel{J}\right) V \cdot \cancel{mol \, e^-}}{2 \, \cancel{mol \, e^-} \times 9.65 \times 10^4 \, \cancel{J}}$$

$$= 2.70 \text{ V}$$

Exercise 19.10 Calculate the equilibrium constant K_c for the following reaction from standard electrode potentials.

$$Fe(s) + Sn^{4+}(aq) \rightleftharpoons Fe^{2+}(aq) + Sn^{2+}(aq)$$

Wanted: K_c.

Given: reaction; standard electrode potentials (text Table 19.1).

Known: $\Delta G° = -nFE°_{cell} = -2.303\ RT \log K$; $K = K_c$;

$F = 9.65 \times 10^4\ \dfrac{J}{V \cdot mol\ e^-}$; $T = 298$ K; $R = 8.31$ J/(K $\cdot$ mol); $E°_{cell} = E°_{cathode} -$

$E°_{anode}$. In order for the units to work out properly, we will use the units of n to be mol e^-/mol, where "mol" in the denominator stands for "mole of reaction." (Refer back to the discussions of Sections 18.3 and 18.4 in this study guide.)

Solution: Since $K_c = K$, and

$$\Delta G° = -2.303\ RT \log K_c = -nFE°$$

we can write

$$2.303\ RT \log K_c = nFE°$$

We see by inspection of the equation that $n = 2$. We obtain the cell emf by the formula

$$E°_{cell} = E°_{cathode} - E°_{anode} = E°_{Sn^{2+}} - E°_{Fe}$$

$$= 0.15\ V - (-0.41\ V) = 0.56\ V$$

Rearranging to solve for $\log K_c$ and putting in values gives

$$\log K_c = \frac{nFE°}{2.303\ RT}$$

$$= \frac{2\ \cancel{mol\ e^-} \times\ 9.65 \times 10^4 \cancel{J}\ \times 0.56\ \cancel{V}(\cancel{K \cdot mol})}{\cancel{mol}\ \times 2.303 \times \cancel{V} \cdot \cancel{mol\ e^-} \times\ 8.31\cancel{J}\ \times 298\cancel{K}}$$

$$= 19.0$$

and

$$K_c = 10^{19}$$

19.7 DEPENDENCE OF emf ON CONCENTRATION

Operational Skill

10. Calculating the cell emf for nonstandard conditions. Given standard electrode potentials and the concentrations of substances in a voltaic cell, calculate the cell emf (Example 19.10).

Not all experiments are carried out under standard conditions, and different conditions will lead to different cell potentials. The Nernst equation will give us the cell potential for a cell under any conditions as long as we know those conditions. The Nernst equation can also be applied to a half-reaction in order to calculate an electrode potential for nonstandard conditions. Check with your instructor to see if he or she wishes you to memorize the equation.

When the Nernst equation is used at 25°C, it can be simplified by multiplying together the constants and the constant units. In this case, we use additional digits for accuracy. Since 2.303 RT/F is constant, and the units of n (mol e⁻/mol, where "mol" in the denominator means "mole of reaction") are constant, we can derive the following value (since n is in the denominator, its units will be inverted):

$$\frac{2.303 \, RT}{F} = \frac{\text{mol} \times 2.303 \times 8.314 \, \text{J} \times 298.15 \, \text{K} \times \text{V} \cdot \text{mol e}^-}{\text{mol e}^- \times (\text{K} \cdot \text{mol}) \times 9.65 \times 10^4 \, \text{J}}$$

$$= 0.05916 \text{ V} = 0.0592 \text{ V}$$

The simplified Nernst equation is

$$E_{\text{cell}} = E_{\text{cell}}^\circ - \frac{0.0592 \text{ V}}{n} \log Q$$

Remember, however, when using the equation in this form, do not write units for n, as they have been included in the constant.

Exercise 19.11 What is the emf of the following voltaic cell at 25°C?

$$\text{Zn}(s) \mid \text{Zn}^{2+}(0.200 \, M) \mid\mid \text{Ag}^+(0.00200 \, M) \mid \text{Ag}(s)$$

Wanted: E_{cell}.

Given: cell notation; $t = 25°C$.

Known: $E_{\text{cell}} = E_{\text{cell}}^\circ - \frac{0.0592 \text{ V}}{n} \log Q$; E° values are in text Table 19.1.

Solution: Write the reaction and solve for $E°_{cell}$

$$Zn(s) \longrightarrow Zn^{2+}(aq) + 2e^- \qquad\qquad -E° = 0.76 \text{ V}$$

$$2Ag^+(aq) + 2e^- \longrightarrow 2Ag(s) \qquad\qquad E° = 0.80 \text{ V}$$

$$2Ag^+(aq) + Zn(s) \longrightarrow Zn^{2+}(aq) + 2Ag(s) \qquad\qquad E° = 1.56 \text{ V}$$

Thus, $n = 2$. To solve the Nernst equation, we need Q from the reaction. Q is

$$\frac{[Zn^{2+}]}{[Ag^+]^2} = \frac{0.200}{(0.00200)^2} = 5.0\underline{00} \times 10^4$$

Then,

$$E_{cell} = E°_{cell} - \frac{0.0592 \text{ V}}{n} \log Q$$

$$= 1.56 - \frac{0.0592}{2} \log (5.0\underline{00} \times 10^4) \text{ V}$$

$$= [1.56 - (0.029\underline{6})(4.69\underline{90})] \text{ V}$$

$$= (1.56 - 0.13\underline{91}) \text{ V}$$

$$= 1.42 \text{ V}$$

Exercise 19.12 What is the nickel(II)-ion concentration in the voltaic cell

$$Zn(s)\,|\,Zn^{2+}(1.00\ M)\,|\,|\,Ni^{2+}(aq)\,|\,Ni(s)$$

if the emf is 0.34 V at 25°C?

Wanted: $[Ni^{2+}]$.

Given: $E = 0.34$ V; cell notation; $[Zn^{2+}] = 1.00\ M$.

Known: We can rearrange the Nernst equation to solve for $[Ni^{2+}]$ as part of Q.

Solution: The reaction and $E°$ are

$$Zn(s) \longrightarrow Zn^{2+}(aq) + 2e^- \qquad\qquad E° = 0.76 \text{ V}$$

$$Ni^{2+}(aq) + 2e^- \longrightarrow Ni(s) \qquad\qquad E° = -0.23 \text{ V}$$

$$Ni^{2+}(aq) + Zn(s) \longrightarrow Zn^{2+}(aq) + Ni(s) \qquad\qquad E° = 0.53 \text{ V}$$

Thus, $n = 2$ and $Q = \dfrac{[Zn^{2+}]}{[Ni^{2+}]}$. Rearrange $E_{cell} = E^{\circ}_{cell} - \dfrac{0.0592\ V}{n} \log Q$ to solve for $\log Q$:

$$E_{cell} - E^{\circ}_{cell} = \frac{-0.0592\ V}{n} \log Q$$

$$\log Q = \frac{(E^{\circ}_{cell} - E_{cell})n}{0.0592\ V} = \frac{(0.53 - 0.34)\ V \times 2}{0.0592\ V} = 6.\underline{4}2$$

$$Q = 2.\underline{6} \times 10^6 = \frac{[Zn^{2+}]}{[Ni^{2+}]} = \frac{1.00}{[Ni^{2+}]}$$

Then,

$$[Ni^{2+}] = \frac{1.00}{2.\underline{6} \times 10^6} = 4 \times 10^{-7}\ M$$

19.8   SOME COMMERCIAL VOLTAIC CELLS

19.9   ELECTROLYSIS OF MOLTEN SALTS

Whether you are dealing with a voltaic or an electrolytic cell, oxidation always occurs at the anode, and reduction always occurs at the cathode. Similarly, anions always move toward the anode, and cations always move toward the cathode. In an electrolytic cell, however, the electrons are pushed toward the more electropositive (less electronegative) metal or electrode.

Exercise 19.13   Write the half-reactions for the electrolysis of the following molten compounds: (a) KCl; (b) KOH.

Known: The most energetically favorable electrode reactions will occur; oxidation occurs at the anode and reduction at the cathode.

Solution: The half-reactions are:

(a) $K^+(l) + e^- \longrightarrow K(l)$ cathode reaction

$Cl^-(l) \longrightarrow 1/2\ Cl_2(g) + e^-$ anode reaction

(b) $K^+(l) + e^- \longrightarrow K(l)$ cathode reaction

$4OH^-(l) \longrightarrow O_2(g) + 2H_2O(g) + 4e^-$ anode reaction

19.10 AQUEOUS ELECTROLYSIS

Operational Skill

11. Predicting the half-reactions in an aqueous electrolysis. Using values of electrode potentials, decide which electrode reactions actually occur in the electrolysis of an aqueous solution   (Example 19.11).

> **Exercise 19.14** Give the half-reactions that occur when aqueous silver nitrate is electrolyzed. Nitrate ion is not oxidized during the electrolysis.
>
> *Known:* The most energetically favorable electrode reactions will occur. Since $\Delta G° = -nFE°_{cell}$; the reaction with the most positive $E°$ is favored.
>
> *Solution:* The possible cathode reactions are

$$Ag^{+}(aq) + e^{-} \longrightarrow Ag(s) \qquad\qquad E° = 0.80 \text{ V}$$

$$2H_2O(l) + 2e^{-} \longrightarrow H_2(g) + 2OH^{-}(aq) \qquad E° = -0.83 \text{ V (pH = 7.00)}$$

Since the electrode potential for the silver ion is higher, it is easier to reduce. Thus, the cathode reaction is the first one above. Since the nitrate ion does not undergo oxidation, the anode reaction is

$$2H_2O(l) \longrightarrow O_2(g) + 4H^{+}(aq) + 4e^{-}$$

19.11 STOICHIOMETRY OF ELECTROLYSIS

Operational Skill

12. Relating the amounts of charge and product in an electrolysis. Given the amount of product obtained by electrolysis, calculate the amount of charge that flowed (Example 19.12). Given the amount of charge that flowed, calculate the amount of product obtained by electrolysis   (Example 19.13).

It is essential that you know the definitions of *current* and *Faraday*, as you will not be able to work the problems without them.

> **Exercise 19.15** A constant electric current deposits 365 mg of silver in 216 min from an aqueous silver nitrate solution. What is the current?
>
> *Wanted:* current, in amperes (A).
>
> *Given:* 365 mg silver deposited, 216 min, $AgNO_3(aq)$.

Known: A = coulomb (C)/s; $F = 9.65 \times 10^4$ C/mol e$^-$; molar mass Ag = 107.9 g/mol. Since the charge on the silver ion is 1$^+$, it takes 1 mol e$^-$ to deposit 1 mol Ag.

Solution: Current is given in amperes, A, which is charge in coulombs, per time, in seconds. First solve for the number of coulombs:

$$0.365 \text{ g Ag} \times \frac{1 \text{ mol Ag}}{107.9 \text{ g Ag}} \times \frac{1 \text{ mol e}^-}{1 \text{ mol Ag}} \times \frac{9.65 \times 10^4 \text{ C}}{1 \text{ mol e}^-} = 326.4 \text{ C}$$

The time the charge flowed is

$$216 \text{ min} \times \frac{60 \text{ s}}{\text{min}} = 1.296 \times 10^4 \text{ s}$$

Thus,

$$\text{Current} = \frac{\text{C}}{\text{s}} = \frac{326.4 \text{ C}}{1.296 \times 10^4 \text{ s}} = 2.52 \times 10^{-2} \text{ A}$$

When you are more familiar with these calculations, you can solve the problem in one line using dimensional analysis, as shown below:

$$\frac{0.365 \text{ g Ag}}{216 \text{ min}} \times \frac{\text{min}}{60 \text{ s}} \times \frac{1 \text{ mol Ag}}{107.9 \text{ g Ag}} \times \frac{1 \text{ mol e}^-}{1 \text{ mol Ag}} \times \frac{9.65 \times 10^4 \text{ C}}{1 \text{ mol e}^-}$$

$$= 2.52 \times 10^{-2} \text{ C/s} = 2.52 \times 10^{-2} \text{ A}$$

Exercise 19.16   How many grams of oxygen are liberated by the electrolysis of water with a current of 0.0565 A after 185 s?

Wanted: g O_2.

Given: 185 s; current = 0.0565 A.

Known: information given in Exercise 19.15; molar mass O_2 = 32.0 g/mol; anode half-reaction for the electrolysis of water is

$$2H_2O(l) \longrightarrow O_2(g) + 4H^+(aq) + 4e^-$$

Thus, 4 mol of electrons are involved in the liberation of 1 mol of O_2. Also, $F = 9.65 \times 10^4$ C/mol e$^-$; ampere = coulomb/s.

Solution: The calculation is

$$185\cancel{s} \times \frac{0.0565 \cancel{C}}{\cancel{s}} \times \frac{1 \cancel{mole}}{9.65 \times 10^4 \cancel{C}} \times \frac{1 \cancel{mol\ O_2}}{4 \cancel{mole}} \times \frac{32.0\ g\ O_2}{1 \cancel{mol\ O_2}}$$

$$= 8.67 \times 10^{-4} g\ O_2$$

A Metal That Matters: ZINC (a Metal for Batteries)

Questions for Study

1. Explain why zinc was discovered long after brass, which is an alloy of zinc.

2. Describe how zinc is prepared from sphalerite, ZnS.

3. Will zinc metal displace nickel from its salts? If so, write the net ionic equation for the reaction. If it does not, give an example of a metal salt from which zinc will displace the metal, and then write the net ionic equation.

4. Describe some important commercial uses of zinc metal.

5. Describe the function of zinc oxide in a photocopier.

Answers to Questions for Study

1. Zinc has a low boiling point for a metal and would have vaporized during the high-temperature reduction processes that were used in ancient times. Not until the zinc vapor was condensed was the metal recovered.

2. Zinc is prepared by roasting the ore sphalerite (ZnS) in air to form its oxide. The zinc oxide (ZnO) is then heated with coke (C) to reduce it to the metal, which is in the vapor state. The vapor is then condensed to give solid zinc.

3. As the reduction potential of Zn^{2+} (–0.76 V) is lower than that of Ni^{2+} (–0.23 V), zinc will displace nickel from its salts. The net ionic equation for the reaction is

$$Zn(s) + Ni^{2+}(aq) \longrightarrow Zn^{2+}(aq) + Ni(s)$$

4. Zinc is used extensively as a protective coating for other metals, because it forms an adherent oxide coating that protects it from further air oxidation. Thus, the metal underneath is protected from oxidation. Since zinc is a strong reducing agent, it keeps other metals from corroding by cathodic protection. Zinc is also used as the anode in batteries such as the zinc–carbon dry cell, the alkaline dry cell, and the mercury(II) oxide cell. It readily alloys with copper to give brass (20% Cu, 80%

Zn), which is used to make castings. In the form of ZnO, it is used to make rubber, used as a paint pigment, and serves as a photoconductive surface.

5. Zinc oxide is used as a photoconductive surface in photocopiers. When this charged surface is exposed to a light image from a printed document, the lighted areas of the photoconducting surface become electrically conducting, so that the electrical charge in these areas is drained away. The dark areas remain electrically charged, so when black toner is spread over the zinc–oxide surface, it sticks to the electrically charged areas. This image is then transferred with heat to a sheet of paper.

ADDITIONAL PROBLEMS

1. A voltaic cell is constructed with a magnesium electrode in a $MgCl_2$ solution and a nickel electrode in a $NiCl_2$ solution. The salt bridge contains a NaCl solution. The Mg electrode is negative and the Ni electrode positive. Draw a diagram of this cell, write each half-cell reaction and the overall cell reaction, label the anode and cathode, describe the migration of ions through the salt bridge, and indicate the flow of electrons through the external circuit.

2. (a) Give the notation for a voltaic cell constructed from a hydrogen electrode (cathode) in 1.0 M HCl at 1.0 atm pressure and a tin electrode (anode) in 1.0 M tin nitrate solution.

    (b) Write the cell reaction for the following notation:

    $$Zn(s)\,|\,Zn^{2+}(aq)\,|\,|\,Fe^{3+}(aq),\,Fe^{2+}(aq)\,|\,Pt$$

3. The value of E°_{cell} for the reaction $3Br_2(l) + 2Cr(s) \longrightarrow 6Br^{-}(aq) + 2Cr^{3+}(aq)$ is 1.81 V. What is the maximum work that could be obtained from the consumption of 1.00 g of chromium?

4. Using standard electrode potentials, calculate E°_{cell} for the following aluminum–nickel cell:

    $$Al(s)\,|\,Al^{3+}(aq)\,|\,|\,Ni^{2+}(aq)\,|\,Ni(s)$$

    Write the half-cell and overall cell reactions. Which electrode is oxidized?

5. Using the standard electrode potential values given in text Table 19.1, rank the following species in order of decreasing oxidizing strength: Sn^{2+}, I_2, Ag^{+}, Mg^{2+}.

6. The voltage, $E°_{cell}$, of a cell composed of a standard bromine half-cell, $Br_2|Br^-|Pt$, connected to a standard hydrogen half-cell is 1.07 V. The Br_2 electrode is positive. Calculate $E°$ for the bromine half-cell. The $E°$ for the standard chlorine cell is 1.36 V. What does this suggest about the reactivity of Br_2 compared with that of Cl_2 in aqueous solution?

7. (a) Calculate the standard free-energy change for the following reaction under standard conditions.

$$Sn(s) + Pb^{2+}(aq) \longrightarrow Sn^{2+}(aq) + Pb(s)$$

The standard potential for $Sn^{2+}|Sn$ is –0.136 V, and for $Pb^{2+}|Pb$ it is –0.126 V.

   (b) Calculate the equilibrium constant for the reaction under standard conditions.

8. What is the emf at 25°C of the following cell?

$$Be(s)|Be^{2+} (0.100 \ M)||Br_2(l)|Br^- (0.500 \ M)|Pt$$

The standard electrode potential for the $Be^{2+}|Be$ cell is –1.85 V.

9. Give the electrode reactions and the overall reaction when each of the following aqueous solutions is electrolyzed: (a) 0.0005 M KBr, (b) 1.0 M KI. Explain your answers.

10. At low voltages, ferric ions are reduced to ferrous ions in the electrolysis of $FeCl_3$ aqueous solutions.

   (a) What amount of current would be needed to reduce ferric ion in 18.2 g $FeCl_3$ to the lower oxidation state in 10.0 min?

   (b) How many liters of chlorine gas measured at STP would be produced at the anode?

ANSWERS TO ADDITIONAL PROBLEMS

If you missed an answer, study the text section and operational skill given in parentheses after the answer.

1. The overall cell reaction is: $Mg(s) + Ni^{2+}(aq) \longrightarrow Mg^{2+}(aq) + Ni(s)$

The cell diagram is as follows:

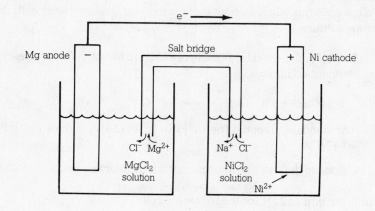

$$Mg(s) \longrightarrow Mg^{2+}(aq) + 2\,e^- \qquad Ni^{2+}(aq) + 2\,e^- \longrightarrow Ni(s)$$

For each Mg^{2+} forming at the anode, two chloride ions migrate from the salt bridge into the anode compartment to maintain electrical neutrality. Similarly, two sodium ions migrate into the cathode chamber for each Ni^{2+} ion that deposits at the cathode surface.   (19.2, Op. Sk. 1)

2.   (a) $Sn(s)\,|\,Sn^{2+}(1.0\ M)\,|\,|\,H^+(1.0\ M)\,|\,H_2(1.0\ atm)\,|\,Pt$

     (b) $Zn(s) + 2Fe^{3+}(aq) \longrightarrow Zn^{2+}(aq) + 2Fe^{2+}(aq)$ (19.3, Op. Sk. 2)

3.   10.1 kJ   (19.4, Op. Sk. 3)

4. $E^\circ_{cell} = +1.43$ V

Cathode half-cell reaction:

   $Ni^{2+}(aq) + 2e^- \longrightarrow Ni(s)$

Anode half-cell reaction:

   $Al(s) \longrightarrow Al^{3+}(aq) + 3e^-$

Overall cell reaction:

$$3Ni^{2+}(aq) + 2Al(s) \longrightarrow 3Ni(s) + 2Al^{3+}(aq)$$

The aluminum electrode is oxidized. (19.5, Op. Sk. 5)

5. decreasing oxidizing strength   (19.5, Op. Sk. 4)
$$\overrightarrow{}$$
$Ag^+ \quad I_2 \quad Sn^{2+} \quad Mg^{2+}$

6. $E° = 1.07$ V. Because $E°$ for the standard chlorine cell is 1.36 V, bromine is less reactive than chlorine in an aqueous solution. (19.5, Op. Sk. 6)

7. (a) –1.9 kJ (19.6, Op. Sk. 7)   (b) 2.2 (19.6, Op. Sk. 9)

8. $E_{cell} = 2.97$ V (19.7, Op. Sk. 10)

9. (a) The expected cathode reaction is

$$2H_2O(l) + 2e^- \longrightarrow H_2(g) + 2OH^-(aq)$$

because $E°$ for this half-reaction is much less negative than that for the reduction of potassium ion. The anode reaction is

$$2H_2O(l) \longrightarrow O_2(g) + 4H^+(aq) + 4e^-$$

because $E°$ for this half-reaction is less negative than for the oxidation of 0.0005 M Br$^-$.

The overall reaction is the electrolysis of water:

$$2H_2O(l) \longrightarrow 2H_2(g) + O_2(g)$$

(b) Since the cation is the same as in part (a), the expected cathode reaction is the same for the same reason. The expected anode reaction is

$$2I^-(aq) \longrightarrow I_2(aq) + 2e^-$$

because $E°$ for the oxidation of iodide ion is less negative than that for the oxidation of water. The overall reaction is

$$2I^-(aq) + 2H_2O(l) \longrightarrow I_2(aq) + H_2(g) + 2OH^-(aq)$$

(19.10, Op. Sk. 11)

10. (a) 18.0 A (b) 1.26 L (19.11, Op. Sk. 12)

CHAPTER POST-TEST

1. Sketch a cell in which a cadmium electrode is in a cadmium nitrate solution and a silver electrode is in a silver nitrate solution. The cells are connected by a salt bridge. Silver is reduced when the cell is operating. Label the anode, cathode, and direction of ion and electron movement, and give the electrode half-reactions.

2. Complete the following table:

Voltaic Cell Notation	Cathode Reaction	Anode Reaction
$Pt \mid H_2(g) \mid H^+(aq) \mid \mid$ $O_2(g) \mid H_2O_2(l) \mid Pt$	(a)	(b)
(c)	$MnO_4^-(aq) + 8H^+(aq) + 5e^-$ $\rightleftharpoons Mn^{2+}(aq) + 4H_2O(l)$	$Ag(s) + Cl^-(aq)$ $\rightleftharpoons AgCl(s) + e^-$
$Zn(s) \mid Zn^{2+}(aq) \mid \mid$ $AuCl_4^-(aq) \mid Au(s)$	(d)	$Zn(s)$ $\rightleftharpoons Zn^{2+}(aq) + 2e^-$
$Pt \mid Cr^{2+}(aq), Cr^{3+}(aq) \mid \mid$ (e)	$Ag(S_2O_3)_2^{3-}(aq) + e^-$ $\rightleftharpoons Ag(s) + 2S_2O_3^{2-}(aq)$	(f)

3. When the cell reaction $Cr_2O_7^{2-}(aq) + 14H^+(aq) + 6Br^-(aq) \longrightarrow 3Br_2(l) + 2Cr^{3+}(aq) + 7H_2O(l)$ takes place with maximum efficiency at 298 K, 5.61 mL of liquid Br_2 (density = 3.119 g/cm^3) are obtained with a maximum electrical work of 5.49×10^3 J. What is the cell emf?

4. Consider the following standard electrode potentials:

   | | $TiO^{2+}(aq)$, $Ti^{3+}(aq)$, $H^+(aq)$ | Pt $E° = 0.1$ V

   | | $Sn^{4+}(aq)$, $Sn^{2+}(aq)$ | Pt $E° = 0.15$ V

   | | $Hg_2^{2+}(aq)$ | $Hg(l)$ $E° = 0.80$ V

   | | $VO^{2+}(aq)$, $V^{3+}(aq)$, $H^+(aq)$ | Pt $E° = 0.361$ V

   | | $Mn^{2+}(aq)$ | $Mn(s)$ $E° = -1.18$ V

   | | $Ag(CN)_2^-(aq)$, $CN^-(aq)$ | $Ag(s)$ $E° = -0.31$ V

   | | $Fe^{3+}(aq)$, $Fe^{2+}(aq)$ | Pt $E° = 0.771$ V

   | | $Pb^{2+}(aq)$ | $Pb(s)$ $E° = -0.13$ V

   | | $I_2(s)$ | $I^-(aq)$ | Pt $E° = 0.54$ V

   | | $Sn^{2+}(aq)$ | $Sn(s)$ $E° = -0.14$ V

   Determine which of the following is not a voltaic cell.

   (a) Pt | $Ti^{3+}(aq)$, $TiO^{2+}(aq)$, $H^+(aq)$ | | $Sn^{4+}(aq)$, $Sn^{2+}(aq)$ | Pt

   (b) $Hg(l)$ | $Hg_2^{2+}(aq)$ | | $H^+(aq)$, $V^{3+}(aq)$, $VO^{2+}(aq)$ | Pt

   (c) $Mn(s)$ | $Mn^{2+}(aq)$ | | $Ag(CN)_2^-(aq)$, $CN^-(aq)$ | $Ag(s)$

   (d) $Pb(s)$ | $Pb^{2+}(aq)$ | | $Fe^{3+}(aq)$, $Fe^{2+}(aq)$ | Pt

   (e) $Sn(s)$ | $Sn^{2+}(aq)$ | | $I_2(s)$ | $I^-(aq)$ | Pt

5. When the standard hydrogen electrode, $H^+(aq)$ | H_2(1 atm) | Pt, is dipped into an aqueous solution and connected to a Cu^{2+}(1.00 M) | $Cu(s)$ cell at 298 K, the emf is 0.36 V. Calculate the pH of this aqueous solution. $E°$ for the $Cu^{2+}(aq)$ | $Cu(s)$ half-cell is 0.34 V.

6. During the discharge of a conventional dry cell,
   (a) the voltage continually drops as current is drawn.
   (b) Zn^{2+} ions are reduced at the cathode.
   (c) the anode reaction is more highly favored than the cathode reaction.

(d)   the graphite cathode reverses polarity.

(e)   none of the above occurs.

7.   The following reaction occurs in a voltaic cell:

$$O_2(g) + 2ClO_3^-(aq) \rightleftharpoons 2ClO_4^-(aq)$$

The standard electrode potentials are: $O_2(g)|OH^-(aq)|Pt$, $E° = 0.40$ V; $ClO_4^-(aq)$, $ClO_3^-(aq)$, $OH^-(aq)|Pt$, $E° = 0.17$ V. Complete the following statement with respect to this cell (more than one choice may be made) and explain your reasoning.

The cell emf can be decreased by

(a)   decreasing $[OH^-]$.

(b)   increasing $[ClO_4^-]$.

(c)   decreasing P_{O_2}.

(d)   increasing $[ClO_3^-]$.

(e)   none of the above.

8.   Calculate K_c at 298 K for the reaction

$$IO^-(aq) + H_2O(l) + Cu(s) \rightleftharpoons Cu^{2+}(aq) + I^-(aq) + 2OH^-(aq)$$

The standard cell emf is 0.15 V.

9.   When a 1 M solution of $CuBr_2$ is electrolyzed, what are the products at the anode and cathode?

10.   How many liters of Cl_2 gas are obtained from a cell that is electrolyzed for 5.0 h at a current of 0.050 A? The gas is collected at STP (assume ideal behavior). The cell half-reaction is $Cl_2(g) + 2e^- \longrightarrow 2Cl^-(aq)$.

ANSWERS TO CHAPTER POST-TEST

If you missed an answer, study the text section and operational skill given in parentheses after the answer.

1.

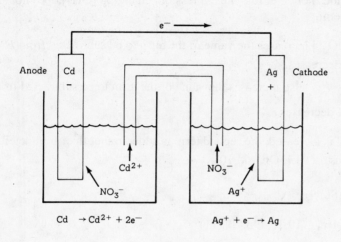

Anode Cd Ag Cathode

$Cd \rightarrow Cd^{2+} + 2e^-$ $Ag^+ + e^- \rightarrow Ag$

(19.2, Op. Sk. 1)

2. (a) $O_2(g) + 2H^+(aq) + 2e^- \rightleftharpoons H_2O_2(l)$

   (b) $H_2(g) \rightleftharpoons 2H^+(aq) + 2e^-$

   (c) $Ag(s), AgCl(s) \mid\mid MnO_4^-(aq), Mn^{2+}(aq) \mid Pt$

   (d) $AuCl_4^-(aq) + 3e^- \rightleftharpoons 4Cl^-(aq) + Au(s)$

   (e) $Ag(S_2O_3)_2^{3-}(aq) \mid Ag(s) \mid Pt$

   (f) $Cr^{2+}(aq) \rightleftharpoons Cr^{3+}(aq) + e^-$ (19.3, Op. Sk. 2)

3. 0.260 V (19.4, 19.11, Op. Sk. 3, 12)

4. b (19.2, 19.5, Op. Sk. 5)

5. 0.3 (19.5, 19.6, 19.7, Op. Sk. 6, 10)

6. a (19.2, 19.8)

7.   b and c. The Nernst equation for this cell is

$$E_{cell} = E°_{cell} - \frac{0.0592 \text{ V}}{2} \log \frac{[\text{ClO}_4^-]^2}{P_{\text{O}_2}[\text{ClO}_3^-]^2}$$

(a)   Since $[\text{OH}^-]$ does not enter into the expression, changing $[\text{OH}^-]$ does not affect the cell emf.

(b)   Increasing $[\text{ClO}_4^-]$ increases the value of the term to be subtracted from $E°$; thus, the emf decreases.

(c)   Decreasing the P_{O_2} also increases the value of the term to be subtracted from $E°$, so the emf decreases.

(d)   When $[\text{ClO}_3^-]$ is increased, the second term is added rather than subtracted and E increases. (19.7, Op. Sk. 10)

8. 1×10^5 (19.6, Op. Sk. 9)

9.   cathode, Cu; anode, Br_2 (19.10, Op. Sk. 11)

10.   0.10 L (19.11, Op. Sk. 12)

CHAPTER 20   NUCLEAR CHEMISTRY

CHAPTER TERMS AND DEFINITIONS

Numbers in parentheses after definitions give the text sections in which the terms are explained. Starred terms are italicized in the text. Where a term does not fall directly under a text section heading, additional information is given for you to locate it.

tekhnetos*   Greek word meaning "artificial"; root of element name technetium   (chapter introduction)

radioactive decay   process in which a nucleus spontaneously disintegrates, giving off radiation   (20.1, introductory section)

nuclear bombardment reaction   rearrangement of reactant nucleus particles after being struck by another nucleus or nuclear particle, to give a product nucleus or nuclei (20.1, introductory section)

nuclide symbol*   symbol that indicates the atomic number, mass number, and identity of a nucleus   (20.1)

nuclear equation   symbolic representation of a nuclear reaction   (20.1)

positron   particle similar to an electron, with the same mass but with a positive charge (20.1)

gamma photon   particle of electromagnetic radiation of short wavelength (1 pm, or 10^{-12} m) and high energy   (20.1)

nucleons*   particles composing an atomic nucleus, i.e., protons and neutrons   (20.1)

nuclear force   strong force of attraction between nucleons; acts only at very short distances (about 10^{-15} m)   (20.1)

shell model of the nucleus   nuclear model in which protons and neutrons exist in levels, or shells, analogous to the shell structure that exists for electrons in an atom   (20.1)

magic number   number of nuclear particles in a completed shell of protons or neutrons, associated with very stable nuclei   (20.1)

band of stability   on a plot of number of protons (Z) against number of neutrons (N), the area or region in which stable nuclides lie   (20.1)

alpha emission (α)   ejection of a $^4_2$He nucleus, or alpha particle, from an unstable nucleus (20.1)

beta emission (β **or** β^-)   ejection of a high-speed electron from an unstable nucleus (20.1)

positron emission (β^+)   ejection of a positron from an unstable nucleus   (20.1)

electron capture (EC)   decay of an unstable nucleus by picking up an electron from an inner obital of an atom   (20.1)

gamma emission (γ)   ejection of a photon of very short wavelength (about 10^{-12} m) from an excited nucleus   (20.1)

metastable nucleus   nucleus in an excited state with a lifetime of at least 10^{-9} s (one nanosecond)   (20.1)

radioactive decay series   sequence of steps in which a naturally radioactive element decays to a stable nuclide   (20.1)

transmutation   change of one element to another element by bombardment of its nucleus with nuclear particles or nuclei   (20.2)

particle accelerator   device used to impart high velocities to electrons, protons, alpha particles, and other ions   (20.2)

electron volt (eV) 1.602×10^{-19} J; energy needed to accelerate an electron (whose charge is 1.602×10^{-19} C) by one volt potential difference   (20.2)

cyclotron   type of particle accelerator consisting of two semicircular metal electrodes in which charged particles are accelerated by stages to higher and higher kinetic energies   (20.2)

dees*   *D*-shaped hollow metal electrodes in a cyclotron   (20.2)

deuterons   nuclei of hydrogen-2 atoms   (20.2)

transuranium elements   elements with atomic numbers greater than 92   (20.2)

ionization counter*   device that uses the production of ions in matter to determine the number of particles emitted in nuclear processes   (20.3)

Geiger counter   ionization counter used to count particles emitted from radioactive nuclei; as the particle enters the metal tube, it ionizes the enclosed gas, causing current to flow to activate a counter or to produce an audible click   (20.3)

scintillation counter   device that detects nuclear radiation from flashes of light generated in a material by the radiation   (20.3)

phosphor*   substance that emits flashes of light when struck by radiation   (20.3)

photomultiplier*   electronic device (tube) that magnifies (amplifies) the effect of the incidence of a photon   (20.3)

activity of a radioactive source   number of nuclear disintegrations per unit time   (20.3)

curie (Ci)   unit measure of radioactivity; equals 3.700×10^{10} disintegrations/s   (20.3)

rad   dosage of radiation that deposits 1×10^{-2} J of energy per kilogram of tissue (<u>ra</u>diation <u>a</u>bsorbed <u>d</u>ose)   (20.3)

rem   unit of radiation dosage used to relate the various kinds of radiation in terms of biological destruction; rems = rads × RBE   (20.3)

relative biological effectiveness (RBE)*   factor for a given type of radiation that, when multiplied by the radiation-absorbed dose, yields the rem, a measure of biological destruction   (20.3)

radioactive decay constant (*k*)   rate constant for radioactive decay   (20.4)

half-life   time required for one-half of the nuclei in a sample to decay   (20.4)

radioactive tracer   radioactive isotope added to a chemical, biological, or physical system to study the system   (20.5)

isotope dilution   technique to determine the quantity of substance in a mixture or the total volume of solution by adding a known amount of a radioactive isotope to it   (20.5)

neutron activation analysis   analysis of elements in a sample based on conversion of stable isotopes to radioactive isotopes by bombardment of the sample with neutrons   (20.5)

radioimmunoassay*   technique for analyzing body fluids for very small quantities of biologically active substances; uses radioactive isotopes and depends on reversible binding of a substance to an antibody   (20.5)

binding energy   energy required to break a nucleus into its individual protons and neutrons   (20.6)

mass defect   total nucleon mass minus the nuclear mass   (20.6)

nuclear fission   nuclear reaction in which a heavy nucleus splits into lighter nuclei and energy is released   (20.6)

nuclear fusion   nuclear reaction in which light nuclei combine to give a more stable, heavier nucleus plus possibly several neutrons, and energy is released   (20.6)

chain reaction   self-sustaining series of successive nuclear fissions caused by absorption of neutrons released from previous nuclear fissions   (20.7)

critical mass   smallest mass of fissionable material in which a chain reaction can be sustained   (20.7)

supercritical (mass)*   mass larger than the critical mass   (20.7)

nuclear fission reactor   device that permits a controlled chain reaction of nuclear fissions   (20.7)

fuel rods   in a nuclear fission reactor, the cylinders that contain fissionable material   (20.7)

control rods   in a nuclear fission reactor, cylinders composed of elements (e.g., boron, cadmium) that absorb neutrons and can therefore slow the chain reaction   (20.7)

moderator   in a nuclear fission reactor, a substance that slows down neutrons   (20.7)

reprocessing plants*   facilities where fuel material is separated from radioactive wastes   (20.7)

plasma   electrically neutral gas of ions and electrons   (20.7)

tokamak nuclear fusion reactor*   nuclear fusion reactor that uses a doughnut-shaped magnetic field to contain the plasma   (20.7)

laser fusion reactor* nuclear fusion reactor that employs a bank of lasers aimed at a
single point (20.7)

CHAPTER DIAGNOSTIC TEST

1.   Determine whether each of the following statements is true or false. If the statement
is false, change it so it is true.
(a)   When a positron is emitted by a radioactive nucleus, the atomic number
remains unchanged but the mass number increases by one unit.   True/False:
_____ .

(b)   Neutron activation analysis is a method of analysis based on the conversion of
stable isotopes to radioactive isotopes through neutron bombardment.
True/False: _____
_____ .

(c)   The biological effect of radiation depends only on the energy of radiation, the
type of radiation, and the length of time of exposure.   True/False: _____
_____ .

(d)   Technetium-99*m* is the radioactive isotope most often used to develop picture
images of internal body organs.   True/False: _____
_____ .

(e)   Nuclides between mass numbers 1 and 10 have the largest binding energies
per nucleon.   True/False: _____
_____ .

(f)   In a nuclear fission reactor, the control rods slow down neutrons for
absorption by other uranium-235 nuclei.   True/False: _____
_____ .

2.   A solution of sodium iodide containing iodine-131 was administered to a patient to
determine the activity of the thyroid gland. What fraction of the iodine-131 would
still be in the patient's body after one day? (The half-life of iodine-131 is 8.04 days.)

3. Complete the following table:

Particle and Identity	Abbreviation	Nuclide or Particle Symbol
beta particle (electron)	(a)	(b)
(c)	β^+	(d)
(e)	(f)	^4_2He
proton	(g)	(h)
(i)	n	(j)
(k)	(l)	^2_1H
gamma photon	(m)	(n)

4. Complete each of the following nuclear reactions by inserting the missing symbols. Use a periodic table.

   (a) $^{34}_{}\text{Cl} \longrightarrow {}^0_1\text{e} + \underline{\hspace{1cm}}$

   (b) $^{235}_{92}\underline{\hspace{0.6cm}} + {}^{}_0\text{n} \longrightarrow {}^{94}_{}\text{Kr} + 3{}^1_0\text{n} + \underline{\hspace{1cm}}$

   (c) $^{93}_{}\text{Zr} \longrightarrow {}^{0}_{-1}\text{e} + {}^{93}_{}\underline{\hspace{1cm}}$

5. Explain how the following would be expected to decay. Use text Figure 20.3.

   (a) $^{39}_{17}\text{Cl}$ (b) $^{84}_{38}\text{Sr}$ (c) $^{232}_{90}\text{Th}$

6. Which one of the following nuclides would be expected to be radioactive, and which two would be stable? Give reasons for your answer.

   (a) $^{37}_{17}\text{Cl}$ (b) $^{51}_{23}\text{V}$ (c) $^{238}_{95}\text{Am}$

7. An 8.6-mg sample of iron-59 registers 125 counts per second on a radiation counter. What is the decay constant?

8. Write the nuclear equation for each of the following:
   (a) $^{27}_{13}\text{Al}(\text{d}, \alpha)^{25}_{12}\text{Mg}$ (b) $^{63}_{29}\text{Cu}(\text{p}, \text{n})^{63}_{30}\text{Zn}$

9. Write the abbreviated notation for each of the following:

    (a) $^{63}_{29}Cu + ^{4}_{2}He \longrightarrow ^{66}_{31}Ga + ^{1}_{0}n$

    (b) $^{12}_{6}C + ^{3}_{1}H \longrightarrow ^{14}_{6}C + ^{1}_{1}H$

    (The symbol for $^{3}_{1}H$ is t.)

10. According to current theory, the elements of which we are composed, and of which all matter is composed, were made in the stars from fusion of simpler nuclei. Calculate the energy released (in joules) when one gram of oxygen is formed in the following fusion reaction: $^{12}_{6}C + ^{4}_{2}He \longrightarrow ^{16}_{8}O$. Use data from text Table 20.3. Compare this with the specific heat of water to get a feel for the size of this value.

11. The Shroud of Turin, once believed by many to have been the burial cloth of Jesus of Nazareth, was subjected to carbon-14 dating in 1988. The results showed that the shroud is of medieval origin, between 1260 and 1390 A.D., with more than 95% certainty. Using 1325 A.D. as the year of origin, calculate the activity of the carbon in the shroud. The half-life of C-14 is 5.73×10^{3} y. The activity of C-14 from living material is 15.3 disintegrations per minute per gram carbon.

12. If 35.0% of a sample of silver-112 decays in 1.99 h, what is the half-life of this isotope (in hours)? What is the activity in Ci of a sample containing 4.8 g of silver-112 (1 Ci = 3.7×10^{10} nuclei/s)?

ANSWERS TO CHAPTER DIAGNOSTIC TEST

If you missed an answer, study the text section and operational skill given in parentheses after the answer.

1. (a) False. When a positron is emitted by a radioactive nucleus, the atomic number decreases by one unit, but the mass number remains unchanged. (20.1)
    (b) True. (20.5)
    (c) False. The biological effect of radiation also depends on the type of tissue involved. (20.3)
    (d) True. (20.5)
    (e) False. Nuclides between mass numbers 56 and 74 have the largest binding energies per nucleon. (20.6, text Figure 20.16)

(f) False. In a nuclear fission reactor, the control rods absorb neutrons to slow the rate of the chain reaction. (20.7)

2. 0.917 (20.4, Op. Sk. 8)

3. (a) β or β^-

(b) $_{-1}^{0}e$ or $_{-1}^{0}\beta$

(c) positron

(d) $_{1}^{0}e$ or $_{1}^{0}\beta$

(e) alpha particle (helium nucleus)

(f) α

(g) p

(h) $_{1}^{1}p$ or $_{1}^{1}H$

(i) neutron

(j) $_{0}^{1}n$

(k) deuterium nucleus

(l) d

(m) γ

(n) $_{0}^{0}\gamma$

(20.1)

4. (a) $_{17}^{34}Cl$, $_{16}^{34}S$

(b) $_{92}^{235}U$, $_{0}^{1}n$, $_{36}^{94}Kr$, $_{56}^{139}Ba$

(c) $_{40}^{93}Zr$, $_{41}^{93}Nb$ (20.1, Op. Sk. 2)

5. (a) β^- emission. Cl-39 has a mass number greater than that of the stable Cl-37 isotope. β^- emission is the expected mode of decay.

(b) Positron emission. Sr-84 has a mass number less than that of the stable Sr-87 isotope. Positron emission is the expected decay, as the nucleus is relatively light.

(c) α emission. The nucleus has more than 83 protons. (20.1, Op. Sk. 4)

6. Nuclides (a) and (b) are stable: Cl-37 has 20 neutrons, a magic number; and V-51 has a magic number of neutrons, 28. Nuclide (c), Am-238, has over 83 protons and would thus be expected to be radioactive. (20.1, Op. Sk. 3)

7. 1.4×10^{-18}/s (20.4, Op. Sk. 6)

8. (a) $_{13}^{27}Al + _{1}^{2}H \longrightarrow _{12}^{25}Mg + _{2}^{4}He$

(b) $_{29}^{63}Cu + _{1}^{1}H \longrightarrow _{30}^{63}Zn + _{0}^{1}n$ (20.2, Op. Sk. 5)

9.   (a) $^{63}_{29}\text{Cu}(\alpha, \text{n})\,^{66}_{31}\text{Ga}$

     (b) $^{12}_{6}\text{C}(\text{t}, \text{p})\,^{14}_{6}\text{C}$ (20.2, Op. Sk. 5)

10. Energy released is 4.3×10^{10} J per gram of oxygen formed. The specific heat of water = 4.184 J to raise 1 g of water 1°C. The calculated energy is approximately 10^{10} times as large!   (20.6, Op. Sk. 10)

11. 14.1 disintegrations per minute per gram carbon. See the article "Radiocarbon Dating of the Shroud of Turin" by P. E. Damon *et al.* in the February 16, 1989, issue of *Nature*, Vol. 337, p. 611.   (20.4, Op. Sk. 9)

12. $t_{1/2} = 3.20$ h; 4.2×10^{7} Ci   (20.4, Op. Sk. 7)

SUMMARY OF CHAPTER TOPICS

20.1   RADIOACTIVITY

Operational Skills

1. Writing a nuclear equation.   Given a word description of a radioactive decay process, write the nuclear equation   (Example 20.1).

2. Deducing a product or reactant in a nuclear equation.   Given all but one of the reactants and products in a nuclear reaction, find that one nuclide   (Examples 20.2 and 20.6).

3. Predicting the relative stabilities of nuclides.   Given a number of nuclides, determine which are most likely to be radioactive and which are most likely to be stable (Example 20.3).

4. Predicting the type of radioactive decay.   Predict the type of radioactive decay that is most likely for given nuclides   (Example 20.4).

In Chapter 2, you were introduced to the term *isotope*, which is one of the terms we use to designate an atomic nucleus. In this chapter, we use the word *nuclides* to refer to atomic nuclei. The difference is that *isotope* refers to the nuclides of one element; *nuclide* is the more general term.

Exercise 20.1 Potassium-40 is a naturally occurring radioactive isotope. It decays to calcium-40 by beta emission. When a potassium-40 nucleus decays by beta emission, it emits one beta particle and gives a calcium-40 nucleus. Write the nuclear equation for this decay.

Wanted: nuclear equation.

Given: Potassium-40 decays to calcium-40 by beta emission.

Known: We write nuclide symbols as follows: Potassium-40 is

$$^{40}_{19}K \qquad \text{Mass number = protons + neutrons} \\ \text{Atomic number = number of protons}$$

Other symbols are $^{40}_{20}Ca$, and $^{0}_{-1}e$ $\left(\text{or } ^{0}_{-1}\beta \right)$ for the beta particle.

Solution: $^{40}_{19}K \longrightarrow ^{40}_{20}Ca + ^{0}_{-1}e$

Exercise 20.2 Plutonium-239 decays by alpha emission, with each nucleus emitting one alpha particle. What is the other product of this decay?

Known: For a nuclear reaction, the sum of the subscripts for the products must equal the sum of the subscripts for the reactants, and the same is true for the superscripts; an alpha particle is $^{4}_{2}He$.

Solution: The transformation written symbolically is

$$^{239}_{94}Pu \longrightarrow ^{4}_{2}He + ?$$

The missing mass number is $239 - 4 = 235$. The missing charge, or atomic number, is $94 - 2 = 92$. The other product is a nuclide with atomic number 92, which we see from the periodic table is U-235, or $^{235}_{92}U$.

To determine whether a nuclide falls above or below the band of stability on text Figure 20.3, a plot of N versus Z, just find the point on the plot that correlates with the number of neutrons and protons and see where that point is in relation to the band.

Exercise 20.3 Of the following nuclides, two are radioactive. Which are radioactive and which is stable? Explain. (a) $^{118}_{50}Sn$; (b) $^{76}_{33}As$; (c) $^{227}_{89}Ac$.

Known: Stability correlates with "magic numbers" of protons (2, 8, 20, 28, 50, 82, and 114) and of neutrons (2, 8, 20, 28, 50, 82, and 126); stability also correlates with paired protons and paired neutrons, with the appropriate N/Z ratio, and with 83 or fewer protons.

Solution: (a) $^{118}_{50}$Sn is expected to be stable. It has a magic number of protons (50) and an appropriate *N/Z* ratio. (b) $^{76}_{33}$As is expected to be radioactive. It has an odd number of protons (33) and an odd number of neutrons (76 − 33 = 43), and the *N/Z* ratio is somewhat high. (c) $^{227}_{89}$Ac is expected to be radioactive since it has more than 83 protons and a very high *N/Z* ratio.

To summarize current theory, nuclides decay for one of four reasons: the *N/Z* ratio is too large or too small, *Z* is too large, or the nuclide is in an excited state. Study guide Table 20.1 lists the first three reasons, with the types of decay that may occur in each case to give greater stability, the equation for the decay, and the effect on the nucleus of the decay. The fourth reason for decay does not affect the number of nuclear particles and thus is not listed. In order to predict the type of decay and to write nuclear decay reactions, you will have to memorize columns 1 and 2 of study guide Table 20.1.

Table 20.1

Reason for Decay	Decay Type and Equation	Effect of Decay on Nucleus
N/Z is too large (above band of stability)	beta emission (more common) $${}_0^1n \longrightarrow {}_1^1p + {}_{-1}^0e$$	A neutron is converted to a proton, decreasing the N/Z ratio.
	neutron emission (less common) $$nuclide_1 \longrightarrow nuclide_2 + {}_0^1n$$	A neutron is lost, decreasing the N/Z ratio.
N/Z is too small (below band of stability)	electron capture ($Z > 80$) $${}_1^1p + {}_{-1}^0e \longrightarrow {}_0^1n$$	A proton is converted to a neutron, increasing the N/Z ratio.
	positron emission ($Z < 20$) $${}_1^1p \longrightarrow {}_0^1n + {}_1^0e$$	A proton is converted to a neutron, increasing the N/Z ratio.
Z is too large (beyond band of stability)	alpha emission ($Z > 83$) $$nuclide_3 \longrightarrow nuclide_4 + {}_2^4He$$	Z decreases by 2.
	spontaneous fission may occur (to be discussed later)	

Exercise 20.4 Predict the type of decay expected for each of the following radioactive nuclides: (a) ${}_7^{13}N$; (b) ${}_{11}^{26}Na$.

Known: reasons for instability and types of decay (study guide Table 20.1); rules for writing nuclear decay reactions.

Solution: (a) The atomic weight of nitrogen is 14.0, so you expect nitrogen-14 to be a stable isotope with an N/Z ratio of 1. Nitrogen-13 has one fewer than in the stable isotope. Thus, you expect it to decay by either positron emission or electron capture. Because positron emission is generally observed with the lighter elements, we predict that nitrogen-13 will undergo positron emission.

$$^{13}_{7}N \longrightarrow {}^{0}_{1}e + {}^{13}_{6}C$$

(b) The atomic weight of sodium is 23.0. Sodium-26 has a mass number greater than that of the stable isotope sodium-23. Thus, we expect sodium-26 to decay by beta emission.

$$^{26}_{11}Na \longrightarrow {}^{0}_{-1}e + {}^{26}_{12}Mg$$

Neutron emission, less commonly observed, would give

$$^{26}_{11}Na \longrightarrow {}^{1}_{0}n + {}^{25}_{11}Na$$

which makes the nucleus even more unstable. We would not expect this.

20.2   NUCLEAR BOMBARDMENT REACTIONS

Operational Skill

5. Using the notation for a bombardment reaction. Given an equation for a nuclear bombardment reaction, write the abbreviated notation, or vice versa (Example 20.5).

Greed for gold led the early alchemists in search of transmutation of the elements. This phenomenon eluded researchers until early in this century. The value of the transmutation process is far greater than the gold the ancients sought. The energies unleashed equal those that light the stars.

In this section you will need to memorize the symbolism used for writing nuclear bombardment reactions.

Exercise 20.5   (a) Write the abbreviated notation for the reaction

$$^{40}_{20}Ca + {}^{2}_{1}H \longrightarrow {}^{41}_{20}Ca + {}^{1}_{1}H$$

(b) Write the nuclear equation for the bombardment reaction ${}^{12}_{6}C(d, p){}^{13}_{6}C$.

Known: Symbolism for writing reactions: n = neutron; p = proton; d = deuteron, ${}^{2}_{1}H$; α = alpha, ${}^{4}_{2}He$.

Solution: (a) $^{40}_{20}Ca(d, p)^{41}_{20}Ca$

(b) $^{12}_{6}C + ^{2}_{1}H \longrightarrow ^{13}_{6}C + ^{1}_{1}H$

Exercise 20.6 Carbon-14 is produced in the upper atmosphere when a particular nucleus is bombarded with neutrons. A proton is ejected for each nucleus that reacts. What is the identity of the nucleus that produces carbon-14 by this reaction?

Wanted: nucleus that reacts, labeled X.

Given: $^{A}_{Z}X(n, p)^{14}_{6}C$.

Known: how to balance a nuclear equation; symbolism for bombardment reactions.

Solution: The equation is

$$^{A}_{Z}X + ^{1}_{0}n \longrightarrow ^{1}_{1}H + ^{14}_{6}C$$

Rewrite the equation to balance mass number (superscripts) and charge (subscripts):

$$A + 1 = 1 + 14 = 15$$

$$Z + 0 = 1 + 6 = 7$$

Solving each of the above, we see that $A = 14$ and $Z = 7$. Thus, the nucleus that reacts is $^{14}_{7}X$, which from the periodic table we see is nitrogen-14.

20.3 RADIATIONS AND MATTER: DETECTION AND BIOLOGICAL EFFECTS

It will be useful for you to refer back to the table ("The Electromagnetic Spectrum") in Section 7.1 of this study guide to look over the various energies and what effect each has on molecules. An enzyme cannot function if its active site (where it binds the molecule whose reaction it catalyzes) is torn apart. A molecule of genetic matter (a DNA molecule) cannot be reproduced properly if part of it is missing. Thus, high-energy radiation plays havoc with living systems.

20.4 RATE OF RADIOACTIVE DECAY

Operational Skills

   6. Calculating the decay constant from the activity. Given the activity (disintegrations per second) of a radioactive isotope, obtain the decay constant   (Example 20.7).

   7. Relating the decay constant, half-life, and activity. Given the decay constant of a radioactive isotope, obtain the half-life   (Example 20.8), or vice versa   (Example 20.9). Given the decay constant and mass of a radioactive isotope, calculate the activity of the sample   (Example 20.9).

   8. Determining the fraction of nuclei remaining after a specified time. Given the half-life of a radioactive isotope, calculate the fraction remaining after a specified time (Example 20.10).

   9. Applying the carbon-14 dating method. Given the disintegrations of carbon-14 nuclei per gram of carbon in a dead organic object, calculate the age of the object — that is, the time since its death   (Example 20.11).

        Before you work the problems in this section, it will be helpful to review text Section 13.4, which covers first-order rate calculations. In our calculations in this section, we simplify the rate law to rate $= kN_t$, where k is the rate constant and N_t is the number of radioactive nuclei at time t. Recall that in the calculations involving first-order kinetics in Section 13.4, you used the concentration of the reactant. We can also use number of nuclei because the number is proportional to the concentration of a solid.

   Exercise 20.7 The nucleus $^{99m}_{43}\text{Tc}$ is a metastable nucleus of technetium-99; it is used in medical diagnostic work. Technetium-99m decays by emitting gamma rays. A 2.5-μg (microgram) sample has an activity of 13 Ci. What is the decay constant (in units of /s)?

   Wanted: k (in the rate law).

   Given: Activity of a 2.5-μg sample of technetium-99m is 13 Ci.

   Known: Nuclear decay is a first-order reaction; the rate law is rate $= kN_t$;

   Tc = 99 g/mol; activity = rate of decay; 1.0 Ci $= 3.7 \times 10^{10}$ nuclei/s;

   1 mol $= 6.02 \times 10^{23}$ nuclei.

Solution: Calculate rate of decay in terms of disintegrations/s:

$$\text{rate} = 13 \, \cancel{Ci} \times \frac{3.7 \times 10^{10} \, \text{nuclei/s}}{1.0 \, \cancel{Ci}} = 4.\underline{8}1 \times 10^{11} \, \text{nuclei/s}$$

Now calculate N_t:

$$N_t = 2.5 \times 10^{-6} \, \cancel{g\,Tc} \times \frac{\cancel{mol\,Tc}}{99 \, \cancel{g\,Tc}} \times \frac{6.02 \times 10^{23} \, \text{nuclei}}{1 \, \cancel{mol\,Tc}}$$

$$= 1.\underline{5}2 \times 10^{16} \, \text{nuclei}$$

Then, solve the rate equation for k and insert values:

$$\text{rate} = kN_t$$

$$k = \frac{\text{rate}}{N_t} = \frac{4.\underline{8}1 \times 10^{11} \, \cancel{\text{nuclei}}/s}{1.\underline{5}2 \times 10^{16} \, \cancel{\text{nuclei}}}$$

$$= 3.2 \times 10^{-5}/s$$

Exercise 20.8 Cobalt-60, used in cancer therapy, decays by beta and gamma emission. The decay constant is $4.18 \times 10^{-9}/s$. What is the half-life in years?

Wanted: $t_{1/2}$ in years.

Given: $k = 4.18 \times 10^{-9}/s$.

Known: For a first-order reaction, $t_{1/2} = \dfrac{0.693}{k}$; conversion factors from s to y.

Solution: The calculation is

$$t_{1/2} = \frac{0.693 \, \cancel{s}}{4.18 \times 10^{-9}} \times \frac{1 \, \cancel{h}}{3600 \, \cancel{s}} \times \frac{1 \, \cancel{day}}{24 \, \cancel{h}} \times \frac{1 \, y}{365 \, \cancel{day}}$$

$$= 5.26 \, y$$

Exercise 20.9 Strontium-90, $^{90}_{38}Sr$, is a radioactive decay product of nuclear fallout from nuclear weapons testing. Because of its chemical similarity to calcium, it is incorporated into the bones if present in food. The half-life of strontium-90 is 28.1 y. What is the decay constant of this isotope? What is the activity of a sample containing 5.2 ng (5.2×10^{-9} g) of strontium-90?

Wanted: k; activity (Ci).

Given: strontium-90; $t_{1/2} = 28.1$ y; 5.2-ng sample.

Known: rate $= kN_t$; $t_{1/2} = 0.693/k$; 1.0 Ci $= 3.7 \times 10^{10}$ nuclei/s;

1 mol $= 6.02 \times 10^{23}$ nuclei; activity = rate in Ci; Sr = 90 g/mol (exact).

Solution: Solve for k:

$$t_{1/2} = \frac{0.693}{k}$$

$$k = \frac{0.693}{t_{1/2}} = \frac{0.693}{28.1 \, \cancel{y}} \times \frac{\cancel{y}}{365 \, \cancel{day}} \times \frac{\cancel{day}}{24 \, \cancel{h}} \times \frac{\cancel{h}}{3600 \, s}$$

$$= 7.8\underline{2}0 \times 10^{-10}/s$$

$$= 7.82 \times 10^{-10}/s$$

To solve for activity, we first determine N_t:

$$N_t = 5.2 \times 10^{-9} \, \cancel{g \, Sr} \times \frac{\cancel{mol \, Sr}}{90 \, \cancel{g \, Sr}} \times \frac{6.02 \times 10^{23} \, nuclei}{\cancel{mol \, Sr}}$$

$$= 3.\underline{4}8 \times 10^{13} \, nuclei$$

Then, activity = rate (in Ci) $= kN_t$

$$= \frac{7.82 \times 10^{-10}}{\cancel{s}} \times 3.\underline{4}8 \times 10^{13} \, \cancel{nuclei} \times \frac{1.0 \, Ci}{3.7 \times 10^{10} \, \cancel{nuclei /s}}$$

$$= 7.4 \times 10^{-7} \, Ci$$

Exercise 20.10   A nuclear power plant emits into the atmosphere a very small amount of krypton-85, a radioactive isotope with a half-life of 10.76 y. What fraction of this krypton-85 remains after 25.0 y?

Wanted: fraction of Kr-85 remaining.

Given: $t_{1/2} = 10.76$ y; $t = 25.0$ y.

Known: $\log \dfrac{N_o}{N_t} = \dfrac{kt}{2.303}$; $t_{1/2} = \dfrac{0.693}{k}$; the fraction remaining is $\dfrac{N_t}{N_o}$.

Solution: To find the fraction of Kr-85 remaining, N_t/N_o, we will first determine k algebraically. Since $t_{1/2} = \dfrac{0.693}{k}$, then $k = \dfrac{0.693}{t_{1/2}}$. Substituting this into

$\log \dfrac{N_o}{N_t} = \dfrac{kt}{2.303}$ gives $\log \dfrac{N_o}{N_t} = \dfrac{0.693\, t}{2.303\, t_{1/2}}$. Putting values into the right side gives

$$\log \frac{N_o}{N_t} = \frac{0.693 \times 25.0\ \cancel{y}}{2.303 \times 10.76\ \cancel{y}} = 0.699\underline{1}$$

$$\frac{N_o}{N_t} = \text{antilog } 0.699\underline{1} = 5.0\underline{0}1$$

The fraction remaining $= \dfrac{N_t}{N_o} = \dfrac{1}{5.0\underline{0}1} = 0.200.$

Exercise 20.11   A jawbone from the archaeological site at Folsom, New Mexico, was dated by analysis of its radioactive carbon. The activity of the carbon from the jawbone was 4.5 disintegrations per minute per gram of total carbon. What was the age of the jawbone? Carbon from living material gives 15.3 disintegrations per minute per gram of carbon.

Wanted: t (age of jawbone).

Given: Activity = 4.5 disintegrations/min per g C;

activity of living material = 15.3 disintegrations/min per g C.

Known: $\log \dfrac{N_o}{N_t} = \dfrac{kt}{2.303}$ and $t_{1/2} = \dfrac{0.693}{k}$; $t_{1/2} = 5{,}730$ y (use three significant figures).

Solution: Solve the $t_{1/2}$ expression above for k and substitute it into the first expression:

$$k = \frac{0.693}{t_{1/2}} \text{ and } \log \frac{N_o}{N_t} = \frac{0.693\, t}{2.303\, t_{1/2}}$$

Solving for t gives

$$t = \log \frac{N_o}{N_t} \times \frac{2.303\, t_{1/2}}{0.693}$$

Assuming that the C-14–to–C-12 atmospheric ratio has been constant, we can say that for one gram of carbon the ratio of disintegrations from living carbon to sample equals the $\dfrac{N_o}{N_t}$ ratio. So,

$$\frac{N_o}{N_t} = \frac{15.3}{4.5} = 3.4$$

Substituting known values into our equation for t gives

$$t = \log 3.4 \times \frac{2.303 \times 5{,}730 \text{ y}}{0.693}$$

$$= 0.531 \times 1.904 \times 10^4 \text{ y} = 1.0 \times 10^4 \text{ y}$$

20.5 APPLICATIONS OF RADIOACTIVE ISOTOPES

20.6 MASS–ENERGY CALCULATIONS

Operational Skill

10. Calculating the energy change for a nuclear reaction. Given nuclear masses, calculate the energy change for a nuclear reaction   (Example 20.12).   Obtain the answer in joules per mole or MeV per particle.

In making calculations in this section, refer to text Table 20.3 for masses in atomic mass units (amu). Be sure that you realize that this is not a table of *atomic* masses but of the masses of some *nuclei* and other atomic particles.

Exercise 20.12   (a) Calculate the energy change in joules when 1.00 g $^{234}_{90}$Th decays to $^{234}_{91}$Pa by beta emission. (b) What is the energy change in MeV when one $^{234}_{90}$Th nucleus decays? Use text Table 20.3 for these calculations.

Known: rules for writing a nuclear reaction; $\Delta E = \Delta mc^2$;

Δm = product mass(es) – reactant mass(es), $c = 3.00 \times 10^8$ m/s; Th = 234 g/mol; masses (in amu's) are in text Table 20.3; 1 J = 1 kg $\bullet$ m^2/s^2;

$N_A = 6.02 \times 10^{23}$ nuclei/mol; 1 MeV = 1.602×10^{-13} J.

Solution: The nuclear reaction is

$$_{90}^{234}\text{Th} \longrightarrow {}_{91}^{234}\text{Pa} + {}_{-1}^{0}\text{e}$$

(a) First find the energy change when 1.00 mol decays:

$$\frac{\Delta m}{\text{Th nuclei}} = (233.9934 + 0.000549) - 233.9942 = -0.000251 \frac{\text{amu}}{\text{Th nuclei}}$$

Since mass in amu's = molar mass in grams,

$$\frac{\Delta m}{\text{mol}} = \text{loss of } 2.5 \times 10^{-4} \text{ g/mol} = \text{loss of } 2.5 \times 10^{-7} \text{ kg/mol}$$

Then,

$$\Delta E = \Delta mc^2 = 2.5 \times 10^{-7} \frac{\text{kg}}{\text{mol}} \left(3.00 \times 10^8 \frac{\text{m}}{\text{s}} \right)^2 = 2.2 \times 10^{10} \frac{\text{kg} \cdot \text{m}^2}{\text{mol} \cdot \text{s}^2}$$

$$= 2.2 \times 10^{10} \text{ J/mol}$$

The energy change for 1.00 g is thus

$$\Delta E = 2.2 \times 10^{10} \frac{\text{J}}{\cancel{\text{mol}}} \times \frac{\cancel{\text{mol}}}{234 \cancel{\text{g}}} \times 1.00 \cancel{\text{g}} = 9 \times 10^7 \text{ J}$$

(b) To find ΔE for decay of one nucleus, we will use the energy change for one mole, and N_A:

$$\Delta E = 2.2 \times 10^{10} \frac{\text{J}}{\cancel{\text{mol}}} \left(\frac{\cancel{\text{mol}}}{6.02 \times 10^{23} \text{ nuclei}} \right) = 3.7 \times 10^{-14} \frac{\text{J}}{\text{nuclei}}$$

$$= \frac{3.7 \times 10^{-14} \cancel{\text{J}} / \text{nuclei}}{1.602 \times 10^{-13} \cancel{\text{J}} / \text{MeV}} = 0.2 \text{ MeV} / \text{nuclei}$$

20.7 NUCLEAR FISSION AND NUCLEAR FUSION

ADDITIONAL PROBLEMS

1. Write the nuclear equation for

    (a)   the decay of Ra-224 by alpha-particle emission.
    (b)   the production of Bi-210 from Pb-210.
    (c)   the beta decay of Po-218.

2. Mendelevium-256 was prepared by alpha-particle bombardment of einsteinium-253. Write a nuclear equation for the synthesis.

3. Which of the following nuclides are radioactive? Explain your choices.

    (a) $^{194}_{78}Pt$ (b) $^{14}_{6}C$ (c) $^{16}_{8}O$ (d) $^{21}_{11}Na$ (e) $^{3}_{1}H$

4. Write the nuclear reaction for the probable mode of decay of each of the following nuclides. Explain how the nucleus is stabilized in each case.

    (a) $^{33}_{17}Cl$ (b) $^{201}_{81}Tl$ (c) $^{246}_{98}Cf$ (d) $^{34}_{15}P$

5. Write the abbreviated notation or the nuclear reaction for each of the following:

    (a) $^{142}_{60}Nd + ^{1}_{0}n \longrightarrow ^{143}_{61}Pm + ^{0}_{-1}e$

    (b) $^{230}_{90}Th + ^{1}_{1}H \longrightarrow ^{223}_{87}Fr + 2^{4}_{2}He$

    (c) $^{239}_{94}Pu(\alpha, n) ^{242}_{96}Cm$

    (d) $^{20}_{10}Ne(\alpha, \gamma) ^{24}_{12}Mg$

6. If 0.143 g of potassium-40 (atomic mass 40.0 amu) has an activity of 1.00 μCi ($10^{-6}Ci$), what is the decay constant for its disintegration?

7. Almost 99% of actinium-227 decays to francium-223. The rate constant for this decay is $3.14 \times 10^{-2}/y$. The other 1% decays to thorium-227, with a rate constant of $3.8 \times 10^{-4}/y$.

    (a)   Given that the overall rate constant for the decay is the sum of the separate rate constants, what is the half-life of actinium-227?
    (b)   Calculate the activity per μg of the nuclide.

8. If we start with a 0.550-mg sample of pure Bi-210, how much of this Bi-210 remains after 63.0 days? ($t_{1/2} = 5.01$ d)

9. The fossil shell discovered by Professor Leakey in the Olduvai Gorge was dated using the Ar-40/K-40 dating method and a mass spectrographic analysis of the isotopes. A small sample of the clay surrounding the fossil contained 6.00×10^{-7} g Ar-40 and 5.64×10^{-4} g K-40. What is the calculated age of the unearthed fossil? The half-life for the electron capture process

$$^{40}_{19}K + ^{0}_{-1}e \longrightarrow ^{40}_{18}Ar$$

is 1.30×10^{9} y.

10. Calculate the energy change in joules per mole of Na-24 (to three significant figures) for the following nuclear reaction:

$$^{24}_{11}Na \longrightarrow ^{0}_{-1}\beta + ^{24}_{12}Mg$$

The nuclidic masses are Na-24, 23.9909623 amu; Mg-24, 23.9850417 amu; and $^{0}_{-1}\beta$, 0.0005486 amu.

ANSWERS TO ADDITIONAL PROBLEMS

If you missed an answer, study the text section and operational skill given in parentheses after the answer.

1. (a) $^{224}_{88}Ra \longrightarrow ^{220}_{86}Rn + ^{4}_{2}He$

    (b) $^{210}_{82}Pb \longrightarrow ^{210}_{83}Bi + ^{0}_{-1}\beta$

    (c) $^{218}_{84}Po \longrightarrow ^{218}_{85}At + ^{0}_{-1}\beta$ (20.1, Op. Sk. 1)

2. $^{253}_{99}Es + ^{4}_{2}He \longrightarrow ^{256}_{101}Md + ^{1}_{0}n$ (20.1, Op. Sk. 1)

3. We would expect nuclides b, d, and e to be radioactive. Nuclide b has paired protons and neutrons, but the N/Z ratio is too high; in nuclide d, protons are unpaired and the N/Z ratio is too low; nuclide e has an unpaired proton and a high N/Z ratio. We would not expect nuclide a to be radioactive because it has paired protons and neutrons and the N/Z ratio is appropriate; nuclide c has a magic number of both protons and neutrons and an appropriate N/Z ratio. (20.1, Op. Sk. 3)

4. (a) $^{33}_{17}\text{Cl} \longrightarrow {}^{33}_{16}\text{S} + {}^{0}_{1}\text{e}$. The loss of a positron converts a proton to a neutron, raising the N/Z ratio.

    (b) $^{201}_{81}\text{Tl} + {}^{0}_{-1}\text{e} \longrightarrow {}^{201}_{80}\text{Hg}$. Electron capture converts a proton to a neutron, increasing the N/Z ratio.

    (c) $^{246}_{98}\text{Cf} \longrightarrow {}^{242}_{96}\text{Cm} + {}^{4}_{2}\text{He}$. The loss of an alpha particle decreases Z by 2. However, the nuclide produced is also unstable because Z is still > 83.

    (d) $^{34}_{15}\text{P} \longrightarrow {}^{34}_{16}\text{S} + {}^{0}_{-1}\text{e}$. Ejection of a β particle from the nucleus converts a neutron to a proton and decreases the N/Z ratio. (20.1, Op. Sk. 4)

5. (a) $^{142}_{60}\text{Nd}(n, \beta)\,{}^{143}_{61}\text{Pm}$ (b) $^{230}_{90}\text{Th}(p, 2\,\alpha)\,{}^{223}_{87}\text{Fr}$

    (c) $^{239}_{94}\text{Pu} + {}^{4}_{2}\text{He} \longrightarrow {}^{242}_{96}\text{Cm} + {}^{1}_{0}\text{n}$

    (d) $^{20}_{10}\text{Ne} + {}^{4}_{2}\text{He} \longrightarrow {}^{24}_{12}\text{Mg} + {}^{0}_{0}\gamma$ (20.2, Op. Sk. 5)

6. 1.72×10^{-17}/s (20.4, Op. Sk. 6)

7. (a) 21.8 y   (b) 7.22×10^{-5} Ci (20.4, Op. Sk. 7)

8. 9.01×10^{-5} mg (20.4, Op. Sk. 8)

9. 1.99×10^{6} y (20.4, Op. Sk. 7, 8)

10. 4.83×10^{11} J (20.6, Op. Sk. 10)

CHAPTER POST-TEST

1. The half-life of rubidium-87 is 4.8×10^{10} y. What is the decay constant?

2. Tell how the following would be expected to decay, and give reasons for your answers. Use text Figure 20.3.

    (a) $^{199}_{82}\text{Pb}$ (b) $^{205}_{80}\text{Hg}$ (c) $^{222}_{89}\text{Ac}$

3. Which two of the following nuclides would you expect to be radioactive, and which one would you expect to be stable? Give reasons.

   (a) $^{209}_{86}Rn$ (b) $^{21}_{11}Na$ (c) $^{48}_{20}Ca$

4. Determine whether each of the following statements is true or false. If the statement is false, change it so it is true.
   (a) There is no mass change during an ordinary chemical reaction. True/False: _____

   (b) Technetium, synthesized in1937 by deuteron bombardment of molybdenum, was the first new element produced in the laboratory from another element.
   True/False: _____

   (c) Gamma radiation from radium-226 is commonly used today in cancer therapy. True/False: _____

   (d) The advantages of using a radioactive tracer are that it behaves chemically as a nonradioactive isotope does and it can be detected in exceedingly small amounts by the measurement of its radiations. True/False: _____

   (e) The critical mass for a particular fissionable material is the smallest mass in which a chain reaction can be sustained. True/False: _____

5. Complete each of the following nuclear reactions by inserting the missing symbols. Use a periodic table.

   (a) $^{252}_{98}\underline{} \longrightarrow \,^{142}_{\underline{}}Ba + \underline{}_{42}Mo + 4\underline{}$

   (b) $^{238}_{\underline{}}U \longrightarrow \,^{234}_{90}Th + \underline{}$

   (c) $^{105}_{42}\underline{} \longrightarrow \,^{105}_{\underline{}}Tc + \underline{}$

6. Potassium-40 occurs naturally as 0.0119% of all potassium atoms. The half-life is 1.28×10^9 y. If the normal human body contains 5.0 g of potassium, how many disintegrations will occur in the normal person each day from the radioactive potassium?

7. Write the nuclear equation for each of the following:

   (a) $^{14}_{7}N(\alpha, p)^{17}_{8}O$ (b) $^{23}_{11}Na(d, p)^{24}_{11}Na$

8.   The trend in elemental abundance in the universe is that as mass increases, abundance decreases. An exception to this trend is iron, of which there is much more present than would be expected by examining the trend. Propose an explanation. (*Hint:* Look at text Figure 20.16.)

9.   Calculate the binding energy per nucleon for iron-56 nuclei. (Use data from text Table 20.3.)

10. Write the abbreviated notations for the following:

     (a) $^{6}_{3}\text{Li} + ^{1}_{0}\text{n} \longrightarrow ^{3}_{1}\text{H} + ^{4}_{2}\text{He}$

     (b) $^{10}_{5}\text{B} + ^{4}_{2}\text{He} \longrightarrow ^{13}_{7}\text{N} + ^{1}_{0}\text{n}$

11. A sample of rock taken from the Black Hills of South Dakota was analyzed for its U-238 and Pb-206 content. The sample showed 1.30×10^{-5} g U-238 and 3.04×10^{-6} g Pb-206. Assuming that all the Pb-206 present came from the decay of the initial U-238, how old are the Black Hills? (The half-life of U-238 is 4.5×10^{9} y.)

12. Uranium-mining operators in Gabon, Africa, discovered in June of 1972 that nearly all of the uranium shipped by one company during the previous 18 months was depleted in the U-235 isotope. Natural uranium is usually 0.720% U-235. The U-235 in one of these shipments was only 0.717%, and one shipment was as low as 0.441%. Account for this depleted percentage of U-235.

ANSWERS TO CHAPTER POST-TEST

If you missed an answer, study the text section and operational skill given in parentheses after the answer.

1. 1.4×10^{-11}/y   (20.4, Op. Sk. 7)

2.   (a)   Electron capture. Pb-199 has a mass number less than that of the stable Pb-207 isotope. The N/Z ratio is 117/82, which Figure 20.3 shows to be low. Nuclides with $Z > 80$ decay by electron capture, which converts a proton in the nucleus to a neutron, thus increasing the N/Z ratio.

(b) β^- emission. Hg-205 has a mass number greater than that of the stable Hg-200 isotope. The N/Z ratio is 125/80, which Figure 20.3 shows to be too high. On β^- emission, a neutron becomes a proton, which is a common way the ratio decreases.

(c) α emission. This is the mode of decay for nuclides with $Z > 83$.

(20.1, Op. Sk. 4)

3.   (a) and (b) would be expected to be radioactive, and (c) stable. Rn-209 has $Z > 83$, and Na-21 has no magic number of nucleons and the N/Z ratio is too small. Ca-48 has all protons paired and a magic number of neutrons, 28.   (20.1, Op. Sk. 3)

4.   (a)   False. There is no detectable mass change during an ordinary chemical reaction.   (20.6)
      (b)   True (Chapter introduction in text)
      (c)   False. Gamma radiation from cobalt-60 is commonly used today in cancer therapy. Radium-226 was used in the past.   (20.5)
      (d)   True   (20.5)   (e)   True   (20.7)

5.   (a) $^{252}_{98}\text{Cf}, \, ^{142}_{56}\text{Ba}, \, ^{106}_{42}\text{Mo}, \, ^{1}_{0}\text{n}$

      (b) $^{238}_{92}\text{U}, \, ^{4}_{2}\text{He}$

      (c) $^{105}_{42}\text{Mo}, \, ^{105}_{43}\text{Tc}, \, ^{0}_{-1}\text{e}$ (20.1, Op. Sk. 2)

6. 1.3×10^7 (20.4, Op. Sk. 7)

7.   (a) $^{14}_{7}\text{N} + ^{4}_{2}\text{He} \longrightarrow ^{17}_{8}\text{O} + ^{1}_{1}\text{H}$

      (b) $^{23}_{11}\text{Na} + ^{2}_{1}\text{H} \longrightarrow ^{24}_{11}\text{Na} + ^{1}_{1}\text{H}$ (20.2, Op. Sk. 5)

8.   Looking at Figure 20.16, we see that Fe-56 is one of the two nuclides with the highest binding energy per nucleon, and thus one of the two most stable. Being at the top of the curve, it is the most stable nuclide that can be formed by either fusion or fission reactions.   (20.6)

9.   8.8 MeV   (20.6, Op. Sk. 10)

10. (a) $^{6}_{3}\text{Li} (n, \alpha) ^{3}_{1}\text{H}$

     (b) $^{10}_{5}\text{B} (\alpha, n) ^{13}_{7}\text{N}$ (20.2, Op. Sk. 5)

11. 1.6×10^{9} y   (20.4)

12. The depletion is postulated to be the result of a naturally occurring nuclear reaction not possible now because the content of fissionable U-235 in natural uranium has become too low to support a chain reaction. See the article "Natural Nuclear Reactors" by Robert West in the June 1976 issue of *J. Chem. Ed.*, Vol. 53, no. 6.
   (20.7)

UNIT EXAM 6

1. The solubility product constant for magnesium arsenate is 2×10^{-20} at 25°C. What is the solubility (in g/L) of $Mg_3(AsO_4)_2$ in pure water?

2. To 0.200 L of a solution containing precipitated silver chromate ($K_{sp} = 1.1 \times 10^{-12}$) was added 0.020 L of a 0.015 M solution of potassium chromate. How did the $[Ag^+]$ change in solution? Comment.

3. If one combines 0.800 L of solution that is 0.0035 M Ca^{2+} with 0.600 L of solution that is 0.01 M SO_4^{2-}, will a precipitate form? (K_{sp} for $CaSO_4 = 2.4 \times 10^{-5}$.)

4. A solution is 0.020 M $BaCl_2$ and 0.020 M $SrCl_2$. If 0.200 L of this solution is mixed with 0.200 L of 0.010 M Na_2SO_4, what fraction of each ion is precipitated? $K_{sp}(BaSO_4) = 1.1 \times 10^{-10}$; $K_{sp}(SrSO_4) = 2.5 \times 10^{-7}$.

5. Determine whether each of the following statements is true or false. If the statement is false, change it so it is true.
    (a) Free energies of formation of compounds are almost always positive numbers. True/False: _____
    _____ .

    (b) A reaction for which the enthalpy of reaction is negative will be spontaneous at any temperature. True/False: _____
    _____ .

    (c) Any system at equilibrium may be harnessed to obtain usable energy. True/False: _____
    _____ .

(d)   In the voltaic cell

$$Fe(s)\,|\,Fe^{2+}(aq)\,|\,|\,Zn^{2+}(aq)\,|\,Zn(s)$$

the anode reaction is $Fe(s) \rightleftharpoons Fe^{2+}(aq) + 2e^-$. True/False: _____

_____ .

(e)   The potential difference across the electrodes of a voltaic cell is usually less than the cell emf, because of nonequilibrium conditions in the cell. True/False: _____

_____

(f)   In a nuclear reactor, a moderator must be used to control the heat output for the production of steam. True/False: _____

_____ .

(g)   Controlled fusion involving hydrogen is not yet feasible because the technology has not been developed to isolate all the tritium released from the reaction. True/False: _____

_____ .

6.   Liquid methanol, $CH_3OH(l)$, at 25°C has an entropy of 127 J/(K • mol). The heat of vaporization, ΔH_{vap}, at this temperature is 37.4 kJ/mol. What is the entropy of the vapor in equilibrium with the liquid at 25°C?

7.   Calculate the free-energy change for the formation of 1.00 g of glucose, $C_6H_{12}O_6$, in a green plant according to the following reaction:

$$6CO_2(g) + 6H_2O(l) \longrightarrow C_6H_{12}O_6(s) + 6O_2(g)$$

ΔG_f°(glucose) = –910.2 kJ/mol. Refer to text Table 18.2 for other values.

8.   A mixture of 3.00 mol Cl_2 and 3.00 mol CO is placed in a 5.00-L flask at 6.00×10^2 °C and allowed to react. $K_p = 8.2 \times 10^{-4}$. Calculate $\Delta G°$ at this temperature.

9.   In a certain electrochemical cell, a lead electrode is in a solution of lead nitrate and is connected by a wire to a zinc electrode in a solution of zinc nitrate. The cells are connected by a salt bridge.
(a)   Calculate $E°$ for this cell. (Use data from text Table 19.1.)
(b)   Sketch the cell, labeling the anode, cathode, and direction of ion and electron movements, and write the electrode half-reactions.
(c)   Calculate $\Delta G°$ for this cell.
(d)   Write the cell notation for the reaction.

10. In an electroplating cell, chromium is plated out from an acidic solution of CrO_3. How many grams of chromium can be plated out in exactly one hour, using a current of 15.0 A?

11. Match each description in the left-hand column with the appropriate term in the right-hand column.

    ___ (a) uses neutron capture by $^{235}_{92}U$ nucleus   1. nuclear binding energy
    ___ (b) mass required for a chain reaction   2. breeder reactor
    ___ (c) needed to absorb neutrons   3. control rod
    ___ (d) required to dissociate the nucleus into   4. conventional fission
          protons and neutrons   reaction
    ___ (e) produces an excess of fissionable fuel   5. critical mass

12. Supply the missing symbol in each of the following nuclear reactions:

    (a) $^{40}_{20}Ca + ^{2}_{1}H \longrightarrow$ ___ $+ ^{1}_{1}H$

    (b) $^{6}_{3}Li + ^{1}_{0}n \longrightarrow ^{3}_{1}H +$ ___

13. Write the abbreviated notation for the following nuclear bombardment reaction:

    $$^{60}_{28}Ni + ^{4}_{2}He \longrightarrow ^{63}_{30}Zn + ^{1}_{0}n$$

14. A sample of paint from a pot shard excavated from what was believed to be an ancient dwelling was burned and the resulting CO_2 was collected. The activity of the CO_2 was measured at 3.31 disintegrations per minute per gram C. If carbon from living material gives 15.3 disintegrations per minute per gram C, what is the probable age of the pot shard? The half-life of C-14 is 5.73×10^3 y.

15. Calculate the nuclear binding energy in J/mol associated with the Li-7 nucleus. The appropriate masses are $^{1}_{1}H$, 1.00728 amu; $^{1}_{0}n$, 1.00867 amu; $^{7}_{3}Li$, 7.01436 amu.

ANSWERS TO UNIT EXAM 6

If you missed an answer, study the text section and operational skill given in parentheses after the answer.

1. 2×10^{-2} g/L   (17.1, Op. Sk. 2)

2. The new concentration (2.8×10^{-5} M) is 1.0×10^{-4} M less than the original (1.3×10^{-4} M). This shows that the common ion depresses the solubility of the salt.
    (17.2, Op. Sk. 3)

3. No. $Q = 8.6 \times 10^{-6} < K_{sp}$ (17.3, Op. Sk. 4)

4. Virtually all of the sulfate precipitates, none of the strontium precipitates, and half of the barium precipitates.   (17.3, Op. Sk. 5)

5. (a) False. Free energies of formation of compounds are almost always negative numbers, as uncombined elements are almost always at a higher energy level than the compound they form.   (18.4, 18.7)
    (b) False. A reaction for which the free energy is negative at a given temperature will be spontaneous at that temperature.   (18.4)
    (c) False. Any system not at equilibrium may be harnessed to obtain usable energy.   (18.5)
    (d) True.   (19.3, Op. Sk. 2)
    (e) False. The potential difference across the electrodes of a voltaic cell is usually less than the cell emf, because of nonstandard conditions in the cell.   (19.7)
    (f) False. In a nuclear reactor, a moderator must be used to control the velocity of the neutrons.   (20.7)
    (g) False. Controlled fusion involving hydrogen is not yet feasible because the high reaction temperature is difficult to maintain in a controlled manner.
    (20.7)

6. 252 J/(K • mol)   (18.2, Op. Sk. 1)

7. 16.0 kJ. Even though this reaction is not spontaneous at 25°C, it does happen at that temperature in green plants. In Section 25.2, you will see that this reaction occurs by the coupling of one unfavorable and one favorable reaction.   (18.4, Op. Sk. 5)

8. 51.6 kJ   (18.6)

9. (a)   0.63 V   (19.5, Op. Sk. 6)

(b)   (19.2, Op. Sk. 1)

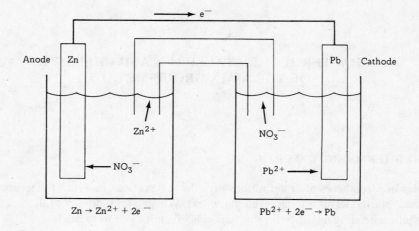

$$\text{Zn} \rightarrow \text{Zn}^{2+} + 2e^-$$

$$\text{Pb}^{2+} + 2e^- \rightarrow \text{Pb}$$

(c) -1.2×10^2 kJ   (19.6, Op. Sk. 7)

(d) $\text{Zn}(s)\,|\,\text{Zn}^{2+}(aq)\,||\,\text{Pb}^{2+}(aq)\,|\,\text{Pb}(s)$ (19.3, Op. Sk. 2)

10.   4.85 g Cr   (19.11, Op. Sk. 12)

11.   (a)   4   (20.7)   (c)   3   (20.7)   (e)   2   (20.7)
      (b)   5   (20.7)   (d)   1   (20.6)

12.   (a) $^{41}_{20}\text{Ca}$ (b) $^{4}_{2}\text{He}$ (20.1, Op. Sk. 2)

13. $^{60}_{28}\text{Ni}\,(\alpha, n)\,^{63}_{30}\text{Zn}$ (20.2, Op. Sk. 5)

14. 1.27×10^4 y   (20.4, Op. Sk. 9)

15. 3.80×10^{12} J/mol   (20.6, Op. Sk. 10)

CHAPTER 21   METALLURGY AND CHEMISTRY
OF THE MAIN-GROUP METALS

CHAPTER TERMS AND DEFINITIONS

Numbers in parentheses after definitions give the text sections in which the terms are explained. Starred terms are italicized in the text. Where a term does not fall directly under a text section heading, additional information is given for you to locate it.

malleable* ability to be pounded into flat sheets   (21.1, introductory section)

ductile* able to be drawn into a wire   (21.1, introductory section)

metal material that is lustrous (shiny), has high electrical and heat conductivities, and is malleable and ductile   (21.1, introductory section)

alloy material with metallic properties that is either a compound or a mixture   (21.1, introductory section)

mineral naturally occurring inorganic solid substance or solid solution with a definite crystalline structure   (21.1)

ore rock or mineral from which a metal or nonmetal can be economically produced   (21.1)

metallurgy scientific study of the production of metals from their ores and the making of alloys having various useful properties   (21.2)

refining* process of purifying a metal   (21.2)

gangue worthless portion of an ore   (21.2)

flotation physical method of separating a mineral from the gangue that depends on differences in their wettabilities by a liquid solution   (21.2)

flotation agents* chemical substances that are used to coat metal-bearing mineral particles selectively so they are wet less easily by water than is the gangue   (21.2)

Bayer process chemical procedure in which purified aluminum oxide, Al_2O_3, is separated from the aluminum ore bauxite   (21.2)

calcined* heated strongly in a furnace   (21.2)

roasting process of heating a mineral in air to obtain the oxide   (21.2)

Hall–Héroult process commercial method for producing aluminum by the electrolysis of a molten mixture of aluminum oxide in cryolite, Na_3AlF_6 (21.2)

blast furnace*   furnace in which metals are obtained by the reduction of their compounds with carbon   (21.2)

Mond process   chemical procedure for the purification of nickel that depends on the formation and later decomposition of a volatile compound of the metal   (nickel tetracarbonyl)   (21.2)

delocalized molecular orbitals*   molecular orbitals that encompass more than two atoms   (21.3)

band theory*   molecular orbital theory of metal bonding, which states that a large number of energy levels are crowded together into "bands"   (21.3)

band*   grouping of energy levels of essentially continuous energies   (21.3)

superconductor*   material that abruptly loses its resistance to an electric current when cooled to a definite characteristic temperature   (A Chemist Looks At: Superconductivity)

salarium*   Latin word for a Roman soldier's salt allowance, from which the word salary is derived   (21.6)

sal*   Latin word for salt   (21.6)

potash*   ashes from burning plant materials in pots; origin of the element name potassium   (21.6)

soda ash*   commercial name for anhydrous sodium carbonate   (21.6)

Solvay process   industrial method for obtaining sodium carbonate from sodium chloride and limestone   (21.6)

Dow process   commercial method for isolating magnesium from seawater   (21.7)

quicklime (lime)*   commercial name for calcium oxide   (21.8)

slag*   molten material formed to carry away impurities in a blast furnace   (21.8)

slaked lime*   commercial name for calcium hydroxide   (21.8)

soften*   to remove certain metal ions from municipal water supplies, usually through the use of CaO or $Ca(OH)_2$ (21.8)

alumina*   commercial name for aluminum oxide   (21.9)

Goldschmidt process   method of preparing a metal by reduction of its oxide with powdered aluminum   (21.9)

thermite*   mixture of iron(III) oxide and aluminum powder used to produce iron for welding   (21.9)

kerimikos*   Greek word from which the word ceramics is derived   (21.9)

gray tin*   nonmetallic form of tin: brittle gray powder; form in which tin exists below 13°C   (21.10)

white tin*   metallic form of tin, existing above 13°C   (21.10)

stannous compounds*   compounds of Sn(II) ion   (21.10)

litharge*   commercial name for Pb(II) oxide   (21.10)

CHAPTER DIAGNOSTIC TEST

1. Complete and balance the following equations.

    (a) $Li(s) + O_2(g) \xrightarrow{\Delta}$

    (d) $Mg(s) + O_2(g) \xrightarrow{\Delta}$

    (b) $Na(s) + Cl_2(g) \xrightarrow{\Delta}$

    (e) $Al(OH)_3(s) \xrightarrow{\Delta}$

    (c) $K(s) + H_2(g) \xrightarrow{\Delta}$

    (f) $Al(s) + Cr_2O_3(s) \longrightarrow$

2. Calculate the mass of NaCl required to prepare 1.0×10^3 kg (metric tonne) NaOH.

3. Write equations for the following reactions:

    (a) $TiCl_4(l)$ is reduced by $Na(l)$ to yield $Ti(s)$.

    (b) Carbon dioxide is added to a solution saturated with ammonia and sodium chloride.

    (c) Calcium is added to water.

    (d) Aluminum hydroxide is added to NaOH solution.

    (e) ZnS is roasted in air.

    (f) Aluminum hydroxide is reacted with dilute HCl.

4. Write the common oxidation number of each of the following:
    (a) Group IIA elements   (c) oxygen family
    (b) carbon family

5. Arrange each of the following rows in order of increasing value of the property listed:

    (a) basicity Al_2O_3 Cl_2O_7 MgO SiO_2

    (b) metallic character   Br Ge K Sc

    (c) metallic character   Bi N P Sb

    (d) multiplicity of oxidation states Cl Na P

    (e) melting point   Cs K Na Rb

6. The solubility of $CaCO_3$ is 6.2×10^{-3} g/L at 25°C. Calculate K_{sp} for $CaCO_3$ at this temperature.

7. Write equations for the preparation of
   (a) sodium from sodium chloride.
   (b) lime from seashells.
   (c) potassium hydroxide from potassium chloride.
   (d) chromium by the Goldschmidt process.

8. Give the differences in behavior of the second-period elements from other members of a periodic table group, and explain the reason(s) for these differences.

9. Give brief explanations for each of the following:
   (a) Sodium is a less reactive metal than potassium, but potassium can be prepared by the reaction

$$Na(l) + KCl(l) \longrightarrow NaCl(l) + K(l)$$

   (b) The barium ion is toxic, yet a suspension of barium sulfate is administered orally by physicians to obtain diagnostic x-ray photographs of the stomach with no harm to the patient.
   (c) Sand, not a CO_2 fire extinguisher, must be used to put out magnesium fires.

10. Using zinc (which occurs as wurtzite, ZnS) as an example, outline the basic steps of metallurgy.

ANSWERS TO CHAPTER DIAGNOSTIC TEST

If you missed an answer, study the text section and operational skill given in parentheses after the answer. Previous text sections and operational skills are included in some cases.

1. (a) $4Li(s) + O_2(g) \xrightarrow{\Delta} 2Li_2O(s)$ (21.5)

   (b) $2Na(s) + Cl_2(g) \xrightarrow{\Delta} 2NaCl(s)$ (21.6)

   (c) $2K(s) + H_2(g) \xrightarrow{\Delta} 2KH(s)$ (21.6)

   (d) $2Mg(s) + O_2(g) \xrightarrow{\Delta} 2MgO(s)$ (21.7)

   (e) $2Al(OH)_3(s) \xrightarrow{\Delta} Al_2O_3(s) + 3H_2O(g)$ (21.2)

   (f) $2Al(s) + Cr_2O_3(s) \longrightarrow Al_2O_3(l) + 2Cr(l)$ (21.9)

2. 1.5×10^3 kg NaCl (21.6; 4.7, Op. Sk. 10)

3. (a) $TiCl_4(l) + 4Na(l) \xrightarrow{\Delta} Ti(s) + 4NaCl(s)$ (21.6)

(b) $CO_2(g) + NaCl(aq) + NH_3(aq) + H_2O(l) \longrightarrow NaHCO_3(s) + NH_4Cl(aq)$
(21.6)

(c) $Ca(s) + 2H_2O(l) \longrightarrow Ca(OH)_2(aq) + H_2(g)$ (21.8)

(d) $Al(OH)_3(s) + NaOH(aq) \longrightarrow NaAl(OH)_4(aq)$ (21.2)

(e) $2ZnS(s) + 3O_2(g) \longrightarrow 2ZnO(s) + 2SO_2(g)$ (21.2)

(f) $Al(OH)_3(s) + 3HCl(aq) \longrightarrow AlCl_3(aq) + 3H_2O(l)$ (21.9)

4. (a)   +2   (b)   +4   (c)   −2 (21.4)

5. (a) Cl_2O_7 SiO_2 Al_2O_3 MgO (21.4)
(b)   Br Ge Sc K (21.4)
(c)   N P Sb Bi (21.4)
(d)   Na P Cl (21.4)
(e)   Cs Rb K Na (21.5, introductory section)

6. 3.8×10^{-9} (17.1, Op. Sk. 2)

7. (a) $2NaCl(l) \xrightarrow{electrolysis} 2Na(l) + Cl_2(g)$ (21.6)

(b) $CaCO_3(s) \xrightarrow{\Delta} CaO(s) + CO_2(g)$ (21.7)

(c) $2KCl(aq) + 2H_2O(l) \xrightarrow{electrolysis} 2KOH(aq) + H_2(g) + Cl_2(g)$
(21.6)

(d) $Cr_2O_3(s) + 2Al(l) \xrightarrow{\Delta} 2Cr(l) + Al_2O_3(l)$ (21.9)

8. Because of the relatively small atoms in the second-period elements, the second-row elements give atoms of relatively high electronegativities. These elements use only *s* and *p* orbitals in bonding. They do not have *d* orbitals available for bonding. This

limits the types of compounds formed. Boron forms BCl_3, but aluminum forms $AlCl_6^-$. (21.4)

9. (a) The reaction is done at 870°C and low pressure. The potassium distills over as a gas and is thus removed from the reaction mixture, forcing the reaction to completion. (21.6)

    (b) Barium sulfate is very insoluble, so there is no appreciable concentration of the barium ion. (3.7)

    (c) Magnesium burns in CO_2, as follows:

$$2Mg(s) + CO_2(g) \longrightarrow 2MgO(s) + C(s) \qquad (21.7)$$

10. Preliminary treatment: concentration of the ore by flotation

Roasting: $\quad 2ZnS(s) + 3O_2(g) \xrightarrow{\Delta} 2ZnO(s) + 2SO_2(g)$

Reduction: $\quad ZnO(s) + C(s) \longrightarrow Zn(g) + CO(g)$

Refining: $\quad$ Zn is purified by distillation. (21.2)

SUMMARY OF CHAPTER TOPICS

21.1   NATURAL SOURCES OF THE METALLIC ELEMENTS

21.2   METALLURGY

21.3   BONDING IN METALS

21.4   GENERAL OBSERVATIONS ABOUT THE MAIN-GROUP ELEMENTS

21.5   LITHIUM

21.6   SODIUM AND POTASSIUM

21.7   MAGNESIUM

21.8   CALCIUM

21.9   ALUMINUM

21.10   TIN AND LEAD

ADDITIONAL PROBLEMS

1.   For each of the following pairs, select the element that is more metallic: (a) Cs or Ba; (b) Ba or Be. Explain your choices on the basis of electron arrangement in the atom.

2.   Tin forms two chlorides, $SnCl_2$ and $SnCl_4$. One chloride is a crystalline substance with a melting point of 246°C. The other is a liquid at room temperature and boils at 114.1°C. Match the properties to the formulas and explain your answer.

3.   Complete and balance the following equations:

(a) $K_2CO_3(s) + HCl(aq) \longrightarrow$

(b) $SiO_2(l) + CaO(s) \longrightarrow$

(c) $K(s) + O_2(g) \longrightarrow$

(d) $Al(s) + NaOH(aq) + H_2O(l) \longrightarrow$

(e) $MgCO_3(s) \xrightarrow{\Delta}$

4.   Write balanced equations for the preparation of the following species, indicating if a catalyst or heat is required:

(a) $CaCO_3$ from CaO

(b) Al_2O_3 from a metal oxide

(c)   pure elemental lead

(d) $Mg(OH)_2$ from $MgCO_3$

5.   List chemical tests that could be used to distinguish between the white solids $MgCO_3$ and KCl.

6.   Match each of the following compounds to its use:

____	(a)	Li	1.	furnace bricks
____	(b)	Na_2CO_3	2.	manufacture of glass
____	(c)	KNO_3	3.	cathode in batteries
____	(d)	MgO	4.	explosives; fireworks
____	(e)	$CaSO_4$	5.	calculator batteries
____	(f)	$Al_2(SO_4)_3 \cdot 18H_2O$	6.	aluminum source
____	(g)	Al_2O_3	7.	plaster; wallboard
____	(h)	PbO_2	8.	making of paper

7.   Strontium is found chiefly as celestite, $SrSO_4$, and strontianite, $SrCO_3$. Which of the two compounds would provide more strontium per kilogram? How much more?

8.   Indicate whether aqueous solutions of the following compounds would be neutral, basic, or acidic.

(a) $Al_2(SO_4)_3$ (d) Al_2Cl_6 (g) KNO_3

(b) Na_2O (e) $B(OH)_3$

(c) K_2CO_3 (f)   CaO

9.   Write a balanced chemical equation that indicates
(a)   an alkali metal acting as a reducing agent.
(b)   the preparation of metallic potassium.
(c)   cesium serving to remove oxygen in a vacuum tube.
(d)   the hydrolysis of the $Al(H_2O)_6^{3+}(aq)$ ion.

10.   Write the equations for the preparation of pure aluminum from an initial mixture of Al_2O_3, $Al(OH)_3$, and $AlO(OH)$.

ANSWERS TO ADDITIONAL PROBLEMS

If you missed an answer, study the text section and operational skill given in parentheses after the answer. Text sections and operational skills from previous chapters are given where appropriate.

1. (a) Cs. Barium has one more electron and one more proton than cesium. The additional electron is at the same energy level as the previous electron. Thus, there is no shielding, giving rise to a tighter hold on the valence electrons by the more positively charged nucleus, which leads to a greater ionization energy.

    (b) Ba. The two *s* electrons in barium are in the sixth energy level. More shielded from the nucleus than they are in beryllium, they have lower ionization energies and are removed more easily. (21.4; 8.6, Op. Sk. 5)

2. $SnCl_4$, with the higher oxidation state of Sn, is more covalently bonded and thus is the liquid. (21.4, 21.10)

3. (a) $K_2CO_3(s) + 2HCl(aq) \longrightarrow 2KCl(aq) + CO_2(g) + H_2O(l)$ (21.6)

    (b) $SiO_2(l) + CaO(s) \longrightarrow CaSiO_3(l)$ (21.8)

    (c) $K(s) + O_2(g) \longrightarrow KO_2(s)$ (21.6)

    (d) $2Al(s) + 2NaOH(aq) + 6H_2O(l) \longrightarrow 2NaAl(OH)_4(aq) + 3H_2(g)$

    $$(21.2, 21.9)$$

    (e) $MgCO_3(s) \xrightarrow{\Delta} MgO(s) + CO_2(g)$ (21.7)

4. (a) $CaO(s) + H_2O(l) \longrightarrow Ca(OH)_2(aq)$

    $Ca(OH)_2(aq) + CO_2(g) \longrightarrow CaCO_3(s) + H_2O(l)$ (21.8)

    (b) $3CaO(s) + 2Al(l) \xrightarrow{\Delta} 3Ca(g) + Al_2O_3(s)$ (21.9)

    (c) $PbO(s) + CO(g) \longrightarrow Pb(l) + CO_2(g)$ (21.10)

    (d) $MgCO_3(s) \xrightarrow{\Delta} MgO(s) + CO_2(g)$

    $MgO(s) + H_2O(l) \longrightarrow Mg(OH)_2(s)$ (21.7)

5. KCl is water-soluble, $MgCO_3$ only slightly soluble. An aqueous solution of $MgCO_3$ would be basic and would turn phenolphthalein red; a KCl solution would be neutral and the indicator would remain colorless. $MgCO_3$ would decompose with moderate heating; KCl would not and would melt with stronger heating. $MgCO_3$ would react with acid to release a gas (CO_2); KCl would not react with acid. (3.3, 3.4, 13.2, 21.6, 21.7)

6. (a) 5   (21.5)   (b) 2   (21.6)   (c) 4   (21.6)
    (d) 1   (21.7)   (e) 7   (21.8)
    (f) 8   (21.9)   (g) 6   (21.9)   (h) 3   (21.10)

7. $SrCO_3$, 116.6 g (4.3, Op. Sk. 5)

8. (a) acidic   (e) acidic
    (b) basic   (f) basic
    (c) basic   (g) neutral (16.4, 21.4, 21.6, 21.8, 21.9)
    (d) acidic

9. (a) $2Na(s) + 2H_2O(l) \longrightarrow 2NaOH(aq) + H_2(g)$ (3.5, 21.6)

    (b) $Na(l) + KCl(l) \xrightarrow{\Delta} NaCl(l) + K(g)$ (21.6)

    (c) $Cs(s) + O_2(g) \longrightarrow CsO_2(s)$ (21.6)

    (d) $Al(H_2O)_6^{3+}(aq) + H_2O(l) \rightleftharpoons Al(H_2O)_5OH^{2+}(aq) + H_3O^+(aq)$
                                                                  (21.9)

10. $AlO(OH)(s) + OH^-(aq) + H_2O(l) \xrightarrow{\Delta} Al(OH)_4^-(aq)$

     $Al(OH)_4^-(aq) \xrightarrow{cool} Al(OH)_3(s) + OH^-(aq)$

     $2Al(OH)_3(s) \xrightarrow{\Delta} Al_2O_3(s) + 3H_2O(g)$

     $2Al_2O_3(s) + 3C(s) \xrightarrow{electrolysis} 4Al(l) + 3CO_2(g)$ (21.2)

CHAPTER POST-TEST

1. Write equations for the following reactions:
   (a) Potassium superoxide reacts with moisture.
   (b) Thorium dioxide is reduced by calcium.
   (c) B_2O_3 is reduced with magnesium.

   (d) $Al^{3+}(aq)$ is reacted with NaOH solution.
   (e) Sodium bicarbonate is strongly heated.
   (f) Calcium hydroxide solution is used as a test for carbon dioxide.

2. Arrange each of the following rows in order of increasing value of the property listed:

   (a) basicity   MgO Na_2O SO_3 SiO_2

   (b) metallic character   As Ca Se Zn

   (c) metallic character   Al B In Tl

   (d) multiplicity of oxidation states Br Ca Ge

   (e) melting point   K Li Na Rb

3. Write the common oxidation state of each of the following:
   (a) Group IA elements   (b) boron family   (c) Group IVA elements

4. Calculate the volume of Cl_2 (25°C, 1.00 atm) produced with 1.00 kg NaOH in the electrolysis of sodium chloride solution.

5. Complete and balance the following equations. Write NR if no reaction occurs.

   (a) $2Na(s) + O_2(g) \xrightarrow{\Delta}$ (d) $Mg(s) + Cl_2(g) \xrightarrow{\Delta}$

   (b) $Li(s) + N_2(g) \xrightarrow{\Delta}$ (e) $Al(s) + O_2(g) \longrightarrow$

   (c) $Li(s) + H_2O(l) + CO_2(g) \longrightarrow$ (f) $Ga(s) + NaOH(s) + H_2O(l) \longrightarrow$

6. Contrast the melting points of the Group IA metals with those of the Group IIA metals. Explain.

7. LiOH is used to remove CO_2 from the air in spacecraft. Explain.

8. Write equations for the preparation of
    (a) NaOH from NaCl.   (b) iron from its oxide, Fe_2O_3, using aluminum.
    (c) aluminum from bauxite.   (d) $SnSO_4$.

9. Calculate the energy evolved when 1.00 kg of chromium is produced according to the reaction

$$Cr_2O_3(s) + 2Al(l) \longrightarrow Al_2O_3(l) + 2Cr(l); \quad \Delta H^\circ = -536 \text{ kJ}$$

10. Many reducing agents are used in the reduction step in metallurgy. By writing equations, cite examples of reduction by an active metal, by hydrogen, and by carbon.

ANSWERS TO CHAPTER POST-TEST

If you missed an answer, study the text section and operational skill given in parentheses after the answer. Previous text sections and operational skills are included in some cases.

1. (a) $4KO_2(s) + 2H_2O(l) \longrightarrow 4KOH(s) + 3O_2(g)$ (21.6)

    (b) $ThO_2(s) + 2Ca(l) \xrightarrow{\Delta} Th(s) + 2CaO(s)$ (21.2)

    (c) $B_2O_3(l) + 3Mg(l) \xrightarrow{\Delta} 2B(s) + 3MgO(s)$ (21.7)

    (d) $Al^{3+}(aq) + 3OH^-(aq) \longrightarrow Al(OH)_3(s)$ (21.9)

    or $Al^{3+}(aq) + 4OH^-(aq) \longrightarrow Al(OH)_4^-(aq)$

    (e) $2NaHCO_3(s) \xrightarrow{\Delta} Na_2CO_3(s) + H_2O(g) + CO_2(g)$ (3.4)

    (f) $Ca(OH)_2(aq) + CO_2(g) \longrightarrow CaCO_3(s) + H_2O(l)$ (21.8)

2. (a) SO_3 SiO_2 MgO Na_2O (15.5, 21.4)

    (b) Se As Zn Ca (21.4)

    (c) B Al In Tl (21.4)

    (d) Ca Ge Br (21.4)

    (e) Rb K Na Li (21.5, introductory section)

3.   (a) + 1   (b) +3   (c) +4   (21.4)

4. 3.06×10^2 L Cl_2 (21.6; 5.4, Op. Sk. 6)

5.   (a) $2Na(s) + O_2(g) \xrightarrow{\Delta} Na_2O_2(s)$ (21.6)

      (b) $6Li(s) + N_2(g) \xrightarrow{\Delta} 2Li_3N(s)$ (21.5)

      (c) $2Li(s) + 2H_2O(l) + 2CO_2(g) \longrightarrow 2LiHCO_3(s) + H_2(g)$ (21.5)

      (d) $Mg(s) + Cl_2(g) \xrightarrow{\Delta} MgCl_2(s)$ (21.7)

      (e) $4Al(s) + 3O_2(g) \longrightarrow 2Al_2O_3(s)$ (21.9)

      (f) $2Ga(s) + 2NaOH(s) + 6H_2O(l) \longrightarrow 2Na[Ga(OH)_4](aq) + 3H_2(g)$

                                                                    (21.2, 21.9)

6.   Group IIA metal melting points are higher due to the increased strength of bonding with the two valence electrons. (21.4)

7.   LiOH readily absorbs CO_2 from air to form lithium carbonate and lithium hydrogen carbonate. (21.5)

8.   (a) $2NaCl(aq) + 2H_2O(l) \xrightarrow{\text{electrolysis}} 2NaOH(aq) + H_2(g) + Cl_2(g)$

                                                                        (21.6)

      (b) $Fe_2O_3(s) + 2Al(s) \xrightarrow{\Delta} 2Fe(l) + Al_2O_3(l)$ (21.9)

      (c) $Al(OH)_3(s) + OH^-(aq) \xrightarrow{\Delta} Al(OH)_4^-(aq)$

             $AlO(OH)(s) + H_2O(l) + OH^-(aq) \xrightarrow{\Delta} Al(OH)_4^-(aq)$

             $Al(OH)_4^-(aq) \xrightarrow{\text{cool}} Al(OH)_3(s) + OH^-(aq)$

             $2Al(OH)_3(s) \xrightarrow{\Delta} Al_2O_3(s) + 3H_2O(g)$

             $2Al_2O_3(s) + 3C(s) \xrightarrow{\text{electrolysis}} 4Al(l) + 3CO_2(g)$ (21.2)
             (a simplified representation)

(d) $Sn(s) + H_2SO_4(aq) \longrightarrow SnSO_4(aq) + H_2(g)$ (21.10)

9. 5.15×10^3 kJ   (21.9; 6.5, Op. Sk. 4)

10.   Examples are:

$$ZrCl_4(g) + 2Mg(l) \longrightarrow Zr(s) + 2MgCl_2(s) \quad (21.7)$$

$$Cr_2O_3(s) + 2Al(s) \xrightarrow{\Delta} 2Cr(l) + Al_2O_3(l) \quad (21.9)$$

$$ZnO(s) + C(s) \xrightarrow{\Delta} Zn(g) + CO(g) \quad (21.2)$$

$$WO_3(s) + 3H_2(g) \xrightarrow{\Delta} W(s) + 3H_2O(g) \quad (21.2)$$

CHAPTER 22   CHEMISTRY OF THE NONMETALS

CHAPTER TERMS AND DEFINITIONS

Numbers in parentheses after definitions give the text sections in which the terms are explained. Starred terms are italicized in the text. Where a term does not fall directly under a text section heading, additional information is given for you to locate it.

brimstone* old English name of sulfur, meaning "a stone that burns" (chapter introduction)

catenation covalent bonding of two or more atoms of the same element to one another (22.1)

carbon black* form of carbon composed of extremely small crystals having an amorphous, or imperfect, graphite structure (22.1)

pyrolyzes* decomposes into gaseous products and solid carbon (22.1)

graphitized* solid carbon transformed to a more or less crystalline graphite structure at high temperatures (22.1)

composite material constructed of two or more different kinds of materials in separate phases (22.1)

abrasive* material used for grinding and polishing (22.1)

synthesis gas* mixture of CO and H useful as the starting material in the production of many organic products (22.1)

silanes* silicon analogues of the hydrocarbons (22.2)

ferrosilicon* silicon alloyed with iron for metallurgical purposes (22.2)

zone refining purification process in which a solid rod of a substance is melted in a small moving band or zone that carries the impurities out of the rod (22.2)

semiconducting element element that exhibits very low electrical conductivity at room temperature when pure, but whose electrical conductivity rises with temperature or with the addition of certain other elements (22.2)

doping* enhancing the electrical conductivity of a semiconductor by adding small amounts of certain elements to it (22.2)

n*-type semiconductor   phosphorus-doped silicon with the electrical current being carried by negative charges (electrons)   (22.2)

p*-type semiconductor   boron-doped silicon with the electrical conductor being carried by positive charges   (22.2)

p*–*n* junction   *p*-type semiconductor joined to *n*-type semiconductor   (22.2)

rectifier*   device that allows current to flow in one direction but not the other   (22.2)

transistor*   device for controlling electrical signals, formed by combining *p*-type and *n*-type semiconductors   (22.2)

silica   covalent network solid in which each silicon atom is covalently bonded in tetrahedral directions to four oxygen atoms; each oxygen atom is in turn bonded to another silicon atom   (22.2)

piezoelectric effect*   property of some crystals, such as quartz, in which compression of the crystal in a particular direction causes an electric voltage to develop across it   (22.2)

silicate   compound of silicon and oxygen (with one or more metals) that may be formally regarded as a derivative of silicic acid, H_4SiO_4 or $Si(OH)_4$ (22.2)

water glass*   soluble material formed by melting silica with sodium carbonate   (22.2)

condensation reaction   reaction in which two molecules or ions are chemically joined by the elimination of a small molecule such as H_2O (22.2)

orthosilicates*   minerals containing SiO_4^{4-} as discrete ions   (22.2)

aluminosilicate mineral   mineral consisting of silicate sheets or three-dimensional networks in which some of the SiO_4 tetrahedra of a silicate structure have been replaced by AlO_4 tetrahedra   (22.2)

ceramics*   nonmetallic, inorganic solids usually produced at elevated temperature (A Chemist Looks At: Ceramics, Ceramics Glaze, and Glass)

plastic*   adjective meaning easily deformable   (A Chemist Looks At: Ceramics, Ceramics Glaze, and Glass)

earthenware*   porous pottery that has been fired at low temperatures (below about 1200°C)   (A Chemist Looks At: Ceramics, Ceramics Glaze, and Glass)

stoneware*   ceramic that has been fired above 1200°C to give a nonporous material   (A Chemist Looks At: Ceramics, Ceramics Glaze, and Glass)

porcelain*   ceramic made from a white clay fired at 1250°C to 1400°C and containing a high proportion of glass   (A Chemist Looks At: Ceramics, Ceramics Glaze, and Glass)

glaze*   nonporous glassy coating that results from firing a ceramic object that is coated with a watery suspension of silica and other oxides (A Chemist Looks At: Ceramics, Ceramics Glaze, and Glass)

supercooled*   describes a liquid cooled below its equilibrium freezing point (A Chemist Looks At: Ceramics, Ceramics Glaze, and Glass)

glass*   supercooled liquid whose viscosity is so high that it has the properties of a solid (A Chemist Looks At: Ceramics, Ceramics Glaze, and Glass)

glass–ceramics* ceramics made by adding materials to a glass in order to introduce many crystal nuclei into it (A Chemist Looks At: Ceramics, Ceramics Glaze, and Glass)

blanketing gas* gas used to protect a material from oxygen during processing or storage (22.3)

Haber process* industrial process for making NH_3 from a gaseous mixture of N_2 and H_2 (22.3)

Ostwald process industrial preparation of nitric acid starting from the catalytic oxidation of ammonia to nitric oxide (22.3)

polyphosphoric acids acids with the general formula $H_{n+2}P_nO_{3n+1}$ formed from linear chains of P—O bonds (22.4)

metaphosphoric acids acids with the general formula $(HPO_3)_n$ (22.4)

polymetaphosphoric acid* metaphosphoric acid with a very large number of repeating units (22.4)

eutrophication* overabundance of plants and algae in lakes, causing oxygen-deficiency from decomposition (22.4)

oxide binary compound with oxygen in the –2 oxidation state (22.5)

peroxide compound with oxygen in the –1 oxidation state (22.5)

superoxide binary compound with oxygen in the $-\frac{1}{2}$ oxidation state (22.5)

rhombic sulfur* stablest form of sulfur under normal conditions, consisting of crown-shaped S_8 molecules (22.6)

monoclinic sulfur* allotrope of sulfur that is unstable below 96°C and differs from rhombic sulfur in the way the S_8 molecules are packed to form crystals (22.6)

plastic sulfur* amorphous mixture of sulfur chains that is rubbery (22.6)

Frasch process mining procedure in which underground deposits of solid sulfur are melted in place with superheated water, and the molten sulfur is forced upward as a froth using air under pressure (22.6)

Claus process method of obtaining free sulfur by the partial burning of hydrogen sulfide (22.6)

sulfurous acid* hydrated species of SO_2, $H_2SO_3(aq)$; has not been isolated (22.6)

contact process industrial method for the manufacture of sulfuric acid that consists of the reaction of sulfur dioxide with oxygen to form sulfur trioxide using a catalyst of vanadium(V) oxide, followed by the reaction of sulfur trioxide with water (22.6)

thio-* prefix indicating that a sulfur atom has replaced an oxygen atom in a compound or ion, the name of which follows the prefix (22.6)

developer* liquid that reduces silver halide on exposed photographic film to bring out the image (22.6)

fixer* solution of sodium thiosulfate used to dissolve unexposed silver halide from exposed film (22.6)

photoconductor* material that is normally a poor conductor of electricity but becomes a good conductor when light falls on it (A Chemist Looks At: Selenium and How Photocopiers Work)

xerography* photocopying technique (A Chemist Looks At: Selenium and How Photocopiers Work)

toner* dry powder, attracted to charged areas of the selenium-coated belt in a photocopier, which is then transferred to paper and fused with heat to form the image (A Chemist Looks At: Selenium and How Photocopiers Work)

pickling* cleaning metal surfaces of oxides with hydrochloric acid (22.7)

inert gases* name used until 1962 for the Group VIIIA elements; given on the basis of the expected lack of reactivity of these elements due to their filled valence shells (22.8, introductory section)

*argos** Greek word meaning "lazy"; root of the element name argon (22.8)

*helios** Greek word meaning "sun"; root of the element name helium (22.8)

CHAPTER DIAGNOSTIC TEST

1. Write the balanced equation for each of the following oxidation–reduction reactions:
   (a) the oxidation of arsenic(III) oxide, As_4O_6, by nitric acid to produce arsenic acid and nitrogen(IV) oxide
   (b) the reaction of silver with hot concentrated sulfuric acid to yield silver sulfate, sulfur dioxide, and water

2. Contrast the properties of carbon dioxide with those of silicon dioxide.

3. Write the positive oxidation states of chlorine, and give a compound exemplifying each state.

4. Calculate the mass of zinc sulfide required to prepare 1.40 L of hydrogen sulfide (782 mmHg, 25°C) by reaction with excess hydrochloric acid.

5. Arrange each row of elements, ions, or compounds in increasing order of the specified property:
   (a) metallic character   B   Be   Li   N
   (b) metallic character   C   Pb   Si   Sn
   (c) stability ClO^- ClO_3^-
   (d) boiling point   HBr   HCl   HF   HI
   (e) oxidizing strength Br_2 Cl_2 F_2 I_2

6.   Write equations for each of the following reactions:
     (a)   Carbon is heated in water vapor.
     (b)   Silicon is reacted with fluorine. (The product is a gas.)
     (c)   Silicon dioxide is heated in concentrated sodium hydroxide solution—a
           metasilicate (SiO_3^{2-}) is formed.
     (d)   Xenon reacts with fluorine in the presence of sunlight.
     (e)   Carbon disulfide is burned in excess oxygen.
     (f)   Phosphorus is reacted with oxygen. (The product is a solid.)

7.   Write equations for each of the following:
     (a)   the preparation of pure silicon from sand
     (b)   the preparation of ammonia from a nitride
     (c)   the preparation of chlorine by electrolysis of a sodium chloride solution
     (d)   the preparation of chlorine from HCl(*aq*)

8.   Sketch and name the expected molecular geometries of the following compounds.
     Also give the bonding orbitals on the central atom.
     (a)   phosphorus(V) fluoride   (c)   sulfate ion
     (b)   phosphoric acid   (d)   chlorate ion

9.   Calculate the mass of sulfuric acid required to react with 31.4 kg of phosphate rock,
     $Ca_3(PO_4)_2$, to make superphosphate fertilizers.

10. Describe the properties and molecular structure of rhombic sulfur as it is slowly
     heated from its melting point at 113°C to its boiling point at 445°C.

11. Account for the electrical conductivity and lubricating properties of graphite.

ANSWERS TO CHAPTER DIAGNOSTIC TEST

If you missed an answer, study the text section and operational skill given in parentheses
after the answer. Previous text sections and operational skills are included in some cases.

1.   (a) $As_4O_6 + 8HNO_3 + 2H_2O \longrightarrow 4H_3AsO_4 + 8NO_2$

                                                   (3.5, 22.3, 22.4, Op. Sk. 5)

     (b) $2H_2SO_4 + 2Ag \longrightarrow Ag_2SO_4 + SO_2 + 2H_2O$ (22.6)

2. CO_2 is a gaseous molecular substance at standard conditions. SiO_2 is a high-melting
     covalent network substance with simplest formula SiO_2.   (22.1, 22.2)

3. +1 Cl_2O or $HClO$

 +3 $HClO_2$

 +4 ClO_2

 +5 $HClO_3$

 +7 $HClO_4$ (10.1, 22.7)

4. 5.74 g ZnS (5.3, 5.4, Op. Sk. 4, 6; 22.6)

5. (a) N B Be Li (chapter introduction, 21.4)

 (b) C Si Sn Pb (22.1, 21.4)

 (c) ClO^- ClO_3^- (22.7)

 (d) HCl HBr HI HF (11.5, Op. Sk. 5)

 (e) I_2 Br_2 Cl_2 F_2 (22.7)

6. (a) $C(s) + H_2O(g) \xrightarrow{\Delta} CO(g) + H_2(g)$ (22.1)

 (b) $Si(s) + 2F_2(g) \longrightarrow SiF_4(g)$ (22.2, 22.8)

 (c) $SiO_2(s) + 2NaOH(aq) \xrightarrow{\Delta} Na_2SiO_3(aq) + H_2O(l)$ (22.2)

 (d) $Xe(g) + 2F_2(g) \longrightarrow XeF_4(s)$ (22.8)

 (e) $CS_2(l) + 3O_2(g) \longrightarrow CO_2(g) + 2SO_2(g)$ (22.1)

 (f) $P_4(s) + 3O_2(g) \xrightarrow{\Delta} P_4O_6(s)$, or $P_4(s) + 5O_2(g) \xrightarrow{\Delta} P_4O_{10}(s)$

 (22.4)

7. (a) $SiO_2(l) + 2C(s) \xrightarrow{\Delta} Si(l) + 2CO(g)$

 $Si(s) + 2Cl_2(g) \xrightarrow{\Delta} SiCl_4(g)$; purify the product

 $SiCl_4(g) + 2H_2(g) \xrightarrow{\Delta} Si(s) + 4HCl(g)$ (22.2)

 (b) $Mg_3N_2(s) + 6H_2O(l) \longrightarrow 3Mg(OH)_2(s) + 2NH_3(g)$ (22.3)

 (c) $2NaCl(aq) + 2H_2O(l) \xrightarrow{\Delta} 2NaOH(aq) + H_2(g) + Cl_2(g)$ (22.7; 19.10)

 (d) $4HCl(aq) + MnO_2(s) \longrightarrow MnCl_2(aq) + Cl_2(g) + 2H_2O(l)$ (22.7)

8.   (a)

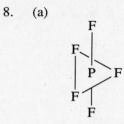

The geometry is trigonal bipyramidal and the bonding orbitals are dsp^3.

(22.4; 10.1, 10.3, Op. Sk. 1, 3)

(b)

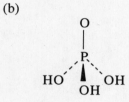

The geometry is tetrahedral and the bonding orbitals are sp^3.

(22.4; 10.1, 10.3, Op. Sk. 1, 3)

(c)

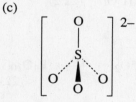

The geometry is tetrahedral and the bonding orbitals are sp^3.

(22.5; 10.1, 10.3, Op. Sk. 1, 3)

(d)

The geometry is trigonal pyramidal and the bonding orbitals are sp^3.

(22.7; 10.1, 10.3, Op. Sk. 1, 3)

9. 29.8 kg H_2SO_4 (22.4; 4.7, Op. Sk. 10)

10. At its melting point, rhombic sulfur is S_8 crown-shaped molecules. The color changes from yellow to straw as the sulfur turns to liquid. On continued heating, the liquid becomes dark reddish-brown and viscous. The S_8 molecules open up and the fragments join to give long spiral chains of sulfur atoms, which intertwine to give a liquid of increased viscosity. At temperatures above 200°C, the chains begin to break apart, resulting in a decreased viscosity. The vapor at the boiling point is composed of S_8, S_6, S_4, and S_2 molecules. (22.6)

11. Graphite is composed of sheets of carbon atoms bonded with sp^2 hybrid orbitals, leaving one electron per carbon to form a pi cloud of delocalized electrons above and below each sheet. This gives rise to the electrical conductivity. Because the sheets are held together only by weak van der Waals forces, the sheets can slip over each other easily, giving rise to the lubricating properties. (11.8, 22.1)

SUMMARY OF CHAPTER TOPICS

22.1 CARBON

22.2 SILICON

22.3 NITROGEN

22.4 PHOSPHORUS

22.5 OXYGEN

22.6 SULFUR

22.7 CHLORINE

22.8 HELIUM AND THE OTHER NOBLE GASES

ADDITIONAL PROBLEMS

1. (a) Draw a portion of a layer of the graphite structure.
    (b) Name the atomic orbitals used in bonding.
    (c) Name the type of solid that graphite is: ionic, molecular, etc.
    (d) Explain the structural basis of the lubricant properties of graphite.

2. Arrange each row of elements, ions, or compounds in increasing order of the specified property:

    (a) metallic character   I Sr In Sb

    (b) metallic character   S Te O Po

    (c) acidity HNO_3 H_3PO_4

    (d) boiling point H_2O H_2S

    (e) stability HNO_3 HNO_2

3. Match each of the following compounds to its use:

    ____ (a) $AgBr$ 1. manufacture of flame retardants

    ____ (b) $Na_2S_2O_3$ 2. manufacture of glass

    ____ (c) $POCl_3$ 3. most manufactured chemical

    ____ (d) $NaClO$ 4. component of photographic film

    ____ (e) H_3PO_4 5. blowing agent for foamed plastics

    ____ (f) N_2H_4 6. photographic fixer

    ____ (g) SiO_2 7. household bleach

    ____ (h) H_2SO_4 8. fertilizers

4. Complete and balance each of the following equations:

    (a) $Cr_2O_3(s) + C(s) \xrightarrow{\Delta}$

    (b) $NO_2(g) + H_2O(l) \longrightarrow$

    (c) $H_2S(g) + SO_2(g) \longrightarrow$

    (d) $Cu(s) + 4HNO_3(aq) \longrightarrow$

5. Write balanced equations for the preparation of each of the following:

   (a) H_2SO_4 from S_8 and concentrated HNO_3

   (b) H_2SO_4 from SO_3 and H_2O

   (c) $NaClO_3$, starting with Cl_2

   (d) $NaClO$

6. Write the balanced chemical equation for the reaction of each of the following species with H_2SO_4. If no reaction occurs, write NR.

   (a) Cu   (b) $Ca_3(PO_4)_2$ (c) NaCl   (d) $KClO_4$

7. Write balanced chemical equations that indicate

   (a) the preparation of $CaCO_3$.

   (b) the decomposition of $NaNO_3$.

   (c) the preparation of white phosphorus from $Ca_3(PO_4)_2$.

   (d) the preparation of $Na_2S_2O_3$.

   (e) the disproportionation of OCl^-.

8. Give the molecular geometry of the following compounds and the bonding orbitals on the central atom:
   (a) phosphorus(III) chloride   (b) silicon hydride
   (c) carbon disulfide   (d) chlorine trifluoride

9. What is the theoretical yield in tons (to three significant figures) of sulfuric acid from each ton of SO_2 that is used in the contact process? The equations are

$$2SO_2(g) + O_2(g) \xrightarrow[\text{V}_2\text{O}_5]{\Delta} 2SO_3(g)$$

$$SO_3(g) + H_2O(l) \longrightarrow H_2SO_4(aq)$$

10. Arrange the halogens in each of the following ways:
    (a) in order of increasing bond energy (see text Table 9.5)
    (b) in order of increasing polarizability of the halide ion
    (c) in order of increasing oxidizing strength of the halogen
    (d) in order of increasing reducing strength of the halide ion

ANSWERS TO ADDITIONAL PROBLEMS

If you missed an answer, study the text section and operational skill given in parentheses after the answer. Text sections and operational skills from previous chapters are given where appropriate.

1.   (a)

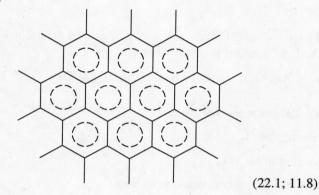

(22.1; 11.8)

   (b) sp^2 (22.1; 10.3)
   (c)   covalent network solid   (22.1; 11.8)
   (d)   The covalently bonded layers are loosely bound by London forces, allowing them to move freely across one another.   (22.1; 11.5)

2.   (a)   I   Sb   In   Sr   (chapter introduction, 21.4)

   (b)   O   S   Te   Po   (22.5)

   (c) H_3PO_4 HNO_3 (22.3, 22.4; 3.4, Op. Sk. 3)

   (d) H_2S H_2O (22.5, 22.6; 11.5, Op. Sk. 5)

   (e) HNO_2 HNO_3 (22.3)

3.   (a)   4   (22.7)   (d)   7   (22.7)   (g)   2   (22.2)
     (b)   6   (22.6)   (e)   8   (22.4)   (h)   3   (22.6)
     (c)   1   (22.3)   (f)   5   (22.3)

4.   (a) $Cr_2O_3(s) + 3C(s) \xrightarrow{\Delta} 2Cr(s) + 3CO(g)$ (22.1)

   (b) $3NO_2(g) + H_2O(l) \longrightarrow 2HNO_3(aq) + NO(g)$ (22.3)

   (c) $16H_2S(g) + 8SO_2(g) \longrightarrow 3S_8(s) + 16H_2O(g)$ (22.6)

(d) $Cu(s) + 4HNO_3(aq) \longrightarrow Cu(NO_3)_2(aq) + 2NO_2(g) + 2H_2O(l)$ (22.3)

5.   (a) $S_8(s) + 48HNO_3(aq) \longrightarrow 8H_2SO_4(aq) + 48NO_2(g) + 16H_2O(l)$ (22.3)

(b) $SO_3(g) + H_2O(l) \longrightarrow H_2SO_4(aq)$ (22.6)

(c) $3Cl_2(g) + 6NaOH(aq) \xrightarrow{\Delta} NaClO_3(aq) + 5NaCl(aq) + 3H_2O(l)$ (22.7)

(d) $Cl_2(g) + 2NaOH(aq) \xrightarrow{\text{cold}} NaCl(aq) + NaClO(aq) + H_2O(l)$ (22.7)

6.   (a) $Cu(s) + 2H_2SO_4(l) \xrightarrow{\Delta} CuSO_4(aq) + 2H_2O(l) + SO_2(g)$ (22.6)

(b) $Ca_3(PO_4)_2(aq) + 3H_2SO_4(aq) \longrightarrow 3CaSO_4(s) + 2H_3PO_4(aq)$ (22.4)

(c) $NaCl(s) + H_2SO_4(aq) \longrightarrow HCl(g) + NaHSO_4(aq)$ (22.7)

(d) $KClO_4(s) + H_2SO_4(aq) \longrightarrow KHSO_4(s) + HClO_4(l)$ (22.7)

7.   (a) $CO_2(g) + Ca(OH)_2(aq) \longrightarrow CaCO_3(s) + 2H_2O(l)$ (22.1)

(b) $2NaNO_3(s) \xrightarrow{\Delta} 2NaNO_2(s) + O_2(g)$ (22.3)

(c) $2Ca_3(PO_4)_2(s) + 6SiO_2(s) + 10C(s) \xrightarrow{\Delta} 6CaSiO_3(l) + 10CO(g) + P_4(q)$

(22.4)

(d) $8Na_2SO_3(aq) + S_8(s) \xrightarrow{\Delta} 8Na_2S_2O_3(aq)$ (22.6)

(e) $3ClO^-(aq) \longrightarrow ClO_3^-(aq) + 2Cl^-(aq)$ (22.7)

8.   (a)   pyramidal, sp^3 orbitals
(b)   tetrahedral, sp^3 orbitals
(c)   linear, sp orbitals
(d)   T-shaped, sp^3d orbitals   (10.1, 10.3, Op. Sk. 1, 3)

9.   1.53 ton   (4.7, Op. Sk. 10)

10.   (a) $I_2 < F_2 < Br_2 < Cl_2$ (9.11)

(b) $F^- < Cl^- < Br^- < I^-$ (11.5)

(c) $I_2 < Br_2 < Cl_2 < F_2$ (22.7)

(d) $F^- < Cl^- < Br^- < I^-$ (3.5, 22.7)

CHAPTER POST-TEST

1. Arrange each of the following rows of elements or compounds in increasing order of the specified property:

    (a)   metallic character   Cl Mg P Si
    (b)   metallic character   As Bi P Sb
    (c)   stability   diamond   graphite
    (d)   melting point Br_2 Cl_2 F_2 I_2

    (e)   acid strength $HClO$ $HClO_2$ $HClO_3$ $HClO_4$

2. Contrast the properties of diamond and graphite. Explain briefly.

3. Contrast the reactions of copper with dilute and concentrated nitric acid.

4. Write equations for each of the following reactions:
    (a)   Potassium is burned in oxygen.
    (b)   Silicon is heated with chlorine.
    (c)   Reaction of CH_4 with steam.
    (d)   Decomposition of $KClO_3$ in the presence of MnO_2.
    (e)   Reaction of $NaHSO_3$ with hydrochloric acid.
    (f)   Production of Cl_2 from HCl.

5. Describe the relationship between the structures of white phosphorus, phosphorus(III) oxide, and phosphorus(V) oxide, using sketches.

6. Write the balanced oxidation–reduction equation for the reaction of hydrogen sulfite ion with iodate ion to produce iodine and sulfate ion.

7. Write equations for each of the following:
    (a)   the preparation of $HClO_4$ from a perchlorate salt
    (b)   the preparation of a metal nitride
    (c)   the preparation of XeF_4
    (d)   the preparation of pure phosphoric acid from phosphate rock

8. Sketch and name the expected geometries of the following species. Also give the bonding orbitals on the central atom.
    (a)   phosphorus acid
    (b)   thiosulfate ion
    (c)   telluric acid, $Te(OH)_6$
    (d)   chlorite ion

9. Calculate the mass of white phosphorus that could be obtained from 1.00×10^3 kg of phosphate rock, which is 92% $Ca_3(PO_4)_2$.

10. Calculate the mass of calcium fluoride required to prepare 19.0 kL of hydrogen fluoride (1.100 atm, and 78.0°C) by reaction with excess sulfuric acid.

11. Write the reaction of chlorine in warm sodium hydroxide solution to produce sodium chloride and sodium chlorate. Calculate $E°$ for this reaction.
    For $ClO_3^-(aq)/Cl_2(g)/$ Pt, $E° = 0.46V$, and for $Cl_2(g)/Cl^-(aq)/Pt$, $E° = 1.36V$.

12. Write balanced equations to show how you can prepare trisodium phosphate, Na_3PO_4, from white phosphorus.

ANSWERS TO CHAPTER POST-TEST

If you missed an answer, study the text section and operational skill given in parentheses after the answer. Previous text sections and operational skills are included in some cases.

1. (a) Cl P Si Mg (chapter introduction, 21.4)
   (b) P As Sb Bi (22.3, 21.4)
   (c) diamond graphite (22.1)
   (d) F_2 Cl_2 Br_2 I_2 (11.5)
   (e) HClO $HClO_2$ $HClO_3$ $HClO_4$ (13.5, 22.7)

2. Diamond is the hardest substance known. It is a brittle solid and does not conduct electricity. Each carbon atom is bonded to four other carbon atoms through sp^3 hybrid orbitals. The color varies depending on impurities, and a pure diamond is colorless and transparent to light. Graphite, on the other hand, is one of the softest substances known. It is a solid composed of carbons bonded in layered sheets through sp^2 orbitals, each sheet being held to the others by weak van der Waals forces. The fourth electron on each carbon atom is involved in an electron cloud above and below each sheet, making graphite a conductor of electricity. The substance is black and opaque to light.   (22.1)

3. The reaction with dilute nitric acid is

$$3Cu(s) + 8HNO_3(aq) \longrightarrow 3Cu(NO_3)_2(aq) + 2NO(g) + 4H_2O(l)$$

and with concentrated nitric acid is

$$Cu(s) + 4HNO_3(aq) \longrightarrow Cu(NO_3)_2(aq) + 2NO_2(g) + 2H_2O(l) \quad (22.3)$$

4. (a) $K(s) + O_2(g) \xrightarrow{\Delta} KO_2(s) \quad (22.5)$

(b) $Si(s) + 2Cl_2(g) \xrightarrow{\Delta} SiCl_4(g) \quad (22.2)$

(c) $CH_4(g) + H_2O(g) \xrightarrow{Ni} CO(g) + 3H_2(g) \quad (22.1)$

(d) $2KClO_3(s) \xrightarrow[MnO_2]{\Delta} 2KCl(s) + 3O_2(g) \quad (22.5)$

(e) $NaHSO_3(aq) + HCl(aq) \longrightarrow NaCl(aq) + H_2O(l) + SO_2(g) \quad (22.6)$

(f) $4HCl(aq) + MnO_2(s) \longrightarrow MnCl_2(aq) + Cl_2(g) + 2H_2O(l) \quad (22.7)$

5. All three structures are based on the P_4 tetrahedron. White phosphorus is the tetrahedron, with a phosphorus atom at each apex. In phosphorus(III) oxide, P_4O_6, there is an oxygen atom between every two phosphorus atoms. In phosphorus(V) oxide, P_4O_{10}, there is also an oxygen atom bonded to the fourth electron pair on each phosphorus atom. The structures are shown below. (The unbonded electron pair on each phosphorus atom is shown in each drawing where appropriate.)

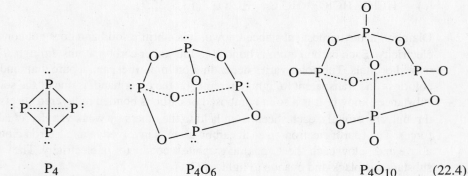

P_4
white phosphorus

P_4O_6

P_4O_{10} (22.4)

6. $5HSO_3^- + 2IO_3^- \longrightarrow 5SO_4^{2-} + I_2 + H_2O + 3H^+ \quad (22.6; 3.6, \text{Op. Sk. } 7)$

7. (a) $KClO_4(s) + H_2SO_4(l) \longrightarrow KHSO_4(s) + HClO_4(l)$ (22.7)

   (b) $3Mg(s) + N_2(g) \longrightarrow Mg_3N_2(s)$ (22.3)

   (c) $Xe(g) + 2F_2(g) \longrightarrow XeF_4(s)$ (22.8)

   (d) $2Ca_3(PO_4)_2(s) + 6SiO_2(s) + 10C(s) \xrightarrow{\Delta} 6CaSiO_3(l) + 10CO(g) + P_4(g)$

$$P_4(g) \longrightarrow P_4(l)$$

$$P_4(l) + 5O_2(g) \xrightarrow{\Delta} P_4O_{10}(s)$$

$$P_4O_{10}(s) + 6H_2O(l) \longrightarrow 4H_3PO_4(l) \quad (22.4)$$

8. (a)

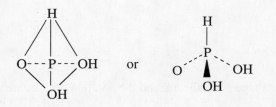

The molecular shape is tetrahedral and the bonding orbitals are sp^3.

(22.4; 10.1, 10.3, Op. Sk. 1, 3)

(b)

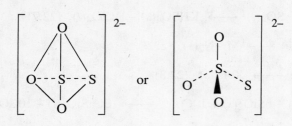

The ionic shape is tetrahedral and the bonding orbitals are sp^3.

(22.6; 10.1, 10.3, Op. Sk. 1, 3)

(c)

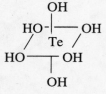

The molecular shape is octahedral and the bonding orbitals are d^2sp^3.

(10.1, 10.3, Op. Sk. 1, 3)

(d)

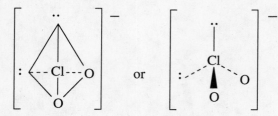

The ionic shape is angular (bent) and the bonding orbitals are sp^3.

(22.7; 10.1, 10.3, Op. Sk. 1, 3)

9. 1.8×10^2 kg P_4 (22.4; 4.7, Op. Sk. 10)

10. 2.83×10^4 g CaF_2 (22.6; 5.4, Op. Sk. 6)

11. $3Cl_2(g) + 6NaOH(aq) \longrightarrow 5NaCl(aq) + NaClO_3(aq) + 3H_2O(l)$
$E° = 0.90$ V   (22.7; 19.4, Op. Sk. 6)

12. $$P_4(s) + 5O_2(g) \xrightarrow{\Delta} P_4O_{10}(s)$$

$$P_4O_{10}(s) + 6H_2O(l) \longrightarrow 4H_3PO_4(aq)$$

$$H_3PO_4(aq) + 3NaOH(aq) \longrightarrow Na_3PO_4(aq) + 3H_2O(l)$$

(22.4; 3.4, Op. Sk. 4)

CHAPTER 23   THE TRANSITION ELEMENTS

CHAPTER TERMS AND DEFINITIONS

Numbers in parentheses after definitions give the text sections in which the terms are explained. Starred terms are italicized in the text. Where a term does not fall directly under a text section heading, additional information is given for you to locate it.

d-block transition elements   elements with an unfilled d subshell in common oxidation states   (23.1, introductory section)

f-block (inner-)transition elements   elements with a partially filled f subshell in common oxidation states   (23.1, introductory section)

lanthanides (rare earths)*   first row of inner-transition elements   (23.1, introductory section)

actinides*   second row of inner-transition elements   (23.1, introductory section)

effective nuclear charge*   positive charge acting on an electron; equals the nuclear charge minus the shielding by intervening electrons   (23.1)

lanthanide contraction*   gradual decrease in atomic radii in the series of elements cerium to lutetium because of the increase in effective nuclear charge   (23.1)

khroma*   Greek word for "color," from which the name for the element chromium is derived   (23.2)

Goldschmidt process*   preparation of pure chromium metal from the exothermic reaction of chromium(III) oxide with aluminum   (23.2)

matte*   intermediate substance, molten copper(I) sulfide, produced in the metallurgy of copper   (23.2)

blister copper*   99% pure copper produced by blowing air through molten Cu_2S (23.2)

complex ion   metal atom or ion with Lewis bases attached to it through coordinate covalent bonds   (23.3)

complex (coordination compound)   compound consisting either of complex ions with other ions of opposite charge or of a neutral complex species   (23.3)

ligands   Lewis bases attached to the metal atom in a complex   (23.3)

coordination number in a complex, the total number of bonds the metal atom forms with ligands (23.3)

monodentate ligand ligand that bonds to the metal atom through one ligand atom (23.3)

bidentate ligand ligand that bonds to the metal atom through two ligand atoms (23.3)

quadridentate ligand* ligand that bonds to the metal atom through four ligand atoms (23.3)

globin* protein that chemically bonds to heme as a quadridentate ligand (23.3)

heme* planar molecule including Fe(II); part of the hemoglobin molecule (23.3)

polydentate ligand (chelating agent) ligand that bonds to the metal atom through two or more ligand atoms (23.3)

chelate complex formed by a metal atom and a polydentate ligand (23.3)

chele* Greek word for "claw"; root of the word chelate (23.3, marginal note)

nomenclature* systematic method of naming compounds (23.4)

mono-* prefix used in nomenclature to denote one ligand; often omitted (23.4)

di-* prefix used in nomenclature to denote two ligands (23.4)

tri-* prefix used in nomenclature to denote three ligands (23.4)

tetra-* prefix used in nomenclature to denote four ligands (23.4)

penta-* prefix used in nomenclature to denote five ligands (23.4)

hexa-* prefix used in nomenclature to denote six ligands (23.4)

bis-* prefix denoting two ligands; used when ligand name contains a Greek number prefix (23.4)

tris-* prefix denoting three ligands; used when ligand name contains a Greek number prefix (23.4)

tetrakis-* prefix denoting four ligands; used when ligand name contains a Greek number prefix (23.4)

-ate* suffix used in nomenclature to denote that a complex is an anion (23.4)

isomerism* phenomenon whereby compounds have the same molecular formula (or the same simplest formula in the case of ionic compounds) but with different arrangements of atoms (23.5)

paramagnetism* phenomenon whereby substances are weakly attracted to a strong magnetic field because of the presence of unpaired electrons in the substance (23.5)

ferromagnetism* aligned magnetism of unpaired electrons, such as in solid iron, having a magnetic attraction about a million times stronger than that of paramagnetism (23.5)

Gouy balance* specially equipped balance used to measure paramagnetism (23.5)

color* property exhibited by a substance because of its absorption of some frequencies of light in the visible region of the electromagnetic spectrum (23.5)

structural isomers isomers that differ in how the atoms are joined together (23.5)

stereoisomers isomers that have the same atoms bonded to each other in the same order but that differ in the precise arrangement of these atoms in space (23.5)

ionization isomers structural isomers of a complex that differ in the anion that is coordinated to the metal atom (23.5)

hydrate isomers structural isomers of a complex that differ in the placement of water molecules in the complex (23.5)

coordination isomers structural isomers of compounds consisting of complex cations and complex anions that differ in the way the ligands are distributed between the metal atoms (23.5)

linkage isomers structural isomers of a complex that differ in the atom of a ligand that is bonded to the metal atom (23.5)

ambidentate* describing a ligand that can bond through either of two ligand atoms (23.5)

geometric isomers isomers in which the atoms are joined to one another in the same way but that differ because some atoms occupy different relative positions in space (23.5)

*cis** isomer that has two A ligands adjacent to each other (23.5)

*trans** isomer that has two A ligands opposite each other (23.5)

*cis–trans** isomers in which the two A ligands are either adjacent to or opposite each other (23.5)

enantiomers (optical isomers) isomers that are nonsuperimposable (or nonsuperposable) mirror images of one another (23.5)

superimposed* placed directly over another so that all like parts coincide; also **superposed** (23.5)

chiral possessing the quality of handedness; an object that is not identical to its mirror image (23.5)

*cheir** Greek word for "hand" (23.5)

achiral* describing an object that is identical to its mirror image (23.5)

plane-polarized light* light whose electromagnetic waves vibrate in only one plane (23.5)

polarizer* device (e.g., Polaroid lens) that selects out one plane in a beam of light (23.5)

optically active able to rotate the plane of polarized light (23.5)

polarimeter* instrument that determines the angular change in a plane of light rotated by an optically active compound (23.5)

dextrorotatory (*d*) able to rotate the plane of polarized light to the right when facing the light source (23.5)

levorotatory (*l*) able to rotate the plane of polarized light to the left when facing the light source (23.5)

racemic mixture mixture of equal amounts of enantiomers (23.5)

resolved* describing a racemic mixture separated into its *d* and *l* isomers (23.5)

valence bond theory* explanation of bonding as overlap of two orbitals, one from each bonding atom (23.6)

high-spin complex ion complex ion in which there is minimum pairing of electrons in the orbitals of the metal atom (23.6)

low-spin complex ion   complex ion in which there is more pairing of electrons in the orbitals of the metal atom than in a corresponding high-spin complex ion   (23.6)

crystal field theory   model of the electronic structure of transition-metal complexes that considers how the energies of the *d* orbitals of a metal ion are affected by the electronic field of the ligands   (23.7)

ligand field theory*   simple extension of the crystal field theory to include covalent character in the bonding   (23.7)

crystal field splitting (Δ)   difference in energy between the two sets of five *d* orbitals on a central metal ion that arises from the interaction of the orbitals with the electric field of the ligands   (23.7)

pairing energy (*P*)   energy required to put two electrons into the same orbital   (23.7)

spectrochemical series   arrangement of ligands according to the relative magnitudes of the crystal field splittings they induce in the *d* orbitals of a metal ion   (23.7)

complementary color*   color observed after absorption of a portion of white light by a sample   (Example 23.9)

tetramer*   structure built of four units   (A Chemist Looks At: The Cooperative Release of Oxygen from Oxyhemoglobin)

cooperative release*   release of O_2 from one heme group in hemoglobin that triggers the release of O_2 from another heme group of the same molecule   (A Chemist Looks At: The Cooperative Release of Oxygen from Oxyhemoglobin)

CHAPTER DIAGNOSTIC TEST

1. List the characteristics of most of the transition elements that set them apart from the main-group elements.

2. Complete the following statements:
   (a) Chromium(III) can be oxidized to chromium(VI) species, by heating chromite strongly with _____ in air.
   (b) The change from chromate ion to dichromate ion on the addition of acid is evidenced by a change in solution color from _____ to _____ .
   (c) In chromite, $FeCr_2O_4$, the oxidation number of iron is +2. The oxidation number of the chromium is _____ .
   (d) The principal commercial use of copper is as _____ .

3. Give the coordination number of the transition-metal atom in each of the following complexes:

   (a) $Ag(S_2O_3)_2^{3-}$ (b) $Cr(en)_3^{3+}$ (c) $Na_2[Ni(CN)_4]$ (d) $Co(CN)_5^{3-}$

4.   Give the IUPAC name for each of the following:

(a) $K_2[Ni(CN)_4]$ (b) $[Ag(NH_3)_2]Cl$

5.   Write the formula for each of the following:
(a)   tetraamminezinc(II) sulfate
(b)   sodium tetrahydroxoaluminate

6.   For each of the following, determine whether the type of isomerism exhibited is ionization, coordination, hydrate, or linkage:

(a) $[Cr(NH_3)_5SO_4]Br$ and $[Cr(NH_3)_5Br]SO_4$

(b) $[Co(en)(C_2O_4)_2][Cr(en)_2(C_2O_4)]$ and $[Co(C_2O_4)_3][Cr(en)_3]$

(c) $[Co(NH_3)_5ONO]Br_2$ and $[Co(NH_3)_5NO_2]Br_2$

7.   Write a balanced equation for the reduction of chromate ion by sufide ion in moderately basic solution. Under these conditions $Cr(OH)_3$ precipitates, as does elemental sulfur.

8.   Using valence bond theory, tell what orbitals would be used for bonding in $Fe(CN)_6{}^{4-}$. Give the number of unpaired electrons, if any, and the resulting magnetic character.

9.   Draw the geometric isomers that exist for any of the following stable complex ions. Name the ones you draw, including the *cis* or *trans* designation.

(a) $Pt(NH_3)_6{}^{4+}$

(b) $Cr(NH_3)_4(SCN)_2{}^+$ (SCN^- is the thiocyanate ion.)

(c) $Co(NH_3)_3(H_2O)_3{}^{3+}$

10.   Tell whether enantiomers (optical isomers) are possible for any of the following stable complexes. Draw them (en = $NH_2CH_2CH_2NH_2$).

(a)   *trans*-$Pt(en)_2Cl_2{}^{2+}$ (c)   *trans*-$[Ir(H_2O)_3Cl_3]$

(b)   *cis*-$Co(en)_2Cl_2{}^+$

11.   Describe the distribution of d electrons in the $Fe(CN)_6{}^{3-}$ ion, using crystal field theory. How many unpaired electrons are there in the ion, and what is the resulting magnetic character?

12. The maximum absorption of the $Co(NH_3)_5H_2O^{3+}$ ion is approximately 500 nm. What color is this ion? If the water is replaced with a chloride ion, would the resulting wavelength of maximum absorption be longer or shorter? Explain. (Use text Table 23.8.)

13. A complex of cobalt has the formula $CoCl_3 \cdot 4NH_3 \cdot 2H_2O$. Conductance measurements show the presence of four ions, and all the chlorides precipitate with $AgNO_3$. Write the formula that corresponds with this information.

ANSWERS TO CHAPTER DIAGNOSTIC TEST

If you missed an answer, study the text section and operational skill given in parentheses after the answer.

1. All transition elements are metals; most have high melting points and high boiling points and are hard solids; most show several oxidation states; compounds are often colored and many are paramagnetic. (23.1)

2. (a) sodium carbonate (23.2)
   (b) yellow, orange (23.2)
   (c) +3 (23.2)
   (d) an electrical conductor (23.2)

3. (a) 2   (b) 6   (c) 4   (d) 5 (23.3)

4. (a) potassium tetracyanonickelate(II)
   (b) diamminesilver(I) chloride (23.4, Op. Sk. 1)

5. (a) $[Zn(NH_3)_4]SO_4$ (b) $Na[Al(OH)_4]$ (23.4, Op. Sk. 1)

6. (a) ionization   (b) coordination   (c) linkage (23.5)

7. $2CrO_4^{2-}(aq) + 8H_2O(l) + 3S^{2-}(aq) \longrightarrow 3S(s) + 2Cr(OH)_3(s) + 10OH^-(aq)$
   (3.6, Op. Sk. 7; 23.2)

8. Two $3d$, the $4s$, and the $4p$ atomic orbitals would combine to form six d^2sp^3 hybrid orbitals to be used for bonding. Because CN^- bonds strongly, the complex would have no unpaired electrons and would thus be diamagnetic. (23.6, Op. Sk. 3)

9. (a)   No geometric isomers exist, as all ligands are the same.

    (b)

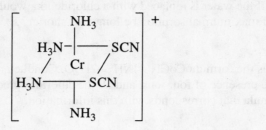

   cis-tetraamminedithiocyanatochromium(III) ion   and

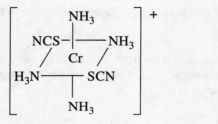

   trans-tetraamminedithiocyanatochromium(III) ion

    (c)

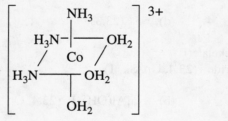

   cis-triamminetriaquacobalt(III) ion   and

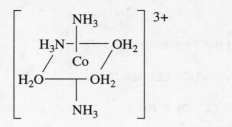

trans-triamminetriaquocobalt(III) ion   (23.5, Op. Sk. 2)

10. (a) No enantiomers exist, as the mirror image is the same compound.
     (b) The enantiomers are:

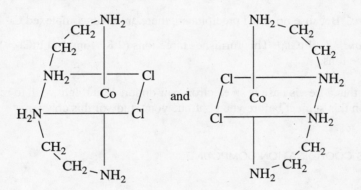   and

     (c) No enantiomers exist, as the mirror image is the same compound.

     (23.5, Op. Sk. 2)

11. Because CN⁻ is a strong-bonding ligand, the complex would be low-spin with the following electron distribution:

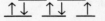

Thus, there is one unpaired electron, making the complex weakly paramagnetic.

     (23.7, Op. Sk. 3)

12. The ion is red. Chloride ion is a weaker-bonding ligand. Therefore, Δ would be smaller, resulting in a longer wavelength of absorption, because in the expression $\Delta = hc/\lambda$, the two variables have an inversely proportional relationship.

     (23.7, Op. Sk. 4)

13. $[Co(NH_3)_4(H_2O)_2]Cl_3$ (23.3)

SUMMARY OF CHAPTER TOPICS

23.1   PERIODIC TRENDS IN THE TRANSITION ELEMENTS

23.2   THE CHEMISTRY OF TWO TRANSITION ELEMENTS

23.3   FORMATION AND STRUCTURE OF COMPLEXES

Exercise 23.1   Another complex studied by Werner had a composition corresponding to the formula $PtCl_4 \cdot 2KCl$. From electrical-conductance measurements, he determined that each formula unit contained three ions. He also found that silver nitrate did not give a precipitate of AgCl with this complex. Write a formula for this complex that agrees with this information.

Known: Because no AgCl precipitated, there are no uncomplexed Cl^- ions.

Solution: $K_2[PtCl_6]$. This furnishes three ions (2 K^+ ions and $PtCl_6^{2-}$) and no chloride ions.

Recall that a Lewis base is an electron-pair donor. You might wish to review text Section 15.3 on this subject before you continue your study of this chapter.

23.4   NAMING COORDINATION COMPOUNDS

Operational Skill

1. Writing the IUPAC name given the structural formula of a coordination compound, and vice versa.   Given the structural formulas of coordination compounds, write the IUPAC names   (Example 23.1). Given the IUPAC names of complexes, write the structural formulas   (Example 23.2).

As with any set of directions, it is best to learn the rules of nomenclature as you use them to name the compounds, rather than to memorize the rules.

Exercise 23.2   Give the IUPAC names of (a) $[Co(NH_3)_5Cl]Cl_2$; (b) $K_2[Co(H_2O)(CN)_5]$; (c) $Fe(H_2O)_5(OH)^{2+}$.

Known: (a) and (b) are neutral compounds, (c) is a cation; rules listed in text.

Solution: (a)   pentaamminechlorocobalt(III) chloride
               (b)   potassium aquapentacyanocobaltate(III)
               (c)   pentaaquahydroxoiron(III) ion

Exercise 23.3 Write structural formulas for each of the following:
(a)   potassium hexacyanoferrate(II)
(b)   tetraamminedichlorocobalt(III) chloride
(c)   tetrachloroplatinate(II) ion

Known: (a) and (b) are neutral compounds, (c) is an anion; rules given in text.

Solution: (a) $K_4[Fe(CN)_6]$ (b) $[Co(NH_3)_4Cl_2]Cl$ (c) $PtCl_4^{2-}$

23.5   STRUCTURE AND ISOMERISM IN COORDINATION COMPOUNDS

Operational Skill

2. Deciding whether isomers are possible. Given the formula of a complex, decide whether geometric isomers are possible and, if so, draw them   (Example 23.3).   Given the structural formula of a complex, decide whether enantiomers (optical isomers) are possible and, if so, draw them   (Example 23.4).

To be able to perform the above operational skill, you will have to memorize the types of geometric and optical isomers and how to identify them. It will also be very helpful to have a model kit so you can make models of the isomers to see the relationships between the ligands. Such kits are available in most college bookstores. However, you can make your own models with toothpicks and colored jellybean candies. Use rubber bands for the bidentate ligands.

Exercise 23.4 Name the type of structural isomerism displayed by each of the following pairs:

(a) $[Co(en)_3][Cr(CN)_6]$ and $[Cr(en)_3][Co(CN)_6]$

(b) $[Mn(CO)_5(SCN)]$ and $[Mn(CO)_5(NCS)]$

(c) $[Co(NH_3)_5(NO_3)]SO_4$ and $[Co(NH_3)_5(SO_4)]NO_3$

(d) $[Co(NH_3)_4(H_2O)Cl]Cl_2$ and $[Co(NH_3)_4Cl_2]Cl \cdot H_2O$

Known: the four types of structural isomerism and how to identify each.

Solution: (a)   coordination isomerism   (c)   ionization isomerism
                (b)   linkage isomerism   (d)   hydrate isomerism

Exercise 23.5 A complex has the composition $Co(NH_3)_4(H_2O)Cl_3$. Conductance measurements show that there are three ions per formula unit, and precipitation of AgCl with silver nitrate shows that there are two Cl^- ions not coordinated to cobalt.

What is the structural formula of the compound? Write the structural formula of an isomer.

Known: Naming rules give the order of specific parts for writing the structural formula; one Cl^- ion would be coordinated to cobalt. The water molecule could either be coordinated to the cobalt or not.

Solution: The structural formula is $[Co(NH_3)_4(H_2O)Cl]Cl_2$. The structural formula of an isomer is $[Co(NH_3)_4Cl]Cl_2 \cdot H_2O$. These are hydrate isomers.

Exercise 23.6 Do any of the following stable octahedral complexes have geometric isomers? If so, draw them.

(a) $Co(NH_3)_5Cl^{2+}$ (b) $Co(NH_3)_4(H_2O)_2^{3+}$

(c) $[Cr(NH_3)_3(SCN)_3]$ (d) $Co(NH_3)_6^{3+}$

Known: definition of geometric isomers and how to draw them.

Solution: (a) No geometric isomers are possible, as there is only one different ligand.

(b) The geometric isomers are

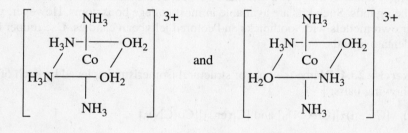

                        cis   *trans*

(c)   The geometric isomers are

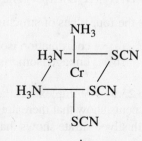

                 trans   *cis*

(d) No geometric isomers are possible, as all ligands are the same.

Exercise 23.7 Do any of the following have optical isomers? If so, draw them.

(a) *trans*-Co(en)$_2$(NO$_2$)$_2$$^+$ (b) *cis*-Co(en)$_2$(NO$_2$)$_2$$^+$

(c) Ir(en)$_3$$^{3+}$ (d) *cis*-[Ir(H$_2$O)$_3$Cl$_3$]

Known: Optical isomers exist if the mirror image of the structure is not identical to the original structure.

Solution: We will draw all complexes, as the determination is easier to make from a drawing.

(a) The structure is

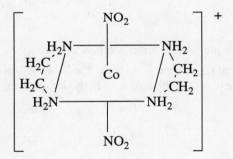

No optical isomers are possible, as this structure is identical to its mirror image.

(b) This complex does have optical isomers. The mirror images are

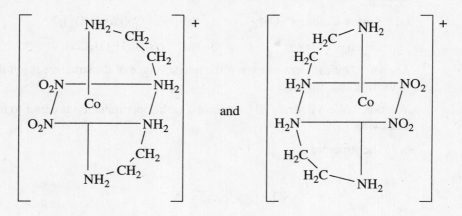

and

We can clearly see that these are not the same compound if we rotate the formula on the right 90° front right 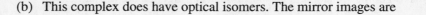 to bring the "bottom" en bidentate ligand to the same place where it is in the structure in the left drawing. We then have

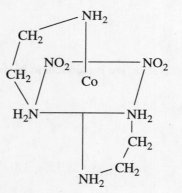

Note particularly that the positions of the NO$_2$ groups on this structure are different in the above left structure. Also note the difference in position of the "top" en bidentate ligand. This mirror image is not the same molecule.

(c) This complex does have optical isomers. They are

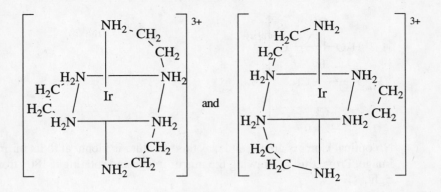

and

To show that these mirror images are not identical, we rotate the structure on the right 90° front right (⌣) to bring the "bottom" en bidentate ligand to the same place where it is in the structure in the left drawing. This gives

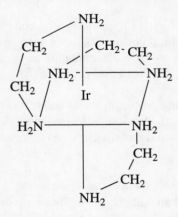

which is not the same structure as in the drawing on the upper left.

(d) The structure is

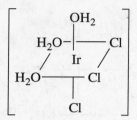

No optical isomers are possible, as this structure is identical to its mirror image. Prove this by drawing the mirror image and rotating it 180° front left or right.

23.6 VALENCE BOND THEORY OF COMPLEXES

Operational Skill

3. Describing the bonding in a complex ion. Given a transition-metal complex ion, describe the bonding types (high-spin and low-spin, if both exist), using valence bond theory for octahedral and four-coordinate complexes. Give the number of unpaired electrons in the complex   (Examples 23.5 and 23.6). (See also Section 23.7.)

An important thing to remember in working problems in this and following sections is that when the transition metals form ions, they lose *s* electrons first. The most common error students make in these problems is to remove *d* electrons first, which yields the wrong number of electrons in the *d* orbitals.

Exercise 23.8 Cobalt(III) forms many stable complex ions, including $Co(NH_3)_6^{3+}$. Most of these are octahedral and diamagnetic. The complex CoF_6^{3-}, however, is paramagnetic. Describe the bonding in $Co(NH_3)_6^{3+}$ and CoF_6^{3-}, using valence bond theory. How many unpaired electrons are there in each complex ion?

Wanted: bonding descriptions; number of unpaired electrons.

Given: $Co(NH_3)_6^{3+}$ is diamagnetic; CoF_6^{3-} is paramagnetic.

Known: correlation between magnetic character and electron pairing; Co is $[Ar]3d^7 4s^2$. The two $4s$ electrons and one $3d$ electron are lost to form Co^{3+}, leaving six electrons in the $3d$ subshell; correlating between octahedral coordination and the hybrid orbitals used for bonding.

Solution: The orbital diagram for the Co^{3+} ion is

[Ar] (⇅)(↑)(↑)(↑)(↑)   (○)   (○)(○)(○)   (○)(○)(○)(○)(○)
       $3d$ $4s$ $4p$ $4d$

$Co(NH_3)_6{}^{3+}$ is diamagnetic; therefore, there are no unpaired electrons and the complex is low-spin. The bonding will thus be six empty hybrid d^2sp^3 orbitals on cobalt overlapping with six doubly filled orbitals from the six NH_3 ligands. We show the following orbital diagram for Co^{3+} with the bonding electrons in the original atomic orbitals and donor electrons shown with dotted arrows:

[Ar] (⇅)(⇅)(⇅)(⇅)(⇅)   (⇅)   (⇅)(⇅)(⇅)   (○)(○)(○)(○)(○)
       $3d$ $4s$ $4p$ $4d$

$$d^2sp^3 \text{ bonds to ligands}$$

$CoF_6{}^{3-}$ is paramagnetic; therefore, there are four unpaired electrons and the complex is high-spin. The bonding will thus be six empty hybrid sp^3d^2 orbitals on cobalt overlapping with six doubly filled orbitals from the six F^- ligands. We show the following orbital diagram for Co^{3+} with the bonding electrons in the original atomic orbitals and donor electrons shown with dotted arrows:

[Ar] (⇅)(↑)(↑)(↑)(↑)   (⇅)   (⇅)(⇅)(⇅)   (⇅)(⇅)(○)(○)(○)
       $3d$ $4s$ $4p$ $4d$

$$sp^3d^2 \text{ bonds to ligands}$$

Exercise 23.9 The complex ion $CoCl_4{}^{2-}$ is paramagnetic, with a magnetism corresponding to three unpaired electrons. If the complex ion indeed has four ligands, as suggested by the formula, what geometry is indicated?

Wanted: geometry.

Given: $CoCl_4{}^{2-}$ is paramagnetic, has three unpaired electrons, and is four-coordinate.

Known: Geometry can be tetrahedral, with sp^3 hybrid orbitals used, or square planar, with dsp^2 hybrid orbitals used; Co in this complex is Co(II) and thus has lost the two $4s$ electrons, leaving seven electrons in the $3d$ subshell.

Solution: There are seven electrons in the $3d$ subshell. The orbital diagrams for the two possible types of coordination follow.

Tetrahedral, with three unpaired electrons in the $3d$ subshell:

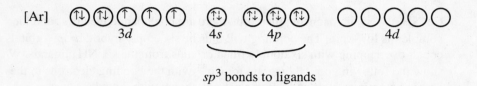

sp^3 bonds to ligands

Square planar, with one unpaired electron in the $3d$ subshell:

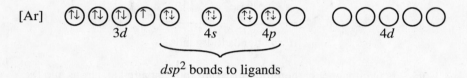

dsp^2 bonds to ligands

The geometry would be tetrahedral, as there are three unpaired electrons in this type of hybridization, correlating with the magnetism corresponding to three unpaired electrons.

23.7   CRYSTAL FIELD THEORY

Operational Skills

3. Describing the bonding in a complex ion. Given a transition-metal complex ion, describe the bonding types (high-spin and low-spin, if both exist), using crystal field theory for octahedral and four-coordinate complexes. Give the number of unpaired electrons in the complex   (Examples 23.7 and 23.8).

4. Predicting the relative wavelengths of absorption of complex ions.   Given two complexes that differ only in the ligands, predict, on the basis of the spectrochemical series, which complex absorbs at higher wavelength. Given the absorption maxima, predict the colors of the complexes   (Example 23.9).

> **Exercise 23.10** Describe the distribution of d electrons in $Ni(H_2O)_6^{2+}$, using crystal field theory. How many unpaired electrons are there in this ion?

Known: Ni^{2+} has lost the two $4s$ electrons, so it is $[Ar]3d^8$; in an octahedral field, the $3d$ orbitals are split into two sets; water is in the middle of the spectrochemical series.

Solution: Because water is not a strong-bonding ligand, a complex with it would be expected to be high-spin. Because Ni^{2+} is octahedral, however, it can have only one electron distribution. The $3d$ orbitals and their electron occupancies are indicated below:

$$\uparrow \ \uparrow$$

$$\uparrow\downarrow \ \uparrow\downarrow \ \uparrow\downarrow \ \uparrow \ \uparrow \qquad\qquad \uparrow\downarrow \ \uparrow\downarrow \ \uparrow\downarrow$$

uncombined Ni^{2+} \qquad\qquad complexed Ni^{2+}

$Ni(H_2O)_6^{2+}$ has two unpaired electrons.

Exercise 23.11 Describe the distribution of d electrons in the $CoCl_4^{2-}$ ion. The ion has a tetrahedral geometry. Assume a high-spin complex.

Known: Cobalt has an oxidation number of +2 in this complex and thus has lost the two $4s$ electrons, so it is $[Ar]3d^7$; the tetrahedral splitting pattern; high-spin means that a maximum number of electrons are unpaired.

Solution: The seven electrons in the $3d$ sublevel of the complexed ion are distributed as follows:

$$\uparrow \ \uparrow \ \uparrow$$

$$\uparrow\downarrow \ \uparrow\downarrow$$

Exercise 23.12 The $Fe(H_2O)_6^{3+}$ ion has a pale purple color, and the $Fe(CN)_6^{3-}$ ion has a ruby-red color. What are the approximate wavelengths of the maximum absorption for each ion? Is the shift of wavelength in the expected direction? Explain.

Wanted: λ_{max} for each ion; is shift in expected direction?

Given: $Fe(H_2O)_6^{3+}$ ion is pale purple; $Fe(CN)_6^{3-}$ is ruby-red.

Known: $\Delta = h\upsilon = hc/\lambda$; the more strongly bonding ligand gives a greater Δ; CN^- is a more strongly bonding ligand than water.

Solution: $Fe(H_2O)_6^{3+}$, being pale purple, absorbs at about 530 nm. $Fe(CN)_6^{3-}$, being ruby-red, absorbs at about 500 nm. Rearranging the expression for Δ to solve for λ gives $\lambda = hc/\Delta$. With the more strongly bonding CN^- ligand, and the resulting greater Δ, the λ of absorption should decrease; so the shift from absorption at about 530 nm to about 500 nm is in the expected direction.

602

23. The Transition Elements

ADDITIONAL PROBLEMS

1. Write the appropriate reaction or reactions for each of the following:
   (a) the preparation of chromium from chromite ore and aluminum
   (b) the preparation of copper by the reduction of copper(I) sulfide

2. Complete and balance the following equations:

   (a) $K_2Cr_2O_7(aq) + H_2SO_4(aq) \longrightarrow$

   (b) $Cr(OH)_3(s) + OH^-(aq) \longrightarrow$

   (c) $Cu(s) + HNO_3(aq) \longrightarrow$
       (conc)

3. The elemental analysis of a purple-red octahedral complex of Co(III) gives the composition as $CoCl_3(NH_3)_5$. One student proposes that the correct formula for the complex is $[CoCl(NH_3)_5]Cl_2$. Describe the qualitative and the quantitative test that would substantiate this formula and rule out $[CoCl_2(NH_3)_4]Cl(NH_3)$ and $[CoCl_3(NH_3)_3](NH_3)_2$ as possibilities.

4. Write the formula for each of the following:
   (a) tetraamminecopper(II) sulfate
   (b) potassium hexacyanochromate(III)
   (c) pentaamminenitrocobalt(III) chloride

5. Draw the geometric isomers of $Co(en)_2Br_2^+$. If any of these isomers are optically active, draw these enantiomers. (en = $NH_2CH_2CH_2NH_2$)

6. Name the following complexes:
   (a) $[Co(NH_3)_5SO_4]Br$
   (b) $[Co(NH_3)_5Br]SO_4$
   (c) $[Cr(H_2O)_5Cl]Cl_2$
   (d) $K_3[Fe(CN)_5CO]$

7. Determine the simplest formula of a compound that is 26.52% chromium, 24.52% sulfur, and 48.96% oxygen.

8. Predict, on the basis of the spectrochemical series, which of the following chromium complexes is green and which is violet. Explain.

    (a) $[Cr(NH_3)_4Cl_2]Cl$ and   (b) $[Cr(OH_2)_4Cl_2]Cl$

9. Discuss the bonding in each of the following complex ions, using valence bond theory:

    (a) CoF_6^{3-} is an octahedral complex with a magnetism corresponding to four unpaired electrons.

    (b) $PtBr_4^{2-}$ is a diamagnetic, square planar complex.

ANSWERS TO ADDITIONAL PROBLEMS

If you missed an answer, study the text section and operational skill given in parentheses after the answer. Text sections and operational skills from previous chapters are given where appropriate.

1. (a) $Cr_2O_3(s) + 2Al(s) \longrightarrow 2Cr(l) + Al_2O_3(s)$

    (b) $Cu_2S(l) + O_2(g) \longrightarrow 2Cu(l) + SO_2(g)$ (23.2)

2. (a) $K_2Cr_2O_7(aq) + 2H_2SO_4(aq) \longrightarrow 2KHSO_4(aq) + 2CrO_3(s) + H_2O(l)$
    $$(23.2)$$

    (b) $Cr(OH)_3(s) + OH^-(aq) \longrightarrow Cr(OH)_4^-(aq)$ (23.2, 23.3)

    (c) $Cu(s) + 4HNO_3(aq) \longrightarrow Cu(NO_3)_2(aq) + 2NO_2(g) + 2H_2O(l)$
        (conc)
    $$(23.2; 3.6, Op. Sk. 7)$$

3. Qualitative test: Dissolve the complex in water and note whether the odor of NH_3 is present. The absence of an NH_3 odor rules out the last two possibilities.

    Quantitative test: Titrate a known quantity of the complex with a standardized $AgNO_3$ solution. Only Cl^- ions outside the first coordination sphere react with Ag^+ ions, so the moles of AgCl that form per mole of complex determine the correct formula.   (23.5)

4.   (a)   [Cu(NH₃)₄]SO₄   (b)   K₃[Cr(CN)₆]   (c)   [Co(NH₃)₅(NO₂)]Cl₂
                                                                  (23.4, Op. Sk. 1)

5.

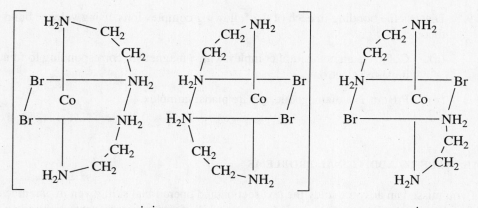

                      cis isomer   *trans* isomer

     The *cis* isomer is optically active.   (23.5, Op. Sk. 2)

6.   (a)   pentaamminesulfatocobalt(III) bromide
     (b)   pentaamminebromocobalt(III) sulfate
     (c)   pentaaquochlorochromium(III) chloride
     (d)   potassium carbonylpentacyanoferrate(II)   (23.4, Op. Sk. 1)

7. $Cr_2S_3O_{12}$ or $Cr_2(SO_4)_3$ (23.2; 4.5, Op. Sk. 8)

8. NH_3 is more strongly bonding than H_2O, so the Δ value for complex (a) will be greater and its wavelength of absorption shorter. Because the complementary color is observed (text Table 23.8), complex (a) is violet.   (23.7, Op. Sk. 4)

9.   (a) $CoF_6{}^{3-}$ is

(b) $PtBr_4^{2-}$ is

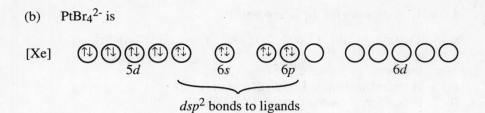

dsp^2 bonds to ligands

In both (a) and (b), donor electrons are shown with dotted arrows.

(23.6, Op. Sk. 3)

CHAPTER POST-TEST

1. Write the balanced molecular equation for the preparation of chromium(VI) oxide by precipitation from a solution of potassium dichromate to which concentrated sulfuric acid is added. Water and bisulfate ion are also formed.

2. Draw the geometric isomers that exist for any of the following linear, square planar, or octahedral complexes. Name those you draw, labeling them *cis* or *trans*.

    (a) $[Pd(NH_3)_2Cl_2]$ (b) $Cr(en)_2Br_2^+$ (c) $Ag(CN)_2^-$

3. A complex studied by Werner had a composition corresponding to the formula $PtCl_4 \cdot 3NH_3$. Electrical-conductance studies showed that each formula unit contained two ions, and precipitation with $AgNO_3$ showed one free chloride ion present. Write a formula for this complex that agrees with this information.

4. Tell whether enantiomers (optical isomers) are possible for any of the following stable complexes. Draw them if so.
    (a)   tris(ethylenediamine)cobalt(III) ion

    (b) ReO_3Cl (tetrahedral geometry)

    (c)   *trans*-$Cr(NH_3)_2(SCN)_4^-$

5. Give the coordination number of the transition-metal atom in the following complexes:

    (a) $MnCl_4^{2-}$ (b) $[Fe(H_2O)_5(NCS)]SO_4$ (c) $[Cu(NH_3)_2(en)]Br_2$

6. Give IUPAC names for the following:

   (a) $PtBr_4^{2-}$ (b) $Cu(C_2O_4)_2^{2-}$ (c) $[Co(NH_3)_5Br]SO_4$

7. Write the formula for each of the following:
   (a) tris(ethylenediamine)cobalt(III) chloride
   (b) triamminetrichloroplatinum(IV) chloride
   (c) hexacyanonickelate(II) anion

8. Answer each of the following questions:
   (a) How do ions of the transition elements exist in aqueous solution?
   (b) What is ferrochrome, how is it made, and what is it used for?
   (c) What is the oxidation number of iron in magnetite, Fe_3O_4? Explain.
   (d) By what mining method is most copper presently obtained? Write the formula of the ore so obtained.

9. Using valence bond theory, tell what orbitals would be used for bonding in $MnCl_4^{2-}$ and give the resulting geometry. Give the number of unpaired electrons, if any, and the resulting magnetic character.

10. Describe the distribution of d electrons in the $Fe(C_2O_4)_3^{3-}$ complex ion, using crystal field theory. Assume it is a high-spin complex. How many unpaired electrons are in the ion, and what is the resulting magnetic character?

11. The $Co(NH_3)_6^{3+}$ ion has a yellow color, whereas the $Co(NH_3)_5Cl^{2+}$ ion is purple. What are the approximate wavelengths of the maximum absorption for each ion? Is the shift of wavelength in the expected direction? Explain. (Use text Table 23.8).

ANSWERS TO CHAPTER POST-TEST

If you missed an answer, study the text section and operational skill given in parentheses after the answer.

1. $K_2Cr_2O_7(aq) + 2H_2SO_4(aq) \longrightarrow 2KHSO_4(aq) + 2CrO_3(s) + H_2O(l)$ (23.2)

2. (a)

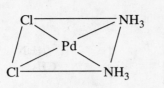 and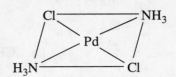

cis-diamminedichloropalladium(II)   *trans*-diamminedichloropalladium(II)

(b)

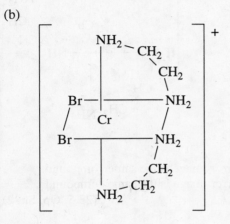

cis-dibromobis(ethylenediamine)chromium(III) ion   and

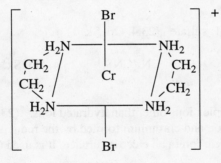

trans-dibromobis(ethylenediamine)chromium(III) ion

(c) No geometric isomers exist. (23.5, Op. Sk. 2)

3. [Pt(NH$_3$)$_3$Cl$_3$]Cl (23.3)

4. (a)   The enantiomers are

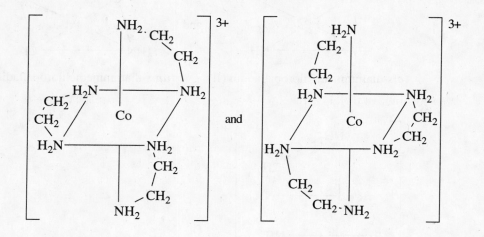

    (b)   No enantiomers exist, as the mirror image is the same compound.
    (c)   No enantiomers exist, as the mirror image is the same compound.

                                                            (23.5, Op. Sk. 2)

5. (a)   4   (b)   6   (c)   4   (23.3)

6. (a)   tetrabromoplatinate(II) anion
    (b)   bis(oxalato)cuprate(II) anion
    (c)   pentaamminebromocobalt(III) sulfate   (23.4, Op. Sk. 1)

7. (a) $[Co(NH_2CH_2CH_2NH_2)_3]Cl_3$ (c) $Ni(CN)_6^{4-}$ (23.4, Op. Sk. 1)
    (b) $[Pt(NH_3)_3Cl_3]Cl$

8. (a)   They frequently exist as complex ions other than hydrated ions.   (23.3)
    (b)   Ferrochrome is an alloy of iron and chromium formed by the reduction of
           chromite ore, $FeCr_2O_4$, with carbon in an electric furnace. It is used to make
           steels.   (23.2)
    (c)   The calculation comes out 2.67. Some of the iron atoms are in the +2 oxidation
           state and some are in the +3 oxidation state.   (23.2)
    (d)   open pit mining; CuS (23.2)

9. Because Cl^- is a weak-bonding ligand, the $4s$ and $4p$ atomic orbitals would be
    combined to form four sp^3 hybrid orbitals to be used for bonding. The geometry

would be tetrahedral. The complex would have five unpaired electrons and would be strongly paramagnetic.   (23.6, Op. Sk. 3)

10.   In a high-spin complex, the electrons would have the following distribution:

showing five unpaired electrons, resulting in highly paramagnetic character.

(23.7, Op. Sk. 3)

11.   The $Co(NH_3)_6^{3+}$ ion absorbs at about 430 nm and the $Co(NH_3)_5Cl^{2+}$ ion absorbs at about 530 nm. Δ should decrease with the weaker-bonding ligand. As energy decreases, λ should increase. This is the shift in wavelength expected.

(23.7, Op. Sk. 4)

UNIT EXAM 7

1. Arrange each of the following rows in increasing order of the property specified:

    (a) basicity CaO Ga_2O_3 K_2O SeO_2
    (b) metallic character In Rb Sn Sr
    (c) metallic character Be Ca Mg Sr
    (d) melting point K Li Na Rb
    (e) boiling point O_2 S Se Te
    (f) oxidizing strength Br_2 Cl_2 F_2 I_2

2. Name the simple picture of bonding in metals.

3. Write equations for the following:
    (a) the dissolving of aluminum hydroxide in concentrated hot sodium hydroxide
    (b) the addition of barium to water
    (c) the preparation of slaked lime from limestone
    (d) the preparation of aluminum from Al_2O_3 (Write one equation.)
    (e) the heating of sand with coke in an electric furnace at 3000°C
    (f) the preparation of lead from galena (PbS)
    (g) the preparation of white phosphorus from phosphate rock

4. Explain briefly why there are no deposits of sodium metal.

5. Contrast the products of the reaction of oxygen with the alkali metals.

6. Sketch and name the expected molecular geometries of the following species:

    (a) phosphorus(III) bromide   (c) SO_2

    (b) hydrogen carbonate ion   (d) boric acid, $B(OH)_3$

7. Write the balanced net ionic equation for the oxidation of lead by concentrated nitric acid. HNO_3 is reduced to NO in this case.

8. Give a major use of each of the following:

    (a) Na_2CO_3 (c) CO_2 (e) $CaSO_4 \cdot 2H_2O$

    (b)   sulfur   (d) H_3PO_4

9. Give compounds exemplifying the +5 and the +7 oxidation states of chlorine.

10. Describe the Frasch process.

11. Give the IUPAC name for each of the following:

    (a) $K[Cr(H_2O)_2(CN)_4]$ (b) $Cr(NH_3)_2(H_2O)_3OH^{2+}$

12. Write the structural formula for trisoxalatochromate(III) anion. Draw one structure for the complex and tell whether enantiomers exist for the compound. The structure of the oxalate anion is

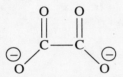

13. Give the theoretical explanation for how partially filled d orbitals are responsible for the following transition-metal properties:
    (a)   high melting points and boiling points of the metals
    (b)   colored complex ions and compounds
    (c)   paramagnetism of complex ions and compounds

14. Write the balanced equation for the preparation of $CuSO_4$ from elemental copper and concentrated sulfuric acid. Sulfuric acid is reduced to SO_2.

15. Using valence bond theory, tell what orbitals would be used for bonding in $AuBr_6^{3-}$. Give the number of unpaired electrons, if any, and the resulting magnetic character.

16. Describe the distribution of d electrons in the square planar $PtCl_4^{2-}$ complex ion, using crystal field theory. How many unpaired electrons are there in the ion, and what is the resulting magnetic character?

ANSWERS TO UNIT EXAM 7

If you missed an answer, study the text section and operational skill given in parentheses after the answer.

1. (a) SeO_2 Ga_2O_3 CaO K_2O (8.7, 13.3, 13.4, 21.4)

   (b) Sn In Sr Rb (21.4)   (e) O_2 S Se Te (22.5)

   (c) Be Mg Ca Sr (21.4)   (f) I_2 Br_2 Cl_2 F_2 (22.7)

   (d) Rb K Na Li (21.5, introductory section)

2. The "electron-sea" model

3. (a) $Al(OH)_3(s) + NaOH(aq) \longrightarrow NaAl(OH)_4(aq)$ (21.2)

   (b) $Ba(s) + 2H_2O(l) \longrightarrow Ba(OH)_2(aq) + H_2(g)$ (21.7, introductory section)

   (c) $CaCO_3(s) \xrightarrow{\Delta} CaO(s) + CO_2(g)$

   $CaO(s) + H_2O(l) \longrightarrow Ca(OH)_2(s)$ (21.8)

   (d) $2Al_2O_3(s) + 3C(s) \xrightarrow{\text{electrolysis}} 4Al(l) + 3CO_2(g)$ (21.2)

   (e) $SiO_2(l) + 2C(s) \xrightarrow{\Delta} Si(l) + 2CO(g)$ (22.2)

   (f) $2PbS(s) + 3O_2(g) \longrightarrow 2PbO(s) + 2SO_2(g)$

   $PbO(s) + CO(g) \longrightarrow Pb(l) + CO_2(g)$ (22.10)

   (g) $2Ca_3(PO_4)_2(s) + 6SiO_2(s) + 10C(s) \longrightarrow 6CaSiO_3(l) + 10CO(g) + P_4(g)$
   (22.4)

4. The metal is very active and reacts with both oxygen and moisture in the atmosphere. (19.8, 21.6)

5. Lithium forms the oxide Li_2O; sodium forms the peroxide Na_2O_2; and potassium, rubidium, and cesium form the superoxides KO_2, RbO_2, and CsO_2.
   (21.5, 21.6, 22.5)

6. (a) PBr$_3$ is pyramidal:

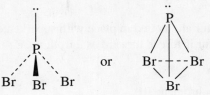

or

(22.4; 10.1, 10.3, Op. Sk. 1, 3)

(b) HCO$_3^-$ is trigonal planar:

(22.1; 10.1, 10.3, Op. Sk. 1, 3)

(c) SO$_2$ is bent (angular):

(22.6; 10.1, 10.3, Op. Sk. 1, 3)

(d) Boric acid is trigonal planar:

OH
|
B
HO   OH

(21.4; 10.1, 10.3, Op. Sk. 1, 3)

7. $3Pb(s) + 8H^+(aq) + 2NO_3^-(aq) \longrightarrow 3Pb^{2+}(aq) + 2NO(g) + 4H_2O(l)$

(21.10, 22.3)

8. (a) manufacture of glass (21.6)

(b) manufacture of H$_2$SO$_4$ (22.6)

(c) carbonation of beverages (22.1)

(d) manufacture of phosphate fertilizers (22.4)

(e) manufacture of plaster and wallboard (21.9)

9.   +5 $HClO_3$ or $KClO_3$
      +7 $HClO_4$ or $NaClO_4$ (22.7; 3.5, Op. Sk. 6)

10.   Underground deposits of solid sulfur are melted in place with superheated water
      piped into the deposit. Molten sulfur is forced upward as a froth via air under
      pressure. (22.6)

11.   (a)   potassium diaquotetracyanochromate(III)
      (b)   diamminetriaquohydroxochromium(III) ion   (23.4, Op. Sk. 1)

12.   The structural formula is $Cr(C_2O_4)_3^{3-}$.   (23.4, Op. Sk. 1)

            One structure is   The mirror image is

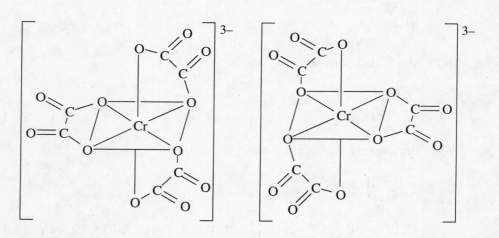

      The mirror image is a different compound, so enantiomers do exist for the
      compound. (23.5, Op. Sk. 2)

13.   (a)   The *d* orbitals and the unpaired electrons in them participate in bonding
             between the metal atoms. (23.1, 23.6)
      (b)   The *d* orbitals are split by incoming ligands of differing electric field strength.
             When there are unfilled orbitals, electrons undergo transition to the higher
             levels and fall back; the energies released (emitted) are in the visual spectral
             range. Differing colors result from differing energy levels of the split *d*
             orbitals. (23.7)
      (c)   Unpaired *d* electrons with unpaired spins make the species paramagnetic.
             (23.6)

14. $Cu(s) + 2H_2SO_4(l) \longrightarrow Cu^{2+}(aq) + SO_4^{2-}(aq) + SO_2(g) + 2H_2O(l)$ (23.2)

15. The $6s$, $6p$, and two of the $6d$ orbitals would combine to form six sp^3d^2 hybrid bonding orbitals. The complex would have two unpaired electrons and would thus be paramagnetic. (23.6, Op. Sk. 3)

16. With eight d electrons, the complex can only be low-spin with the following electron distribution:

The complex would have no unpaired electrons and would thus be diamagnetic.

(23.7, Op. Sk. 3)

CHAPTER 24   ORGANIC CHEMISTRY

CHAPTER TERMS AND DEFINITIONS

Numbers in parentheses after definitions give the text sections in which the terms are explained. Starred terms are italicized in the text. Where a term does not fall directly under a text section heading, additional information is given for you to locate it.

hydrocarbons compounds containing only carbon and hydrogen (24.1, introductory section)

organic compounds compounds that are hydrocarbons or derived from hydrocarbons (24.1, introductory section)

aromatic hydrocarbons hydrocarbons that contain benzene rings or similar structural features (24.1)

aliphatic hydrocarbons hydrocarbons that do not contain benzene rings (24.1)

molecular formula* group of symbols that gives the number and kind of atoms in a molecule (24.1)

structural formula* drawing that gives the number and kind of atoms in a molecule and shows how the atoms are bonded to one another (24.1)

saturated hydrocarbons hydrocarbons in which all carbon atoms are bonded to the maximum number of hydrogen atoms (24.1)

alkanes saturated hydrocarbons with the general formula C_nH_{2n+2} (24.1)

paraffin* from the Latin *parum affinus*, meaning "little affinity"; another term for alkane (24.1, marginal note)

condensed structural formulas* structural formulas using certain conventional abbreviations, e.g., CH_3 for

$$-\overset{\displaystyle H}{\underset{\displaystyle H}{\overset{\displaystyle |}{\underset{\displaystyle |}{C}}}}-H \quad (24.1)$$

homologous series   series of compounds in which one compound differs from a preceding one by a — CH_2 — group   (24.1)

straight-chain (normal) alkanes*   alkanes that have all carbon atoms bonded to one another to give a single chain   (24.1)

branched-chain (alkanes)*   alkanes that have at least one carbon atom bonded to at least three other carbon atoms   (24.1)

structural isomers*   compounds with the same molecular formula but different structural formulas   (24.1)

-ane*   in nomenclature of organic compounds, a suffix that indicates a saturated hydrocarbon, e.g., pentane   (24.1)

alkyl group   alkane less one hydrogen atom   (24.1)

-yl*   in nomenclature of organic compounds, a suffix on a hydrocarbon name indicating that the group is a branch on a larger compound   (24.1)

di-*   prefix meaning "two"   (24.1)

tri-*   prefix meaning "three"   (24.1)

tetra-*   prefix meaning "four"   (24.1)

cycloalkanes   saturated hydrocarbons with the general formula C_nH_{2n}, in which carbon atoms form a ring   (24.1)

petroleum refining*   process of separating petroleum into various hydrocarbon fractions by distillation   (24.1)

unsaturated hydrocarbons   hydrocarbons in which not all carbon atoms are bonded to the maximum number of hydrogen atoms; such compounds contain carbon–carbon multiple bonds   (24.2)

alkenes (olefins)   hydrocarbons with the general formula C_nH_{2n} that contain a carbon–carbon double bond   (24.2)

-ene*   in nomenclature of organic compounds, a suffix that indicates the presence of one or more carbon–carbon double bonds, e.g., pentene   (24.2)

acetylene   the common name for C_2H_2, the *-ene* ending of which does not follow IUPAC rules   (24.2, marginal note)

geometric isomers   isomers in which the atoms are joined to one another in the same way but that differ because some atoms occupy different relative positions in space   (24.2)

cis **(isomer)***   in alkenes, the isomer with two identical groups attached to the same side of the double bond, one group on each double-bonded carbon   (24.2)

trans **(isomer)***   in alkenes, the isomer with two identical groups attached to opposite sides of the double bond, one group on each double-bonded carbon   (24.2)

alkynes   hydrocarbons with the general formula C_nH_{2n-2} that contain a carbon–carbon triple bond   (24.2)

-yne*   in nomenclature of organic compounds, a suffix that indicates the presence of one or more carbon–carbon triple bonds, e.g., propyne   (24.2)

polycyclic aromatic hydrocarbons   aromatic hydrocarbons in which two or more rings share carbon atoms   (24.3)

ortho- (*o*)*   in nomenclature of benzene derivatives, a prefix indicating that two groups on the benzene ring are in adjacent positions   (24.3)

meta- (*m*)*   in nomenclature of benzene derivatives, a prefix indicating that two groups on the benzene ring are in alternate positions   (24.3)

para- (*p*)*   in nomenclature of benzene derivatives, a prefix indicating that two groups on the benzene ring are directly opposite each other   (24.3)

Bayer test*   test using $KMnO_4$ solution to distinguish unsaturated hydrocarbons, which react with it, from saturated hydrocarbons that do not   (24.4)

substitution reaction   reaction in which part of a reagent molecule replaces a hydrogen atom on a hydrocarbon or hydrocarbon group   (24.4)

addition reaction   reaction in which parts of a reactant are added to each carbon atom of a carbon–carbon double bond   (24.4)

Markownikoff's rule   when an unsymmetrical reagent such as HCl is added to an alkene, the major product formed is the one obtained when the hydrogen atom of the reagent adds to the carbon of the double bond that already has more hydrogen atoms attached to it   (24.4)

petroleum refining*   industrial separation of petroleum into its constituent hydrocarbons   (24.4)

catalytic cracking*   process wherein a vaporized hydrocarbon molecule, in the presence of a heated catalyst, breaks up into hydrocarbons of lower molecular weight   (24.4)

octane-number scale*   scale that rates the antiknock characteristics of gasoline   (24.4)

functional group   reactive portion of a molecule that undergoes predictable reactions   (24.5, introductory section)

alcohol   organic compound in which a hydroxyl group (— OH) takes the place of an — H atom on an aliphatic hydrocarbon   (24.5)

hydroxyl group*   — OH group   (24.5)

-ol*   in IUPAC nomenclature of organic compounds, a suffix that denotes an alcohol   (24.5)

primary alcohol*   alcohol in which the hydroxyl group is attached to a carbon atom bonded to only one other carbon atom   (24.5)

secondary alcohol*   alcohol in which the hydroxyl group is attached to a carbon atom bonded to two carbon atoms   (24.5)

tertiary alcohol*   alcohol in which the hydroxyl group is attached to a carbon atom bonded to three carbon atoms   (24.5)

ether   compound formed by the replacement of both atoms of hydrogen in water by hydrocarbon groups   (24.5)

carbonyl group* $\diagdown C {=\!=} O$ group   (24.5)

aldehyde   organic compound containing a carbonyl group that has at least one hydrogen atom attached to it   (24.5)

ketone   organic compound containing a carbonyl group that has two hydrocarbon groups attached to it   (24.5)

-al*   in IUPAC nomenclature of organic compounds, a suffix that denotes an aldehyde   (24.5)

-one*   in IUPAC nomenclature of organic compounds, a suffix that denotes a ketone   (24.5)

carboxylic acid   organic compound containing the carboxyl group, — COOH   (24.5)

carboxyl group*
$$\begin{array}{c} O \\ \| \\ {-}{-}C{-}{-}O{-}{-}H \end{array}$$
group   (24.5)

-oic acid*   in IUPAC nomenclature of organic compounds, a suffix and word that denote an acid   (24.5)

ester   organic compound formed from a carboxylic acid, RCOOH, and an alcohol, R′OH,
with the general structure
$$\begin{array}{c} :O: \\ \| \quad .. \\ RC{-}{-}O{-}{-}R' \\ .. \end{array}$$
   (24.5)

oxidation*   in organic chemistry, the addition of oyxgen atoms to, or the removal of hydrogen atoms from, an organic compound   (24.6)

reduction*   in organic chemistry, the addition of hydrogen atoms to, or the removal of oxygen atoms from, an organic compound   (24.6)

Tollen's test*   test used to distinguish between aldehydes, which reduce silver ion in ammonia solution to metallic silver, and ketones, which do not   (24.6)

condensation reaction   reaction in which two molecules are joined by the elimination of a small molecule such as water   (24.6)

-ate*   in nomenclature of organic compounds, a suffix that denotes an ester   (24.6)

hydrolysis*   reaction of a compound with water, such as the reaction of an ester with water in the presence of H^+ to give a carboxylic acid and an alcohol   (24.6)

saponification   hydrolysis of an ester in the presence of a base   (24.6)

*sapon**   Latin word for "soap"   (24.6)

fatty acids*   long-chain carboxylic acids   (24.6)

amines   organic compounds in which one or more hydrogen atoms of ammonia are replaced by hydrocarbon groups   (24.7)

amides   organic compounds derived from the reaction of ammonia, or of a primary or secondary amine, with a carboxylic acid   (24.7)

polymer   chemical species of very high molecular weight that is made up from many repeating units of low molecular weight   (24.8)

monomer   compound used to make a polymer (and from which the polymer's repeating unit arises)   (24.8)

poly-*   prefix meaning "many"   (24.8)

addition polymer   polymer formed by linking together many molecules by addition reactions   (24.8)

initiator* compound used to induce the formation of an addition polymer by producing free radicals (species having an unpaired electron) (24.8)

vulcanization* heating crude rubber with sulfur to promote cross-linking of the polymer chains (24.8)

homopolymer polymer whose monomer units are all alike (24.8)

copolymer polymer consisting of two or more different monomer units (24.8)

condensation polymer polymer formed by linking together many molecules by condensation reactions (24.8)

polyester* polymer whose repeating units are joined by ester groups (24.8)

polyamide* condensation polymer whose repeating units are joined by amide groups (24.8)

CHAPTER DIAGNOSTIC TEST

1. Give the IUPAC name for each of the following compounds.

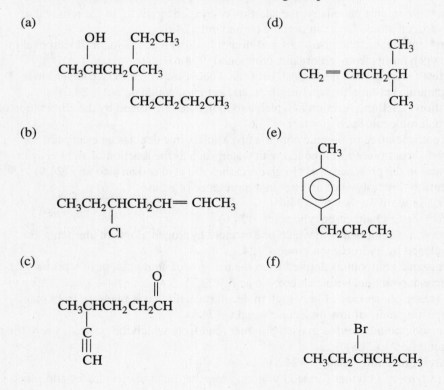

(a)

OH   CH$_2$CH$_3$
 | |
CH$_3$CHCH$_2$CCH$_3$
             |
          CH$_2$CH$_2$CH$_2$CH$_3$

(b)

CH$_3$CH$_2$CHCH$_2$CH═CHCH$_3$
           |
           Cl

(c)

                O
                ‖
CH$_3$CHCH$_2$CH$_2$CH
   |
   C
  ‖‖
   CH

(d)

                    CH$_3$
                     |
CH$_2$═CHCH$_2$CH
                     |
                    CH$_3$

(e)

CH$_3$ (attached to benzene ring)

CH$_2$CH$_2$CH$_3$ (attached to benzene ring)

(f)

                Br
                 |
CH$_3$CH$_2$CHCH$_2$CH$_3$

2. Write the condensed structural formula for each of the following compounds.
   (a) methyl propyl ether   (d) 3-hexanone
   (b) phenyl 2-pentynoate   (e) 1,3-diphenylbutane
   (c) 3-methyl-3-butenoic acid   (f) 2-heptyne

3. Determine whether each of the following statements is true or false. If the statement is false, change it so it is true.
   (a) The general formula for an alkane is C_nH_{2n}. True/False: _____ .
   (b) Unsaturated hydrocarbons form the bulk of petroleum and are not readily reactive with most reagents at normal temperatures. True/False: _____ .
   (c) Markownikoff's rule states that the major product formed by the addition of an unsymmetrical reagent, such as HCl, to a multiple bond of an alkene or alkyne is the one obtained by addition of the H atom to the carbon having more H atoms bonded to it. True/False: _____ .
   (d) Fossil fuels (natural gas, petroleum, and coal) are the principal sources of hydrocarbons. True/False: _____ .
   (e) The antiknock characteristics of a gasoline can be improved by changing straight-chain hydrocarbons to branched-chain isomers. True/False: _____ .

4. For each of the following compounds, decide if *cis–trans* isomers are possible. If they are, draw the condensed structural formula and give the IUPAC name, and/or label the *cis* or *trans* isomers.

   (a) 2-butene   (d) $CH_3CH_2C = CHCH_3$
   (b) propene |
   (c) $CH_3C \equiv CCH_2CH_3$ $CH_2CH_2CH_3$

5. Polystyrene is an addition polymer of styrene, $C_6H_5CH = CH_2$. Write the structure of the polymer.

6. Write the balanced equation for the oxidation of 2-propanol to propanone by chromic oxide (CrO_3) in acid solution. The reduction product of chromic oxide is Cr^{3+}.

7. Nylon-66 is a condensation polymer of 1,6-diaminohexane and adipic acid. The repeating structure is

$$\left(- \overset{\overset{\displaystyle O}{\|}}{C}CH_2CH_2CH_2CH_2\overset{\overset{\displaystyle O}{\|}}{C} - NHCH_2CH_2CH_2CH_2CH_2CH_2NH - \right)$$

Give the structures of 1,6-diaminohexane and adipic acid.

8.   Match the terms on the left with the letter of the appropriate formula on the right.

_____ (1) amine
_____ (2) amide
_____ (3) ester
_____ (4) carboxylic acid
_____ (5) alcohol
_____ (6) aldehyde

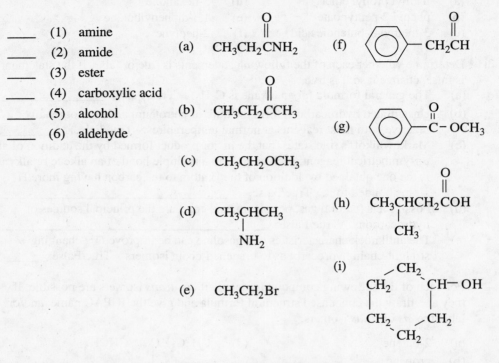

9.   Complete and balance the following reactions. Write NR if no reaction occurs.

(a) $C_6H_6(l) + O_2(g) \xrightarrow{\Delta}$

(b) $CH_3CH_2CH_2CH = CH_2(l) + HBr(g) \longrightarrow$ (major product is a liquid)

(c) $C_4H_9CH = CHCH_3(l) + Br_2(l) \longrightarrow$ (product is a liquid)

(d) ⬡ $(l) + Br_2(l) \xrightarrow{FeBr_3}$ (products are a liquid and a gas)

(e) ⬡ $(l) + HO - NO_2(l) \xrightarrow{H_2SO_4}$ (organic product is a liquid)

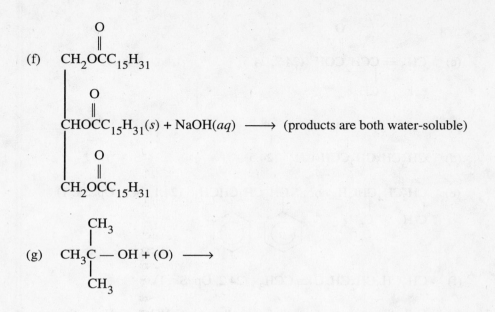

(f)

$$CH_2OCC_{15}H_{31}$$
$$O$$
$$\|$$

$$CHOCC_{15}H_{31}(s) + NaOH(aq) \longrightarrow \text{(products are both water-soluble)}$$

$$CH_2OCC_{15}H_{31}$$

(g)

$$\underset{\underset{CH_3}{|}}{\overset{\overset{CH_3}{|}}{CH_3C}} — OH + (O) \longrightarrow$$

ANSWERS TO CHAPTER DIAGNOSTIC TEST

If you missed an answer, study the text section and operational skill given in parentheses after the answer.

1. (a) 4-ethyl-4-methyl-2-octanol (24.5)
   (b) 5-chloro-2-heptene (24.2, 24.4, Op. Sk. 1)
   (c) 4-methyl-5-hexynal (24.5)
   (d) 4-methyl-1-pentene (24.2, Op. Sk. 1)
   (e) 4-methyl-1-propylbenzene or *p*-methylpropylbenzene (24.3, Op. Sk. 1)
   (f) 3-bromopentane (24.1, 24.4, Op. Sk. 1)

2. (a) $CH_3OCH_2CH_2CH_3$ (24.5)

   (b) $CH_3CH_2C \equiv CCOOC_6H_5$ or $CH_3CH_2C \equiv CCO{-}\bigcirc$

   (24.2, 24.3, 24.5)

(c) $\underset{\overset{|}{CH_3}}{CH_2 = C}CH_2\overset{\overset{O}{\|}}{C}OH$ (24.2, 24.5)

(d) $CH_3CH_2CH_2\overset{\overset{O}{\|}}{C}CH_2CH_3$ (24.5)

(e) $\underset{\overset{|}{C_6H_5}}{CH_2}\underset{}{CH_2}\underset{\overset{|}{C_6H_5}}{CH}CH_3$ or   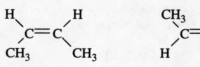   (24.1, 24.3, Op. Sk. 1)

(f) $CH_3CH_2CH_2CH_2C \equiv CCH_3$ (24.2, Op. Sk. 1)

3.   (a)   False. The general formula for an alkane is C_nH_{2n+2}. C_nH_{2n} is the general
         formula for an alkene.   (24.1, 24.2)
     (b)   False. Saturated hydrocarbons form the bulk of petroleum and are not readily
         reactive with most reagents at normal temperatures.   (24.4)
     (c)   True. (24.4)   (d)   True. (24.1)   (e)   True. (24.4)

4.   (a)   Yes.

cis-2-butene   *trans*-2-butene

     (b)   No
     (c)   No

(d)   Yes

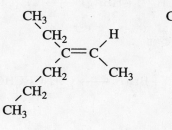 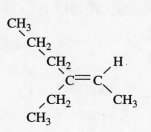

cis-3-ethyl-2-hexene   *trans*-3-ethyl-2-hexene

(24.1, 24.2, Op. Sk. 1, 2)

5.   — $CHCH_2CHCH_2$ —   (24.8)

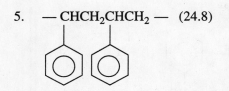

6. $3CH_3CHCH_3(aq) + 2CrO_3(aq) + 6H^+(aq) \longrightarrow$

      | OH

$$3CH_3\overset{O}{\overset{\|}{C}}CH_3(aq) + 2Cr^{3+}(aq) + 6H_2O(l)$$

(24.6, Op. Sk. 4)

7.   1,6-diaminohexane is $H_2NCH_2CH_2CH_2CH_2CH_2CH_2NH_2$; adipic acid is

$$\overset{O}{\overset{\|}{HOC}}CH_2CH_2CH_2CH_2\overset{O}{\overset{\|}{C}}OH \quad (24.8)$$

8.   (1)   d (24.7)   (2)   a (24.7)   (3)   g (24.5)   (4)   h (24.5)
     (5)   i (24.5)   (6)   f (24.5)

9.   (a) $2C_6H_6(l) + 15O_2(g) \overset{\Delta}{\longrightarrow} 12CO_2(g) + 6H_2O(g)$ (24.4)

(b) $CH_3CH_2CH_2CH = CH_2(l) + HBr(g) \longrightarrow CH_3CH_2CH_2CHCH_3(l)$

$$\underset{Br}{|}$$

(24.4, Op. Sk. 3)

(c) $C_4H_9CH = CHCH_3(l) + Br_2(l) \longrightarrow C_4H_9CHCHCH_3(l)$ (24.4)

$$\underset{Br\ Br}{|\ |}$$

(d) 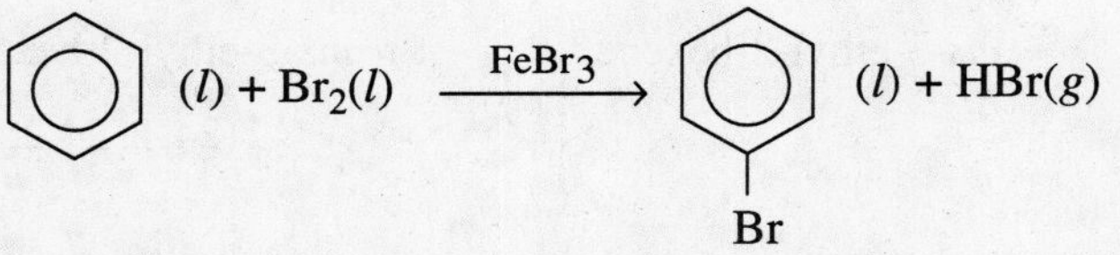 benzene $(l) + Br_2(l) \xrightarrow{FeBr_3}$ bromobenzene $(l) + HBr(g)$ (24.4)

(e) 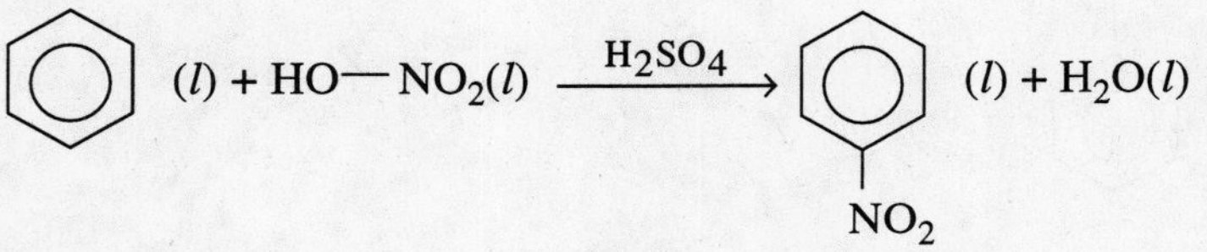 benzene $(l) + HO-NO_2(l) \xrightarrow{H_2SO_4}$ nitrobenzene $(l) + H_2O(l)$ (24.4)

(f)

$$CH_2OCC_{15}H_{31}$$
$$O$$
$$CHOCC_{15}H_{31}(s) + 3NaOH(aq) \longrightarrow \begin{matrix} CH_2OH \\ CHOH(aq) \\ CH_2OH \end{matrix} + 3C_{15}H_{31}CO^-Na^+(aq)$$
$$CH_2OCC_{15}H_{31}$$

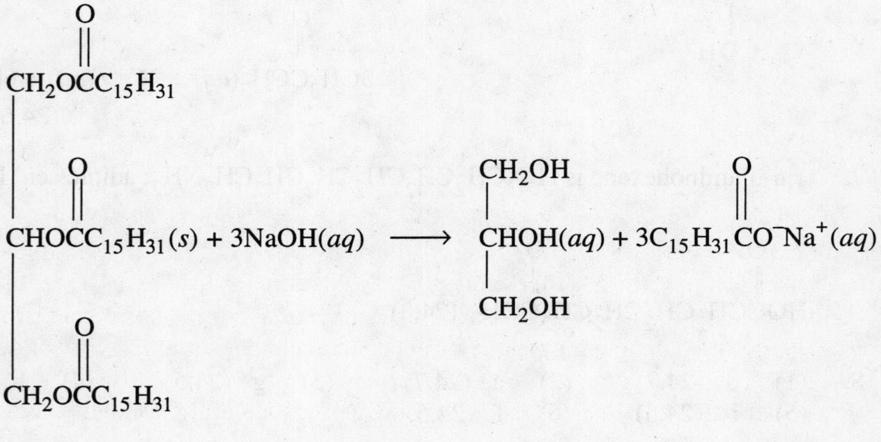

(g) $CH_3 - \underset{\underset{CH_3}{|}}{\overset{\overset{CH_3}{|}}{C}} - OH + (O) \longrightarrow NR$ (24.6)

SUMMARY OF CHAPTER TOPICS

24.1   ALKANES AND CYCLOALKANES

Operational Skill

1.   Writing the IUPAC name of a hydrocarbon given the structural formula, and vice versa.   Given the structure of a hydrocarbon, state the IUPAC name   (Example 24.1).   Given the IUPAC name of a hydrocarbon, write the structural formula   (Example 24.2).

It is important to understand that although we use the term *straight-chain* to denote the simplest alkanes, the three-dimensional structure is not a straight chain, due to the 109.5° bond angles between adjacent carbons. We use the term merely to indicate the linkage of one carbon atom to another. Also, because the atoms are continually spinning, any structure we draw "freezes" the molecule in space just as a camera catches a ballet dancer at the height of a leap. The drawing below exemplifies one "frozen" atomic arrangement of butane. Recall that the wedged bonds are to atoms in front of the plane of the paper, and the dotted bonds are to atoms behind the plane of the paper.

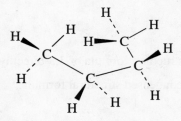

To name the compounds, you must know the prefixes that denote the number of carbon atoms in the straight-chain hydrocarbons, given in text Table 24.1: *meth-* for one, *eth-* for two, etc. These are also the alkyl group names. Commit them to memory as quickly as possible. Also pay particular attention to the placement of commas, dashes, and spaces in writing compound names. Correct punctuation is part of the correct name.

Exercise 24.1   What is the IUPAC name for each of the following hydrocarbons?

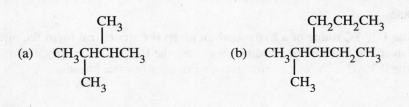

Known: naming rules given in text; prefix names indicating numbers of carbon atoms are in text Table 24.1.

Solution:

(a)   2,3-dimethylbutane. The carbons are numbered below. Note that no matter where we start or end, the longest chain has four carbons:

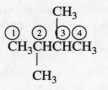

(b)   3-ethyl-2-methylhexane. The carbons are numbered as follows:

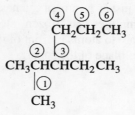

Exercise 24.2 Write the condensed structural formula of 3,3-dimethyloctane.

Known: naming rules; definition of condensed structural formula.

Solution: The formula is

$$CH_3CH_2\underset{\underset{CH_3}{|}}{\overset{\overset{CH_3}{|}}{C}}CH_2CH_2CH_2CH_2CH_3$$

24.2   ALKENES AND ALKYNES

Operational Skills

1. Writing the IUPAC name of a hydrocarbon given the structural formula, and vice versa. Given the structure of a hydrocarbon, state the IUPAC name (Example 24.3). Given the IUPAC name of a hydrocarbon, write the structural formula.

2. Predicting *cis–trans* isomers. Given a condensed structural formula of an alkene, decide whether *cis* and *trans* isomers are possible, and, if so, draw the structural formulas (Example 24.4).

Exercise 24.3 Give the IUPAC name for each of the following compounds.

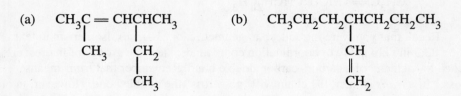

(a) $CH_3C = CHCHCH_3$
          | |
         CH_3 CH_2
                      |
                    CH_3

(b) $CH_3CH_2CH_2CHCH_2CH_2CH_3$
                          |
                         CH
                         ‖
                        CH_2

Known: rules for naming compounds; alkyl-group names.

Solution:

(a)   2,4-dimethyl-2-hexene. The carbons are numbered as follows:

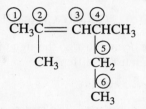

(b)   3-propyl-1-hexene. The numbering of the carbon atoms is as follows:

Note that the double-bonded carbon must be in the chain that determines the name.

Exercise 24.4 Write the condensed structural formula of 2,5-dimethyl-2-heptene.

Solution: The formula is

$$\underset{\displaystyle CH_3C = CHCH_2CHCH_2CH_3}{\overset{\displaystyle CH_3 \qquad\qquad CH_3}{|\qquad\qquad\quad |}}$$

Recall that you were introduced to geometric, or *cis–trans*, isomerism in text Section 23.5 in your study of coordination compounds. Here the groups of interest are attached to each end of a carbon–carbon double bond. Remember that *trans* means "across." In a *trans* isomer, the chain will "go across" the double bond. However, in order for these geometric isomers to exist, the two groups attached to each carbon in the double bond must be different.

Exercise 24.5 Decide whether *cis–trans* isomers are possible for each of the following compounds. If isomers are possible, draw the structural formulas and give the IUPAC names, labeling *cis* or *trans*.

(a) $CH_3CH = CHCH_2CH_2CH_3$ (b) $CH_3CH_2CH = CH_2$

Known: For geometric isomers to exist, the chain must go through the double bond and the groups attached to each carbon must be different.

Solution:

(a)   Structural isomers are possible for this compound. The structural formulas and names are

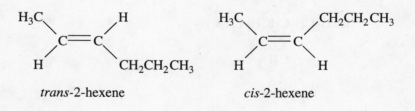

        trans-2-hexene   *cis*-2-hexene

(b)   Geometric isomers do not exist for this compound, as both "groups" on the right-hand carbon in the double bond are the same: H.

With the discussion of alkynes, you have been introduced to the three major groups of nonaromatic hydrocarbons. You should memorize the suffixes for each group: *-ane*, for no multiple bonds; *-ene*, for double bond(s); and *-yne*, for triple bond(s).

Exercise 24.6   Give the IUPAC name for each of the following alkynes.

(a) $CH_3C \equiv CH$

(b) $CH \equiv CCHCH_3$
                    $|$
                    CH_2CH_3

Solution:
(a)   1-propyne
(b)   3-methyl-1-pentyne. The numbering is as follows:

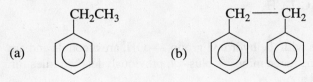

24.3   AROMATIC HYDROCARBONS

Although the aromatic hydrocarbons are fragrant, the term *aromatic* refers to the special properties these molecules exhibit because of their delocalized bonding. Be sure that when you use the abbreviated structure for benzene and other aromatic compounds, you do not forget that there are hydrogens on the ring.

Exercise 24.7   Write the structural formula of (a) ethylbenzene;
(b) 1,2-diphenylethane.

Solution:

(a)   (b)

24.4   REACTIONS OF HYDROCARBONS

Operational Skill

3. Predicting the major product of an addition reaction.   Predict the major product in the addition of an unsymmetrical reagent to an unsymmetrical alkene   (Example 24.5).

Exercise 24.8 Predict the main product when HBr adds to 1-butene.

Known: Markownikoff's rule, stating in effect that H adds to the carbon of a double bond that has more hydrogens; how to write the structure for 1-butene.

Solution: The reaction is

$$CH_2 = CHCH_2CH_3 + HBr \longrightarrow CH_3\underset{\underset{Br}{|}}{C}HCH_2CH_3$$

The H of HBr adds to carbon 1, as it is the carbon with the greater number of hydrogens. The product name is 2-bromobutane.

24.5 ORGANIC COMPOUNDS CONTAINING OXYGEN

It is important to memorize the name and formula of each functional group and the general name of the compound with each functional group. These are found in text Table 24.4.

Exercise 24.9 Give the IUPAC name of the following compound.

$$CH_3CH_2CH_2\underset{\underset{CH_2CH_3}{|}}{\overset{\overset{OH}{|}}{C}}CH_2CH_3$$

Known: Compounds with the hydroxyl group, — OH, are alcohols and are named using the alkane name minus -*e* plus -*ol*; previously learned rules for naming hydrocarbons apply.

Solution: 3-ethyl-3-hexanol. The numbering is as follows:

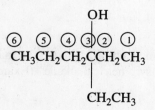

Exercise 24.10 Give the common name of each of the following compounds: (a) CH_3OCH_3; (b) $CH_3OCH_2CH_3$.

Known: Compounds of the type R — O — R are ethers; common names are formed using the alkyl group names in alphabetical order plus *ether*.

Solution: (a) dimethyl ether or methyl ether. (b) ethyl methyl ether.

Exercise 24.11 Name the following compounds by IUPAC rules:

(a)
$$CH_3CH_2CH_2 \overset{\displaystyle O}{\underset{\displaystyle \|}{—C}} — CH_3$$
(b)
$$H \overset{\displaystyle O}{\underset{\displaystyle \|}{—C}} — CH_2CH_2CH_3$$

Known: Compounds of the type $R \overset{\displaystyle O}{\underset{\displaystyle \|}{—C}} R$ are ketones and are named with the alkane name minus *-e* plus *-one* and an initial number to designate the carbonyl

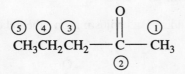 carbon. Compounds of the type $R \overset{\displaystyle O}{\underset{\displaystyle \|}{—C}} H$ are aldehydes and are named with the alkane name minus *-e* plus *-al*.

Solution:
(a)   2-pentanone. The numbering is as follows:

$$\overset{⑤}{CH_3}\overset{④}{CH_2}\overset{③}{CH_2} \overset{\displaystyle O}{\underset{\displaystyle \underset{②}{\|}}{—C—}} \overset{①}{CH_3}$$

(b)   butanal.

24.6   REACTIONS OF OXYGEN-CONTAINING ORGANIC COMPOUNDS

Operational Skill

4. Balancing oxidation–reduction equations involving organic compounds. Write a complete balanced equation for the oxidation or reduction of an organic compound (Example 24.6).

Note that when discussing organic reactions, we define oxidation as the addition of oxygen atoms to, or the removal of hydrogen atoms from, an organic compound. Reduction is defined as the addition of hydrogen atoms to, or the removal of oxygen atoms from, an organic compound. When you work at balancing these equations, you will see that the definitions you learned back in text Section 3.5 — in terms of what is happening to the oxidation state of carbon — still apply.

Exercise 24.12 Write the balanced equation for the oxidation of ethanol to acetaldehyde by permanganate ion in acidic solution. Permanganate is reduced to Mn^{2+} in acidic solution.

Wanted: balanced oxidation–reduction equation.

Given: reactants and products.

Known: how to balance oxidation–reduction reactions, text Section 3.6. Acetaldehyde is the common name for ethanal (24.5).

Solution: The oxidation half-reaction is

$$CH_3CH_2OH \longrightarrow CH_3\overset{\overset{\displaystyle O}{\|}}{C}H + 2H^+ + 2e^-$$

The reduction half-reaction is

$$5e^- + 8H^+ + MnO_4^- \longrightarrow Mn^{2+} + 4H_2O$$

Multiplying by appropriate factors, adding, and adjusting give the balanced equation

$$5CH_3CH_2OH \longrightarrow 5CH_3\overset{\overset{\displaystyle O}{\|}}{C}H + 10H^+ + 10e^-$$

$$\underline{10e^- + 16H^+ + 2MnO_4^- \longrightarrow 2Mn^{2+} + 8H_2O}$$

$$6H^+ + 5CH_3CH_2OH + 2MnO_4^- \longrightarrow 5CH_3\overset{\overset{\displaystyle O}{\|}}{C}H + 2Mn^{2+} + 8H_2O$$

Exercise 24.13 Complete each of the following equations. If no reaction occurs, write NR. Name any organic products.

(a) $(CH_3)_3COH + (O) \longrightarrow$

(b) $CH_3\overset{\displaystyle |}{\underset{\displaystyle CHO}{C}}HCH_2CH_3 + (O) \longrightarrow$

(c) $CH_3\overset{\displaystyle |}{\underset{\displaystyle CHO}{C}}HCH_2CH_3 + (H) \longrightarrow$

(d) $CH_3\overset{\displaystyle |}{\underset{\displaystyle OH}{C}}HCH_2CH_3 + (O) \longrightarrow$

Known: Primary alcohols are oxidized to aldehydes, then to acids if the aldehyde is not removed; secondary alcohols go to ketones; tertiary alcohols don't react; aldehydes are oxidized to acids and are reduced to alcohols; rules for naming compounds.

Solution:

(a)   The alcohol is tertiary: $(CH_3)_3COH + (O) \longrightarrow$ NR.

(b) $CH_3\overset{\displaystyle |}{\underset{\displaystyle CHO}{C}}HCH_2CH_3 + (O) \longrightarrow CH_3\overset{\displaystyle |}{\underset{\displaystyle \underset{\displaystyle OH}{|}}{\underset{\displaystyle C=O}{C}}}HCH_2CH_3$

2-methylbutanoic acid

(c) $CH_3\overset{\displaystyle |}{\underset{\displaystyle CHO}{C}}HCH_2CH_3 + 2(H) \longrightarrow CH_3\overset{\displaystyle |}{\underset{\displaystyle CH_2OH}{C}}HCH_2CH_3$

2-methyl-1-butanol

(d) $CH_3\overset{\displaystyle |}{\underset{\displaystyle OH}{C}}HCH_2CH_3 + (O) \longrightarrow CH_3\overset{\displaystyle \|}{\underset{\displaystyle O}{C}}CH_2CH_3 + H_2O$

2-butanone

Exercise 24.14 Write an equation for the preparation of methyl propionate. Note any catalyst used.

Known: The desired compound is an ester, the alkyl-group name being from the alcohol that is reacted with the organic acid to form the compound; an inorganic acid, such as H_2SO_4, is needed; rules for writing compound structures.

Solution: The equation is

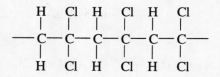

$$CH_3CH_2\overset{\displaystyle \overset{O}{\|}}{C}OH + CH_3OH \; \underset{}{\rightleftharpoons}^{H^+} \; CH_3CH_2\overset{\displaystyle \overset{O}{\|}}{C} - OCH_3 + H_2O$$

24.7 ORGANIC COMPOUNDS CONTAINING NITROGEN

24.8 ORGANIC POLYMERS

Exercise 24.15 An addition polymer is prepared from vinylidene chloride, $CH_2{=\!=}CCl_2$. Write the structure of the addition polymer.

Known: An addition polymer is formed by linking monomers.

Solution: The structure is

$$
\begin{array}{ccccccccccc}
 & H & & Cl & & H & & Cl & & H & & Cl \\
 & | & & | & & | & & | & & | & & | \\
-\!\!\!& C & -\!\!& C & -\!\!& C & -\!\!& C & -\!\!& C & -\!\!& C & \!\!\!-\\
 & | & & | & & | & & | & & | & & | \\
 & H & & Cl & & H & & Cl & & H & & Cl \\
\end{array}
$$

Note: poly(vinylidene chloride) is commonly known as Saran.

ADDITIONAL PROBLEMS

1. Match the terms on the left with the letter of the appropriate formula on the right.

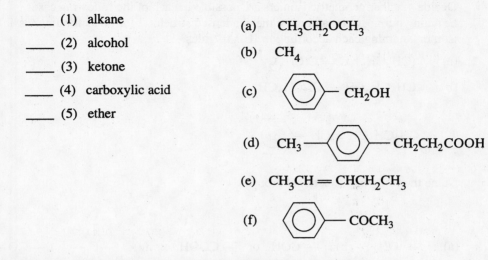

    ____ (1) alkane

    ____ (2) alcohol

    ____ (3) ketone

    ____ (4) carboxylic acid

    ____ (5) ether

(a) $CH_3CH_2OCH_3$

(b) CH_4

(c) ⬡— CH_2OH

(d) CH_3—⬡— CH_2CH_2COOH

(e) $CH_3CH = CHCH_2CH_3$

(f) ⬡— $COCH_3$

2. Give the IUPAC name for each of the following compounds.

(a)
$$\begin{array}{ccc} & CH_3 & CH_3 \\ & | & | \\ CH_3 & CHCHCHCH_2CH_3 \\ & | & \\ & CH_3 & \end{array}$$

(b) $CH_3C \equiv CCH_2CH_3$

(c)
$$\begin{array}{cc} & CH_3 \quad O \\ & | \quad\; \| \\ CH_3CH_2 & CHCHCH \\ & | \\ & CH_2 \\ & | \\ & CH_3 \end{array}$$

(d)
$$\begin{array}{c} O \\ \| \\ CH_3CH_2CCOH \\ \| \\ CH_2 \end{array}$$

(e)
$$\begin{array}{c} O \\ \| \\ CH_3CH_2C - OCH_2CH_3 \end{array}$$

3. Write the condensed structural formula for each of the following compounds.
    (a) 2,3-dimethyl-2-hexene
    (b) 2-chloro-2-methylpropane
    (c) 2-phenylethanol

    (d)   3-heptanone

    (e)   methyl pentyl ether

4.   Decide whether geometric isomers are possible in each of the following cases. Explain your answers, and write the condensed structural formulas for any possible isomers, naming each according to IUPAC rules.

    (a) $(CH_3CH_2)_2C = C(CH_3)_2$

    (b) $CH_3CH_2CH = CHCH_2CH_3$

    (c) $CH_3CH_2CH = CH -$

5.   Name the following functional groups.

    (a) $- OH$ (b) $- COH$ or $- COOH$ (c) $- \overset{\overset{\displaystyle O}{\|}}{C} -$ or $- CO -$

6.   Write the structure of the product you would expect in each of the following reactions.

    (a) $CH_3\underset{\underset{\displaystyle CH_3}{|}}{C} = CH_2 + H_2O \xrightarrow{H^+}$

    (b) $CH_3CH_2CH_2C \equiv CH + 2HBr \longrightarrow$

    (c) $CH_2 = CHCH_2CH = CH_2 + 2HCl \longrightarrow$

7.   Complete and balance each of the following equations.

    (a) $CH_3CH_2CH_2CH_2OH + CH_3COOH \underset{\displaystyle \rightleftharpoons}{\overset{\displaystyle H^+}{}}$

    (b) $CH_3CH_2CHO + (H) \xrightarrow{LiAlH_4}$

    (c) $CH_3CH_2CH_2 - \overset{\overset{\displaystyle OH}{|}}{CH} - CH_3 + Cr_2O_7^{2-} + H^+ \longrightarrow$

8. When C_6H_5Cl reacts with Cl_2, three isomers of $C_6H_4Cl_2$ are formed. Draw and name these three isomers.

9. Write the equation for each of the following. Note any catalyst used.
    (a) Preparation of ethyl acetate, the major component of nail polish remover. Name the reactants according to the IUPAC rules.
    (b) Saponification of butyl formate. Give the IUPAC names of the products.

10. Teflon, used to coat "nonstick" cooking surfaces, is an addition polymer of tetrafluoroethene.
    (a) Draw the structure of tetrafluoroethene.
    (b) Draw a portion of the polymer.

ANSWERS TO ADDITIONAL PROBLEMS

If you missed an answer, study the text section and operational skill given in parentheses after the answer.

1. (1) b   (2) c   (3) f   (4) d   (5) a (24.1, 24.5)

2. (a) 2,3,4-trimethylhexane   (d) 2-ethyl-2-propenoic acid
    (b) 2-pentyne   (e) ethyl propanoate
    (c) 2-ethyl-3-methylpentanal   (24.1, 24.2, 24.5, Op. Sk. 1)

3. (a) $CH_3CH_2CH_2C=CCH_3$
            $\quad\quad\quad\quad\ |\quad\ |$
            $\quad\quad\quad\quad H_3C\quad CH_3$

    (c)  $-CH_2CH_2OH$

    (b) $\quad\quad\ CH_3$
         $\quad\quad\quad |$
         CH_3C-CH_3
         $\quad\quad\quad |$
         $\quad\quad\ Cl$

    (d) $\quad\quad\quad\quad\quad\quad\quad\quad\quad O$
         $\quad\quad\quad\quad\quad\quad\quad\quad\quad ||$
         $CH_3CH_2CH_2CH_2CCH_2CH_3$

    (e) $CH_3OCH_2CH_2CH_2CH_2CH_3$
         (24.2, 24.3, 24.5, Op. Sk. 1)

4. (a) No; groups A and B are the same, as are groups C and D.

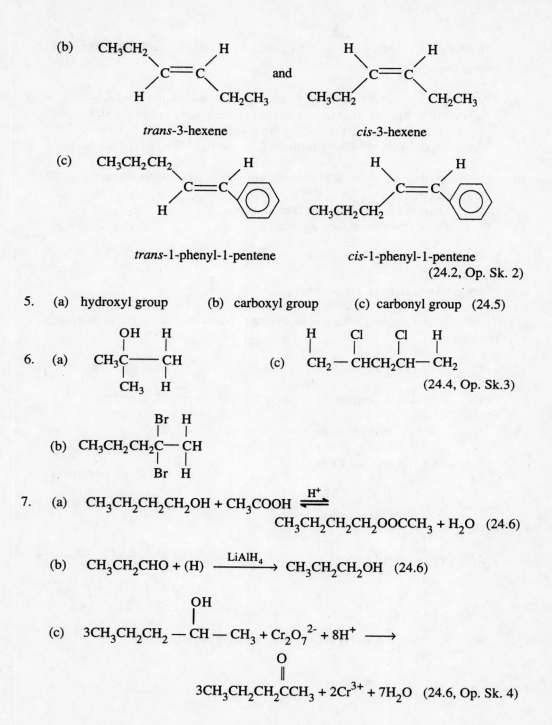

(b)

CH$_3$CH$_2$ C=C   H / CH$_2$CH$_3$ and   H / CH$_3$CH$_2$ C=C   H / CH$_2$CH$_3$

trans-3-hexene   *cis*-3-hexene

(c)

CH$_3$CH$_2$CH$_2$ C=C   H / (phenyl)   and   H / CH$_3$CH$_2$CH$_2$ C=C   H / (phenyl)

trans-1-phenyl-1-pentene   *cis*-1-phenyl-1-pentene
                                 (24.2, Op. Sk. 2)

5.   (a)   hydroxyl group   (b)   carboxyl group   (c)   carbonyl group   (24.5)

6.   (a)

OH   H
| |
CH$_3$C——CH
| |
CH$_3$ H

(c)

H   Cl   Cl   H
| | | |
CH$_2$—CHCH$_2$CH—CH$_2$
                              (24.4, Op. Sk.3)

(b)

Br   H
| |
CH$_3$CH$_2$CH$_2$C—CH
| |
Br   H

7.   (a)   CH$_3$CH$_2$CH$_2$CH$_2$OH + CH$_3$COOH $\overset{H^+}{\rightleftharpoons}$

CH$_3$CH$_2$CH$_2$CH$_2$OOCCH$_3$ + H$_2$O   (24.6)

(b)   CH$_3$CH$_2$CHO + (H) $\xrightarrow{\text{LiAlH}_4}$ CH$_3$CH$_2$CH$_2$OH   (24.6)

(c)

OH
|
3CH$_3$CH$_2$CH$_2$ — CH — CH$_3$ + Cr$_2$O$_7^{2-}$ + 8H$^+$ $\longrightarrow$

O
‖
3CH$_3$CH$_2$CH$_2$CCH$_3$ + 2Cr^{3+} + 7H$_2$O   (24.6, Op. Sk. 4)

8.

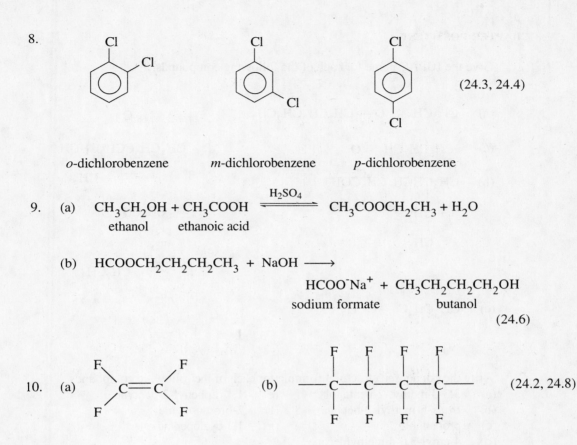

o-dichlorobenzene   *m*-dichlorobenzene   *p*-dichlorobenzene

(24.3, 24.4)

9. (a) $CH_3CH_2OH + CH_3COOH \xrightleftharpoons{H_2SO_4} CH_3COOCH_2CH_3 + H_2O$
        ethanol   ethanoic acid

(b) $HCOOCH_2CH_2CH_2CH_3 + NaOH \longrightarrow$

$HCOO^-Na^+ + CH_3CH_2CH_2CH_2OH$
sodium formate   butanol

(24.6)

10. (a)   (b)   (24.2, 24.8)

CHAPTER POST-TEST

1. Give the IUPAC name for each of the following compounds.

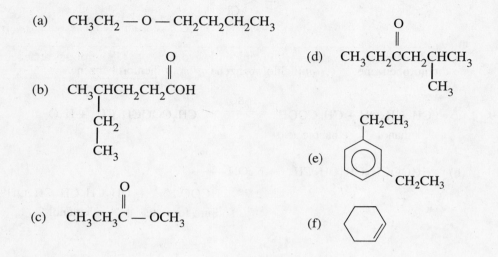

    (a) $CH_3CH_2 — O — CH_2CH_2CH_2CH_3$

    (b)
    $$CH_3CHCH_2CH_2COH$$
    (with O double bonded to terminal C, and branch CH_2 then CH_3)

    (c)
    $$CH_3CH_2C — OCH_3$$
    (with O double bonded to C)

    (d)
    $$CH_3CH_2CCH_2CHCH_3$$
    (with O double bonded to third C, and branch CH_3)

    (e) (benzene ring with CH_2CH_3 substituents)

    (f) (cyclohexene)

2. Write the condensed structural formula for each of the following compounds.
    (a) 3,3-dimethyl-2-pentanol
    (b) *trans*-5-methyl-2-heptene
    (c) propenal
    (d) 4-ethyl-3,5-dimethyldecane
    (e) 1,4-diphenyl-2-pentyne
    (f) 2-bromobutane
    (g) 1,3-cyclopentadiene

3. Dacron polyester is a condensation polymer of terephthalic acid,

    $$HO—C—\text{(benzene ring)}—C—OH$$
    (each C double bonded to O)

    and ethylene glycol, $HOCH_2CH_2OH$. Draw the repeating structure of this polymer.

4. Determine whether each of the following statements is true or false. If the statement is false, change it so it is true.
    (a) The first four paraffins in the homologous series are gases under normal conditions. True/False: _____

    _____ .

(b) Natural gas and petroleum are the major sources of organic chemicals.
True/False: _____
_____ .

(c) The process of petroleum refining is passing hydrocarbon vapor over a heated
catalyst so that each hydrocarbon molecule breaks up to give hydrocarbons of
lower molecular weight. True/False: _____
_____ .

(d) A secondary alcohol is one in which the hydroxyl (— OH) group is attached
to the second carbon in the chain. True/False: _____
_____ .

(e) Natural flavors are generally complex mixtures of amines and other
constituents. True/False: _____
_____ .

5. In a laboratory preparation of chrome alum, potassium chromium(III) sulfate
24-hydrate, ethanol is used to reduce dichromate, $Cr_2O_7^{2-}$, to Cr^{3+} in acid solution.
The oxidation product of the ethanol is ethanal, which escapes as a gas from the
reaction mixture. Write the balanced equation for this reaction.

6. For each of the following compounds, decide if *cis–trans* isomers are possible. If
they are, draw the condensed structural formula and give the IUPAC name, and/or
label the *cis* or *trans* isomers.

(a) 1,2-dichloro-1-butene

(c) $CH_3CHCH = C(CH_3)_2$
$|$
CH_3

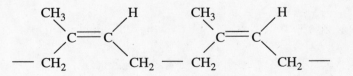

(b)

(d) 3-methyl-2-hexene

7. Natural rubber is essentially an addition polymer of isoprene and has the following
repeating structure:

$$\begin{array}{c}
CH_3 \qquad\quad H \qquad\qquad CH_3 \qquad\quad H \\
\diagdown C = C \diagup \qquad \diagdown C = C \diagup \\
\diagup \qquad\quad \diagdown \qquad\qquad \diagup \qquad\quad \diagdown \\
— CH_2 \qquad CH_2 — CH_2 \qquad CH_2 —
\end{array}$$

Draw the structure of isoprene and give its IUPAC name.

8.   Complete and balance the following reactions.

(a) $C_4H_{10}(g) + O_2(g) \xrightarrow{\Delta}$

(b) $CH_3CH_2CH_2C \equiv CH(l) + 2HCl(g) \longrightarrow$ (give major product, a liquid)

(c) $CHCl_3(l) + Cl_2(g) \xrightarrow{h\upsilon}$ (products are gases at the reaction temperature)

(d) $C_3H_7CH = CHCH_2CH_3(l) + Br_2(l) \longrightarrow$ (product is a liquid)

(e) 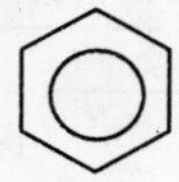   ⬡ $(l) + Cl_2(g) \xrightarrow{AlCl_3}$ (products are a liquid and a gas)

(f) $CH_2 = CH - \overset{\overset{\displaystyle O}{\|}}{C}OH(l) + CH_3\underset{\underset{\displaystyle CH_3}{|}}{C}HOH(l) \underset{\longleftarrow}{\overset{H^+}{\rightleftharpoons}}$ (products are liquids)

(g) $CH_3CH = CHCH_2CH_3(l) + H_2(g) \xrightarrow{Ni}$ (product is a liquid)

(h) $CH_3CH_2CH_2\overset{\overset{\displaystyle O}{\|}}{C}CH_3(l) + 2(H) \longrightarrow$ (product is a liquid)

9. Match the terms on the left with the letter of the appropriate formula on the right.

_____ (1) amine

_____ (2) amide

_____ (3) ketone

_____ (4) ether

_____ (5) organic halide

_____ (6) substituted benzene

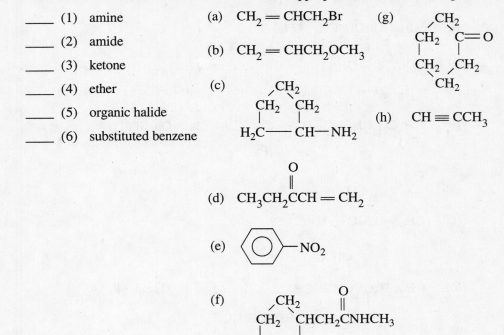

ANSWERS TO CHAPTER POST-TEST

If you missed an answer, study the text section and operational skill given in parentheses after the answer.

1.   (a)   1-ethoxybutane (24.5)
     (b)   4-methylhexanoic acid (24.5)
     (c)   methyl propanoate (24.5)
     (d)   5-methyl-3-hexanone (24.5)
     (e)   1,3-diethylbenzene (24.3, Op. Sk. 1)
     (f)   cyclohexene (24.1, 24.2, Op. Sk. 1)

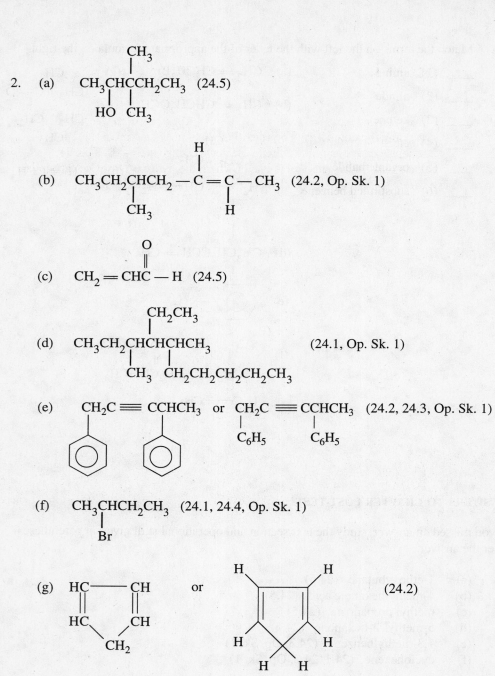

2. (a) CH$_3$CHCCH$_2$CH$_3$ (24.5)
         with CH$_3$ above and HO, CH$_3$ below

    (b) CH$_3$CH$_2$CHCH$_2$—C=C—CH$_3$ (24.2, Op. Sk. 1)
         with H above, CH$_3$ and H below

    (c) CH$_2$=CHC—H (24.5) with O double-bonded

    (d) CH$_3$CH$_2$CHCHCHCH$_3$ (24.1, Op. Sk. 1)
         with CH$_2$CH$_3$ above, CH$_3$ and CH$_2$CH$_2$CH$_2$CH$_2$CH$_3$ below

    (e) CH$_2$C≡CCHCH$_3$ or CH$_2$C≡CCHCH$_3$ (24.2, 24.3, Op. Sk. 1)
         with C$_6$H$_5$ below each

    (f) CH$_3$CHCH$_2$CH$_3$ (24.1, 24.4, Op. Sk. 1)
         with Br below

    (g) HC—CH or (cyclopentene structure) (24.2)
         HC CH
           CH$_2$

3. 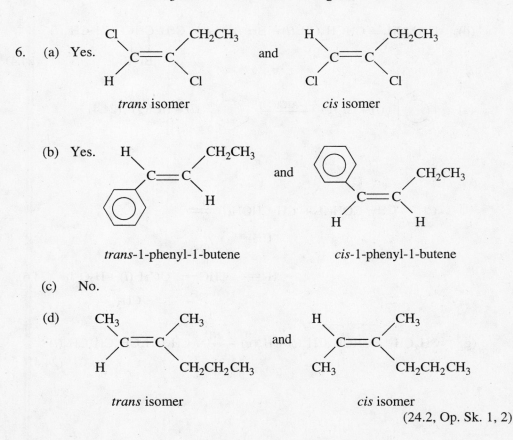 (24.8)

4. (a) True. (24.1) (b) True. (24.4)

   (c) False. The process described is catalytic cracking. Petroleum refining is separating the components by fractional distillation and processing the fractions to achieve products with the desired characteristics. (24.1)

   (d) False. A secondary alcohol is one in which the hydroxyl (— OH) group is attached to a carbon to which two other carbons are attached. (24.5)

   (e) False. Natural flavors are generally mixtures of esters and other constituents. (24.5)

5. $8H^+(aq) + Cr_2O_7^{2-}(aq) + 3CH_3CH_2OH(aq) \longrightarrow$
   $3CH_3CHO(g) + 2Cr^{3+}(aq) + 7H_2O(l)$ (24.5, 24.6, Op. Sk. 1, 4)

6. (a) Yes.

   trans isomer   *cis* isomer

   (b) Yes.

   trans-1-phenyl-1-butene   *cis*-1-phenyl-1-butene

   (c) No.

   (d)

   trans isomer   *cis* isomer

   (24.2, Op. Sk. 1, 2)

7.

$$CH_3 \quad H$$
$$\underset{CH_2}{C} - \underset{CH_2}{C}$$

2-methyl-1,3-butadiene   (24.2, 24.8, Op. Sk. 1)

8.   (a) $2C_4H_{10}(g) + 13O_2(g) \xrightarrow{\Delta} 8CO_2(g) + 10H_2O(g)$ (24.4)

(b) $CH_3CH_2CH_2C \equiv CH(l) + 2HCl(g) \longrightarrow CH_3CH_2CH_2\underset{Cl}{\overset{Cl}{C}}CH_3(l)$

(24.4, Op. Sk. 3)

(c) $CHCl_3(l) + Cl_2(g) \xrightarrow{h\upsilon} CCl_4(g) + HCl(g)$ (24.4)

(d) $C_3H_7CH = CHCH_2CH_3(l) + Br_2(l) \longrightarrow C_3H_7\underset{Br}{\overset{}{C}}H\underset{Br}{\overset{}{C}}HCH_2CH_3(l)$ (24.4)

(e) $\bigcirc (l) + Cl_2(g) \xrightarrow{AlCl_3} \bigcirc\!\!-Cl (l) + HCl(g)$ (24.4)

(f) $CH_2 = CH - \overset{O}{\overset{\|}{C}}OH(l) + CH_3\underset{CH_3}{\overset{}{C}}HOH(l) \rightleftharpoons$

$CH_2 = CHC - O\underset{CH_3}{\overset{CH_3}{C}}H (l) + H_2O(l)$ (24.6)

(g) $CH_3CH = CHCH_2CH_3(l) + H_2(g) \xrightarrow{Ni} CH_3CH_2CH_2CH_2CH_3(l)$ (24.4) (h)　$\overset{\displaystyle O}{\underset{\displaystyle \|}{}}$

(h)　$CH_3CH_2CH_2\overset{\overset{\textstyle O}{\|}}{C}CH_3(l) + 2(H) \longrightarrow CH_3CH_2CH_2\overset{\overset{\textstyle OH}{|}}{C}HCH_3(l)$　(24.6)

9.　(1)　c　(24.7)　　(2)　f　(24.7)　　　　(3)　d and g　(24.5)

　　(4)　b　(24.5)　　(5)　a　(text Table 24.4)　(6)　e　(24.3, 24.4)

CHAPTER 25   BIOCHEMISTRY

CHAPTER TERMS AND DEFINITIONS

Numbers in parentheses after definitions give the text sections in which the terms are explained. Starred terms are italicized in the text. Where a term does not fall directly under a text section heading, additional information is given for you to locate it.

metabolism   process of building up and breaking down organic molecules in cells   (25.1)

enzyme   protein that catalyzes a biochemical reaction   (25.1)

metabolic pathway*   series of interconnected enzyme-catalyzed reactions   (25.1)

regulation*   slowing down or speeding up of metabolic reactions by which the cell exerts control over them through enzyme activity   (25.1)

coupling*   combination of a process requiring free energy with a process releasing free energy so that the overall free-energy change for the two processes is favorable and the unfavorable reaction can occur   (25.2)

chlorophylls*   green pigments in plant cells that trap radiant energy for conversion to chemical energy   (25.2, marginal note)

macromolecules   very large molecules having molecular weights ranging into the millions of amu's   (25.3, introductory section)

condensation reactions*   reactions in which molecules join to each other with the splitting out of small molecules   (25.3, introductory section)

proteins   biological polymers of amino acids with a very wide range of molecular weights   (25.3)

insulin*   protein hormone made in the pancreas and secreted to regulate the body's blood-sugar level   (25.3)

glucagon*   protein hormone made in the pancreas and secreted to regulate the body's blood-sugar level   (25.3)

amino acid   molecule containing an amino group ($—NH_2$) and a carboxyl group ($—COOH$)   (25.3)

side chain*   in amino acid–protein chemistry, the hydrocarbon group (R group) of an amino acid   (25.3)

zwitterion* amino acid in the doubly ionized form, with $- NH_3^+$ and $- COO^-$ (25.3)

enantiomers (D- and L-isomers, optical isomers)* isomers that are nonsuperimposable mirror images of one another (25.3)

D-isomer one of a pair of enantiomers; the D-amino acid has the structural formula

$$\begin{array}{c} COOH \\ | \\ H - C - NH_2 \\ | \\ R \end{array} \quad (25.3)$$

L-isomer one of a pair of enantiomers; the L-amino acid has the structural formula

$$\begin{array}{c} COOH \\ | \\ H_2N - C - H \\ | \\ R \end{array} \quad (25.3)$$

peptide (amide) bond $C - N$ bond resulting from a condensation reaction between the carboxyl group of one amino acid and the amino group of a second amino acid (25.3)

dipeptide* molecule formed by linking together two amino acids (25.3)

tripeptide* molecule formed by linking together three amino acids (25.3)

polypeptide polymer formed by the linking of many amino acids by peptide bonds (25.3)

primary structure (of a protein) order, or sequence, of the amino-acid units in a protein (25.3)

disulfide linkage* in polypeptide chemistry, $- S - S -$ divalent linkage between two cysteine side chains which helps anchor the folded polypeptide into position (25.3)

fibrous proteins polypeptides that form long coils or align themselves in parallel to form long, water-insoluble fibers (25.3)

secondary structure (of a protein) in polypeptide chemistry, a simple coiled or parallel arrangement of a protein molecule (25.3)

globular proteins polypeptides in which long coils fold into compact, roughly spherical shapes (25.3)

tertiary structure (of a protein) in polypeptide chemistry, the structure associated with the way the protein coil is folded (25.3)

active site* surface feature of an enzyme where catalysis occurs (25.3)

carbohydrates polyhydroxy aldehydes or polyhydroxy ketones, or substances that yield such products when they react with water (25.4)

monosaccharides simple sugars, each containing three to nine carbon atoms, all but one of which bear a hydroxyl group, the remaining one being a carbonyl carbon (25.4)

oligosaccharides short polymers of two to ten simple sugar units (25.4)

polysaccharides polymers consisting of more than ten simple sugar units (25.4)

hemiacetals* organic compounds in which an $- OH$ group, an $- OR$ group, and an H

atom are attached to the same carbon:
$$R' - \underset{\underset{OR}{|}}{\overset{\overset{OH}{|}}{C}} - H \quad (25.4)$$

hemiketal*   organic compound in which an — OH group, an — OR group, and an

— R'' group are attached to the same carbon:
$$R' - \underset{\underset{OR}{|}}{\overset{\overset{OH}{|}}{C}} - R'' \quad (25.4)$$

pyranoses*   six-membered ring forms of simple sugars   (25.4)

furanoses*   five-membered ring forms of simple sugars   (25.4)

maltose*   common disaccharide composed of two α-D-glucose units   (25.4)

lactose*   milk sugar; common disaccharide   (25.4)

cellulose*   structural polysaccharide in plants; a linear polymer of β-D-glucopyranose units   (25.4)

turgor*   water pressure in cells that gives nonwoody plants their shapes   (25.4)

amylose*   type of starch; contains long, unbranched chains of α-D-glucopyranose units (25.4)

amylopectin*   type of starch; contains long, branched chains of α-D-glucopyranose units (25.4)

glycogen*   storage polysaccharide in animals; similar in structure to amylopectin but more branched   (25.4)

carbohydrate loading*   process of trying to increase the amount of glycogen stored in muscle cells; includes a long run, minimization of carbohydrates, then eating food high in starch   (25.4, marginal note)

genetic engineering*   manipulation of nucleic acids to change the characteristics of organisms   (25.5, marginal note)

nucleotides   building blocks of nucleic acids; composed of an organic base linked to a 5-carbon sugar, which in turn is linked to a phosphate group   (25.5)

ribonucleotides*   nucleotides found in RNA; the 5-carbon sugar is β-D-ribose   (25.5)

deoxyribonucleotides*   nucleotides found in DNA; the 5-carbon sugar is 2-deoxy-β-D-ribose   (25.5)

nucleoside*   in nucleotides, the base plus the sugar   (25.5)

adenine* organic base found in nucleotides; structural formula:

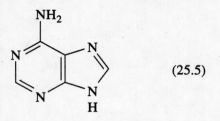 (25.5)

guanine* organic base found in nucleotides; structural formula:

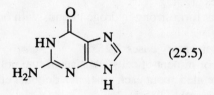 (25.5)

cytosine* organic base found in nucleotides; structural formula:

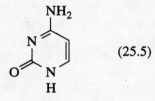 (25.5)

uracil* organic base found in nucleotides; structural formula:

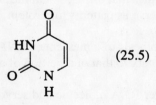 (25.5)

thymine*   organic base found in nucleotides; structural formula:

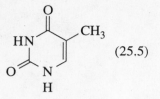

$$(25.5)$$

polynucleotide   linear polymer of nucleotide units   (25.5)

nucleic acids   carriers of the cell and species inheritance; polynucleotides folded or coiled into specific three-dimensional shapes   (25.5)

complementary bases   nucleotide bases that form strong hydrogen bonds with one another   (25.5)

base pairing*   hydrogen bonding of complementary bases   (25.5)

deoxyribonucleic acid (DNA)   hereditary constituent of cells; consists of two polymer strands of deoxyribonucleotide units coiled about each other in a double helix, with base pairing along their entire strand lengths   (25.5)

ribonucleic acid (RNA)   constituent of cells that is used to manufacture proteins from genetic information; a polymer of ribonucleotide units   (25.5)

chromosomes*   cell structures that contain proteins and DNA   (25.5)

gene   sequence of nucleotides in a DNA molecule that codes for a given protein   (25.5)

polymerase*   enzyme involved in DNA replication   (25.5)

polymerase chain reaction (PCR)*   basis of technique used in the laboratory to amplify the quantity of DNA in a sample   (25.5)

mutation*   change in the genetic information, or a genetic error   (25.5)

ribosomes   tiny cellular particles on which protein synthesis takes place   (25.5)

ribosomal RNA   RNA contained in ribosomes   (25.5)

messenger RNA (mRNA)   relatively small RNA molecules that diffuse about the cell and attach themselves to ribosomes, where they serve as patterns for protein biosynthesis   (25.5)

transcription*   first step in protein biosynthesis, the synthesis of a messenger-RNA molecule that has a sequence of bases complementary to that of a portion of a DNA strand coresponding to a single protein   (25.5)

codon   sequence of three bases that occurs in a messenger-RNA molecule and serves as the code for a particular amino acid   (25.5)

translation*   biosynthesis of protein using messenger-RNA codons   (25.5)

termination codons*   codons that signify the end of a genetic message, i.e., the end of protein biosynthesis   (25.5)

transfer RNA (tRNA)   smallest RNA molecules, which bond to amino acids and carry them to the ribosomes, then attach themselves, through base pairing, to messenger-RNA codons   (25.5)

anticodon* in a transfer-RNA molecule, a triplet sequence complementary to the codon on a messenger-RNA molecule (25.5)

lipids biological substances that are soluble in nonpolar organic solvents, such as chloroform and carbon tetrachloride, and insoluble in water (25.6)

fats* solid esters formed from glycerol and three fatty acids (25.6)

oils* liquid esters formed from glycerol and three fatty acids (25.6)

triacylglycerols esters formed from glycerol and three fatty acids; called triglycerides (25.6)

glycerol* trihydroxy alcohol: $CH_2OHCHOHCH_2OH$ (25.6)

fatty acids* long-chain carboxylic acids (25.6)

adipose tissue* specialized fat-storage tissue in animals (25.6)

membranes* proteins inserted into a phospholipid matrix that surround and define a biological cell (25.6)

phospholipid diacylglycerol of two fatty acids, with the third — OH group bonded to a phosphate group that is bonded in turn to an alcohol (25.6)

polar head* hydrophilic end of a phospholipid, which consists of the phosphate group and its bonded alcohol (25.6)

phospholipid bilayer* sheet of phospholipids, two molecules thick, which presents a barrier to charged or polar substances (25.6)

CHAPTER DIAGNOSTIC TEST

1. Determine whether each of the following statements is true or false. If the statement is false, change it so it is true.

    (a) Metabolism is the breakdown of organic molecules in cells. True/False: _____ .

    (b) Lipids have polymeric structural units of individual triacylglycerols. True/False: _____ .

    (c) Primary structure refers to the order or sequence of the building-block units in a polymer. True/False: _____ .

    (d) The basic structural units of proteins are α-amino acids. True/False: _____ .

    (e) Cellulose is a linear polymer of α-D-glucose units. True/False: _____ .

    (f) The building blocks of nucleic acids are nucleosides, each composed of a sugar and a base. True/False: _____ .

    (g) The genetic code is the relationship between the nucleotide sequence in DNA and the amino-acid sequence in proteins. True/False: _____ .

    (h) The DNA sequence that is complementary to AAGCUGAUUCG is TTCGACTAAGC. True/False: _____ .

2.   Which of the following two structures is an L-amino acid?

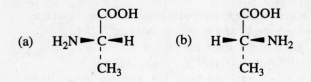

(a) $H_2N \blacktriangleright C \blacktriangleleft H$ (b) $H \blacktriangleright C \blacktriangleleft NH_2$

with COOH above and CH_3 below each central C.

3.   The structure of L-valine is

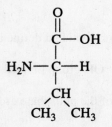

Draw the zwitterion.

4.   Draw the two peptides that could form from L-tyrosine and L-phenylalanine.

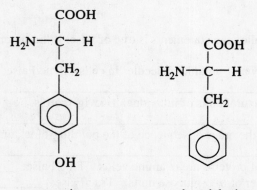

L-tyrosine   L-phenylalanine

5. Draw a diagram showing how a hydrogen bond could form between the side chains of L-threonine and L-aspartate.

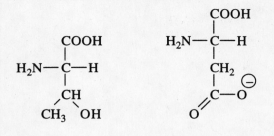

L-threonine   L-aspartate

6. D-galactose has the straight-chain formula

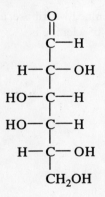

Write the reaction for the equilibria between this form and the pyranose forms of the sugar that exist in solution. Number the carbons in each structure, and label the ring forms α or β.

7. Write the structural formula for cellobiose, which consists of two β-D-glucose units linked from carbons 1 to 4. The structure of D-glucose is

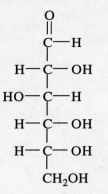

8. Draw the structural formula of the nucleotide containing adenine, β-D-ribose, and three phosphate groups. Name the compound. The structure of adenine is

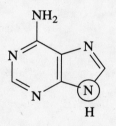

It links through the circled nitrogen. β-D-ribose is

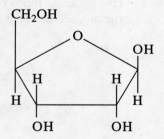

9. The following messenger-RNA sequence codes for the amino acids in the human female reproductive hormone ocytocin. Write the amino-acid sequence of ocytocin. Use text Table 25.2.

UGCUAUAUACAGAAUUGUCCACUAGGA

10. Write the letter of each of the steps below to give the proper chronological sequence of protein biosynthesis. A letter may be used more than once.
    (a) Amino acid enters the ribosome bonded to a transfer-RNA anticodon.
    (b) Transfer-RNA anticodon base pairs with the messenger-RNA codon.
    (c) Enzyme binds the amino acid.
    (d) Messenger-RNA molecule is synthesized in the cell nucleus.
    (e) Peptide bond is formed between the two amino acids.
    (f) Messenger-RNA molecule moves out of the nucleus and into a ribosome.
    (g) Termination codon is processed.
    (h) Amino acids continue to be brought into the ribosome and bonded.
    (i) Finished protein is released from the ribosome.

11. Using text Table 25.3, write the structure of the triacylglycerol formed from lauric acid, palmitic acid, and oleic acid.

ANSWERS TO CHAPTER DIAGNOSTIC TEST

If you missed an answer, study the text section and operational skill given in parentheses after the answer.

1. (a) False. Metabolism is the synthesis and breakdown of organic molecules in cells. (25.1)
    (b) False. Lipids differ from the other classes of biological molecules in that there are no lipid polymers. (25.6)
    (c) True. (25.3)   (d) True. (25.3)
    (e) False. Starch is a linear polymer of α-D-glucose units. Cellulose is a linear polymer of β-D-glucose units. (25.4)
    (f) False. The building blocks of nucleic acids are nucleotides, each composed of a sugar, base, and phosphate group. (25.5)
    (g) True. (25.5)   (h) True. (25.5)

2. a (25.3)

3. 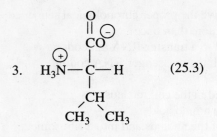   (25.3)

4.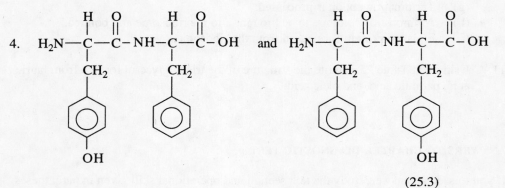

(25.3)

5.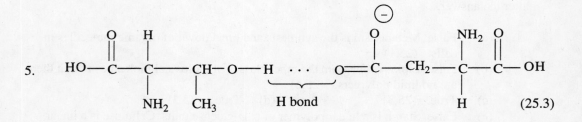

(25.3)

6.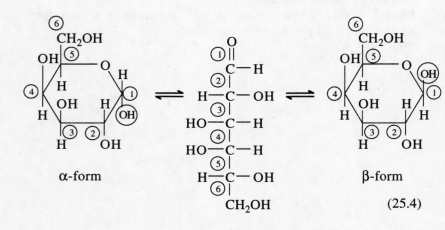

(25.4)

7.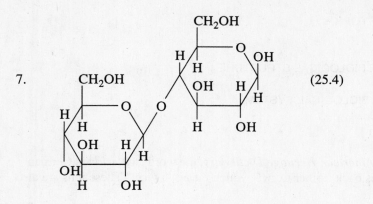

(25.4)

8.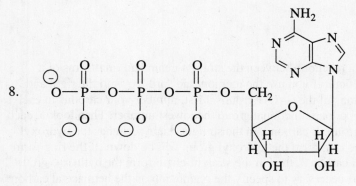

adenosine triphosphate (ATP)　　　　　　　(25.5)

9.　　cys–tyr–ile–gln–asn–cys–pro–leu–gly　(25.5)

10.　　d, f, a, b, c, a, b, c, e, h, g, i　(25.5)

11.

$$CH_2-O-\overset{\overset{\displaystyle O}{\|}}{C}-(CH_2)_{10}CH_3$$

$$CH-O-\overset{\overset{\displaystyle O}{\|}}{C}-(CH_2)_{14}CH_3$$

$$CH_2-O-\overset{\overset{\displaystyle O}{\|}}{C}-(CH_2)_7CH=CH(CH_2)_7CH_3 \qquad (25.6)$$

SUMMARY OF CHAPTER TOPICS

25.1   THE CELL: UNIT OF BIOLOGICAL STRUCTURE

25.2   ENERGY AND THE BIOLOGICAL SYSTEM

25.3   PROTEINS

According to *The American Heritage Dictionary*, the word *zwitterion* is German for "mongrel ion" and goes back further to *zwi-*, which means "twice." These ions are also called inner salts.

25.4   CARBOHYDRATES

Writing the equilibria reactions between the straight-chain and ring forms of a D-monosaccharide is not difficult if you use the memory aid "ard": α is to the right and down, referring to the positions of the — OH groups. First, number your carbons in each form, giving the aldehyde or ketone functional group the lowest number. Then look at each carbon that has a hydroxyl group. If carbon 2 in the straight-chain form has the hydroxyl group to the right, then in the ring form the hydroxyl group will be down. If the ring form has the hydroxyl group up at carbon 3, then in the straight-chain form the hydroxyl group will be written to the left. It is necessary to specify the orientation on the hemiacetal carbon as α or β. Again, looking at the position of the hydroxyl group and using the memory aid, give the answer. If the hemiacetal form is written as a straight chain and the hydroxyl group is to the right, it is the α form; if in the ring form that hydroxyl group is up, it is the β form. Practice by writing the straight-chain form for an equilibrium given in the text; then close your text and write both ring forms.

25.5   NUCLEIC ACIDS

You will have to know the base content of each of the nucleic acids in order to draw structures of nucleotides or write the complementary sequence for a DNA sequence or the amino-acid sequence for an RNA sequence. The best way to commit these to memory is to construct a memory aid using the first letters of the nucleic acids and bases. DNA contains adenine (A), guanine (G), cytosine (C), and thymine (T). A possible memory aid would be DAGCT. RNA contains adenine (A), guanine (G), cytosine (C), and uracil (U). An aid might be RAGCU.

You will also need to know which bases pair. Again, a memory aid of first letters will be helpful. AT, AU, and CG are complementary bases.

25.6 LIPIDS

ADDITIONAL PROBLEMS

1. Aspartic acid has the structural formula

$$HOOCCH_2CHCOOH$$
$$|$$
$$NH_2$$

Draw the zwitterion of aspartic acid.

2. Two common amino acids are

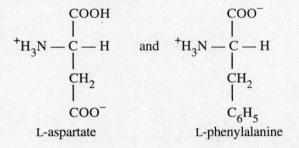

L-aspartate   and   L-phenylalanine

(a) Draw the dipeptides that the two amino acids can form.
(b) One of these dipeptides is the basis for the sweetener aspartame, L-aspartyl-L-phenylalanine methyl ester (asp-phe-OMe), sold under the brand name "NutraSweet." Draw the structure of aspartame.

3. The formulas of the two amino acids L-serine and L-glutamate are

$$^-OOC - \underset{\underset{NH^+_3}{|}}{\overset{\overset{H}{|}}{C}} - CH_2 - OH \quad and \quad ^-OOC - CH_2 - CH_2 - \underset{\underset{H}{|}}{\overset{\overset{NH^+_3}{|}}{C}} - COO^-$$

Show by sketching structural formulas of these two molecules how their side chains could form a hydrogen bond between them.

4. The straight-chain carbohydrate D-gulose is a flexible molecule. Write the reaction showing the equilibria between the straight-chain form and its hemiacetal forms. The formula for D-gulose is

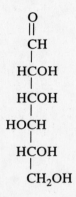

5. Diagram the reaction to give a disaccharide that could form between

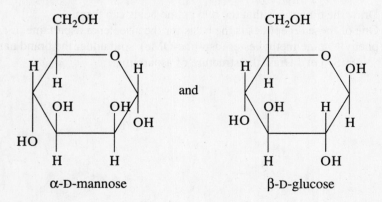

α-D-mannose   β-D-glucose

linked from carbon 1 of glucose to carbon 4 of mannose.

6. Draw the structural formula for the nucleotide containing cytosine, 2-deoxy-β-D-ribose, and phosphate group at the $5'$ position.

7. Write the DNA nucleotide sequence of bases that is complementary to ACCGTCG.

8. Write the letter of the response that matches the numbered terms on the left.

____ (1) protein units		(a)	nucleosides
____ (2) enzyme		(b)	mannose, glucose, fructose
		(c)	monosaccharides
____ (3) primary structure		(d)	nucleotides
____ (4) disaccharides		(e)	A — T, A — U
		(f)	sequence of polymer units
____ (5) complementary bases		(g)	U — T, A — G
____ (6) translation		(h)	messenger-RNA synthesis
		(i)	biological catalyst
____ (7) nucleic-acid units		(j)	α-amino acids
		(k)	sucrose, maltose, lactose
		(l)	coiling of polymer chain
		(m)	protein synthesis using codons

9. The tissue hormone bradykinin, which dilates the blood vessels and lowers blood pressure, is synthesized according to the following codon sequence:

    AGGCCACCCGGGUUUUCACCUUUCAGAUGA

    (a) How many amino acids are represented? Explain.
    (b) Give the amino acid sequence of bradykinin.

10. Write a structural formula for the triacylglycerol containing lauric, oleic, and linolenic acids.

ANSWERS TO ADDITIONAL PROBLEMS

If you missed an answer, study the text section and operational skill given in parentheses after the answer.

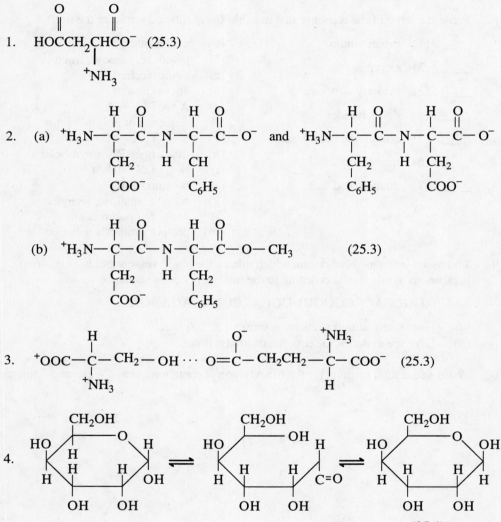

1. $\underset{\text{O}}{\overset{\text{O}}{\|}}\quad\underset{\text{O}}{\overset{\text{O}}{\|}}$ HOCCH$_2$CHCO$^-$ (25.3)
      |
      $^+$NH$_3$

2. (a) $^+$H$_3$N—C—C—N—C—C—O$^-$ and $^+$H$_3$N—C—C—N—C—C—O$^-$

(b) $^+$H$_3$N—C—C—N—C—C—O—CH$_3$ (25.3)

3. $^+$OOC—C—CH$_2$—OH $\cdots$ O=C—CH$_2$CH$_2$—C—COO$^-$ (25.3)

4.
                                                                                        (25.4)

5.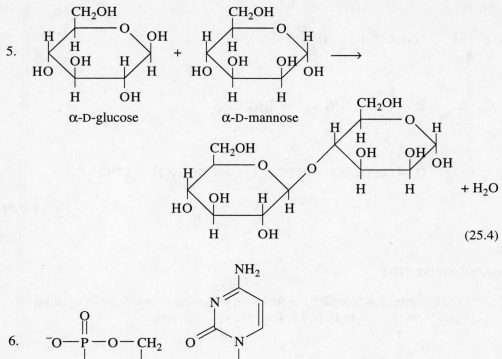

α-D-glucose   α-D-mannose

$+ H_2O$

(25.4)

6.

(25.4)

7. TGGCAGC (25.5)

8. (1) j   (2) i   (3) f (25.3)
    (4) k (25.4)
    (5) e   (6) m   (7) d (25.5)

9. (a) 9; each amino acid is coded for by three bases; the final codon (three bases)
         signifies the end of the peptide.
    (b) arg–pro–pro–gly–phe–ser–pro–phe–arg (25.5)

10.
$$CH_2-O-\overset{\overset{\displaystyle O}{\|}}{C}(CH_2)_{10}CH_3$$

$$CH_2-O-\overset{\overset{\displaystyle O}{\|}}{C}(CH_2)_7CH=CH(CH_2)_7CH_3$$

$$CH_2-O-\overset{\overset{\displaystyle O}{\|}}{C}(CH_2)_7CH=CHCH_2CH=CHCH_2CH=CHCH_2CH_3$$

(25.6)

CHAPTER POST-TEST

1. Write the structural formula of gentiobiose, consisting of two β-D-glucose units linked from carbon 1 to carbon 6. The structure of glucose is

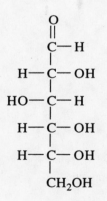

2. The pyranose form of α-D-mannose is

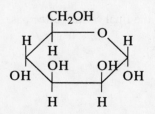

Write the structural formula for the straight-chain form in equilibrium with this hemiacetal.

3. The structure of L-proline is

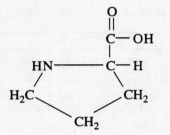

Draw the zwitterion.

4. Which of the following has a polar side chain? Explain.

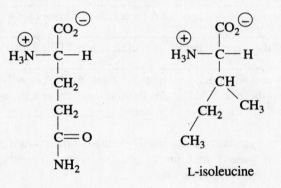

L-glutamine   L-isoleucine

5. Draw the two peptides that could form from L-valine and L-proline.

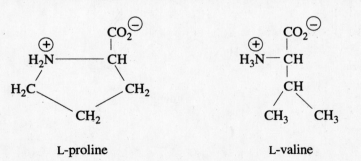

L-proline   L-valine

6. Determine whether each of the following statements is true or false. If the statement is false, change it so it is true.

    (a) The fact that energetically unfavorable reactions occur in metabolism is explained through the concept of coupling, which holds that the unfavorable process occurs together with a process releasing energy. True/False: _____

    _____

    _____ .

    (b) The three-dimensional conformations of polymers depend on covalent bonds holding them in place. True/False: _____

    _____ .

    (c) All of the amino acids of known, naturally occurring proteins are D-amino acids. True/False: _____

    _____ .

    (d) The structural units of a protein are linked by hydrogen bonds. True/False: _____

    _____ .

    (e) The structural units of polymeric carbohydrates are monosaccharides. True/False: _____ .

    (f) A gene is a sequence of nucleotides in a DNA molecule that codes for a given protein. True/False: _____ .

    (g) Nucleotides function as the structural units of nucleic acids. True/False: _____

    _____ .

    (h) The structures of DNA and RNA differ in that the sugar in DNA is 2-β-D-deoxyribose rather than ribose, DNA contains thymine and not uracil, and RNA contains uracil and not thymine. True/False: _____ .

7. Draw the structural formulas of cytosine and guanine in their base-paired form. Guanine uses only atoms on the 6-membered ring to form the hydrogen bonds. The bases bond covalently through the circled nitrogens.

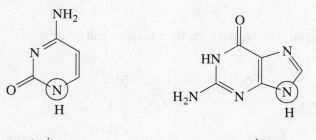

cytosine   guanine

8. Write the structural formula of deoxyuridine monophosphate (dUMP). The structures are

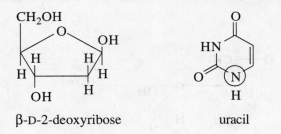

β-D-2-deoxyribose   uracil

Uracil bonds through the circled nitrogen.

9. Write the RNA sequence complementary to AAGCUGAUUCG.

10. The polypeptide hormone vasopressin elevates blood pressure. The peptide sequence is cys–tyr–phe–gln–asn–cys–pro–lys–gly. Write a possible messenger-RNA sequence that would code for this peptide.

11. Write structures of the four molecules from which the following compound was formed, and name each. Use text Table 25.3.

$$CH_2 - O - \overset{\overset{\displaystyle O}{\|}}{C}(CH_2)_{18}CH_3$$

$$CH - O - \overset{\overset{\displaystyle O}{\|}}{C}(CH_2)_7CH = CHCH_2CH = CH(CH_2)_4CH_3$$

$$CH_2 - O - \overset{\overset{\displaystyle O}{\|}}{C}(CH_2)_7CH = CH(CH_2)_5CH_3$$

ANSWERS TO CHAPTER POST-TEST

If you missed an answer, study the text section and operational skill given in parentheses after the answer.

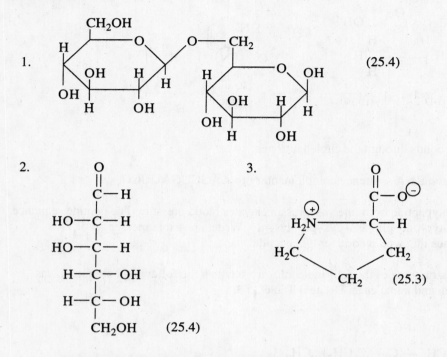

1.   (25.4)

2.   3.

                                                                                      (25.3)

(25.4)

4. L-glutamine has the polar side chain. The side chain is the hydrocarbon group attached to the α-carbon. In L-glutamine, the side chain includes the polar carbonyl and amine groups. In L-isoleucine, the side chain is the nonpolar 2-butyl group. (25.4)

5.

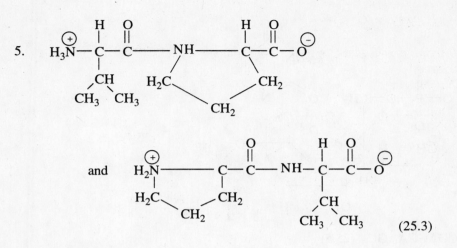

(25.3)

6. (a) True. (25.2)
   (b) False. The three-dimensional conformations of polymers depend on covalent bonds and noncovalent bonds, including hydrogen bonds and ionic bonds. (25.3)
   (c) False. All of the amino acids of known, naturally occurring proteins are L-amino acids. (25.3)
   (d) False. The structural units of a protein are linked by peptide (amide) bonds. (25.3)
   (e) True. (25.4)   (f) True. (25.5)   (g) True. (25.5)
   (h) True. (25.5)

7.

to sugar

(25.5)

to sugar

8.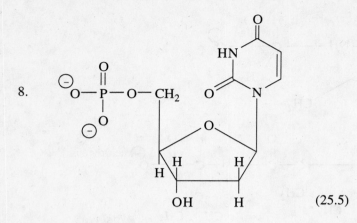

(25.5)

9.   UUCGACUAAGC   (25.5)

10.   The top line below would be a possible sequence. Individual codons could be
      replaced by any below them.   (25.5)

UGU UAU UUU CAA AAU UGU CCU AAA GGU UAA
or   or   or   or   or   or   or   or   or   or
UGC UAC UUC CAG AAC UGC CCC AAG GGC UAG
                              or   or   or
                              CCA   GGA UGA
                              or   or
                              CCG   GGG

11.   CH$_2$OH
       |
       CHOH
       |
       CH$_2$OH   glycerol

$$\underset{\text{HO}}{}\;-\;\overset{\displaystyle\text{O}}{\underset{\displaystyle\|}{\text{C}}}\;-(\text{CH}_2)_{18}\text{CH}_3 \qquad\qquad \text{arachidic acid}$$

$$\text{HO} - \overset{\displaystyle\text{O}}{\underset{\displaystyle\|}{\text{C}}}(\text{CH}_2)_7\text{CH} = \text{CHCH}_2\text{CH} = \text{CH(CH}_2)_4\text{CH}_3 \qquad \text{linoleic acid}$$

$$\text{HO} - \overset{\displaystyle\text{O}}{\underset{\displaystyle\|}{\text{C}}}(\text{CH}_2)_7\text{CH} = \text{CH(CH}_2)_5\text{CH}_3 \qquad\qquad \text{palmitoleic acid}$$

(25.6)

UNIT EXAM 8

1. Write the IUPAC name for each of the following compounds.

(a) 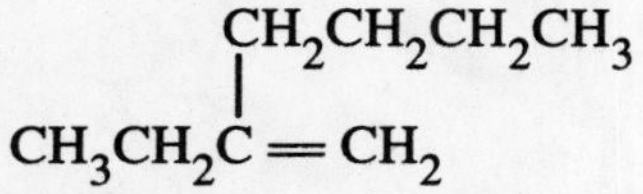
$$\underset{\displaystyle CH_3CH_2\overset{\textstyle |}{C}=CH_2}{\overset{\textstyle CH_2CH_2CH_2CH_3}{}}$$

(b)
$$CH_3CH_2CH_2 - \underset{\displaystyle CH_2CH_2CH_2CH_3}{\overset{\displaystyle OH}{\underset{|}{\overset{|}{C}}CH_3}}$$

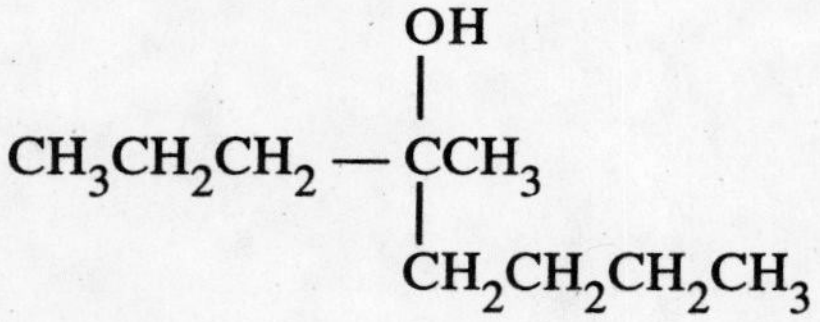

(c)
$$CH_3\overset{\displaystyle O}{\overset{|}{C}} - OCH_3$$

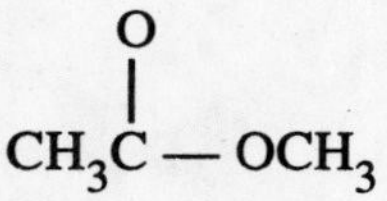

(d) 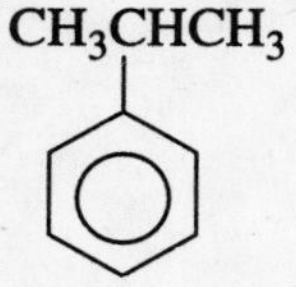 CH_3CHCH_3

2. Write the structural formula for each of the following compounds.
   (a) 3-ethyl-2-methylpentane   (c) 2-pentanone
   (b) 2-methyl-butanal   (d) propanoic acid

676

3. Tell whether geometric isomers exist for each of the following compounds. If they do, draw and name the isomers, including the *cis* or *trans* designation.

   (a) $CH_2 = CHCH_2CH_3$ (c) $CH_3C \equiv CCH_3$

   (b) $CH_3CH_2CH = CHCH_3$

4. Complete and balance the following equation:

   $$C_6H_{14}(l) + O_2(g) \xrightarrow{\Delta}$$

5. Write the major organic product resulting from each of the following reactions. Write NR if no reaction occurs.

   (a) $CH_3CH = CH_2 + Br_2 \longrightarrow$

   (b) 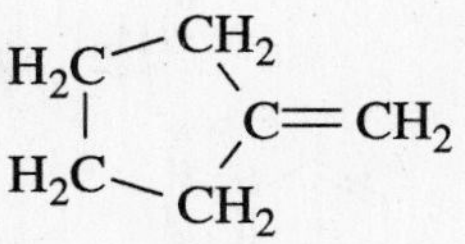 $+ \; HBr \longrightarrow$

   (c) 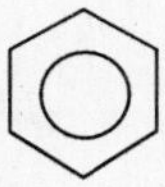 $+ \; HONO_2 \xrightarrow{H_2SO_4}$

   (d) 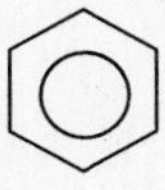 $+ \; Cl_2 \xrightarrow{AlCl_3}$

   (e)
   $$H_3C - \overset{\overset{\displaystyle CH_3}{|}}{\underset{\underset{\displaystyle CH_3}{|}}{C}} - OH + (O) \longrightarrow$$

   (f)
   $$\overset{\overset{\displaystyle OH}{|}}{CH_3CHCH_3} + (O) \longrightarrow$$

   (g)
   $$\overset{\overset{\displaystyle O}{\|}}{CH_3CH} + 2(H) \longrightarrow$$

(h) $CH_3CH_2\overset{O}{\overset{\|}{C}}OH + C_2H_5OH \overset{H^+}{\rightleftharpoons}$

6. Explain the meaning of the term *octane number.*

7. Match each term in the left-hand column with the most appropriate item in the right-hand column.

____ (a) addition polymers

____ (b) condensation polymers

____ (c) amide

____ (d) cycloalkene

____ (e) aromatic compound

____ (f) alcohol

____ (g) amine

1. 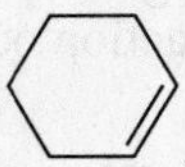

2. $CH_3CH_2\underset{NH_2}{\underset{|}{C}}HCH_3$

3. 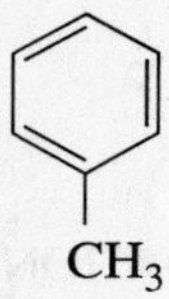

4. saran, polypropylene

5. $CH_3\underset{OH}{\underset{|}{C}}HCH_3$

6. 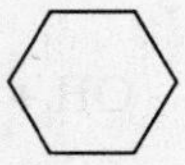

7. $CH_3CH_2\overset{O}{\overset{\|}{C}}OCH_3$

8. $CH_3\overset{O}{\overset{\|}{C}}NHCH_3$

9. nylon-66, dacron polyester

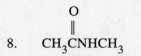

8. The structure of L-methionine is

$$
\begin{array}{c}
O \\
\| \\
C-OH \\
| \\
H_2N-C-H \\
| \\
CH_2 \\
| \\
CH_2 \\
| \\
S \\
| \\
CH_3
\end{array}
$$

Draw the zwitterion.

9. Draw the two peptides that could form from L-methionine (met) and L-asparagine (asn). The structure of L-methionine is given in Problem 8, and L-asparagine is

$$
\begin{array}{c}
O \\
\| \\
COH \\
| \\
H_2N-C-H \\
| \\
CH_2 \\
| \\
C=O \\
| \\
NH_2
\end{array}
$$

10. Write all possible sequences of all possible tripeptides with amino-acid composition histidine (his), arginine (arg), and isoleucine (ile).

11.   D-allose has the straight-chain formula

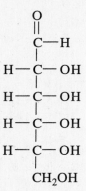

Write the reaction for the equilibria that exist in solution between this form and the two pyranose forms. Name the pyranose forms.

12.   Draw two linked monosaccharide units of the starch amylose. The linkage is $1 \rightarrow 4$.

13.   Determine whether each of the following statements is true or false. If the statement is false, change it so it is true.
    (a)   An example of a sugar without an oxygen on all carbons is D-ribose. True/False: _____

    _____ .

    (b)   A dipeptide of two different amino-acid structural units could have two possible sequences. True/False: _____

    _____ .

    (c)   Energy-storage polysaccharides are amylose and amylopectin in plants and glycogen in animals. True/False: _____

    _____ .

    (d)   Complementary base pairs are adenine–guanine, adenine–uracil, and thymine–cytosine. True/False: _____

    _____ .

14. Draw the structural formula for cytidine diphosphate. The following structural formulas are provided for you:

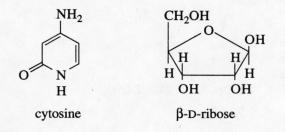

cytosine   β-D-ribose

15. Write the DNA sequence complementary to the following RNA-base sequence:

GUUGUCGGGCAUACAAGAUGGUGA

16. Use text Table 25.2 to write the amino-acid sequence for which the RNA-base sequence in Problem 15 would code.

17. Write the structural formulas of the products of the following reaction:

$$
\begin{array}{l}
\text{CH}_2 - \text{O} - \overset{\displaystyle \overset{\text{O}}{\|}}{\text{C}}(\text{CH}_2)_{10}\text{CH}_3 \\[2em]
\text{CH} - \text{O} - \overset{\displaystyle \overset{\text{O}}{\|}}{\text{C}}(\text{CH}_2)_7\text{CH} = \text{CH}(\text{CH}_2)_7\text{CH}_3 + 3\text{OH}^- \longrightarrow \\[2em]
\text{CH}_2 - \text{O} - \overset{\displaystyle \overset{\text{O}}{\|}}{\text{C}}(\text{CH}_2)_7\text{CH} = \text{CH}(\text{CH}_2)_7\text{CH}_3
\end{array}
$$

ANSWERS TO UNIT EXAM 8

If you missed an answer, study the text section and operational skill given in parentheses after the answer.

1. (a)   2-ethyl-1-hexene
    (b)   4-methyl-4-octanol
    (c)   methyl ethanoate
    (d)   2-phenylpropane   (24.2, Op. Sk. 1; 24.3, 24.5)

2. (a)   (c)

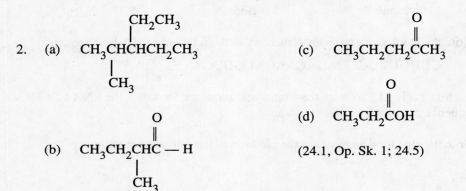

    (b)   (d)

                                                                 (24.1, Op. Sk. 1; 24.5)

3. (a)   No geometric isomers exist.
    (b)   Yes. The geometric isomers are

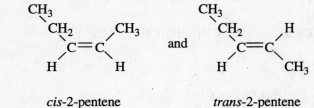

        cis-2-pentene   *trans*-2-pentene

    (c)   No geometric isomers exist.   (24.2, Op. Sk. 2)

4. $2C_6H_{14}(l) + 19O_2(g) \xrightarrow{\Delta} 12CO_2(g) + 14H_2O(g)$ (24.4)

5. (a) CH₃CHCH₂   (24.4)
     | |
     Br   Br

(e) NR   (24.6)

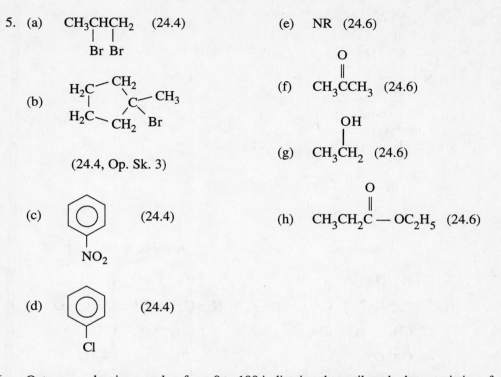

(b)

(24.4, Op. Sk. 3)

(f) CH₃CCH₃   (24.6)
         O (double bond)

(g) CH₃CH₂   (24.6) with OH

(c) [benzene ring with NO₂]   (24.4)

(h) CH₃CH₂C—OC₂H₅   (24.6) with O double bond

(d) [benzene ring with Cl]   (24.4)

6. Octane number is a number from 0 to 100 indicating the antiknock characteristics of a gasoline. The scale is based on heptane, with octane number 0, and 2,2,4-trimethylpentane (an octane isomer), with octane number 100. The higher the number, the better are the antiknock characteristics. (24.4)

7. (a) 4 (24.8)   (b) 9 (24.8)   (c) 8 (24.7)   (d) 1 (24.2)
    (e) 3 (24.3)   (f) 5 (24.5)   (g) 2 (24.7)

8.

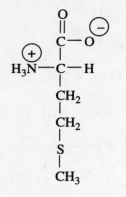

(25.3)

9.   Asn–met is   Met–asn is

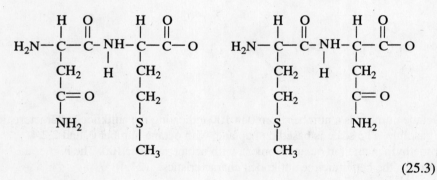

(25.3)

10.   his–arg–ile   arg–his–ile   ile–arg–his
       his–ile–arg   arg–ile–his   ile–his–arg   (25.3)

11.

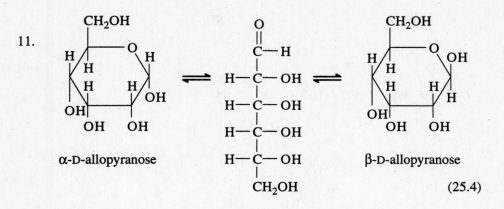

α-D-allopyranose   β-D-allopyranose

(25.4)

12. Amylose is a polymer of α-D-glucose linked through carbons 1 and 4:

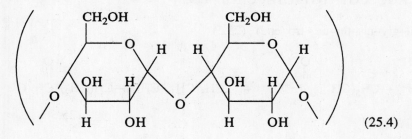

(25.4)

13. (a) False. An example of a sugar without an oxygen on all carbons is 2-deoxy-D-ribose. (25.4)

    (b) True. (25.3)

    (c) True. (25.4)

    (d) False. Complementary base pairs are adenine–thymine, adenine–uracil, and guanine–cytosine. (25.5)

14. 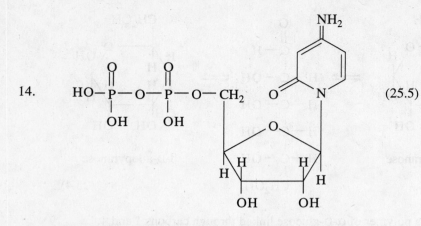   (25.5)

15.   CAACAGCCCGTATGTTCTACCACT   (25.5)

16.   val–val–gly–his–thr–arg–trp (end)   (25.5)

17. CH_2OH
       |
       $CHOH$ + $CH_3(CH_2)_{10}\overset{O}{\overset{\|}{C}}\!-O^-$ + $2CH_3(CH_2)_7CH=CH(CH_2)_7\overset{O}{\overset{\|}{C}}\!-O^-$
       |
       CH_2OH (25.6)